Microbiology in Dairy Processing

IFT | Press

The *IFT Press* series reflects the mission of the Institute of Food Technologists—to advance the science of food contributing to healthier people everywhere. Developed in partnership with Wiley, *IFT Press* books serve as leading - edge handbooks for industrial application and reference and as essential texts for academic programs. Crafted through rigorous peer review and meticulous research, *IFT Press* publications represent the latest, most significant resources available to food scientists and related agriculture professionals worldwide. Founded in 1939, the Institute of Food Technologists is a nonprofit scientific society with 18,000 individual members working in food science, food technology, and related professions in industry, academia, and government. IFT serves as a conduit for multidisciplinary science thought leadership, championing the use of sound science across the food value chain through knowledge sharing, education, and advocacy.

WILEY Blackwell

Titles in the IFT Press series

- *Accelerating New Food Product Design and Development, second edition* (Jacqueline H. Beckley, Leslie J. Herzog and M. Michele Foley)
- *Advances in Dairy Ingredients* (Geoffrey W. Smithers and Mary Ann Augustin)
- *Anti-Ageing Nutrients: Evidence-based Prevention of Age-Associated Diseases* (Deliminda Neves)
- *Bioactive Compounds from Marine Foods: Plant and Animal Sources* (Blanca Hernández-Ledesma and Miguel Herrero)
- *Bioactive Proteins and Peptides as Functional Foods and Nutraceuticals* (Yoshinori Mine, Eunice Li - Chan, and Bo Jiang)
- *Biofilms in the Food Environment, second Edition* (Anthony L. Pometti III and Ali Demicri)
- *Bitterness: Perception, Chemistry and Food Processing* (Michel Aliani and Michael N.A. Eskin)
- *Calorimetry in Food Processing: Analysis and Design of Food Systems* (Gönü l Kaletunç)
- *Coffee: Emerging Health Effects and Disease Prevention* (YiFang Chu)
- *Flavor, Satiety and Food Intake* (Beverly Tepper, Martin Yeomans)
- *Food Carbohydrate Chemistry* (Ronald E. Wrolstad)
- *Food Carotenoids: Chemistry, Biology and Technology* (Delia B. Rodriguez-Amaya)
- *Food Industry Design, Technology & Innovation* (Helmut Traitler, Birgit Coleman and Karen Hofmann)
- *Food Ingredients for the Global Market* (Yao - Wen Huang and Claire L. Kruger)
- *Food Irradiation Research and Technology, second edition* (Christoper H. Sommers and Xuetong Fan)*Foodborne Pathogens in the Food Processing Environment: Sources, Detection and Control* (Sadhana Ravishankar, Vijay K. Juneja, and Divya Jaroni)
- *Food Oligosaccharides: Production, Analysis and Bioactivity* (F. Javier Moreno and Maria Luz Sanz
- *Food Texture Design and Optimization* (Yadunandan Dar and Joseph Light)
- *High Pressure Processing of Foods* (Christopher J. Doona and Florence E. Feeherry)
- *Hydrocolloids in Food Processing* (Thomas R. Laaman)
- *Improving Import Food Safety* (Wayne C. Ellefson, Lorna Zach, and Darryl Sullivan)
- *Innovative Food Processing Technologies: Advances in Multiphysics Simulation* (Kai Knoerzer, Pablo Juliano, Peter Roupas, and Cornelis Versteeg)
- *Mathematical and Statistical Methods in Food Science and Technology* (Daniel Granato and Gastón Ares)
- *Membrane Processes for Dairy Ingredient Separation*(Kang Hu and James Dickson)
- *Microbial Safety of Fresh Produce* (Xuetong Fan, Brendan A. Niemira, Christopher J. Doona, Florence E. Feeherry, and Robert B. Gravani)
- *Microbiology and Technology of Fermented Foods* (Robert W. Hutkins)
- *Microbiology in Dairy Processing: Challenges and Opportunities* (Palmiro Poltronieri)
- *Multiphysics Simulation of Emerging Food Processing Technologies* (Kai Knoerzer, Pablo Juliano, Peter Roupas and Cornelis Versteeg)

- *Multivariate and Probabilistic Analyses of Sensory Science Problems* (Jean - Fran ç ois Meullenet, Rui Xiong, and Christopher J. Findlay
- *Nanoscience and Nanotechnology in Food Systems* (Hongda Chen)
- *Nanotechnology and Functional Foods: Effective Delivery of Bioactive Ingredients* (Cristina Sabliov, Hongda Chen and Rickey Yada)
- *Natural Food Flavors and Colorants, second edition* (Mathew Attokaran)
- *Nondestructive Testing of Food Quality* (Joseph Irudayaraj and Christoph Reh)
- *Nondigestible Carbohydrates and Digestive Health* (Teresa M. Paeschke and William R. Aimutis)
- *Nonthermal Processing Technologies for Food* (Howard Q. Zhang, Gustavo V. Barbosa - C á novas, V.M. Balasubramaniam, C. Patrick Dunne, Daniel F. Farkas, and James T.C. Yuan)
- *Nutraceuticals, Glycemic Health and Type 2 Diabetes* (Vijai K. Pasupuleti and James W. Anderson)
- *Organic Meat Production and Processing* (Steven C. Ricke, Ellen J. Van Loo, Michael G. Johnson, and Corliss A. O ' Bryan)
- *Packaging for Nonthermal Processing of Food* (Jung H. Han)
- *Practical Ethics for the Food Professional: Ethics in Research, Education and the Workplace* (J. Peter Clark and Christopher Ritson)
- *Preharvest and Postharvest Food Safety: Contemporary Issues and Future Directions* (Ross C. Beier, Suresh D. Pillai, and Timothy D. Phillips, Editors; Richard L. Ziprin, Associate Editor)
- *Processing and Nutrition of Fats and Oils* (Ernesto M. Hernandez and Afaf Kamal - Eldin)
- *Processing Organic Foods for the Global Market* (Gwendolyn V. Wyard, Anne Plotto, Jessica Walden, and Kathryn Schuett)
- *Regulation of Functional Foods and Nutraceuticals: A Global Perspective* (Clare M. Hasler)
- *Resistant Starch: Sources, Applications and Health Benefits* (Yong - Cheng Shi and Clodualdo Maningat)
- *Sensory and Consumer Research in Food Product Design and Development* (Howard R. Moskowitz, Jacqueline H. Beckley, and Anna V.A. Resurreccion)
- *Spray Drying Techniques for Food Ingredient Encapsulation* (C. Anandharamakrishnan and Padma Ishwarya S.)
- *Sustainability in the Food Industry* (Cheryl J. Baldwin)
- *Thermal Processing of Foods: Control and Automation* (K.P. Sandeep)
- *Trait - Modified Oils in Foods* (Frank T. Orthoefer and Gary R. List)
- *Water Activity in Foods: Fundamentals and Applications* (Gustavo V. Barbosa - Cánovas, Anthony J. Fontana Jr., Shelly J. Schmidt, and Theodore P. Labuza)
- *Whey Processing, Functionality and Health Benefits* (Charles I. Onwulata and Peter J. Huth)

WILEY Blackwell

Microbiology in Dairy Processing

Challenges and Opportunities

Edited by Palmiro Poltronieri
Institute of Sciences of Food Productions (CNR-ISPA)

IFT Press

WILEY Blackwell

This edition first published 2018

Registered Offices
John Wiley & Sons, Inc., 111 River Street, Hoboken, NJ 07030, USA
John Wiley & Sons Ltd, The Atrium, Southern Gate, Chichester, West Sussex, PO19 8SQ, UK

Editorial Office
The Atrium, Southern Gate, Chichester, West Sussex, PO19 8SQ, UK
For details of our global editorial offices, customer services, and more information about Wiley products visit us at www.wiley.com.

Wiley also publishes its books in a variety of electronic formats and by print-on-demand. Some content that appears in standard print versions of this book may not be available in other formats.

Library of Congress Cataloging-in-Publication Data

Names: Poltronieri, Palmiro, editor.
Title: Microbiology in dairy processing : challenges and opportunities / edited by Palmiro Poltronieri.
Description: Hoboken, NJ : John Wiley & Sons, 2017. | Series: IFT Press series | Includes index. |
Identifiers: LCCN 2017016747 (print) | LCCN 2017035565 (ebook) | ISBN 9781119114970 (pdf) |
ISBN 9781119114987 (epub) | ISBN 9781119114802 (cloth)
Subjects: LCSH: Dairy microbiology. | Dairy processing.
Classification: LCC QR121 (ebook) | LCC QR121 .M54 2017 (print) | DDC 579.3/7–dc23
LC record available at https://lccn.loc.gov/2017016747

Cover Design: Wiley
Cover Images: (Background Image) © karandaev/Gettyimages; (Inset Images: from top to bottom)
© nicolas_/Gettyimages; © WitthayaP/Shutterstock; © Maurizio Milanesio/Shutterstock

Set in 10.5/12.5pt Times by SPi Global, Pondicherry, India
Printed and bound in Malaysia by Vivar Printing Sdn Bhd

10 9 8 7 6 5 4 3 2 1

Contents

List of contributors

Iolanda Altomonte
Department of Veterinary Sciences, University of Pisa, Pisa, Italy

Maria Aponte
Department of Agricultural Sciences, Division of Microbiology, University of Naples Federico II, Portici, Italy

Stefania Arioli
Department of Food Environmental Nutritional Sciences (DeFENS), University of Milan, Milan, Italy

Marta Ávila Arribas
Departamento de Tecnología de Alimentos, Instituto Nacional de Investigación y Tecnología Agraria y Alimentaria (INIA), Madrid, Spain

François Baglinière
Department of Microbiology, Federal University of Viçosa, Minas Gerais, Brazil

Giovanna Battelli
ISPA-CNR, Institute of Sciences of Food Productions, Milano, Italy

Giuseppe Blaiotta
Department of Agricultural Sciences, Division of Grape and Wine Sciences, University of Naples Federico II, Avellino, Italy

Milena Brasca
ISPA-CNR, Institute of Sciences of Food Productions, Milano, Italy

Cinzia Caggia
Department of Agriculture, Food and Environment (Di3A), University of Catania, Catania, Italy

Cesare Cammà
Istituto Zooprofilattico Sperimentale 'G. Caporale', Teramo, Italy

Domenico Carminati
CREA-ZA, Centro di ricerca Zootecnia e Acquacoltura, sede di Lodi, Italy

Laura Cavallarin
CNR-ISPA, Institute of Sciences of Food Productions, Dipartimento di Scienze Veterinarie dell'Università di Torino, Grugliasco, TO, Italy

Luca Simone Cocolin
Settore di Microbiologia agraria e Tecnologie alimentari, Dipartimento di Scienze Agrarie, Forestali e Alimentari, Grugliasco, Italy

Fabio Dal Bello
Sacco System, Sacco srl, Cadonago (CO), Italy

Maria Cristina Dantas Vanetti
Department of Microbiology, Federal University of Viçosa, Minas Gerais, Brazil

Marilù Decimo
ISPA-CNR, Institute of Sciences of Food Productions, Milano, Italy

Elisabetta Di Giannatale
Istituto Zooprofilattico Sperimentale, "G. Caporale", Teramo, Italy

Paola Dolci
Settore di Microbiologia agraria e Tecnologie alimentari, Dipartimento di Scienze Agrarie, Forestali e Alimentari, Grugliasco, Italy

Sonia Garde Lopez-Brea
Departamento de Tecnología de Alimentos, Instituto Nacional de Investigación y Tecnología Agraria y Alimentaria (INIA), Madrid, Spain

Giorgio Giraffa
CREA-ZA, Centro di ricerca Zootecnia e Acquacoltura, sede di Lodi, Italy

Marzia Giribaldi
CREA-IT, Research Centre for Engineering and Agro-Food Processing, Turin, Italy

Maria Gabriella Giuffrida
CNR-ISPA, Institute of Sciences of Food Productions, Dipartimento di Scienze Veterinarie dell'Università di Torino, Grugliasco, TO, Italy

Natalia Gómez-Torres
Departamento de Tecnología de Alimentos, Instituto Nacional de Investigación y Tecnología Agraria y Alimentaria (INIA), Madrid, Spain

Solimar Gonçalves Machado
Federal Institute of Education, Science and Technology of North of Minas Gerais, Campus Salinas, Minas Gerais, Brazil

Nicoletta Pasqualina Mangia
Department of Agriculture, University of Sassari, Italy

Mina Martini
Department of Veterinary Sciences, University of Pisa, Pisa, Italy

Giacomo Migliorati
Istituto Zooprofilattico Sperimentale, "G. Caporale", Teramo, Italy

Diego Mora
Department of Food Environmental Nutritional Sciences (DeFENS), University of Milan, Milano, Italy

Stefano Morandi
ISPA-CNR, Institute of Sciences of Food Productions, Milano, Italy

Alessandra Pino
Department of Agriculture, Food and Environment (Di3A), University of Catania, Catania, Italy

Palmiro Poltronieri
Institute of Sciences of food Productions (CNR-ISPA), CNR, National Research Council of Italy, Lecce, Italy

Francesco Pomilio
Istituto Zooprofilattico Sperimentale, "G. Caporale", Teramo, Italy

Cinzia Randazzo
Department of Agriculture, Food and Environment (Di3A), University of Catania, Catania, Italy

Franca Rossi
Istituto Zooprofilattico Sperimentale, "G. Caporale", Teramo, Italy

Lorena Sacchini
Istituto Zooprofilattico Sperimentale, "G. Caporale", Teramo, Italy

Tiziana Silvetti
ISPA-CNR, Institute of Sciences of Food Productions, Milano, Italy

Gianluigi Scolari
Istituto di Microbiologia degli alimenti, Università Cattolica del Sacro Cuore, Piacenza, Italy

Panagiotis Sfakianakis
Food Chemistry and Technology, School of Chemical Engineering, National Technical University of Athens, Zografou, Greece

Federica Salari
Department of Veterinary Sciences, University of Pisa, Pisa, Italy

Constantina Tzia
Food Chemistry and Technology, School of Chemical Engineering, National Technical University of Athens, Zografou, Greece

Koenraad van Hoorde
Laboratory of Brewing and Biochemistry, Faculty of Bioscience Engineering, Ghent University, Gent, Belgium

Miriam Zago
CREA-ZA, Centro di ricerca Zootecnia e Acquacoltura, sede di Lodi, Italy

Foreword

Microbiology in Dairy Processing: Challenges and Opportunities is directed to the following: dairy scientists; dairy professionals in industry and academia; those in food science, dairy science and microbiology; intermediate course and post-graduate students; trained laboratory personnel; and R&D and production personnel in dairy industry companies of all sizes. The idea to write this book came from the section "Questions" in the Researchgate community. I realised that there is a need to introduce lactic acid bacteria (LAB) growth media at various levels of expertise, from young researchers starting their laboratory work to food technologists devoted to microbiological analyses. Therefore, from this starting point, I searched the recent literature and produced a list of exceptionally interesting publications on how far the genomics field has advanced in its knowledge of LAB species in recent years. The chapters in this book reflect these advancements and offer a panoramic view of the research fields in which to apply these advancements in knowledge, either for LAB and dairy-associated species and their applications in dairy productions and for the technologies to maintain the milk products safe and devoid of undesired pathogens and milk spoilage bacteria. The challenges of dairy microbiology are either to maintain the product safety devoid of undesired bacteria that may spoil the quality and change the taste or to the further advancement in the microbiota and the interaction among bacteria at community level. The opportunities remain in the exploration of the biodiversity of LAB and dairy-associated species, either at genome rearrangements and horizontal gene transfer or at the biochemistry level, to produce novel dairy products that are low fat, low salt, or with beneficial properties for human health.

Preface

Microbiology in Dairy Processing: Challenges and Opportunities introduces and reviews the knowledge regarding dairy technologies and lactic acid bacteria (LAB) and dairy-associated species in the fermentation of dairy products for laboratory technicians and researchers and students learning the protocols for LAB isolation and characterisation. It provides application notes useful in laboratories of food technology departments and for students and researchers studying all aspects of the milk-processing industry, from microbiology to food productions.

The chapters deal with the industrial processing of milk – the problems solved and those still affecting the processes, from microfiltration to deterioration of stored milk in cold by psychrotrophic bacteria (such as *Pseudomonas fragi*) and by spore-forming bacteria – and cheese-manufacturing technologies. The book introduces culture methods and species-selective growth media to grow, separate and characterise LAB and dairy-associated species, molecular methods for species identification and strain characterization, Next Generation Sequencing for genome characterization, comparative genomics, phenotyping, and current applications in dairy and non-dairy productions, as well as the potential future exploitation of the culture of novel strains with useful traits (probiotics, fermentation of sugars, metabolites produced, bacteriocins).

Chapter 1 introduces the quality and properties of milk fats and differences in milks of various origin. Chapter 2 overviews the spore-forming bacteria associated with milk. Chapter 3 discusses the problem of psychrotrophic bacteria in milk deterioration. Chapter 4 presents the various types of industrial milk according to the freshness and quality. Chapter 5 presents the advancements in LAB and dairy-associated species genomics and strain differences, related to gene content and their applications. Chapter 6 presents very broadly the biochemistry of LAB and dairy-associated species. Chapter 7 reviews selective growth media for different species of LAB and non-LAB dairy-associated bacteria. Chapter 8 introduces the molecular tools for strain identification and characterization. Chapter 9 discusses the bacteriocin-producing LAB species and their potential applications in food products. Chapter 10 analyses in detail the complex interactions among starter and non-starter strains. Chapter 11 reviews the physical-chemical properties of milk cream products and technological processes involving milk fats and cream-derived products. Chapter 12 analyses technological traits of lactic acid bacteria, their industrial relevance and new perspectives. Chapter 13 overviews LAB bacteriophages in dairy products, their problems and solutions. Chapter 14 details the application of LAB as a cell factory for delivering functional biomolecules in dairy products. Chapter 15 reviews the dairy technologies applied to yogurt production. Finally, Chapter 16 introduces properties of milk proteins, the differences in amino acids of protein variants, and the potential to originate bioactive peptides and the proteolysis by co-fermenting LAB species, a process that may ensure the safety and healthiness of the fermented products, as assessed by EFSA authority. Last, the potential for milks of different origin to be administered to individuals suffering of milk allergies or intolerance is discussed.

Palmiro Poltronieri
Institute of Sciences of Foods Productions (CNR-ISPA)

Acknowledgements

The contributions of all members of staff at Wiley–IFT who were involved with writing and reviewing the draft of this book are thankfully acknowledged. This book was made possible thanks to the support of colleagues and professors from Italy, Spain, Belgium, Greece, and Brazil. Among others, I would like to thank Dr. Bruno Battistotti, whose course in dairy technologies opened my mind and my working perspectives during my first years of research; Dr. Pier Sandro Cocconcelli, who introduced me to LAB molecular methods; Drs. Giuseppe Zacheo, Franco Dellaglio and Marco Gobbetti, for supervising microbial technologies and project proposals of my Institute; and Dr. Maria Morea for including me in her projects on "Microbiology for food quality and safety". A special appreciation goes to the contributors of the book chapters and to AITeL, the Italian Association of Milk Technologists, providing the opportunity to meet and include some of the last-minute contributors. Finally, I would like to thank the scientific community, the European Research Council (previously ESF), for the opportunity to work and collaborate with high-level scientists in several international projects, such as Ferbev, Riboreg, TransBio, and COST 853 "Agricultural biomarkers for array technology". This book was made possible thanks to the scientific vision of many other colleagues who, because of their schedule, could not be directly involved at this time, but whose work has enlightened the process of writing the chapters and the organization of the book.

1 Milk fat components and milk quality

Iolanda Altomonte, Federica Salari and Mina Martini

Department of Veterinary Sciences, University of Pisa, Pisa, Italy

1.1 INTRODUCTION

From a physico-chemical point of view, milk is an emulsion of lipid globules and a colloidal suspension of protein and mineral aggregates in a solution of carbohydrates (mainly lactose). In Western countries, milk and dairy products, and in general food of animal origin, are often accused of causing adverse health effects, especially with regard their food lipid intake, since lipids have been implicated in several diseases such as obesity, insulin resistance and atherosclerosis (Olofsson et al., 2009). For these reasons, the number of studies on the physical and chemical structure of fat in several edible products of animal origin have increased. Although milk and dairy products contain saturated fatty acids, they also provide specific beneficial components for human health and also lipid components (phospholipids, some individual fatty acids (FAs) and fat-soluble vitamins) that have a role in health maintenance. In addition, milk is a major source of dietary energy, especially in developing countries, where there is shortage of animal-source food (FAO, 2013), and in childhood.

Milks of different origins have long been used, and they have been processed to dairy products for their longer shelf life. Due to the wide natural variability from species to species in the proportion of milk macronutrients and to variations along lactation, milk represents a flexible source of nutrients that may be exploited to produce a variety of dairy products.

Ruminant milk is the main source available for humans to use to manufacture dairy products and fermented milk. Besides cow's milk and milk from other ruminants (such as buffalo, goat and sheep), research on milk from other species is still poorly exploited (FAO, 2013). More recently, equine milks have been suggested for use in children with severe IgE-mediated cow milk protein allergy (CMPA) (Monti et al., 2007, 2012; Sarti et al., 2016), and local producers have established a niche for the application of donkey products with well-characterised profile of its constituents (Martini et al., 2014a).

Microbiology in Dairy Processing: Challenges and Opportunities, First Edition.
Edited by Palmiro Poltronieri.

Published 2018 by John Wiley & Sons Ltd and the Institute of Food Technologists.

1.1.1 Milk fat globules

Milk lipids are composed of milk fat globules (MFGs) made up of triglycerides enveloped by a biological membrane. MFGs are responsible and/or contribute to some properties and phenomena in milk and dairy products and may affect milk fatty acid composition and the way in which fat is digested (Baars et al., 2016; Huppertz and Kelly, 2006; Martini et al., 2017). For the dairy industry it is of interest that changes in the morphometry of the MFGs lead to changes in milk quality, yields, and ripening and the nutritional quality of cheeses (Martini et al., 2004).

In milk of different species there are MFGs of various sizes, ranging from a diameter smaller than 0.2 μm to a maximum of about 15 μm, with an average diameter that varies as a function of endogenous (species, breed), physiological (parity, stage of lactation), and exogenous factors (feeding) (Martini et al., 2010a).

Different average diameters have been reported in the literature for ruminant species (3.5–5.5 μm for cows; 2.79–4.95 μm for sheep; 2.2 and 2.5–2.8 μm for goats and 2.96–5.0 μm for buffalos) (Table 1.1) (Martini et al., 2016b). However average diameter of globules in equids is considerably lower than other dairy species (about 2 μm in donkey) (Martini et al., 2014b), while regarding human MFGs, larger dimensions have also been found (4 μm) (Lopez and Ménard, 2011).

The MFG membrane (MFGM) is a triple membrane resulting from the mammary secretory cell that surrounds a core of triglycerides distributed in a lamellar way (Heid and Keenan, 2005).

The MFGM consists of different classes of lipids (phospholipids, triglycerides and cholesterol) and of several proteins and enzymes. Phospholipids, in the form of mixtures of fatty acid esters of glycerol and sphingosine, possibly containing phosphoric acid, and a nitrogen-based compound (choline, ethanolamine or serine). These are natural emulsifiers able to maintain the milk lipids as discrete globules, ensuring high stability. MFGM contains about 1% of the total milk proteins. Most of them are present in very low amounts and are enzymes and proteins involved in milk synthesis. The principal proteins in the MFGM include mucins (MUC) 1 and 5, adipophilin (ADPH), butyrophilin (BTN), periodic acid-Schiff glycoproteins (PAS) 6 and 7, fatty acid binding protein (FABP), and xanthine oxidoreductase (XOR), a metal (Mo, Fe) binding protein (Spertino et al., 2012). In the last few years, research on the composition and structure of the milk membranes have been increased and have improved the knowledge of the MFGM from species other than the bovine (Saadaoui et al., 2013; Pisanu et al., 2012; Lu et al., 2016; Martini et al., 2013).

These studies have increased also due to the fact that MFGM is a dietary source of functional substances and is considered a nutraceutical (Rosqvist et al., 2014; Timby et al., 2015; Hernell et al., 2016). The functionality of the MFGM seems to be provided by its content of phospholipids, sphingolipids, fatty acids and proteins with an antibacterial effect (such as xanthine oxidoreductase and mucins) and/or health benefits.

MFGM conveys fat in an aqueous environment and is damaged by some treatment, such as homogenization, whipping and freezing, affecting milk physicochemical properties, for example producing hydrolytic activity, rancidity, and oiling off, and low wettability of milk powders. MFGM composition also affects the creaming rate on the milk surface (Martini et al., 2017); in bovine milk this phenomenon is due to the effect

Table 1.1 Average values in literature for fat content, milk fat globules characteristics and fatty acid composition of milk from different species.

		Cow	**Buffalo**	**Goat**	**Sheep**	**Donkey**	**Horse**	**Human**
Fat	%	3.70	8.14	3.90	6.50	0.36	1.48	3.34
Average diameter of the fat globules	µm	3.5–5.5	2.96–5.0	2.2–2.8	2.79–4.95	2	2–3	3.3
SFA	g/100g fat	71.24	65.9	70.42	71.85	55.55	45.18	41.77
MUFA	g/100g fat	25.56	31.4	25.67	26.04	22.21	31.88	38.73
PUFA	g/100g fat	3.20	2.70	4.08	2.10	21.08	22.93	16.96
UFA	g/100g fat	28.76	34.1	29.75	28.14	43.29	54.81	55.29
UFA:SFA ratio		0.40	0.52	0.42	0.39	0.78	1.20	1.32
SCFA	g/100g fat	10.52	9.72	17.51	17.13	12.29	10.79	1.87
MCFA	g/100g fat	52.81	53.70	48.28	45.87	40.08	42.47	37.94
LCFA	g/100g fat	34.38	32.73	32.64	35.87	47.64	46.75	57.72
CLA c9, t11	g/100g fat	0.65	0.45	0.70	1.00	–	0.09	0.19
C18:2 n6 (LA)	g/100g fat	2.42	1.71	2.72	1.20	9.5	16.17	12.96
C18:3 n3 (ALA)	g/100g fat	0.25	0.51	0.53	0.77	7.25	5.96	1.15
C18:2 n6: C18:3 n3 ratio	—	9.68	3.35	5.13	1.56	1.31	2.71	11.26
C20:4 (AA)	g/100g fat	0.13	0.10	0.16	0.10	0.09	0.10	0.4
C20:5 (EPA)	g/100g fat	0.05	0.03	nd	nd	0.26	–	0.11
C22:6 (DHA)	g/100g fat	nd	–	0.05	0.04	0.28	–	0.51

SFA: saturated fatty acids; MUFA: monounsaturated fatty acids; PUFA: polyunsaturated fatty acids; UFA: unsaturated fatty acids; SCFA: short chain fatty acids (≤10C); MCFA: medium chain fatty acids (≤11C, ≥17C); LCFA: long chain fatty acids (≥18C); nd: no data.

of cryoglobulins, an M-type immunoglobulin that aggregates globules during cold storage. Other types of milk are lacking these cryoglobulins and do not agglutinate. Homogenization reduces globule diameter, making globules insensitive to the action of cryoglobulins and prevents agglutination. During butter production, extensive agitation and kneading causes the MFGM to form the water-in-oil emulsion. The partitioning in the aqueous phase produces the loss of MFGM in the buttermilk.

1.1.2 Milk fat and fatty acid composition

Milk lipid content and fatty acid composition vary by virtue of various endogenous and exogenous factors. Among endogenous factors, the species, breed and stage of lactation are the main factors.

Regarding the species, buffalo and sheep milk contains higher fat percentages and are particularly suitable for processing, such as cheese making. Fat percentages vary in a range between 7 and 9% for buffalo, but can reach 15% under favourable conditions (Altomonte et al., 2013; Varricchio et al., 2007), whereas in sheep the range is between 6.5 and 9% depending on the breed (Haenlein, 2007; Martini et al., 2012). Regarding cow and goat milk, fat content are comparable; in fact cow total lipid ranges from 3.4% in Holstein to about 6% in Jersey breeds (Nantapo et al., 2014; Pegolo et al., 2016; Sanz Ceballos et al., 2009), and goat range from a minimum of 3.5% to a maximum of 5.6% in some native goats (Haenlein, 2007; Martini et al., 2010b). Equid milk has lower fat percentages compared to ruminant milk; the average values reported in literature are 0.30–0.53% in donkey and 1.5% in horse milk (Pikul and Wójtowski, 2008; Martemucci and D'Alessandro, 2012; Martini et al., 2014b; Salimei et al., 2004). Furthermore, some authors stated lower contents (1.04%, 0.92%, 0.8%) in the milk of Halfinger, Hucul and Wielkopolski mares, respectively (Salamon et al., 2009; Pieszka Huszczyński and Szeptalin, 2011). The low fat content in equid milk could be a limiting factor in its use in infant nutrition in a diet exclusively based on milk, thus an appropriate lipid integration should be introduced. On the other hand it is encouraging for studies on the possible use of donkey milk in dietotherapy.

Regarding human milk, fat content is more similar to cow milk, varying between 2.8 and 3.8% (Antonakou et al., 2013).

From a nutritional point of view, donkey milk leads to lower saturated fatty acid (SFA) intake, about 2.00 g/l (Table 1.2), than the other milks commonly used for human feeding. Despite being rich in unsaturated fatty acids (UFAs) and having a UFA:SFA ratio intermediate between ruminant and human milk, donkey provides a limited amount of fat; thus, the total intake of UFA per 1 l of milk is lower (1.56 g) than milk of other species (Martini et al., 2016a).

In milk from ruminants, especially sheep and goats, triglycerides contain short chain fatty acids (SCFAs) such as butyric acid and hexanoic, octanoic and decanoic acid. On the contrary, human (Yuhas, Pramuk and Lien, 2006) and donkey milk (Martini et al., 2014b) are characterized by low amounts of SCFA—especially the chains shorter than C8—and high quantities of long chain fatty acids (LCFAs).

SCFAs are synthesized by the fermentation of dietary fibre, are water soluble and volatile, and contribute to the typical flavour of ovine and caprine milk. When freed by endogenous lipase or bacterial enzymes, SCFA can also give rancidity and quality

Table 1.2 Calculated average values for fat content and some fatty acids (g/l) in milk from different species and Dietary Reference Values.

		Dietary Reference Values	Cow	Buffalo	Goat	Sheep	Donkey	Horse	Human
Fat	g/l	Adults: 20–30% of the energy of the diet (E); Infants (6–12 months): % 40 E%; Children (2–3 years): 35–40 E%.	37.0	81.4	39.0	65.0	3.60	14.8	33.4
SFA	g/l	as low as possible	26.36	53.64	27.46	46.70	2.00	6.68	13.95
MUFA	g/l	Not set	9.45	25.56	10.01	16.93	0.80	4.71	12.94
PUFA	g/l	Not set	1.18	2.20	1.59	1.36	0.76	3.39	5.66
UFA	g/l	Not set	10.63	27.76	11.60	18.29	1.56	8.1	18.60
C18:2 n6 (LA)	g/l	Adequate Intake (AI): 4 E%	0.89	1.39	1.06	0.78	0.34	2.39	4.33
C18:3 n3 (ALA)	g/l	AI: 0.5% E	0.09	0.41	0.21	0.50	0.26	0.88	0.38
C20:4 (AA)	g/l	Not set	0.05	0.08	0.06	0.06	0.006	0.01	0.03
C20:5 (EPA)	g/l	Not set	0.02	0.02	–	nd	0.009	–	0.04
C22:6 (DHA)	g/l	Infants and young children (between 6 and 24 months) AI: 0.10 g DHA	–	–	0.02 (20% of AI)	0.03 (30% of AI)	0.010 (9% of AI)	–	0.17 (170.34% of AI)
C20:5 + C22:6 (EPA + DHA)	g/l	Adults. AI: 0.25 g	–	–	–	–	0.017 (6.80% AI)	–	0.21 (84% of AI)

SFA: saturated fatty acids; MUFA: monounsaturated fatty acids; PUFA: polyunsaturated fatty acids; UFA: unsaturated fatty acids; nd: no data.
Source: Data from EFSA (2010).

deterioration. SCFAs and MCFAs, which are a source of rapidly available energy, are particularly relevant for people suffering from malnutrition or fat absorption syndromes and for elderly people (Raynal-Ljutovac et al., 2008).

Recent studies have highlighted effects of SCFA at cellular and molecular levels in the organism; their presence or their deficiency may affect pathogenesis of some diseases (autoimmune, inflammatory diseases). In addition, SCFAs have antimicrobial activity and anti-inflammatory effects in the gut (Tan et al., 2014).

Ruminant milk, in particular milk from sheep that feed in pastures, is the richest natural source of conjugated linoleic acids (CLA) and of C18:1 trans-11 (vaccenic acid) (Bauman and Lock, 2005; Lim et al., 2014). The CLA content in milk varies depending on species, breed and individual, farming system, feeding and season. In sheep milk CLA varies from 1.2 to 2.9 g/100 g of fat; in goat between 0.5 and 1 g/100 g of fat (Parodi, 2003). Cow milk is generally reported to vary from 0.1 to 2.2 g/100 g total FA (Elgersma, Tamminga and Ellen, 2006), whereas human and equid milk are poor sources of CLA (Table 1.1).

Ninety percent of CLA isomers in milk is made up of *cis*-9, *trans*-11C18:2 (rumenic acid) produced mainly by stearoyl Co-A desaturase (SCD) *o*-Δ9–desaturase enzyme in the mammary gland using vaccenic acid as precursor, but also by the rumen bacterium *Butyrivibrio fibrisolvens* as intermediate of biohydrogenation of linoleic and linolenic acids ingested with feed (Bauman and Lock, 2005). Rumenic acid vary between 0.29 and 0.71% of total human milk fatty acids, while in the horse it is between 0.07 and 0.10%. Moreover, in equids, cecum seems to contribute little to CLA synthesis (Markiewicz-Keszycka et al., 2014).

Anticarcinogenic properties and modulation of immunological functions have been demonstrated for rumenic acid in animal models and cell cultures (Field and Schley, 2004; O'Shea et al., 2004). However, the most documented effects of CLA in humans are the gain of muscle mass at the expense of body fat, whereas in vivo studies on the effects on atherosclerosis and cholesterol have shown conflicting results in humans (Crumb, 2011).

Vaccenic acid has shown anticancer properties in human mammary adenocarcinoma cells (Lim et al., 2014).

Regarding the omega-3 FAs, milk is not a good source of this family of FAs. However, among the mammalian species reared for milk production, horse, sheep and donkey are richest sources of C18:3 n3 (α-linolenic acid (ALA))(g/l) (Table 1.2), in particular donkey and horse milk provide a good ALA intake (0.22–0.88 g/l) although they have low fat content. In adults minimum intake levels for ALA are recommended to prevent deficiency symptoms (0.5% of energy) (FAO-WHO, 2010).

Linoleic acid (LA) and ALA are precursors of omega 6 and omega-3 families, respectively, and their ratio is generally considered as indicative of their balanced intake in the diet. The interest in the LA:ALA ratio derives from the antagonistic effects between the two families of FAs observed in human body. In fact, the higher intake of *n*-6 fatty acids may reduce the formation of anti-inflammatory mediators from omega-3 fatty acids. Observations on animal models suggest that raising the *n*-6 to *n*-3 fatty acids ratio (*n*6:*n*3) acts on adipogenesis and the risk of obesity in the offspring later in life (Rudolph et al., 2015). However, research is yet not supported by studies in humans, and an optimal ratio of these fatty acids in the diet has not yet been established (EFSA, 2010).

Furthermore, the prevalence of *n*-6 in human diets has increased over the decades while *n*-3 fatty acids remain unchanged, thus increasing the n-6/n-3 milk fatty acid ratio (Rudolph et al., 2015). Thus, a reduction of omega-6 in the diet is desirable, and donkey's milk appears to have a balanced rapport of these two families (about 1) compared to other milks (Martini et al., 2014b).

Arachidonic acid (AA) C20:4 is essential component of cellular membranes and also of MFGM, where it may have an essential role (Fong et al., 2007; Martini et al., 2013).

AA is present in almost similar amounts in the milk of ruminants (Table 1.2), while it shows lower values in equids.

Despite the importance of AA for membrane integrity (Fong, Norris and MacGibbon, et al., 2007), it has been described as an adipogenetic-, pro-inflammatory- and hypertension-promoting factor (Vannice and Rasmussen, 2014), and recommended intake levels have not been established.

C20:5 (EPA) and C22:6 (DHA) have showed evidence of both independent and shared effects in neuroprotection and in the treatment for a variety of neurodegenerative and neurological disorders. In particular, DHA is an important constituent of the retina and the nervous system, and it has unique and indispensable roles in neuronal membranes (Dyall, 2015).

There is still insufficient evidence to support beneficial effects of EPA and DHA in foetal life or early childhood on obesity, blood pressure, or blood lipids (Voortman et al., 2015).

Overall levels of DHA and EPA in milk are quite low, and in human milk DHA content is highly variable; values from 0.17 to 0.99 % have been reported, depending on the diets and on different countries (Yuhas, Pramuk and Lien, 2006). The recommended daily intake of EPA plus DHA is 0.25 g in adults (EFSA, 2010).

1.2 CONCLUSIONS

The transformation of milks of different origin may be the source of dairy products with different and peculiar characteristics. Since a role in health maintenance has been reported for several lipid components of milk, a deep knowledge of milk lipid constituents from different dairy species is of utmost relevance for both the nutritional uptake and effects on human health.

References

Altomonte I, Mannari I, Martini M, Salari F. 2013. Il latte di bufala: studio di alcuni parametri produttivi. *Large Anim Rev* 1:17–20.

Antonakou A, Skenderi KP, Chiou A, Anastasiou CA, Bakoula C, Matalas AL. 2013. Breast milk fat concentration and fatty acid pattern during the first six months in exclusively breastfeeding Greek women. *Eur J Nutr* 52:963–73.

Baars A, Oosting A, Engels E, Kegler D, Kodde A, Schipper L, Verkade HJ, van der Beek EM. 2016. Milk fat globule membrane coating of large lipid droplets in the diet of young mice prevents body fat accumulation in adulthood. *Br J Nutr* 115, 1930–37.

Bauman DE, Lock AL. 2005. Conjugated linoleic acid. In: Pond WG, Bell AW. editors. *Encyclopedia of animal science*. New York: Marcel Dekker, Inc., pp 235–38.

Crumb DJ. 2011. Conjugated linoleic acid (CLA)–An overview. *Int J Applied Res Nat Prod* 4:12–18.

Dyall SC. 2015. Long-chain omega-3 fatty acids and the brain: a review of the independent and shared effects of EPA, DPA and DHA. *Front Aging Neurosci* 7:52.

EFSA [European Food Safety Authority] – Panel on Dietetic Products, Nutrition, and Allergies (NDA). 2010. Scientific opinion on Dietary Reference Values for fats, including saturated fatty acids, polyunsaturated fatty acids, monounsaturated fatty acids, trans fatty acids, and cholesterol. *EFSA J* 8:1461.

Elgersma A, Tamminga S, Ellen G. 2006. Modifying milk composition through forage. *Anim Feed Sci Tech* 131:207–25.

FAO. 2013. *Milk and dairy products in human nutrition*. Muehlhoff E, Bennett A, McMahon D. Rome. Food and Agriculture Organization. Retrieved from www.fao.org/docrep/018/i3396e/i3396e.pdf.

FAO-WHO. 2010. *Fats and fatty acids in human nutrition*. Report of an expert consultation.10 – 14 November 2008. Geneva: Food and Agriculture Organization of the United Nations Rome.

Field CJ, Schley PD. 2004. Evidence for potential mechanisms for the effect of conjugated linoleic acid on tumor metabolism and immune function: lessons from n-3 fatty acids. *Am J Clin Nutr* 79:1190–98.

Fong, BY, Norris, CS, MacGibbon, AKH. 2007. Protein and lipid composition of bovine milk-fat globule membrane. *Int Dairy J* 17:275–88.

Haenlein GFW. 2007. About the evolution of goat and sheep milk production. *Small Rumin Res* 68:3–6.

Heid HW, Keenan TW. 2005. Intracellular origin and secretion of milk fat globules. *Eur J Cell Biol* 84:245–58.

Hernell O, Timby N, Domellöf M, Lönnerdal B. 2016. Clinical benefits of milk fat globule membranes for infants and children. *J Pediatr* 173:S60–65.

Hupperz T, Kelly AL. 2006. Physical chemistry of milk fat globules. In: Fox PF, McSweeney PLH. editors. *Advanced dairy chemistry*, vol 2. *Lipids*, 3rd ed. New York: Springer, pp. 173–204.

Lim J-N, Oh J-J, Wang T, Lee J-S, Kim S-H, Kim Y-J, Lee H-G. 2014. Trans-11 18:1 vaccenic Acid (TVA) has a direct anti-carcinogenic effect on MCF-7 human mammary adenocarcinoma cells. *Nutrients* 6:627–36.

Lopez C, Ménard O. 2011. Human milk fat globules: polar lipid composition and in situ structural investigations revealing the heterogeneous distribution of proteins and the lateral segregation of sphingomyelin in the biological membrane. *Colloids Surf B Biointerfaces* 83:29–41.

Lu J, Liu L, Pang X, Zhang S, Jia Z, Ma C, Zhao L, Lv J. 2016. Comparative proteomics of milk fat globule membrane in goat colostrum and mature milk. *Food Chem* 209:10–16.

Markiewicz-Kszycka M, Wójtowski J, Czyzak-Runowska G, Kuczynska B, Puppel K, Krzyzewski J, Strzakowska N, Józwik A, Bagnicka E. 2014. Concentration of selected fatty acids, fat-soluble vitamins and b-carotene in late lactation mares' milk. *Int Dairy J* 38:31–16.

Martemucci G, D'Alessandro AG. 2012. Fat content, energy value and fatty acid profile of donkey milk during lactation and implications on human nutrition. *Lipids Health Dis* 11:113–26.

Martini M, Altomonte I, Salari F. 2012. Relationship between the nutritional value of fatty acid profile and the morphometric characteristics of milk fat globules in ewe's milk. *Small Rumin Res* 105:33–37.

Martini M, Altomonte I, Salari F. 2013. Evaluation of the fatty acid profile from the core and membrane of fat globules in ewe's milk during lactation. *LWT-Food Sci Technol* 50:253–58.

Martini M, Altomonte I, Salari F. 2014b. Amiata donkeys: fat globule characteristics, milk gross composition and fatty acids. *Ital J Anim Sci* 13:123–26.

Martini M, Altomonte I, Salari F, Caroli AM. 2014a. Monitoring nutritional quality of Amiata donkey milk: Effects of lactation and productive season. *J Dairy Sci* 97:6819–22.

Martini M, Altomonte I, Sant'Ana da Silva AM, Salari F. 2017. Fatty acid composition of the bovine milk fat globules obtained by gravity separation. *Int Food Res J* (in press).

Martini M, Liponi GB, Salari F. 2010a. Effect of forage:concentrate ratio on the quality of ewe's milk, especially on milk fat globules characteristics and fatty acids composition. *J Dairy Res* 77:239–44.

Martini M, Salari F, Altomonte I, Rignanese D, Chessa S, Gigliotti C, Caroli A., 2010b. The Garfagnina goat: a zootechnical overview of a local dairy population. *J Dairy Sci* 93:4659–67.

Martini M, Salari F, Altomonte I. 2016b. The macrostructure of milk lipids: the fat globules. *Crit Rev Food Sci Nutr* 56:1209–21.

Martini M, Salari F, Altomonte I, Ragona G, Casati D, Brajon G. 2016a. Conservazione del latte d'asina: aspetti nutrizionali e sanitari. AITEL 5th workshop, Bari, 9 September 2016.

Monti G, Bertino E, Muratore MC, Coscia A, Cresi F, Silvestro L, Fabris C, Fortunato D, Giuffrida MG, Conti A. 2007. Efficacy of donkey's milk in treating highly problematic cow's milk allergic children: an in vivo and in vitro study. *Pediatr Allergy Immunol* 18:258–64.

Monti G, Viola S, Baro C, Cresi F, Tovo PA, Moro G, Ferrero MP, Conti A, Bertino E. 2012. Tolerability of donkey's milk in 92 highly-problematic cow's milk allergic children. *J Biol Regul Homeost Agents* 26:75–82.

Nantapo CTW, Muchenje V, Hugo A. 2014. Atherogenicity index and health-related fatty acids in different stages of lactation from Friesian, Jersey and Friesian × Jersey cross cow milk under a pasture-based dairy system. *Food Chem* 146:127–33.

Olofsson SO, Boström P, Andersson L, Rutberg M, Perman J, Borén J., 2009. Lipid droplets as dynamic organelles connecting storage and efflux of lipids. *Biochim Biophys Acta* 1791:448.

O'Shea M, Bassaganya-Riera J, Mohede ICM. 2004. Immunomodulatory properties of conjugated linoleic acid. *Am J Clin Nutr* 79:1199–206.

Parodi, PW. 2003. Conjugated linoleic acid in food. In: Sébédio J-L, Christie WW, Adlof R. editors. *Advances in conjugated linoleic acid research*, vol 2. Champaign, IL: The American Oil Chemists Society. pp 101–122.

Pegolo S, Cecchinato A, Mele M, Conte G, Schiavon S, Bittante G. 2016. Effects of candidate gene polymorphisms on the detailed fatty acids profile determined by gas chromatography in bovine milk. *J Dairy Sci* 99:4558–73.

Pieszka M, Huszczyński J, Szeptalin A. 2011. Comparison of mare's milk composition of different breeds. *Nauka Przyroda Technologie* 5:112.

Pikul J, Wójtowski J. 2008. Fat and cholesterol content and fatty acid composition of mares' colostrum and milk during five lactation months. *Livest Sci* 113:285–90.

Pisanu S, Ghisaura S, Pagnozzi D, Falchi G, Biosa G, Tanca A, Roggio T, Uzzau S, Addis MF. 2012. Characterization of sheep milk fat globule proteins by two-dimensional polyacrylamide gel electrophoresis/mass spectrometry and generation of a reference map. *Int Dairy J* 24:78–86.

Ragona G, Corrias F, Benedetti M, Paladini I, Salari F, Altomonte I, Martini M. 2016. Amiata donkey milk chain: animal health evaluation and milk quality. *IJFS* 5:173–78.

Raynal-Ljutovac K, Lagriffoul G, Paccard P, Guillet I, Chilliard Y. 2008. Composition of goat and sheep milk products: an update. *Small Ruminant Res* 79:57–72.

Rosqvist F, Paulsson, M, Lindmark-Månsson, H, Smedman, A, Risérus U. 2014. Role of milk fat globule membrane in the regulation of blood lipids in humans: a randomized trial (642.3). *FASEB J* 28:642–43.

Rudolph MC, Houck JA, Aikens RM, Erickson CB, Lewis AS, Friedman JE, MacLean PS. 2015. Neonates consuming milk with a high n-6 to n-3 fatty acid ratio have larger adipocytes but smaller subcutaneous adipose depot by 14 days of life. Endocrine Society's 97th Annual Meeting and Expo, San Diego, March 5–8, 2015.

Saadaoui B, Henry C, Khorchani T, Mars M, Martin P, Cebo C. 2013. Proteomics of the milk fat globule membrane from Camelus dromedarius. *Proteomics* 13:1180–84.

Salamon R, Csapo J, Salamon S, Csapo-Kiss Z. 2009. Composition of mare's colostrum and milk. I. Fat content, fatty acid composition and vitamin contents. *Acta Univ Sapientiae Alimentaria* 2:119–31.

Salimei E, Fantuz F, Coppola R, Chiofalo B, Polidori P, Varisco G. 2004. Composition and characteristics of ass's milk. *Anim Res* 53:67–78.

Sanz Ceballos L, Ramos Morales E, de la Torre Adarve G, Díaz Castro J, Pérez Martínez L, Sanz Sampelayo MR. 2009. Composition of goat and cow milk produced under similar conditions and analyzed by identical methodology. *J Food Compos Anal* 22:322–29.

Sarti L, Martini M, Ragona G, Casati D, Belli F, Salari F, Altomonte I, Barni S, Mori F, Pucci N, Novembre E., 2016. Il latte d'asina di razza Amiatina nella gestione del bambino con allergia alle proteine del latte vaccine. Acta of Xth Forum of Pratical Nutrition, NutriMi, Milano. April 21, 2016–April 22, 2016.

Spertino S, Cipriani V, De Angelis C, Giuffrida MG, Marsano F, Cavaletto M. 2012. Proteome profile and biological activity of caprine, bovine and human milk fat globules. *Mol BioSyst* 8:967–74.

Tan J, McKenzie C, Potamitis M, Thorburn AN, Mackay CR, Macia L. 2014. The role of short-chain fatty acids in health and disease. *Advances Immunol* 121:91–119.

Timby N, Hernell O, Vaarala O, Melin M, Lönnerdal B, Domellöf M., 2015. Infections in infants fed formula supplemented with bovine milk fat globule membranes. *J Pediatr Gastroenterol Nutr* 60:384–89.

Vannice G, Rasmussen H. 2014. Position of the academy of nutrition and dietetics: dietary fatty acids for healthy adults. *J Acad Nutr Diet* 114:136–53.

Varricchio ML, Di Francia A, Masucci F, Romano R, Proto V. 2007. Fatty acid composition of Mediterranean buffalo milk fat. *Ital J Anim Sci* 6:509–11.

Voortman T, van den Hooven EH, Braun KVE, van den Broek M, Bramer WM, Chowdhurry R, Franco OH. 2015. Effects of polyunsaturated fatty acid intake and status during pregnancy, lactation, and early childhood on cardiometabolic health: a systematic review. *Prog Lipid Res* 59:67–87.

Yuhas R, Pramuk K, Lien EL., 2006. Human milk fatty acid composition from nine countries varies most in DHA. *Lipids* 41:851–57.

2 Spore-forming bacteria in dairy products

Sonia Garde Lopez-Brea, Natalia Gómez-Torres and Marta Ávila Arribas

Departamento de Tecnología de Alimentos, Instituto Nacional de Investigación y Tecnología Agraria y Alimentaria (INIA), Madrid, Spain

2.1 INTRODUCTION

Spore-forming bacteria are gram-positive microorganisms with low G+C content, aerobic or anaerobic, ubiquitous in nature, and belong to the phylum Firmicutes, which includes the classes Bacilli, Clostridia, Erysipelotrichia, Negativicutes and Thermolithobacteria. Bacterial spores are common contaminants of food products, and their outgrowth may cause food spoilage or foodborne diseases. Bacilli and Clostridia remain the most dominant classes within the Firmicutes phylum, consisting of 16 and 21 families, respectively, and are arguably the most important classes relevant to the dairy industry (Gopal et al., 2015). They are a primary cause of concern for the international dairy industry because of the pervasive and resistant nature of their spores in comparison to vegetative cells, surviving environmental challenges, such as heat, desiccation, freezing, thawing, presence of organic solvents and oxidizing agents, and UV irradiation, as well as predation by protozoa (Setlow et al., 2014b). Spore formers survive heat manufacturing/processing conditions of milk and dairy products, form biofilms in the manufacturing equipment and contaminate final products. Heat treatments may even activate the germination of spores, which then start the metabolically active phase of growth and multiplication, in some cases even at refrigeration temperatures, and can negatively affect the quality and safety of dairy products. Certain species represent a hazard due to the production of toxins.

Spore-forming bacteria pose the greatest spoilage threat to dairy products, causing severe economic losses, equipment impairment and/or reputational damage of food companies. As will be further addressed in this chapter, the main genera implicated in spoilage of milk and dairy products are *Bacillus* spp. and *Paenibacillus* spp. in milk spoilage, *Geobacillus* spp. and *Anoxybacillus* spp. in contamination of dairy powders, and *Clostridium tyrobutyricum* and related *Clostridium* spp. in cheese spoilage. In addition, they are also implicated in the formation of biofilms on stainless steel surfaces of processing equipment in dairy manufacturing plants. Bacilli and related genera are responsible for spoilage problems in milk and dairy products such as bitty cream, sweet

Microbiology in Dairy Processing: Challenges and Opportunities, First Edition.
Edited by Palmiro Poltronieri.

Published 2018 by John Wiley & Sons Ltd and the Institute of Food Technologists.

curdling, off flavour, flat sour, nonsterility, bitterness, ropiness, interference with cheese production, and cheese blowing (Ternström et al., 1993; Heyndrickx and Scheldeman, 2002; Quiberoni et al., 2008; Burgess et al., 2010), and clostridia are responsible for the late blowing defect (LBD) of cheeses (Garde et al., 2013). On the other side, the main spore-forming bacteria implicated in poisoning of milk and dairy products are *Bacillus cereus*, *Clostridium perfringens* and *Clostridium botulinum* (a, b). In recent years, the European Food Safety Authority (EFSA) and the European Centre for Disease Prevention and Control (ECDC) have reported that *Bacillus* and *Clostridium* toxins were implicated in 5.5 and 6.0% (mean values from 2010 to 2014) of total strong-evidence foodborne outbreaks, respectively (EFSA 2012, 2013, 2014, 2015a,b). As can be seen in Table 2.1, there was an increase in the percentage of strong-evidence foodborne outbreaks caused by *Bacillus* and *Clostridium* toxins, mainly from 2010 to 2011. Milk or dairy products represented from 1.9 to 5.3% of food vehicles implicated in strong-evidence foodborne outbreaks caused by *Bacillus* toxins for the period 2010–2013, but milk or dairy products were not associated with these outbreaks caused by *Bacillus* toxins in 2014 or caused by *Clostridium* toxins during all the period.

Table 2.1 Strong-evidence foodborne outbreaks caused by *Bacillus* and *Clostridium* toxins reported by the EFSA and the ECDC from 2010 to 2014.

		2010	**2011**	**2012**	**2013**	**2014**
***Bacillus* toxins**	SE outbreaks, %	3.7	6.7	5.0	6.4	5.9
	Milk or dairy products as food vehicle in SE outbreaks caused by *Bacillus*, %	3.8 (milk)	2.1 (milk)	5.3 (cheese)	1.9 (milk)	none
	Hospitalisations associated with SE outbreaks, %	0.2	0.7	1.1	4.6	4.4
	Deaths associated with SE outbreaks, %	None	None	None	None	None
***Clostridium* toxins**	SE outbreaks, %	3.3	5.3	7.1	7.2	7.1
	Milk or dairy products as food vehicle in SE outbreaks caused by *Clostridium*, %	None	None	None	None	None
	Hospitalisations associated with SE outbreaks, %	1.6	0.8	1.1	1.5	2.4
	Deaths associated with SE outbreaks, %	6.7	1.5	8.3	11.1	20.0
	Causative agent of deaths	*Clostridium botulinum*	*Clostridium botulinum*	*Clostridium perfringens*	*Clostridium perfringens*	*Clostridium perfringens*

SE: strong evidence.

Contamination of milk and dairy products with bacterial spores includes two routes of contamination: the raw milk route and the post-pasteurization route (Heyndrickx, 2011). Raw milk in the farm tank is contaminated via the exterior of the cattle's teats and through improperly cleaned milking equipment contaminated with soil, faeces and bedding material with spores. Soil can be considered as the initial and a direct contamination source for spore-forming bacteria into foods, since it is a major habitat of these microorganisms. Spores present in faeces probably originate from feed (indirectly, after ingestion and subsequent excretion of spores), from soil (directly during grazing on the pasture when soil is also taken up by the ruminant), and when the bedding material is contaminated with faeces (Heyndrickx, 2011). On the other hand, the post-pasteurization contamination of milk with spores is related to the dairy plant equipment (foam in processing equipment, milk foulant, worn gaskets, pitted metal) or to the biofilms of spore-forming bacteria growing in processing equipment that can subsequently be dispersed by release into the milk production system (Heyndrickx, 2011). To limit the contamination with spore formers, it is possible to act at a preventive level at the dairy farm by reducing the bacterial load of raw milk (i.e. controlling feed quality, changing the feed, improving hygiene at milking and storage) and at the dairy plant (cleaning of milking equipment, transport tankers and manufacturing plant, avoiding product recycling or holding, controlling temperature and avoiding growth temperatures, and limiting manufacturing run lengths) or at a palliative level at the industry by means of technological procedures (physically removing the spores, preventing spore germination) (Garde et al., 2013).

In this chapter, we review early and recent research on structure and composition of bacterial spores, spore resistance and the life cycle of spore-forming bacteria, and focus on the problems caused by those spore-forming bacteria relevant for the dairy industry and the available control strategies.

2.2 THE BACTERIAL SPORE

Bacterial endospores (hereafter referred as spores in this chapter) are dormant cellular structures in the life cycle of spore-forming bacteria and are arguably the hardiest life forms on the Earth. The first studies about the bacterial spores were carried out by Cohn and independently by Koch in 1876, although the presence of refractile bodies into bacterial vegetative cells had earlier been reported by Ehrenberg in 1838 (Gould, 2006). Cohn established the basis for the much later discovery of specific germinants, and Koch described for the first time the complete sporulation–germination–multiplication–resporulation life cycle of spore-forming bacteria.

Sporulation is a bacteria strategy for surviving environmental challenges such as nutrient starvation, whereby bacteria cease vegetative growth and form a spore inside the mother cell cytoplasm. Bacterial spores are metabolically inactive and extremely resistant to environmental stress conditions, including heat, salinity, acidity, radiation, oxygen and/or water depletion, and low availability of nutrients (McKenney et al., 2013), and they can survive in their dormant state for many years (Gould, 2006). Spore resistance and longevity are related to spore structure and composition.

2.2.1 Structure and chemical composition of bacterial spores

The structure of a bacterial spore is very different from that of a vegetative cell, including several layers and many constituents that are unique to spores (Figure 2.1). These differences permit their survival in environmental stresses that would be lethal to vegetative cells. The structure of bacterial spores comprises, from outermost to innermost, an exosporium, spore coat, outer membrane, cortex, inner membrane and spore core (Leggett et al., 2012).

2.2.1.1 Exosporium

It has been found in spores of members belonging to classes Bacilli and Clostridia, but it is not present in all species, that is, in *Bacillus subtilis* spores (Stewart, 2015). Exosporium from *Bacillus anthracis* and *Bacillus cereus* are the best studied, consisting of a protein basal layer surrounded by an external nap of glycoprotein hairlike projections. Functions attributed to the exosporium include roles in germination, outgrowth, and attachment (Brunt et al., 2015; Stewart, 2015). The exosporium layer contribute to the overall hydrophobicity of the spore and, therefore, to its binding properties, which are important in the formation of biofilms in food-processing surfaces. It has been suggested that the exosporium is involved in the pathogenicity of some spores (Stewart, 2015).

2.2.1.2 Spore coat

Underlying the exosporium is the spore coat, which comprises a series of thin, concentric layers, the numbers of which vary among species and which is composed mainly of proteins (McKenney et al., 2013). Three layers have been described by electron microscopy for *B. subtilis* spore coat (inner coat, outer coat and the crust), and at least 70 different proteins build this multilayered structure. The crust is the outermost layer of the coat,

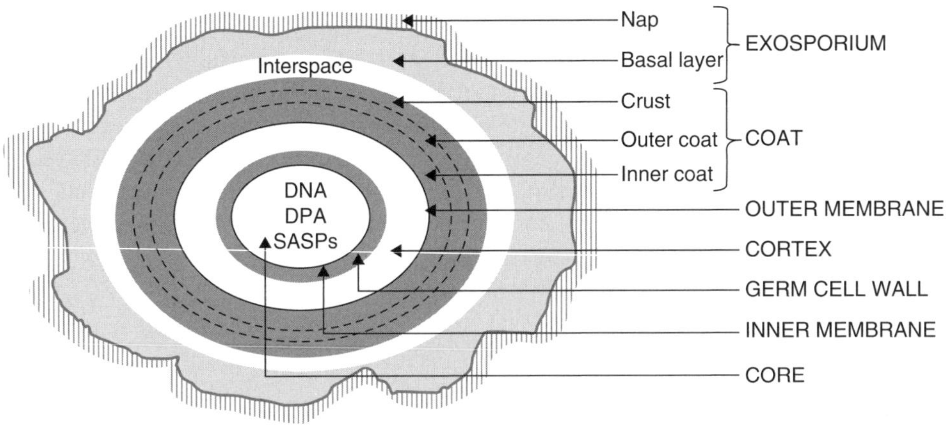

Figure 2.1 Scheme of bacterial spore structure (layers are not drawn to scale). DNA: deoxyribonucleic acid; DPA: dipicolinic acid; SASPs: small acid-soluble spore proteins.

and it has been suggested that it would be functionally equivalent to the exosporium in bacterial spores lacking this structure (Stewart, 2015). The major known function of the coat is spore protection from environmental stress (McKenney et al., 2013), acting as a barrier and passively excluding degradative enzymes such as lysozyme and toxic molecules. In addition, it has been suggested that some enzymes associated with the spore coat may serve to detoxify potentially damaging chemicals. In some species, the coats may contain pigments that absorb strongly in the UV region and might be involved in spore UV resistance (McKenney et al., 2013). Another coat function is the regulation of germination, acting as a molecular sieve that excludes large molecules while allowing the passage of small-molecule germinants, sequestering enzymes required for degradation of the cortex peptidoglycan or modifying or degrading germinants by enzymes present in the coat (Moir and Cooper, 2015).

2.2.1.3 Outer spore membrane

The outer spore membrane is under the spore coat and is essential for spore formation, but the importance of this membrane in the mature spore remains unclear (Leggett et al., 2012).

2.2.1.4 Cortex and germ cell wall

The cortex and germ cell wall is the next layer and is composed of spore-specific peptidoglycan (PG), characterised by the complete absence of teichoic acids from the *N*-acetylmuramic acid, the muramic-δ-lactam moiety and low peptide cross-linking (Leggett et al., 2012). These cortex-specific modifications appear crucial in attaining spore dormancy and/or resistance properties. There is a second layer of PG under the cortex, termed the *germ cell wall*, which becomes the cell wall as the spore undergoes germination and outgrowth. The PG in this layer has a structure similar to that of vegetative cell PG.

2.2.1.5 Inner spore membrane

Under the germ cell wall is the inner spore membrane. Despite a not unusual fatty acid and phospholipid composition, this membrane is relatively impermeable to small molecules (Leggett et al., 2012). It has been suggested that this property may result from the largely immobile lipids located in the inner spore membrane, perhaps by the compression of the inner membrane by the spore cortex (Cowan et al., 2004). The low permeability of inner spore membrane seems to be important in spore resistance to some biocidal chemicals by restricting its access to targets in the spore core (Setlow, 2014b).

2.2.1.6 The core spore

At the centre of the spore is the core, which contains DNA, RNA, ribosomes, most of its enzymes and small molecules (Setlow, 2014a). The core has a number of characteristic that play many roles in spore resistance: (i) a low water content (25–55% of wet weight), a factor important in both spores' enzymatic dormancy and their resistance to heat and some chemicals; (ii) a high level of pyridine-2,6-dicarboxylic acid (dipicolinic acid, or DPA) in a 1:1 complex with various divalent cations, generally Ca^{2+}, and important in resistance to some DNA-damaging agents and in maintaining spore dormancy; and (iii) a high levels of small acid-soluble spore proteins (SASPs) that saturate spore DNA and protect it from damage due to many genotoxic chemicals, desiccation, dry and wet heat, and UV and γ-radiation.

2.2.2 Spore resistance

As mentioned before, spore resistance is due to a variety of factors related to spore structure such as properties of spore coat, spore inner membrane impermeability, low core hydration, and high levels of DPA and SASPs, both components implicated in mechanisms for protecting and repairing spore DNA. Spores from different strains, species and genera can exhibit quite large differences in their resistance. The spore resistance properties were recently reviewed by Setlow (2014b), thus only the main aspects will be mentioned in this chapter.

The saturation of spore DNA with SASPs is the main mechanism that protects spores from dry heat, although DNA repair by spore enzymes during outgrowth and the mineralization of the core with DPA and divalent cations also play a role (Setlow, 2014b). However, the low water content in the spore core is the major factor in wet heat resistance since the low water content results in reduced molecular mobility of core proteins and thus elevated protein resistance. Other factors such as DNA saturation by SASPs, DPA content and sporulation conditions (including temperature, divalent metal ion content and optimum sporulation temperature) are also implicated in wet heat resistance. DNA saturation by SASPs and DPA content are also implicated in desiccation spore resistance. Spore mechanisms involved in radiation resistance are DNA saturation by SASPs, DNA repair during spore outgrowth, low core water content and carotenoid production in the spore coat. Detoxifying enzymes in the spore coat and/or exosporium, nonspecific detoxification by spore coat components, low permeability of the inner spore membrane, protection of DNA by SASP binding and repair of chemically DNA damage during spore outgrowth are the factors implicated in chemical spore resistance.

Bacterial spores are also more resistant than vegetative cells to high pressure, abrasion and freeze–thawing, but resistance mechanisms against these treatments remain unknown. In addition, spores are resistant to predation by bacteriovores (protozoa and nematodes), probably because the spore coat limits the access of bacteriovores' PG hydrolases to the cortex (McKenney et al., 2013).

On the other hand, and despite their extreme resistance, bacterial spores can be killed by several mechanisms, depending on the treatment applied: DNA damage (i.e. dry heat, radiation, formaldehyde or nitrous acid), inner membrane damage (oxidizing agents), core enzymes damage (wet heat, small oxidizing agents as H_2O_2), damage of components of the spore germination apparatus (NaOH), breaching all spore permeability barriers (strong acids) and also unknown mechanisms to date (high pressure, gas dynamic heating, plasma, supercritical fluids) (Setlow, 2014b).

2.2.3 Life cycle of spore-forming bacteria

Bacterial spores are formed during sporulation, and this process is very similar for Bacilli and Clostridia (Legget et al., 2012). It generally begins with an unequal vegetative cell division generating two compartments, the larger mother cell and the smaller prespore (Figure 2.2). In the next stage, the prespore is engulfed by the mother cell in a process resembling phagocytosis to form the forespore, which is a double membrane-bound cell within the mother cell. The spore then matures through a series of biochemical and morphological changes, and eventually the mother cell lyses, releasing the mature spore into the environment. Those changes include the synthesis of spore cortex (composed of peptidoglycan), which is assembled between the inner and outer forespore membranes, the formation of proteinaceous spore coat, the synthesis of DPA, which accumulates in the spore core and is accompanied by a reduction in the water content, the synthesis of SASPs and of exosporium (if any). Precursors of spore structures and components are synthesized by mother cell and transferred to forespore.

The mature spore structure protects the dormant microorganism from environmental stress until the conditions become favourable for vegetative cell growth. Spores are continually sensing their environment for the presence of nutrients using a group of receptors located in the spore inner membrane (Setlow 2014a; Moir and Cooper, 2015). The transition from dormant spore to vegetative cell can be divided into four main stages: activation, stages I and II of germination, and outgrowth (Moir and Cooper, 2015).

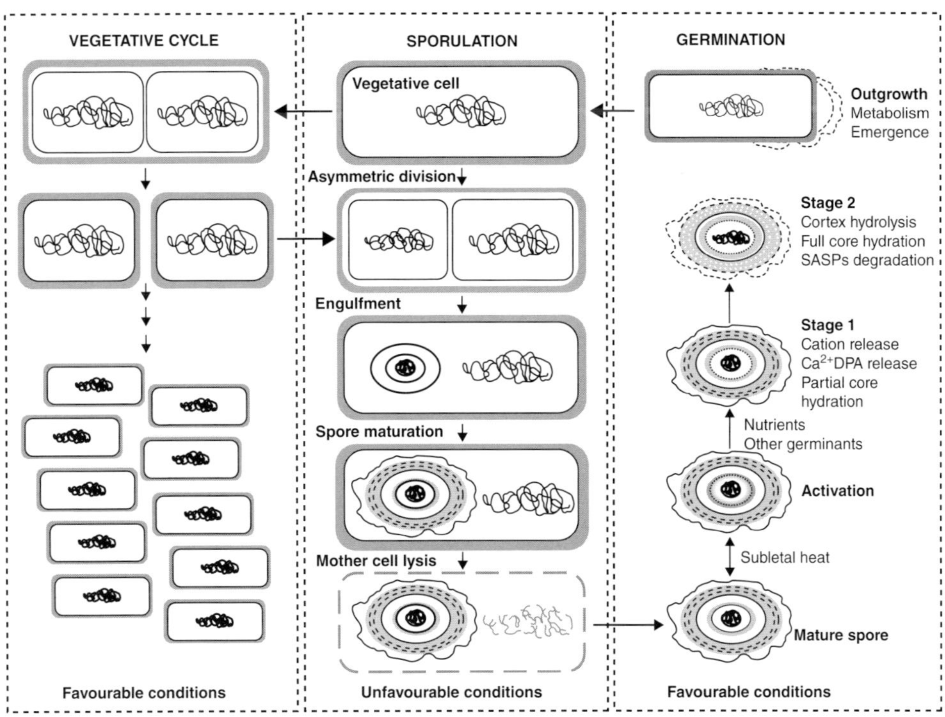

Figure 2.2 Life cycle of spore-forming bacteria. DPA: dipicolinic acid; SASPs: small acid-soluble spore proteins.

Activation is a reversible process that can be triggered by exposure to sublethal heating that makes germinant receptors more accessible or receptive to nutrient germinants, probably due to reversible conformational changes in receptors (Setlow, 2014a). In this germination stage, activated spores retain most properties of the dormant spores. In stage I of germination, germinants interact with membrane-located receptors, which transduce the stimulus, activating the membrane-associated changes and hydrolytic reactions that occur in germination (Moir and Cooper, 2015). Germinants include "nutrient" germinants such as amino acids, sugars, purine nucleosides and inorganic salts, and "non-nutrient" germinants such as Ca^{2+}-DPA, dodecylamine, lysozyme, high pressure or peptidoglycan fragments (Moir and Cooper, 2015). In this stage, release of monovalent cations (including H^+, Na^+, and K^+) occurs, followed by the release of all Ca^{2+}-DPA, which is replaced by water. These events are associated with a major change in inner membrane permeability and perhaps inner membrane structure (Setlow, 2014a). After stage I of germination, the extreme heat resistance of the spore has been lost, but protein mobility in the core and lipid mobility in the inner membrane have not yet increased, and there is no detectable metabolism (Moir and Cooper, 2015). In stage II of germination, the cortex is hydrolysed by cortex lytic enzymes located in spore coat, and inner and outer membranes, leading to full rehydration and expansion of the core, the inner membrane and the germ cell wall, and the germinated spore regains more vegetative cellular properties. SASPs are also degraded, releasing the DNA for transcription and providing a source of amino acids for biosynthesis during outgrowth. DNA repair proteins present in the spore are now active to repair damage incurred during spore dormancy, and ATP is generated from 3-phosphoglycerate. Spore coat breakdown is also initiated during this germination stage (Moir and Cooper, 2015). Finally, metabolism and macromolecules synthesis in the cell are resumed in the outgrowth. This latter stage also includes the swelling of the spore, emergence (where the outer spore layers, coat and exosporium, are shed), replication of DNA and the first cell division that originate the new vegetative cell which re-enters the vegetative cycle.

In food products, spore germination is a critical step regarding spoilage and foodborne disease caused by spore formers, generally triggered by "nutrient germinants" that are sensed by specific germinant receptors. Although spore germination has been extensively studied, most germination studies have been performed with mutants of *Bacillus* and *Clostridium* spp. in laboratory media because it is difficult to establish the behaviour of wild spore-forming bacteria in actual food products.

2.3 SPORE-FORMING BACTERIA IMPORTANT FOR THE DAIRY INDUSTRY

2.3.1 Class *Bacilli*

The class *Bacilli* includes two orders, Bacillales and Lactobacillales, that include spore-forming and non-spore-forming representatives. Among the spore formers, *Bacillus, Paenibacillus, Geobacillus* and *Anoxybacillus* are the most relevant for the dairy industry due to their spoilage or pathogenic potential (Scott et al., 2007; Ivy et al., 2012; Gopal et al., 2015). The harmful effects concerning food safety and product quality caused by these spore formers are threefold: (i) production of toxins, (ii) production of spoilage enzymes and (iii) interference with cheese making (De Jonghe at al., 2010).

2.3.1.1 *Bacillus* genus

Bacillus is a diverse bacterial genus belonging to the family Bacillaceae, endospore-forming, aerobic or facultative anaerobic, rod-shaped, motile and gram-positive bacterium (on occasion displays a gram-negative or variable reaction). Their taxonomy is quite complex and has been subject to considerable revision in recent years. At the time of writing this chapter, this genus included 318 species with validly published names (LPSN: www.bacterio.net).

Bacillus spp. are characterised by different nutritional requirements, the ability to grow in a wide range of temperatures (mesophilic and thermophilic psychrotrophic strains) and pH values. They are the most common isolated gram-positive bacteria among the microbial species that can survive the usual heat treatment of milk, and they are also able to grow in biofilms and contaminate the product as it flows through the processing line. Although the initial concentration of thermophiles and spores in raw milk is low (<10 cfu/ml), *Bacillus* has been determined in several steps of the dairy production chain, which selects for spore formers through several heating processes, and for psychrophiles during cold storage (Heyndrickx and Scheldeman, 2002).

2.3.1.1.1 Bacillus cereus

Among *Bacillus* spp., *B. cereus* is one of the most important spoilage microorganisms in the dairy environment, and it is also a safety concern as it is associated with incidences of food poisoning (Kumari and Sarkar, 2016). The *B. cereus* group, also known as *B. cereus sensu lato*, is a species complex that includes 11 closely related species according to current taxonomy: *B. cereus* (*sensu stricto*), *Bacillus thuringiensis*, *B. anthracis*, *Bacillus mycoides*, *Bacillus pseudomycoides*, *Bacillus weihenstephanensis, Bacillus cytotoxicus* (Kumari and Sarkar, 2016), *Bacillus toyonensis* (Jiménez et al., 2013), and the recently proposed species "*Bacillus gaemokensis*" (Jung et al., 2010), "*Bacillus manliponensis*" (Jung et al., 2011) and "*Bacillus bingmayongensis*" (Liu et al., 2014). Growth and survival characteristics of *B. cereus* vary widely between strains and depend upon a complex series of interacting factors such as temperature, pH, water activity (NaCl concentration), nutrients and presence of competitive microbiota (Spanu et al., 2016). A very broad thermal range for growth temperature in the *B. cereus* group strains has been recorded (ranging from 4 °C to 50 °C), and strains seem to adapt to very different habitats, including refrigerated to dehydrated foods (Guinebretière and Nguyen-The, 2003).

In the dairy industry, *B. cereus* is frequently isolated from raw, pasteurised and ultra-high temperature (UHT) milk, and dairy products such as fermented milk, dried milk product, milk powder, powdered infant formulae, ice cream, cheeses and processed cheeses (EFSA, 2005a; Svensson et al., 2006; Samaržija et al., 2012; Kumari and Sarkar, 2016; Spanu et al., 2016). The dairy heating processes may select this noncompetitive bacteria, whose vegetative cells grow well in the absence of competing microorganisms (Granum and Lund, 1997). Psychrotrophic strains would grow below 10 °C, at temperatures as low as 4 °C, and have optimal growth temperatures of 30–37 °C and maximum growth temperatures of 37–42 °C (EFSA, 2005a). The development of psychrotolerant strains has led to increasing surveillance of the *B. cereus* group since they are a recurrent problem in perishable dairy products during refrigerated storage and distribution, affecting the product shelf life (Granum and Lund, 1997). On the other hand, mesophilic strains of *B. cereus* can grow at between 10 and 42 °C, some strains being able to grow at 50–55 °C, with an optimal growth temperature of 30–37 °C (EFSA, 2005a), allowing these strains to adhere to and germinate in hot dairy equipment, what may limit the keeping quality of pasteurised milk (Svensson et al., 2006). The hydrophobic properties of *B. cereus* spores and presence of appendages, and their resistance towards heat, desiccation and disinfectants, allow them to attach to processing equipment and survive cleaning procedures of dairy plants, resulting in biofilm formation, which can be an important reservoir for recurrent contamination of dairy products (Kumari and Sarkar, 2016). They can also survive the drastic processing of dehydrated foods like dry milk powder and subsequently contaminate diverse foodstuffs via dehydrated ingredients (Guinebretière and Nguyen-The, 2003).

2.3.1.1.2 Other *Bacillus* species

Some species of the *B. subtilis* group are usual contaminants of dairy products. Along with *B. subtilis sensu stricto* other closely related species with high genetic and/or biochemical similarities, and thus not easily distinguishable, are now clustered in the *B. subtilis* group: *Bacillus amyloliquefaciens*, *Bacillus atrophaeus*, *Bacillus axarquiensis*, *Bacillus licheniformis*, *Bacillus malacitensis*, *Bacillus mojavensis*, *Bacillus pumilus*, *Bacillus sonorensis*, *Bacillus tequilensis*, *Bacillus vallismortis* and *Bacillus velezensis* (Jeyaram et al., 2011). These species are in general mesophilic with regard to temperature and neutrophilic with respect to pH for growth, while often being tolerant to higher pH levels. *B. licheniformis*, *B. subtilis* and *B. pumilus*, the predominant mesophilic species in milk, are also facultative thermophiles, depending on the strain (Burgess et al., 2010).

B. licheniformis is widespread in the environment and across the dairy farms, being the predominant mesophilic *Bacillus* species in raw milk, despite some regional, seasonal and methodological differences (Phillips and Griffiths, 1986). It is facultatively anaerobic, grows at temperatures up to 55 °C, and can reduce nitrate to nitrite (Burgess et al., 2010). Thermotolerant isolates of *B. licheniformis* are also some of the most common thermotolerant species in milk, predominating in milk powders (Yuan et al., 2012; Sadiq et al., 2016). *B. licheniformis* is known to cause spoilage of milk and dairy products, raise specification compliance problems and have adverse effects on milk and functional properties (Gopal et al., 2015). *B. subtilis* grows at temperatures from 5 to 20 °C up to 45–55 °C, is strictly aerobic, and can reduce nitrate (Burgess et al., 2010). It has been associated with spoilage of raw and pasteurised milk and UHT and canned milk products (Heyndrickx and Scheldeman, 2002). *B. pumilus* is strictly aerobic, and grows at temperatures from 5 to 15 °C up to 50–55 °C (Burgess et al., 2010).

Bacillus sporothermodurans and *Bacillus coagulans* are other examples of facultative thermophiles causing undesired growth in consumer milk products. *B. sporothermodurans* is encountered globally in milk products, producing mesophilic vegetative cells but highly heat-resistant spores that may survive UHT treatment or industrial sterilization. It is strictly aerobic and grows at temperatures from 20 °C up to 45–55 °C

(Burgess et al., 2010). It is detected at the farm in low levels, but propagates in the plant, and has been reported to contaminate different kinds of UHT products such as evaporated or reconstituted milk, cream, chocolate milk and milk powder (Scheldeman et al., 2006). *B. coagulans* is facultatively anaerobic and grows at temperatures from 5 to 15 °C up to 57–61 °C. Some strains are able to reduce nitrate (Burgess et al., 2010). It has been associated with spoilage of evaporated milk (Kalogridou-Vassiliadou, 1992).

2.3.1.1.3 Importance of *Bacillus* spp. in the dairy industry

Product quality is the main issue posed by *Bacillus* spp., and members of the *B. cereus* and *B. subtilis* groups are the most important spoilage microorganisms in the dairy environment (Lücking et al., 2013). *Bacillus* spp. are able to produce serious off-flavours in dairy products such as milk, cream and cheese (McClure, 2006), caused by protease and/or lipase activities. The majority of *Bacillus* species produce various thermostable extracellular and intracellular enzymes (e.g. proteinases, lipases, phospholipases) that are not sufficiently inactivated by pasteurisation or even UHT treatment and that may therefore cause spoilage of the final products during storage, decreasing their organoleptic quality and reducing their shelf life. Maximum production of *Bacillus* extracellular enzymes normally occurred in the late exponential and early stationary phases of growth, before sporulation (Priest, 1977). Matta and Punj (1999) reported that 48 out of 100 raw milk samples contained diverse lipolytic psychrotrophic *Bacillus* species, with *B. cereus* being the predominant species. De Jonghe and colleagues (2010) showed strains of *B. subtilis*, the *B. cereus* group and *B. amyloliquefaciens* to be strongly proteolytic, along with *B. licheniformis* and *B. pumilus* to a lesser extent, and demonstrated lipolytic activity in strains of *B. subtilis*, *B. pumilus* and *B. amyloliquefaciens* isolated from raw milk. In 40–84% of cases, *Bacillus* spp. isolated from milk expressed both proteolytic and lipolytic activities, and in approximately 80% of cases also a phospholipolytic activity, and certain species can produce more than one type of proteinase simultaneously (Samaržija et al., 2012).

The presence of proteases can lead to uncontrolled or unwanted proteolysis, adversely affecting food quality and flavour through the formation of bitter peptides from milk proteins (McClure, 2006). The action of these enzymes can also lead to clotting and gelation of milk (Furtado, 2005; De Jonghe et al., 2010) at a relatively high pH. In pasteurised milk, sweet curdling (coagulation of the milk without acidification) is a rather common defect, mainly caused by *B. cereus* proteolytic activity (De Jonghe et al., 2010), although *B. subtilis* strains have been also linked to this defect. Proteolysis by *B. subtilis* in canned, evaporated milk results in thickening and a bitter taste followed by digestion to a brown liquid (Varnam and Sutherland, 1994). In cheese production, the presence of *Bacillus* proteases in milk may cause lower yield, shorter coagulation time due to higher concentration of amino acids that stimulate the growth of starter culture (Kumari and Sarkar, 2016).

Bacterial lipases also contribute to unpleasant flavour, such as rancid, butyric, buttery, bitter, unclean, soapy and astringency in milk and dairy products resulting from extensive free fatty acid release and subsequent reactions (Furtado, 2005). In cream, *Bacillus* species are important in spoilage at temperatures above 10 °C (Varnam and Sutherland, 1994).

Another common example of spoilage of milk is the bitty cream phenomenon caused by the lecithinase activity of *B. cereus* that hydrolyses the phospholipids associated with the milk fat globule membrane to produce small proteinaceous fat particles that float on the surface of hot drinks and adhere to surfaces of crockery and glasses (Adams and Moss, 2008). Bitty cream is associated mostly with milk that has been subject to temperature abuse, although psychrotrophic *Bacillus* species are becoming increasingly associated with the spoilage of refrigerated milk. Yogurt can be defective (lumpy structure, separation of whey, off-flavours) in exceptional cases where *B. cereus* develops to 10^6 cfu/ml before the pH is reduced to 6.1 by the starter bacteria (Heyndrickx and Scheldeman, 2002). *B. cereus* or *B. subtilis* may also induce bitterness in yogurt (Mistry, 2001).

Spoilage of UHT and sterilized milk occurs only occasionally, usually as a result of recontamination after the heat treatment, during filling, and is caused mostly by proteolytic activity of some *Bacillus* species (Heyndrickx and Scheldeman, 2002). Proteolytic and lipolytic activity of *Bacillus* spp. may increase the concentration of free tyrosine (indicator of proteinase activity and general organoleptic quality) and free fatty acids after storage of UHT milk (Janštová et al., 2006). However, as mentioned before, it has been reported the massive occurrence of nonsterility in all kind of UHT products by highly heat-resistant spores of *Bacillus sporothermodurans* present in the initial milk (Scheldeman et al. 2006). Nevertheless, real spoilage defects in consumer milk are only rarely noticed as a slight pink colour change, off-flavours and coagulation, since *B. sporothermodurans* grows poorly in milk (only up to 10^5 cfu/ml) and does not affect the pH of the product (Heyndrickx and Scheldeman, 2002). Even when there are no negative effects on the products, the colony count at 30 °C of unopened packages of heat-treated milk-based products after 15 days of incubation at 30 °C must be below 10 cfu per 0.1 ml to meet the legal requirements established by the European Union (EU, Council Directive 92/46/EEC).

In the powdered dairy ingredients industry, the presence of high numbers ($>10^4$ cfu/g) of thermophiles in finished dairy products is an indicator of poor hygiene during processing (Burgess et al., 2010). Although *Bacillus* growth in these products is minimal because of their low water content, once introduced to a more favourable environment during reconstitution, the spores can germinate, grow, and produce enzymes and acid and consequential off-flavour development and spoilage in the products containing dairy powders (Scott et al., 2007; Lücking et al., 2013). Thus, the US Dairy Export Council has set strict tolerances for mesophilic (<1000 cfu/g) and thermophilic spores (<500 cfu/g) in dairy powders.

The fermentative growth of *Bacillus* spp. may also be involved in the spoilage of dairy products, causing flat sour spoilage of evaporated milk due to acid production during fermentation of carbohydrates without formation of gas (Kalogridou-Vassiliadou, 1992). *Bacillus coagulans* can cause firm coagulation in evaporated milk due to the production of high concentrations of lactic acid, and a slight cheesy odour and flavour, but it is normally only a problem when storage temperatures are high (Varnam and Sutherland, 1994; McClure, 2006). *Bacillus megaterium* causes an acid formation accompanied by gas and a cheesy odour in canned, evaporated milk (Varnam and Sutherland, 1994). Furthermore, some strains of *B. licheniformis* are capable of producing a slimy extracellular substance that can affect the quality of pasteurised milk and cream, and *B. subtilis* has been associated with ropiness in raw and pasteurised milk as well as the spoilage of UHT and canned milk products, as reviewed by Burgess et al. (2010).

A further concern regarding *Bacillus* spp. spoilage is the interference with cheese production. Nitrate is often added as preservative to the milk to prevent growth and germination of *Clostridium* species causing the LBD of cheese, and *Bacillus* species able to reduce nitrate to nitrite may limit its effectiveness (Ternström et al., 1993).

Formation of bacterial biofilms by *Bacillus* and related genera within dairy processing plants is also a matter of great concern to processors, as bacteria within biofilms are more difficult to eliminate than planktonic cells and can act as a source of recurrent contamination to plant, product and personnel; biofilm renders its inhabitants resistant to antimicrobial agents and cleaning (Kumari and Sarkar, 2016).

Although the overwhelming majority of *Bacillus* species are nonpathogenic, the role of *B. cereus* as a potential food-poisoning agent is acknowledged, being the predominant pathogenic *Bacillus* species found in raw milk and at all stages of dairy processing (Kalogridou-Vassiliadou 1992; Scheldeman et al. 2006). Within the *B. cereus* group, *B. cereus sensu stricto* is known to cause two kinds of foodborne diseases if conditions are favourable: an emetic (vomiting) intoxication due to the ingestion of a toxin preformed in the food, and a diarrhoeal infection due to the ingestion of bacterial cells/spores that produce enterotoxins in the small intestine (EFSA, 2005a). *B. cereus* food poisoning is underreported as both types of illness are relatively mild and usually last for less than 24 h (Granum and Lund, 1997). A dose of 10^5–10^8 cells or spores per gram is generally considered necessary to cause illness (Granum and Lund, 1997). The emetic syndrome is caused by a heat-stable, low molecular weight substance, a ring formed, modified peptide, cereulide, similar in structure and function to the potassium ionophore valinomycin (Agata et al., 1994). The vast majority of food-poisoning cases have been attributed to rice and rice dishes. Diarrhoeal food poisoning is caused by different heat-labile protein toxins like haemolysin BL, nonhaemolytic enterotoxin and cytotoxin K. A wide range of proteinaceous foods has been implicated in diarrhoeal food poisoning (Kumari and Sarkar, 2016). The occurrence of emetic and diarrhoeal toxin-producing strains of *B. cereus* in the dairy production chain has been reported (Kumari and Sarkar, 2016). However, dairy products have seldom been associated with human illness despite the frequent contamination with *B. cereus* (EFSA, 2005a), and no food safety criteria for *B. cereus* are applicable to foodstuffs placed on the market during their shelf life (EC Regulation No. 2073/2005), with the exception of dried infant formulae intended for infants below 6 months of age (EC Regulation No. 1441/2007). In these products, five samples have to be analysed and presumptive *B. cereus* in four samples must be below 50 cfu/g, while the remaining sample can be between 50 and 500 cfu/g. Although *Bacillus* species other than *B. cereus* have generally been considered to be of little significance in food-poisoning incidents, there is evidence that some strains of other species, including *B. licheniformis*, may cause foodborne illness (McClure, 2006). As summarized by De Jonghe et al. (2010), cellular assays confirm both production and functionality of heat-labile and heat-stable toxins by strains of several *Bacillus* spp. such as *B. subtilis*, *B. licheniformis* and *B. pumilus*. However, the production of toxins has been reported only at mesophilic conditions (Burgess et al., 2010).

Gopal et al. (2015) suggested a possible association of *Bacillus* with another cheese safety concern, the production of biogenic amines, since amine-producing capabilities have been reported for a few *Bacillus* strains.

2.3.1.2 *Geobacillus* and *Anoxybacillus* genera

Members of the genus *Geobacillus,* belonging to the family Bacillaceae, are rod-shaped and facultatively anaerobic. Most are thermophilic, motile, produce stain gram-positive or negative and have optimum growth temperatures above 50 °C (McClure, 2006; Burgess et al., 2010). At the time of writing, this genus included 20 species with validly published names (LPSN: www.bacterio.net). The most important species is *Geobacillus stearothermophilus*, formerly known as *Bacillus stearothermophilus*, with some of the most heat-resistant bacterial spores known.

The genus *Anoxybacillus,* also belonging to the family Bacillaceae, was separated from the genus *Bacillus* by Pikuta et al. (2000), and species of this genus are defined as rod-shaped (often with rounded ends), gram-positive, aerotolerant or facultatively anaerobic, thermophilic and alkaliphilic or alkalitolerant bacteria (McClure, 2006; Burgess et al., 2010). At the time of writing, this genus included 21 species with validly published names (LPSN: www.bacterio.net). The organism of relevance for the food industry is *Anoxybacillus flavithermus*, formerly known as *Bacillus flavothermus* (Pikuta et al., 2000), which is described as a facultatively anaerobic thermophile, with terminal endospores and motility (Heinen et al., 1982).

Geobacillus spp. and *A. flavithermus* are important contaminants in the dairy industry, most of them being obligate thermophiles, which grow at elevated temperatures. Spores of these thermophilic bacilli are usually found at very low levels in raw milk, but since they survive the harsh conditions during dairy manufacturing, they are common in finished products like milk powders and other dried dairy products. *A. flavithermus* has an optimum growth temperature at 60 and 65 °C under aerobic and anaerobic conditions, respectively. *G. stearothermophilus* resists and grows at temperatures up to 75 °C, and its spores have the potential to survive UHT treatments (Khanal et al., 2014). Both exhibit a fast growth rate and tend to form biofilm on the stainless steel surfaces of processing equipment (i.e. heat exchangers and evaporators), from which cells and spores can slough off and contaminate the product flowing past, resulting in numbers of up to 10^6 cfu/g of bacteria and spores released in the final products (Scott et al., 2007). Germination of spores and survival, and even growth, of *G. stearothermophilus* was reported to occur in concentrated milk between the pasteuriser and the drying tower during the manufacture of skim and whole milk powders (Murphy et al., 1999).

A. flavithermus and *G. stearothermophilus*, along with *B. licheniformis*, have been reported as the prevalent microorganisms in milk powders from different countries (Yuan et al., 2012; Sadiq et al., 2016). These bacilli are generally nonpathogenic and, although their growth may result in milk product defects caused by the production of acids or enzymes after the powder is recombined and if the conditions are suitable, their real potential to spoil dairy products is thought to be low, as dairy products are usually stored at temperatures below 37 °C, at which obligate thermophiles will not grow (Burgess et al., 2010). *G. stearothermophilus* produces acid but no gas, hence causing the flat sour defect in canned milk products (Kalogridou-Vassiliadou, 1992).

2.3.1.3 *Paenibacillus* genus

Members of the genus *Paenibacillus* belong to the family Paenibacillaceae. In 1993, Ash et al. proposed that members of 'group 3' within the genus *Bacillus* should be transferred to the genus *Paenibacillus*, for which they proposed *Paenibacillus polymyxa* as the type species. The genus *Paenibacillus* is a diverse group of organisms that appear to be predominantly psychrotolerant, with an ability to grow in milk and possibly other foods at temperatures as low as 6 °C (Ivy et al., 2012). At the time of writing, this genus included 182 species with validly published names (LPSN: www.bacterio.net) and contains rod-shaped, gram-positive or gram-negative, motile, facultatively anaerobic or aerobic bacteria and produce ellipsoidal spores (McClure, 2006). Members of this genus appear to occupy diverse ecological niches and have been isolated from farm and processing environments, raw milk, various pasteurised foodstuffs and even commercial UHT-treated milk, suggesting that at least some isolates can survive short-time heat treatments over 100 °C (Ivy et al., 2012). *Paenibacillus* spp. are generally present in low numbers in raw milk and early in pasteurised milk shelf life, yet they can reproduce to high numbers during refrigerated storage, becoming the predominant spoilage organism (Ivy et al., 2012) and limiting the further extension of fluid milk's shelf life. Thus, *P. polymyxa* was found as the main gram-positive spoilage organism of Swedish and Norwegian pasteurised milk stored at 5 or 7 °C (Ternström et al., 1993). Although *Paenibacillus* persistence on processing equipment has not been established, certain species have been shown to produce exopolysaccharide or to form biofilms, which, if present in appropriate locations, may lead to post-pasteurization contamination of fluid milk (Ivy et al., 2012).

P. polymyxa may cause a sour, yeasty and gassy defect in milk (Heyndrickx and Scheldeman, 2002). De Jonghe et al. (2010) have reported strains of *P. polymyxa* isolated from raw milk to be strongly proteolytic, producing lecithinase and able to reduce nitrate, and found a *P. polymyxa* strain capable of gas production during lactose fermentation. In Argentinian Cremoso, Mozzarella, and semihard cheeses, gas-producing *P. polymyxa* and *Paenibacillus macerans* have been associated with blowing spoilage (Quiberoni et al., 2008). Another possible problem for the cheese industry caused by *P. polymyxa* is the denitrification activity of this species, which may diminish the anticlostridial effect of added nitrate, alongside with *Bacillus* spp. (Heyndrickx and Scheldeman, 2002). Furthermore, bitter cream defect in pasteurised milk may also be caused by lecithinase of *P. polymyxa* (De Jonghe et al., 2010).

2.3.2 Class Clostridia

The class Clostridia includes four orders, Clostridiales, Halanaerobiales, Natranaerobiales and Thermoanaerobacterales, which in turn include 17 families. As mentioned before, spore-forming bacteria of the class Clostridia that are problematic for the dairy industry belong to the genus *Clostridium* (family Clostridiaceae) with 210 species with validly published names (LPSN: www.bacterio.net) at the time of writing this chapter. *Clostridium* spp. are gram-positive rods, motile or nonmotile, that form oval or spherical spores that usually deform the vegetative cell. Most of species are obligate anaerobes, although some species such as *C. perfringens* are facultative anaerobes and can tolerate the presence of oxygen (Charlebois et al., 2014). Several species of this genus can grow in food, causing food spoilage and/or food poisoning in humans. The majority of the numerous species of the genus *Clostridium* are nonpathogenic, but this genus also includes pathogens such as *Clostridium botulinum*, *Clostridium tetani*, *Clostridium difficile* and *Clostridium perfringens*. Although rare, *C. perfringens* and *C. botulinum* are the main *Clostridium* species involved in poisoning of dairy products. However, *C. tyrobutyricum* and related species are one of the most common groups of bacteria responsible for cheese spoilage.

2.3.2.1 *Clostridium botulinum*

C. botulinum produces a highly potent neurotoxin that causes botulism, which affects humans and animals and may lead to death due to paralysis of the respiratory muscles since botulinum toxin causes progressive flaccid paralysis by binding to nerve endings. The neurotoxin is completely odourless and tasteless; therefore, food containing preformed neurotoxin may not transmit any warning signs to the consumer. Nevertheless, the neurotoxin is destroyed by heating at 80 °C for 20 min or at 85 °C for 5 min (Siegel, 1993). *C. botulinum* can produce eight types of neurotoxins (A–H), based on the serological properties of these ones, and some *C. botulinum* strains producing two or even three neurotoxins types have been described (Dover et al., 2014). In addition, some strains of *Clostridium baratii* and *Clostridium butyricum* can also produce botulinum toxins (Hatheway, 1993). *C. botulinum* strains can be categorized into four phenotypically and genetically diverse groups (Carter and Peck, 2015). *C. botulinum* groups I and II are associated with botulism in humans, strains of group III with botulism in animals, and group IV strains have not been associated with illness. *C. botulinum* group I strains are mesophilic, proteolytic, produce toxin types A, B, F and H, and are generally of terrestrial origin. Group II strains are psychrotrophic and nonproteolytic, produce toxin types B, E, or F, and are frequently found in aquatic environments.

C. botulinum can cause two types of foodborne illness: classical botulism and infant botulism. Classical botulism is an intoxication caused by consumption of neurotoxin preformed in the food, while infant botulism is a toxicoinfection, which evolves by ingestion of spores followed by subsequent germination and toxin production in the gut of infants (Carter and Peck, 2015). Adults with heavy antibiotic treatment or intestinal lesions can also suffer this latter botulism disease, but it is very rare. Foodborne botulism in humans is a rare but severe disease. The reporting rate of foodborne outbreaks caused by *Clostridium* toxins (including toxins produced by *C. botulinum*, *C. perfringens* and unspecified *Clostridium* spp.) was 0.04 per 100,000 population in the EU in 2014, and outbreaks caused by *C. botulinum* represented 0.17% of all outbreaks (EFSA, 2015b). However, the estimated cost per case of botulism associated with commercial food products is very high. For example, a milk product that was incorrectly related to a *C. botulinum* contamination in New Zealand in 2013 cost hundreds of millions of dollars (Carter and Peck, 2015). On the other hand, botulism outbreaks in cattle, which have increased in the past decades and which are often large, affecting hundreds of animals, cause enormous economic losses to the dairy industry due to death and euthanasia of intoxicated animals, in addition to reduction in milk production (Lindström et al., 2010).

Although in recent years any foodborne outbreak of botulism has been associated with milk or dairy products in the EU (EFSA 2012, 2013, 2014, 2015a, 2015b), at least 20 human botulism outbreaks have previously been associated with the consumption of contaminated dairy products, with a mean fatality rate of 17.9%, and type A and B neurotoxins as the main implicated toxins (Lindström et al., 2010). As reviewed by these authors, the limited studies on the prevalence and contamination level of *C. botulinum* in milk and dairy products would indicate that low numbers of *C. botulinum* spores may occasionally be present in milk and other dairy foods. Dairy food vehicles for botulinum toxins or *C. botulinum* were cheeses or cheese products (70% of outbreaks), milk (15%), yogurts (10%) and infant formula milk (5%). All the outbreaks were intoxications, except for the case of infant botulism associated with infant formula milk. The majority of the outbreaks were caused by *C. botulinum* group I. It was suspected that an abused storage temperature of dairy products could be responsible for clostridial growth and neurotoxin production. Botulism outbreaks related to the consumption of commercial milk were the first cases of botulism linked to dairy products (Meyer and Eddie, 1965). In addition, worldwide botulism outbreaks due to homemade camel sour (Smith et al., 1979), Brie cheese (Sébald et al., 1974; 1980), Mascarpone cheese or tiramisu made with this cheese (Aureli et al., 1996), oil-preserved cheese (Pourshafie et al., 1998) and yogurt (O'Mahony et al., 1990, Akdeniz et al., 2007) have been reported. Cheese sauces were also associated with botulism outbreaks, but investigations revealed that were not the source of *C. botulinum* (Lindström et al., 2010).

Besides those products, a dried infant formula milk powder was initially associated to an infant botulism outbreak, but further investigations did not establish a clear and definite relation (Brett et al., 2005; Johnson et al. 2005). *C. botulinum* was neither detected in samples of infant formula from the home of infants with botulism nor from retail outlets (Barash et al., 2010). With respect to hazardous microorganisms associated with infant formula, the Food and Agriculture Organization/World Health Organization (FAO/WHO) categorized *C. botulinum* as "causality less plausible or not yet demonstrated" because, although having been identified in powdered infant formula, the microorganisms had not been implicated as causing illness in infants (FAO/WHO, 2004). Consequently, *C. botulinum* is not considered a hazard in the Codex international hygiene standard for infant formula (CAC, 2008). The International Commission on Microbiological Specifications for Foods (ICMSF) recently concluded that there is not sufficient scientific support to initiate any specifications for *C. botulinum* spores in dried dairy products and does not recommend routine testing for *C. botulinum* in dried dairy powders, but rather it recommends determining the number of sulfite-reducing clostridia in dried dairy ingredients as indicator of soil and/or faecal contamination (ICMSF, 2014). Although in the EU there are no regulations regarding the number of spore formers allowed in milk or dairy products, milk quality payment schemes exist whereby payments to farmers are based on how their milk scores on a number of criteria such as somatic cell count, butyric acid spores, total bacteria count, sulfite-reducing clostridia, the presence of antibiotic residues and milk cleanliness (Velthuis and van Asseldonk, 2010).

2.3.2.2 *Clostridium perfringens*

C. perfringens can be considered a major pathogen since it is the most widely distributed pathogenic microorganism in nature (its spores are highly prevalent in soil and the intestinal tract of humans and animals), has a fast growth (a generation time of less than 10 min under optimal conditions), produces over 15 toxins causing numerous human and animal diseases and has been ranked among the most common foodborne illnesses worldwide (Lindström et al., 2011). *C. perfringens* strains are classified into five toxigenic types (A–E) based on the repertoire of four major toxins (alpha, beta, epsilon and iota) that each individual strain produces. The type A and C strains have been associated with human diseases, being *C. perfringens* type A food poisoning the second most common foodborne illness reported in the United States, causing one million illnesses each year (Grass et al., 2013). In addition, *C. perfringens* can also produce a diversity of other toxins, such as enterotoxin, heat and pH labile, which is the cause of *C. perfringens* food poisoning. Illness occurs after ingestion of large numbers of enterotoxin-producing vegetative cells of *C. perfringens* that survive the acid conditions of the stomach and subsequently form spores in the intestine with concomitant enterotoxin production. Enterotoxin causes abdominal cramps and diarrhoea about 8–12 h after eating contaminated food with *C. perfringens* vegetative cells. The disease is mostly self-limiting, lasting for about 24 h. Deaths are rare (but possible) and may occur due to dehydration, mainly in elderly, very young or debilitated humans.

In the EU in 2014, outbreaks caused by *C. perfringens* represented 2.4% of all outbreaks and 20% of fatalities related to strong-evidence outbreaks (EFSA, 2015b), but as mentioned previously, any strong-evidence outbreak was associate with milk or dairy products for the period 2010–2014 (Table 2.1) (EFSA 2012, 2013, 2014, 2015a, 2015b). *C. perfringens* has been detected in a wide range of foods as a result of contamination by soil or with faecal matter such as meat, poultry, fish, vegetables, dairy products and dehydrated foods (EFSA, 2005b), being cooked meat and poultry the foods most commonly involved in *C. perfringens* outbreaks. Despite this, a few foodborne outbreaks associated with dairy products have been reported due to the consumption of contaminated milkshakes, cheese sauce and an unspecified dairy product (Bennett et al., 2013; McAuley et al. 2014; Doyle et al., 2015). *C. perfringens* has been detected in infant formula (Barash et al., 2010), but no illness has ever been attributed to infant formula consumption.

2.3.2.3 *Clostridium tyrobutyricum* and related species

The butyric acid bacteria, a group of mainly spoilage bacteria such as *C. tyrobutyricum, C. butyricum* and *Clostridium beijerinckii,* are responsible for the late blowing defect (LBD) of cheese as a consequence of their ability to metabolize lactate to acetate, butyrate, CO_2 and H_2 (Klijn et al., 1995;Cocolin et al., 2004; Le Bourhis et al., 2007; Garde et al., 2011a; Bassi et al., 2015; Gómez-Torres et al., 2015), resulting in abnormal cheese flavour and texture (Figure 2.3). Also, *Clostridium sporogenes* can produce gas due to proteolysis in the anaerobic cheese environment (Le Bourhis et al., 2007; Garde et al., 2011a; Gómez-Torres et al., 2015). Blowing is the most serious microbial spoilage occurring in cheese because (i) a great variety of cheeses can be affected, (ii) once the

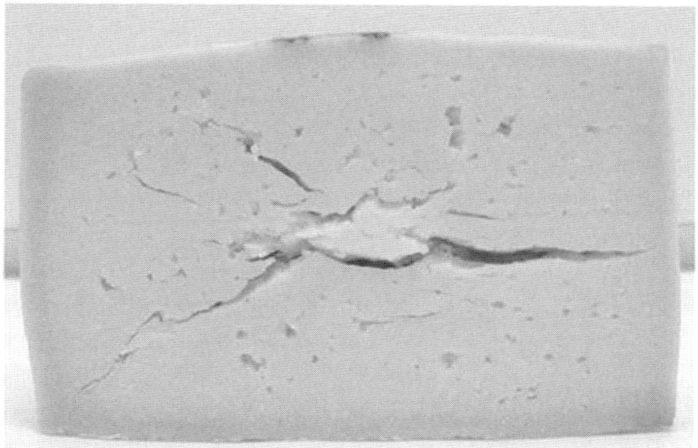

Figure 2.3 Hard ovine milk cheese with LBD.

defect emerges in cheese it is impossible to correct, (ii) it has a high frequency of occurrence and (iv) it has an important economic impact, as blowing usually affects large-volume productions (Quiberoni et al., 2008). LBD is a major cause of spoilage in semihard and hard cheeses (Garde et al., 2013), with *C. tyrobutyricum* considered the primary cause of LBD in cheese (Klijn et al., 1995). More recently, Gómez-Torres et al. (2015) reported that the degree of LBD in cheese was species-dependent, and *C. tyrobutyricum* strains potent cheese spoilers, noticeably affecting cheese pH and colour and leading to intense blown-packaging, cheese cracking and accumulation of volatile compounds associated with rancid and pungent off-flavours at the end of ripening.

A correlation between *Clostridium* spore counts in milk and the incidence of LBD in cheeses has been reported (Goudkov and Sharpe, 1965; Garde et al., 2011b). Although the number of *Clostridium* spores found in raw milk is generally low and it typically shows a seasonal dependence, the reported incidence of clostridial spores was often high, ranging from 16 to 100%, with a mean value of 58% (Goudkov and Sharpe, 1965; Garde et al., 2011b; Arias et al., 2013; Reindl et al., 2014; Chaturvedi et al., 2015; Feligini et al, 2014). The most common *Clostridium* spp. identified in raw milk varied according to each study, but *C. sporogenes, C. beijerinkii, C. tyrobutyricum* and *C. perfringens* have been identified (Arias et al., 2013; Reindl et al., 2014; Chaturvedi et al., 2015; Feligini et al, 2014).

With regard to cheeses with LBD, 47.5% isolates from Grana cheeses with LBD were identified as *C. tyrobutyricum*, 45% as *C. sporogenes* and 7.5% as *C. butyricum* (Matteuzzi et al., 1977). In Manchego cheeses with LBD, 78.9%, 10.3%, 9.0% and 1.8% of isolates were identified as *C. sporogenes*, *C. beijerinckii*, *C. tyrobutyricum* and *C. butyricum*, respectively, and more than one species was isolated from 40% of the LBD cheeses (Garde et al., 2011b). These authors found a high *Clostridium* intraspecies genomic diversity, and isolates could be grouped according to the factory and the date of cheese manufacture (Garde et al., 2012). On the other hand, *Clostridium* spp. have been detected and identified directly in LBD cheeses by the development of PCR-based techniques, without previous strain isolation (Klijn et al., 1995; Cocolin et al., 2004; Le Bourhis et al., 2005). DNA from *C. tyrobutyricum, C. sporogenes, C. butyricum* and *C. beijerinckii* has been detected in these studies.

The presence of more than one *Clostridium* species and/or strains in cheeses with LBD, suggests that it may not be the result of the metabolic activity of a single species but depend on a synergistic action of several *Clostridium* species and/or strains. The association of *C. beijerinckii* or *C. sporogenes* to *C. tyrobutyricum* has been reported to enhance butyric fermentation and cheese defects (Le Bourhis et al., 2007). Garde et al. (2011b) also reported a synergistic effect on the production of butyric acid and gas in milk when a *C. beijerinkii* strain was combined with a *C. tyrobutyricum* or *C. sporogenes* strain. In addition, as reported by Bassi et al. (2015), the interaction of *Clostridium* spp. with the rest of cheese microbiota can be crucial for the appearance of LBD in cheese. It is known that the growth of *Clostridium* is favoured by the presence of other microorganisms in cheese, such as coliforms, lactococci, leuconostocs, lactobacilli and propionibacteria (Goudkov and Sharpe, 1965; Gómez-Torres et al., 2014).

2.4 CONTROL STRATEGIES TO PREVENT POISONING AND SPOILAGE OF MILK AND DAIRY PRODUCTS BY SPORE-FORMING BACTERIA

As can be seen from the discussions above, it is necessary that the dairy industry adopt control strategies in order to prevent poisoning and spoilage of milk and dairy products by spore-forming bacteria to ensure safety and quality of these foods. The first control measure is minimizing the presence of spores in raw milk by the application of good farming practices (Vissers et al., 2007a, 2007b) since it is impossible to eliminate the risk of contamination of milk with spores due to their ubiquitous presence in nature, including environmental niches on the dairy farm. The good farming practices include the use of good quality silage, parlour/milking equipment cleaning routines and maintenance, and udder cleaning and teat preparation prior to milking.

Other strategies to reduce the number of spores in raw milk are bactofugation and microfiltration (Garde et al., 2013). Bactofugation and microfiltration eliminate about 94–98% and 99% of spores from milk, respectively, but these techniques are expensive, time-consuming and labour intensive. In addition, microfiltration can be applied only to skim milk because the milk fat globules are too large to pass through the microfiltration membrane.

Destruction of spores and/or toxins is another control strategy to avoid poisoning and/or spoilage of milk and dairy products. However, this strategy is limited in the case of dairy products, since the treatments required for this purpose (i.e. heating at 121 °C for 3 min is targeted to eliminate the risk of *C. botulinum* group I spores by a factor of 10^{12}) cannot be applied to many dairy products (Lindström et al., 2010). The UHT treatment (135 °C or above for a few seconds) and the sterilization process (115 °C or above for 20 min) used commonly in the dairy industry are likely to destroy most spores and toxins, but some thermophilic spore-forming bacteria may survive. Besides, these treatments cannot be applied to all dairy products. For this reason, other strategies are based on the control of spore-former growth and the production of toxins and spoilage enzymes. Proper refrigeration combined with limited storage times is a first important measure to control spore-forming bacteria growth (Lindström et al., 2010). In addition, some dairy foods inhibit bacteria growth through intrinsic factors such as low water activity in milk powder or the low pH in some fermented product. The use of food additives is another strategy to prevent growth of spore-forming bacteria. Nitrate and/or lysozyme are the most common additives used to control *C. tyrobutyricum* growth and to prevent LBD in cheese, but nitrate could lead to the formation of potentially carcinogenic nitrosamines and lysozyme could pose risks for allergic subjects (Garde et al., 2013). Nisin, a bacteriocin produced by *Lactococcus lactis*, is used as additive in pasteurised processed cheese spreads to control *C. botulinum* growth and neurotoxin production (Lindström et al., 2010). The efficacy of nisin against *B. cereus* and related species has also been tested in milk alone or in combination with high pressure (Egan et al., 2016). Continuing with bacteriocins, the inoculation of milk with bacteriocinogenic lactic acid bacteria in cheese manufacture has been successfully investigated to prevent LBD (Rilla et al., 2003; Bogovic-Matijasic et al., 2007; Anastasiou et al., 2009; Martínez-Cuesta et al., 2010; Garde et al., 2011a).

More recently, LBD has been prevented by the use of *Lactobacillus reuteri* INIA P572, a reuterin (β-hydroxypropionaldehyde) producer strain, as an adjunct culture, together with glycerol (required for reuterin production) in cheese manufacture (Gómez-Torres et al., 2014). Other approaches investigated in order to diminish the appearance of LBD in cheese include irradiation (Velasco et al., 2011), the addition of aromatic plant extracts to cheese milk (Moro et al., 2015), the use of jenny milk, which is characterised by a high lysozyme content (Cosentino et al., 2015), and the application of high pressure (Ávila et al., 2016). High-pressure treatments also have been assayed to inactivate *B. subtilis* and *B. cereus* spores in cheese (Capellas et al., 2000; López et al., 2003). A combination of ultrasonication and pasteurisation resulted in complete inactivation of the thermally resistant vegetative cells of *B. coagulans* and *A. flavithermus* in milk (Khanal et al., 2014).

With regard to the control of biofilms formed by *Bacillus* and related genera, they are more resistant to antimicrobials and cleaning regimes compared to planktonic cells, and this makes their elimination from dairy industry a big challenge (Srey et al., 2013). With regard to their control, strategies used in the dairy industry have been recently reviewed by Gopal et al. (2015). These approaches include the alteration of the surface chemistry to prevent cell attachment, treatment of surfaces with antimicrobial agents, optimization of the process and equipment design or the use of intensive cleaning regimens. Antimicrobial agents can be physical (cleaning-in-place processes and ultrasonication), biological (biological biocides such as enzymes, bacteriocins and reuterin, and bacteriophages) and chemical (chemical biocides, ozone). Current methods for controlling thermophilic bacilli and their biofilm growth in dairy manufacturing plants include shorter production lengths (therefore increasing the cleaning frequency), the use of sanitisers (disinfectants), altering temperatures, reducing the surface area in the optimal temperature growth zone and the use of dual equipment (Burgess et al., 2010).

2.5 CONCLUSIONS

Spore-forming bacteria from *Bacillus* spp. and related genera and *Clostridium* spp. persist as a major cause of spoilage in the dairy industry, and a few of these bacteria are also a concern for food safety as they can produce different types of toxins, being potential food-poisoning agents. Spore-forming bacteria are adapted to endure usual commercial heat treatments followed by refrigeration carried out to reduce the microbial load of milk and to limit the number of spoilage microorganisms and to prevent foodborne disease, and some spore formers are even able to germinate and reproduce in those harsh conditions. Despite the preventive and palliative measures adopted by farmers and at the dairy industry, effective control of spore-forming bacteria in farms, in the processing environment and in dairy products is still a difficult issue because these microorganisms are ubiquitous in nature and because of the high number of critical factors involved. For these reasons, the research and development of new control strategies throughout the dairy chain are necessary.

References

Adams MR, Moss MO. 2008. Microbiology of primary food commodities. In: Adam MR, Moss MO. editors. *Food microbiology*, 3rd ed. Cambridge, UK: RSC Publishing. pp 119–53.

Agata N, Mori M, Ohta M, Suwan S, Ohtani I, Isobe M. 1994. A novel dodecadepsipeptide, cereulide, isolated from *Bacillus cereus* causes vacuole formation in HEp-2 cells. *FEMS Microbiol Lett* 121:31–34.

Akdeniz H, Buzgan T, Tekin M, Karsen H, Karahocagil MK. 2007. An outbreak of botulism in a family in Eastern Anatolia associated with eating süzme yogurt buried under soil. *Scand J Infect Dis* 39:108–14.

Anastasiou R, Aktypis A, Georgalaki M, Papadelli M, De Vuyst L, Tsakalidou E. 2009. Inhibition of *Clostridium tyrobutyricum* by *Streptococcus macedonicus* ACA-DC 198 under conditions mimicking Kasseri cheese production and ripening. *Int Dairy J* 19:330–35.

Arias C, Oliete B, Seseña S, Jimenez L, Pérez-Guzmán M, Arias R. 2013. Importance of on-farm management practices on lactate-fermenting *Clostridium* spp. spore contamination of Manchega ewe milk: determination of risk factors and characterisation of *Clostridium* population. *Small Rumin Res* 111:120–28.

Ash C, Priest FG, Collins MD. 1993. Molecular identification of rRNA group 3 bacilli (Ash, Farrow, Wallbanks and Collins) using a PCR probe test. Proposal for the creation of a new genus *Paenibacillus*. *Antonie van Leeuwenhoek* 64:253–60.

Aureli P, Franciosa G, Pourshaban M. 1996. Foodborne botulism in Italy. *Lancet* 348:1594.

Ávila M, Gómez-Torres N, Delgado D, Gaya P, Garde S. 2016. Application of high pressure processing for controlling *Clostridium tyrobutyricum* and late blowing defect on semi-hard cheese. *Food Microbiol* 60:165–73.

Barash JR, Hsia JK, Arnon SS. 2010. Presence of soil-dwelling clostridia in commercial powdered infant formulas. *J Pediatr* 156:402–408.

Bassi D, Puglisi E, Cocconcelli PS. 2015. Understanding the bacterial communities of hard cheese with blowing defect. *Food Microbiol* 52:106–118.

Bennett SD, Walsh KA, Gould LH. 2013. Foodborne disease outbreaks caused by *Bacillus cereus, Clostridium perfringens*, and *Staphylococcus aureus*—United States, 1998–2008. *Clin Infect Dis* 57:425–33.

Bogovic-Matijasic B, Rajsp MK, Perko B, Rogelj I. 2007. Inhibition of *Clostridium tyrobutyricum* in cheese by *Lactobacillus gasseri*. *Int Dairy J* 17:157–66.

Brett M, McLauchlin J, Harris A, O'Brien S, Black N, Forsyth R, Roberts D, Bolton F. 2005. A case of infant botulism with a possible link to infant formula milk powder: evidence for the presence of more than one strain of *Clostridium botulinum* in clinical specimens and food. *J Med Microbiol* 54:769–76.

Brunt J, Cross KL, Peck MW. 2015. Apertures in the *Clostridium sporogenes* spore coat and exosporium align to facilitate emergence of the vegetative cell. *Food Microbiol* 51:45:50.

Burgess SA, Lindsay D, Flint SH. 2010. Thermophilic bacilli and their importance in dairy processing. *Int J Food Microbiol* 144:215–25.

CAC (Codex Alimentarius Commission). 2008. Code of hygienic practice for powdered infant formulae for infants and young children. CAC/RCP 66–2008.

Capellas M, Mor-Mur M, Gervilla R, Yuste J, Guamis B. 2000. Effect of high pressure combined with mild heat or nisin on inoculated bacteria and mesophiles of goat's milk fresh cheese. *Food Microbiol* 17:633–41.

Carter AT, Peck MW. 2015. Genomes, neurotoxins and biology of *Clostridium botulinum* group I and group II. *Res Microbiol* 166:303–317.

Charlebois A, Jacques M, Archambault M. 2014. Biofilm formation of *Clostridium perfringens* and its exposure to low-dose antimicrobials. *Front Microbiol* 5:183.

Chaturvedi A, Shukla S. 2015. Occurrence of *Clostridium* species in different dairy products and its associated health risk. *Int J Recent Sci Res* 6:2827–29.

Cocolin L, Innocente N, Biasutti M, Comi G. 2004. The late blowing in cheese: a new molecular approach based on PCR and DGGE to study the microbial ecology of the alteration process. *Int J Food Microbiol* 90:83–91.

Cosentino C, Paolino R, Valentini V, Musto M, Ricciardi A, Adduci F, D'Adamo C, Pecora G, Freschi P. 2015. Effect of jenny milk addition on the inhibition of late blowing in semihard cheese. *J Dairy Sci* 98:5133–42.

Council Directive 92/46/EEC. 1992. Council of the European Communities of 16 July 1992. Health rules for the production and the trade of raw milk, heat-treated milk, and products based on milk. *Off J Eur Community* L268:1–32.

Cowan AE, Olivastro EM, Koppel DE, Loshon CA, Setlow B, Setlow P. 2004. Lipids in the inner membrane of dormant spores of *Bacillus* species are largely immobile. *Proc Natl Acad Sci USA* 101:7733–38.

De Jonghe V, Coorevits A, De Block J, Van Coillie E, Grijspeerdt K, Herman L, De Vos P, Heyndrickx M. 2010. Toxinogenic and spoilage potential of aerobic spore-formers isolated from raw milk. *Int J Food Microbiol* 136:318–25.

Dover N, Barash JR, Hill KK, Xie G, Arnon SS. 2014. Molecular characterization of a novel botulinum neurotoxin type H gene. *J Infect Dis* 209:192–202.

Doyle CJ, Gleeson D, Jordan K, Beresford TP, Ross RP, Fitzgerald GF, Cotter PD. 2015. Anaerobic sporeformers and their significance with respect to milk and dairy products. *Int J Food Microbiol* 197:77–87.

EFSA. 2005a. Opinion of the scientific panel on biological hazards on *Bacillus cereus* and other *Bacillus* spp. in foodstuffs. *EFSA J* 175:1–48.

EFSA. 2005b. Opinion of the scientific panel on biological hazards on the request from the commission related to *Clostridium* spp in foodstuffs. *EFSA J* 169:1–65.

EFSA. 2012. The European Union summary report on trends and sources of zoonoses, zoonotic agents and food-borne outbreaks in 2010. *EFSA J* 10:2597.

EFSA. 2013. The European Union summary report on trends and sources of zoonoses, zoonotic agents and food-borne outbreaks in 2011. *EFSA J* 11:3129.

EFSA. 2014. The European Union summary report on trends and sources of zoonoses, zoonotic agents and food-borne outbreaks in 2012. *EFSA J* 12:3547.

EFSA. 2015a. The European Union summary report on trends and sources of zoonoses, zoonotic agents and food-borne outbreaks in 2013. *EFSA J* 13:3991.

EFSA. 2015b. The European Union summary report on trends and sources of zoonoses, zoonotic agents and food-borne outbreaks in 2014. *EFSA J* 13:4329.

Egan K, Field D, Rea MC, Ross RP, Hill C, Cotter PD. 2016. Bacteriocins: novel solutions to age old spore-related problems? *Front Microbiol* 7:461.

FAO/WHO. 2004. *Enterobacter sakazakii and other microorganisms in powdered infant formula: meeting report*. Microbiological Risk Assessment Series 6. Retrieved from http://www.fao.org/docrep/007/y5502e/y5502e00.htm.

Feligini M, Brambati E, Panelli S, Ghitti M, Sacchi R, Capelli E, Bonacina C. 2014. One-year investigation of *Clostridium* spp. occurrence in raw milk and curd of Grana Padano cheese by the automated ribosomal intergenic spacer analysis. *Food Control* 42:71–77.

Furtado MM. 2005. *Main problems of cheeses: causes and prevention*. São Paulo, Brazil: FontePress.

Garde S, Arias R, Gaya P, Nuñez M. 2011b. Occurrence of *Clostridium* spp. in ovine milk and Manchego cheese with late blowing defect: Identification and characterization of isolates. *Int Dairy J* 21:272–78.

Garde S, Ávila M, Arias R, Gaya P, Nuñez M. 2011a. Outgrowth inhibition of *Clostridium beijerinckii* spores by a bacteriocin-producing lactic culture in ovine milk cheese. *Int J Food Microbiol* 150:59–65.

Garde S, Ávila M, Gómez N, Nuñez M. 2013. *Clostridium* in late blowing defect of cheese: detection, prevalence, effects and control strategies. In: Castelli H, du Vale L. editors. *Cheese: production, chemistry and sensory properties*. New York: Nova Science Publishers, pp 503–18.

Garde S, Gaya P, Arias R, Nuñez M. 2012. Enhanced PFGE protocol to study the genomic diversity of *Clostridium* spp. isolated from Manchego cheeses with late blowing defect. *Food Control* 28:392–99.

Gómez-Torres N, Ávila M, Gaya P, Garde S. 2014. Prevention of late blowing defect by reuterin produced in cheese by a *Lactobacillus reuteri* adjunct. *Food Microbiol* 42:82–88.

Gómez-Torres N, Garde S, Peirotén A, Ávila M. 2015. Impact of *Clostridium* spp. on cheese characteristics: microbiology, color, formation of volatile compounds and off-flavors. *Food Control* 56:186–94.

Gopal N, Hill C, Ross PR, Beresford TP, Fenelon MA, Cotter PD. 2015. The prevalence and control of *Bacillus* and related spore-forming bacteria in the dairy industry. *Front Microbiol* 6:1418.

Goudkov AV, Sharpe ME. 1965. Clostridia in dairying. *J Appl Bacteriol* 28:63–73.

Gould GW. 2006. History of science spores. *J Appl Microbiol* 101:507–13.

Granum PE, Lund T. 1997. *Bacillus cereus* and its food poisoning toxins. *FEMS Microbiol Lett* 157:223–28.

Grass JE, Gould LH, Mahon BE. 2013. Epidemiology of foodborne disease outbreaks caused by *Clostridium perfringens*, United States, 1998–2010. *Foodborne Pathog Dis* 10131–36.

Guinebretière MH, Nguyen-The C. 2003. Sources of *Bacillus cereus* contamination in a pasteurized zucchini puree processing line, differentiated by two PCR-based methods. *FEMS Microbiol Ecol* 43:207–215.

Hatheway CL. 1993. *Clostridium botulinum* and other clostridia that produce botulinum neurotoxin. In: Hauschild AHW, Dodds KL. editors. *Clostridium botulinum: ecology and control in foods*. New York: Marcel Dekker, pp 3–21.

Heinen W, Lauwers AM, Mulders JWM. 1982. *Bacillus flavothermus*, a newly isolated facultative thermophile. *A Van Leeuwenhoek* 48:265–72.

Heyndrickx M. 2011. The importance of endospore-forming bacteria originating from soil for contamination of industrial food processing. *Appl Environ Soil Sci*. Retrieved from http://dx.doi.org/10.1155/2011/561975.

Heyndrickx M, Scheldeman P. 2002. Bacilli associated with spoilage in dairy products and other food. In: Berkeley R, Heyndrickx MN, de Vos P. editors. *Applications and systematics of Bacillus and relatives*. 1st ed. Oxford: Blackwell Science, pp 64–82.

ICMSF. 2014. Usefulness of testing for *Clostridium botulinum* in powdered infant formula and dairy-based ingredients for infant formula. Retrieved from http://www.icmsf.org/pdf/ICMSF_Infant_Formula_Testing_Revision1-20140117.pdf.

Ivy RA, Ranieri M, Martin NH, den Bakker HC, Xavier BM, Wiedmann M, Boor KJ. 2012. Identification and characterization of psychrotolerant spore formers associated with fluid milk production and processing. *Appl Environ Microbiol* 78:1853–64.

Janštová B, Dračková M, Vorlová L. 2006. Effect of *Bacillus cereus* enzymes on the milk quality following ultra high temperature processing. *Acta Vet Brno* 75:601–609.

Jeyaram K, Romi W, Singh TA, Adewumi GA, Basanti K, Oguntoyinbo FA. 2011. Distinct differentiation of closely related species of *Bacillus subtilis* group with industrial importance. *J Microbiol Methods* 87:161–64.

Jiménez G, Urdiain M, Cifuentes A, López-López A, Blanch AR, Tamames J, Kämpfer P, Kolstø AB, Ramón D, Martínez JF, Codoñer FM, Rosselló-Móra R. 2013. Description of *Bacillus toyonensis* sp. nov., a novel species of the *Bacillus cereus* group, and pairwise genome comparisons of the species of the group by means of ANI calculations. *Syst Appl Microbiol* 36:383–91.

Johnson EA, Tepp WH, Bradshaw M, Gilber RJ, Cook PE, McIntosh EDG. 2005. Characterization of *Clostridium botulinum* strains associated with an infant botulism case in the United Kingdom. *J Clin Microbiol.* 43:2602–2607.

Jung MY, Kim JS, Paek WK, Lim J, Lee H, Kim PI, Ma JY, Kim W, Chang YH. 2011. *Bacillus manliponensis* sp. nov., a new member of the *Bacillus cereus* group isolated from foreshore tidal flat sediment. *J Microbiol* 49:1027–1032.

Jung MY, Paek WK, Park IS, Han JR, Sin Y, Paek J, Rhee MS, Kim H, Song HS, Chang YH. 2010. *Bacillus gaemokensis* sp. nov., isolated from foreshore tidal flat sediment from the Yellow Sea. *J Microbiol* 48:867–71.

Kalogridou-Vassiliadou D. 1992. Biochemical activities of *Bacillus* species isolated from flat sour evaporated milk. *J Dairy Sci* 75:2681–86.

Khanal SN, Anand S, Muthukumarappan K, Huegli M. 2014. Inactivation of thermoduric aerobic spore formers in milk by ultrasonication. *Food Control* 37:232–239.

Klijn N, Nieuwenhof FFJ, Hoolwerf JD, Van Der Waals CB, Weerkamp AH. 1995. Identification of *Clostridium tyrobutyricum* as the causative agent of late blowing in cheese by species-specific PCR amplification. *Appl Environ Microbiol* 61:2919–24.

Kumari S, Sarkar PK. 2016. *Bacillus cereus* hazard and control in industrial dairy processing environment. *Food Control* 69:20–29.

Le Bourhis AG, Dore J, Carlier JP, Chamba JF, Popoff MR, Tholozan JL. 2007. Contribution of *C. beijerinckii* and *C. sporogenes* in association with *C. tyrobutyricum* to the butyric fermentation in Emmental type cheese. *Int J Food Microbiol* 113:154–63.

Le Bourhis AG, Saunier K, Dore J, Carlier JP, Chamba JF, Popoff MR, Tholozan JL. 2005. Development and validation of PCR primers to assess the diversity of *Clostridium* spp. in cheese by temporal temperature gradient gel electrophoresis. *Appl Environ Microbiol* 71:29–38.

Leggett MJ, McDonnell G, Denyer SP, Setlow P, Maillard JY. 2012. Bacterial spore structures and their protective role in biocide resistance. *J Appl Microbiol.* 113:485–498.

Lindström M, Heikinheimo A, Lahti P, Korkeala H. 2011. Novel insights into the epidemiology of *Clostridium perfringens* type A food poisoning. *Food Microbiol* 28:192–198.

Lindström M, Myllykoski J, Sivela S, Korkeala H. 2010. *Clostridium botulinum* in cattle and dairy products. *Crit Rev Food Sci Nutr* 50:281–304.

Liu B, Liu GH, Hu GP, Sengonca C, Lin NQ, Tang JY, Tang WQ, Lin YZ. 2014. *Bacillus bingmayongensis* sp. nov., isolated from the pit soil of emperor Qin's terra-cotta warriors in China. *A. Van Leeuwenhoek* 105:501–510.

López TJ, Roig AX, Capellas M, Trujillo AJ, Hernández M, Guamis B. 2003. Evaluation of the importance of germinative cycles for destruction of spores of *Bacillus cereus* in miniature cheeses. *High Pressure Res* 23:81–85.

Lücking G, Stoeckel M, Atamer Z, Hinrichs J, Ehling-Schulz M. 2013. Characterization of aerobic spore-forming bacteria associated with industrial dairy processing environments and product spoilage. *Int J Food Microbiol* 166:270–79.

Martínez-Cuesta MC, Bengoechea J, Bustos I, Rodríguez B, Requena T, Pelaez C. 2010. Control of late blowing in cheese by adding lacticin 3147-producing *Lactococcus lactis* IFPL 3593 to the starter. *Int Dairy J* 20:18–24.

Matta H, Punj V. 1999. Isolation and identification of lipolytic, psychrotrophic spore forming bacteria from raw milk. *Int J Dairy Technol* 52:59–62.

Matteuzzi D, Trovatelli LD, Biavati B, Zani G. 1977. Clostridia from Grana cheese. *J Appl Bacteriol* 43:375–82

McAuley CM, McMillan K, Moore SC, Fegan N, Fox EM. 2014. Prevalence and characterization of foodborne pathogens from Australian dairy farm environments. *I Dairy Sci* 97:7402–412.

McClure PJ. 2006. Spore-forming bacteria. In: Blackburn C.d.W. editor. *Food spoilage micro-organisms.* Cambridge: Woodhead Publishing, pp 579–23.

McKenney PT, Driks A, Eichenberger P. 2013. The *Bacillus subtilis* endospore: assembly and functions of the multilayered coat. *Nat Rev Microbiol* 11:33–44.

Meyer KF, Eddie B. 1965. Sixty-five years of human botulism in the United States and Canada: Epidemiology and tabulations of reported cases 1899 through 1964. San Francisco: University of California Printing Department.

Mistry VV. 2001. Fermented milks and cream. In: Marthand EH, Steele JL. editors. *Applied dairy microbiology*, 2nd ed. New York: Marcel Dekker, pp 301–326.

Moir A, Cooper G. 2015. Spore germination. *Microbiol Spectr* 3: TBS-0014-2012.

Moro A, Librán CM, Berruga MI, Carmona M, Zalacain A. 2015. Dairy matrix effect on the transference of rosemary (*Rosmarinus officinalis*) essential oil compounds during cheese making. *J Sci Food Agricul* 95:1507–1513.

Murphy PM, Lynch D, Kelly PM. 1999. Growth of thermophilic spore forming bacilli in milk during the manufacture of low heat powders. *Int J Dairy Technol* 52:45–50.

O'Mahony M, Mitchell E, Gilbert RJ, Hutchinson DN, Begg NT, Rodhouse JC, Morris JE. 1990. An outbreak of foodborne botulism associated with contaminated hazelnut yogurt. *Epidemiol Infect* 104:389–95.

Phillips JD, Griffiths MW. 1986. Factors contributing to the seasonal variation of *Bacillus* spp. in pasteurized dairy products. *J Appl Bacteriol* 61:275–85.

Pikuta E, Lysenko AM, Chuvilskaya N, Mendrock U, Hippe H, Suzina N, Nikitin D, Osipov GA, Laurinavichius K. 2000. *Anoxybacillus pushcinensis* gen. nov., sp nov., a novel anaerobic, alkaliphilic, moderately thermophilic bacterium from manure, and description of *Anoxybacillus flavithermus* comb. nov. *Int J Syst Evol Microbiol* 50:2109–117.

Pourshafie MR, Saifie M, Shafiee A, Vahdani P, Aslani M, Salemian J. 1998. An outbreak of food-borne botulism associated with contaminated locally made cheese in Iran. *Scandinavian J Infect Dis* 30:92–94.

Priest FG. 1977. Extracellular enzyme synthesis in the genus *Bacillus*. *Bacteriolog Rev* 41:711–53.
Quiberoni A, Guglielmotti D, Reinheimer J. 2008. New and classical spoilage bacteria causing widespread blowing in Argentinean soft and semihard cheeses. *Int J Dairy Technol* 61:358–63.
Reindl A, Dzieciol M, Hein I, Wagner M, Zangerl P. 2014. Enumeration of clostridia in goat milk using an optimized membrane filtration technique. *J Dairy Sci* 97:6036–45.
Rilla N, Martínez B, Delgado T, Rodríguez A. 2003. Inhibition of *Clostridium tyrobutyricum* in Vidiago cheese by *Lactococcus lactis* ssp. *lactis* IPLA 729, a nisin Z producer. *Int J Food Microbiol* 85:23–33.
Sadiq FA, Li Y, Liu T, Flint S, Zhang G, He G. 2016. A RAPD based study revealing a previously unreported wide range of mesophilic and thermophilic spore formers associated with milk powders in China. *Int J Food Microbiol* 217:200–208.
Samaržija D, Zamberlin Š, Pogačić T. 2012. Psychrotrophic bacteria and milk and dairy products quality. *Mljekarstvo* 62:77–95.
Scheldeman P, Herman L, Foster S, Heyndrickx M. 2006. *Bacillus sporothermodurans* and other highly heat-resistant spore formers in milk. *J Appl Microbiol* 101:542–55.
Scott SA, Brooks JD, Rakonjac J, Walker KMR, Flint SH. 2007. The formation of thermophilic spores during the manufacture of whole milk powder. *Int J Dairy Technol* 60:109–117.
Sébald M, Billon J, Cassaigne R, Rosset R, Poumeyrol G. 1980. Le botulisme en France – incidence, mortalite, aliments responsables avec etude des foyers dus a un aliment qui n'est pas de preparation familiale. *Méd Nutr* 16:262–68.
Sébald M, Jouglard J, Gilles G. 1974. Botulisme human de type B aprés ingestion du fromage. *Ann Microbiol* 125A:349–57.
Setlow P. 2014a. Germination of spores of *Bacillus* species: what we know and do not know. *J Bacteriol* 196:1297–305.
Setlow P. 2014b. Spore resistance properties. *Microbiol Spectr* 2(5):TBS-0003-2012.
Siegel LS. 1993. Destruction of botulinum toxins in food and water. In: Hauschild AHW, Dodds KL. editors. *Clostridium botulinum. Ecology and control in foods*. New York: Marcel Dekker, pp 323–32.
Smith DH, Timms GL, Refai M. 1979. Outbreak of botulism in Kenyan nomads. *Ann Trop Med Parasit* 73:145–48.
Spanu C, Scarano C, Spanu V, Pala C, Casti D, Lamon S, Cossu F, Ibba M, Nieddu G, De Santis EPL. 2016. Occurrence and behavior of *Bacillus cereus* in naturally contaminated ricotta salata cheese during refrigerated storage. *Food Microbiol* 58:135–38.
Srey S, Jahid IK, Ha SD. 2013. Biofilm formation in food industries: a food safety concern. *Food Control* 31:572–85.
Stewart GC. 2015. The exosporium layer of bacterial spores: a connection to the environment and the infected host. *Microbiol Mol Biol Rev* 79:437–57.
Svensson B, Monthan A, Shaheen R, Andersson MA, Salkinoja-Salonen M, Christiansson A. 2006. Occurrence of emetic toxin producing *Bacillus cereus* in the dairy production chain. *Int Dairy J* 16:740–49.
Ternström A, Lindberg AM, Molin G. 1993. Classification of the spoilage flora of raw and pasteurized bovine milk, with special reference to *Pseudomonas* and *Bacillus*. *J Appl Bacteriol* 75:25–34.
Varnam AH, Sutherland JP. 1994. Milk and milk products: technology, chemistry and microbiology. London: Chapman & Hall.
Velasco R, Ordóñez JA, Cabeza MC, de la Hoz L, Cambero MI. 2011. Use of the E-beam radiation to diminish the late blowing of cheese. *Int Dairy J* 21:493–500.
Velthuis A, van Asseldonk M. 2010. *Association between product quality control and process quality control of bulk milk*. Ninth Wageningen International Conference on Chain and Network Management. Wageningen, The Netherlands.
Vissers MMM, Driehuis F, Te Giffel MC, De Jong P, Lankveld JMG. 2007a. Minimizing the level of butyric acid bacteria spores in farm tank milk. *J Dairy Sci* 90:3278–85.
Vissers MMM, Te Giffel MC, Driehuis F, De Jong P, Lankveld JMG. 2007b. Minimizing the level of *Bacillus cereus* spores in farm tank milk. *J Dairy Sci* 90:3286–93.
Yuan DD, Liu GC, Ren DY, Zhang D, Zhao L, Kan CP, Yang YZ, Ma W, Li Y, Zhang LB. 2012. A survey on occurrence of thermophilic bacilli in commercial milk powders in China. *Food Control* 25:752–57.

3 Psychrotrophic bacteria

Milena Brasca[1], Marilù Decimo[1], Stefano Morandi[1], Solimar Gonçalves Machado[2], François Baglinière[3] and Maria Cristina Dantas Vanetti[3]

[1]*ISPA-CNR, Institute of Sciences of Food Productions, Milano, Italy*
[2]*Federal Institute of Education, Science and Technology of North of Minas Gerais, Campus Salinas, Minas Gerais, Brazil*
[3]*Department of Microbiology, Federal University of Viçosa, Minas Gerais, Brazil*

3.1 INTRODUCTION

Bacterial food spoilage is a serious global problem that causes significant economic losses for the food industry because of inadequate processing and refrigeration facilities (Bhunia, 2008).

The contamination of raw milk after collection (at the time of processing) and the premature deterioration of market-processed milk and dairy products continue to be a problem for the dairy industry. For dairy companies, there is a real need for high-quality, extended shelf-life products to gain and maintain consumer loyalty. The wide array of available dairy food challenges the microbiologist, engineer, and technologist to find the best ways to prevent the entry of microorganisms, inactivate those that get in along with their enzymes, and prevent the growth and activities of those that escape processing treatments (Ledenbach and Marshall, 2009).

The highly nutritious nature of dairy products makes them especially good media for the growth of microorganisms. Milk contains abundant water and nutrients and has a nearly neutral pH, providing a physicochemical environment that is favorable for the multiplication of a broad spectrum of microorganisms (Ray, 2004). The microbial population of raw milk is complex and variable according to a variety of sources, including the health status of the cattle, the nature of their feed, teat apex, milking equipment, air, water, soil and other environments (LeJeune and Rajala-Schultz, 2009; Bava et al., 2011; Vacheyrou et al., 2011). The microbial community of raw milk can include bacteria of technological relevance such as lactic acid bacteria that are fundamental if the milk is employed in cheese making, contributing to the sensory richness and variety of traditional cheese (Morandi et al., 2011). Nevertheless, troublesome spoilage bacteria, including both Gram-positive and Gram-negative microorganisms, may enter in raw milk by unsatisfactory milk production practices, failures in processing

Microbiology in Dairy Processing: Challenges and Opportunities, First Edition.
Edited by Palmiro Poltronieri.

Published 2018 by John Wiley & Sons Ltd and the Institute of Food Technologists.

systems or unsanitary practices, affecting the technological and organoleptic properties of the finished dairy products (Cousin, 1982).

Whereas milk contamination is diverse during milking process, the conditions under which the raw milk is stored and transported and the length of storage time before it is processed define the dominant microbiota. The refrigeration is universally acceptable for prolonging shelf life and eliminating spoilage by mesophilic bacteria in raw milk; however, changes in the bacterial communities occur, and this procedure creates selective conditions for growth of psychrotrophic bacteria.

Psychrotrophic microorganisms are defined as those that can grow at 7 °C or below, within 7–10 days incubation, regardless of their optimal growth temperature (Cousin, 1982; Frank and Yousef, 2004). Psychrotrophic microorganisms are ubiquitous in nature primarily in water and soil, including vegetation, and a small number can also be present in the air; they become an important part of the microbiota of raw milk since the introduction of bulk refrigerated storage. The psychrotroph counts, which are approximately 10% of the total count of mesophilic aerobes immediately after milking performed under hygienic conditions, may reach, on average, 90% after cold storage (Sørhaug and Stepaniak, 1997; Catanio et al., 2012). However, this group of bacteria, that are able to grow at cold temperatures, can represent more than 75% of the initial microbiota of raw milk in situations of poor hygiene (Hantsis-Zacharov and Halpern, 2007; Malacarne et al., 2013).Ubiquitous distribution of these bacteria indicates a remarkable degree of their physiological and genetic adaptability (Spiers et al., 2000).

3.2 SOURCES OF PSYCHROTROPHIC BACTERIA CONTAMINATION OF MILK

Psychrotrophic bacteria are not part of the natural microbial population of the udder, and milk freshly drawn from the udder often does not contain detectable populations of culturable psychrotrophic bacteria. Therefore, the presence of psychrotrophs in raw milk is exclusively the result of milk contamination from teat, water, milking and storage equipment and environment, contamination during transportation, and refrigerated storage at the processing facility. Poor cow and bedding hygiene strongly affect bacterial counts, coliforms, and psychrotrophic bacteria content in bulk milk (Elmoslemany et al., 2010; Zucali et al., 2011). Milking operation routine strongly influences the microbial quality of raw milk at farm level. Forestripping (the removal of the first streams of milk), pre-milking teat sanitation, and post-dipping (the disinfection of the teat after the milking) effectively reduce teat contamination; in particular, pre-milking teat sanitation can reduce test surface bacteria by 75% (Bade et al., 2008).

The mammary glands of very young cows yield no bacteria in aseptically collected milk samples, but as numbers of lactations increase, so do the chances of isolating bacteria in milk drawn aseptically from the teats. The stresses placed on the cow's teats and mammary glands by the very large amounts of milk produced and the actions of the milking machine cause teat canals to become more open, and teat ends to become misshapen as time passes. These stresses may open the teat canal for the entry of bacteria capable of infecting the glands and, consequently, to enter in milk. Environmental contaminants, like psychrotrophic bacteria, represent a significant percentage of spoilage microorganisms. They are ubiquitous in the environment from which they contaminate the cow, equipment, water, and milkers' hands. Since milking machines exert about 38 cm (15 in.) of vacuum on the teats during milking, and since air often leaks into the system, psychrotrophic bacteria on the surfaces of the cow or in water retained from pre-milking preparation can be drawn into the milk. In addition, when inflation clusters drop to the floor, they pick up microorganisms that can be enter into the milk (Ledenbach and Marshall, 2009).

Contamination of milk by psychrotrophic microorganisms also occurs due to inadequately sanitized surfaces of milking and milk storage equipment (residual water in milking machines, milk pipelines or coolers, dirty udders and teats, and inadequate cleaning of surfaces of dairy equipment for reception). The psychrotrophic microorganisms grow in milk residues present in crevices, joints, rubber gaskets, and dead ends of badly cleaned milking plants and adhere to stainless-steel milk transfer pipelines. Ancillary equipment such as agitators, dipsticks, outlet plugs and cocks can be difficult to clean, and these may be a possible source of contamination (McPhee and Griffiths, 2011). Under sanitary conditions, <10% of the total microbiota are psychrotrophs, compared to >75% under unsanitary conditions (Suhren, 1989; Cempìrkovà 2002). A study of Bava et al. (2011) evidenced that milk samples from farms applying higher water temperature during the washing phase of the milking equipment (>35 °C) had significantly lower total and psychrotrophic bacteria counts. Moreover, the effect of the temperature reached during washing cycle is strictly correlated to the presence of psychrotrophic bacteria on the surface of the milking equipment, where biofilm can easily grow. A proper setting of the whole washing system (water temperature, turbulence, and water filling, detergent concentration) contributes significantly to obtaining lower psychrotrophic bacteria count in bulk milk. Liner swabs represent effective tools to highlight microbiological problems in the milking equipment.

An important characteristic of psychrotrophic bacteria is their ability to grow at low temperatures (0–7 °C), regardless of their optimal temperature (30 °C) (Sørhaug and Stepaniak 1997). This characteristic makes these bacteria especially significant with regard to dairy spoilage and safety, given that the storage of milk and derived products at cold temperatures is a routine practice during production, transportation, processing, and post-purchase in order to control the growth of mesophilic organisms (Beales, 2004; Russell, 2002). The combination of longer storage times and lower temperatures of raw milk throughout the dairy chain prior to heat treatments effectively controls the development of populations of mesophilic spoilage organisms while at the same time providing a selective advantage for the growth of psychrotrophic bacteria (Chen et al., 2003; Lafarge et al., 2004). Milk may be collected from farms on alternate days, or even longer in some instances. The cooling temperature and time of milk storage may vary from one country or region to another. In some countries, part of the milk in the farm bulk tank may be held for 48 h more. Storage of raw milk at 2 °C supports significantly lower growth, enzymatic activities of psychrotrophs, and better sensory qualities compared to 4 and 7 °C of storage (Kumaresen et al., 2007).

Although alternate-day collection may have little effect on the bacteriological quality of milk rapidly cooled to 4 °C or below before addition to the tank, the growth potential of the raw milk microbiota is significantly affected. Thus, milk collected on alternate days will contain a greater number of bacteria that are entering the exponential phase of growth when the milk arrives at the processing site, and the amount of time that this milk can be subsequently stored will be reduced (Tekinson and Tothwell, 1974). For example, it has been shown that *Pseudomonas* spp. isolated from milk that had been stored at 7 °C for 3 days grew 10 times faster at this temperature, had 1000-fold more proteolytic activity, and were 280-fold more lipolytic than pseudomonads isolated from freshly drawn milk (McPhee and Griffiths, 2011).

Milk is usually transported in insulated tanks or in refrigerated tankers and may be transferred to larger vehicles for longer journeys. During transportation, the main cause of increased bacterial count is inadequately cleaned vehicles and growth of bacteria already present in the milk. The latter is dependent on the milk temperature and journey time. A twofold increase in count is common during transportation of milk from the farm to the processing site, and this is due primarily to the growth of psychrotrophic bacteria, including pseudomonads. Critical sites in the milk tanker for cleaning have been identified as the air separator, the milk meter, the milk sieve and the suction hose, and factors that contribute to inadequate cleaning include blockage of the cleaning-in-place (CIP) spray system and low water pressure and flow rate. These can lead to buildup of milk stone on the inner surface of the tanker (McPhee and Griffiths, 2011). The current trend in the dairy industries is to reduce the frequency of milk collection; thus, the refrigerated storage of milk has been lengthened from 2 to 5 days prior to heat treatment (O'Brien and Guinee, 2011). This practice has been stimulated in part by a desire for a 5-day work week and in response to a decreased milk supply at certain times of the year (McPhee and Griffiths, 2011). Thus, the temperature at which milk is stored becomes critical. It has been recommended that milk be cooled to, and maintained at, 3 °C on receipt at the processing plant before storage. The average counts of psychrotrophic aerobic bacteria in milk silos at several dairies in southwest Scotland were reported to be 1.3×10^5 CFU ml^{-1}. The majority of the bacteria present were pseudomonads (70.2%), but Enterobacteriaceae (7.7%), Gram-positive bacteria (6.9%), and other Gram-negative, rod-shaped organisms were also isolated (McPhee and Griffiths, 2011). Analogously, Decimo et al. (2014) isolated psychrotrophs from bovine bulk tank milk samples in Italy, and among the 80 identified as Gram-negative, *Pseudomonas* were the most commonly occurring contaminants (78.75%), with *Pseudomonas fluorescens* the predominant isolated species (30.16%), along with Enterobacteriaceae (21.25%), primarily *Serratia marcescens* (52.94%). Following storage for a further 48 h at 6 °C, the psychrotrophic counts increased by two log cycles to 1.3×10^7 CFU ml^{-1}. Growth rates of psychrotrophic bacteria were highest during filling of the silos, due possibly to temperature fluctuation, and final bacterial numbers for the evident appearance of spoilage were dependent on initial counts immediately after milking and on cold storage time, the latter being the most significant factor affecting milk quality. It has been shown that raw milk stored at low temperatures is spoilt exclusively by Gram-negative bacteria.

Although Gram-negative psychrotrophic bacteria present in raw milk do not survive pasteurization, these organisms are commonly isolated from pasteurized milk and cream, again with *Pseudomonas* spp. being the most frequently encountered. Thus, the shelf life of pasteurized products is limited by post-pasteurization contamination. A likely cause of post-pasteurization contamination in the processing plant is shedding of psychrotrophic bacteria from biofilms formed on gaskets in pasteurized milk pipelines. Pseudomonads utilize biofilm formation for colonization and environmental persistence.

Biofilms are surface-associated bacterial communities that are embedded in a matrix of self-produced extracellular polymeric substance consisting of nucleic acids, exopolysaccharides, lipids, and proteins resulting from the successful attachment and subsequent growth of microorganisms on a surface (Mann et al., 2012). In nature, biofilms can be composed of a single species, but more commonly, they comprise a consortium of species (Skandamis and Nychas, 2012). The nature of the extracellular three-dimensional matrix, the ratio of proliferation, and interaction between cells within the biofilm are determined by the available conditions for growth, the medium, and substrate (Constantin, 2009). Bacteria in biofilms (sessile form) are more resistant to chemical sanitizers and the majority of antibiotics than are the same bacteria in suspension (planktonic form) (Mosteller and Bishop, 1993). Yet, when population densities in the biofilms become high, bacteria are released into the environment, providing a continuous source of planktonic bacteria capable of replication within milk (Bai and Rai, 2011). Gram-negative psychrotrophic bacteria, like *Pseudomonas*, are largely recognized by their ability to produce large amounts of exopolysaccharides, which contribute to adhesion and biofilm growth (Drenkard and Ausubel, 2002).

Bacterial adhesion and biofilm formation are also influenced by the chemical–physical properties of the surface, temperature, nutrients, and the presence of other microorganisms. *Pseudomonas* biofilms can develop on the sides of gaskets, despite operation of CIP systems, and represent a long-lasting source of permanent product contamination. An intriguing feature of milk-spoiling *Pseudomonas* recovered from biofilms is their ability to alter phenotypes via the process of phase variation. Through this process, high-frequency phenotypic switching is mediated by mutation, reorganization, or modification of the genome (Van Den Broek *et al.*, 2005), contributing to the survival of the biofilm population during environmental stresses such as temperature fluctuations and frequent exposure to sanitizers during the cleaning of dairy processing and storage equipment (Marchand et al., 2012).

A biofilm can build up in a few hours, thus in a food processing plant, the period for biofilm development is strictly dependent on the frequency of sanitation practices. Cleaning is the key procedure to control biofilm formation since, generally, disinfectants do not penetrate the biofilm matrix and thus do not destroy all the living cells. In addition, biofilm cells can tolerate much higher concentrations of biocides, thus the sanitation procedures used in the food industry are sometimes ineffective for killing them. A mixture of enzymes can be used to degrade the biofilm matrix and a mixture of enzymes, surfactants, and dispersing and chelating agents can reduce the use of chemical agents (Cappitelli et al., 2014).

There is substantial evidence that the filling operation has the greatest influence on the potential shelf life of pasteurized milk, and filling machines in the dairy industry represent the most common site for product recontamination by psychrotrophs. The efficiency of the process and the resulting quality of dairy products are directly influenced by the microbiological quality of raw milk (Nörnberg et al., 2010).

In the production of sterilized milk, fermented milks, butter, and cheese, the presence and subsequent replication of populations of psychrotrophs are consequences of the growth of these bacteria in raw milk stored at low temperatures for a longer period of time prior to processing or heat treatments. According to Sørhaug and Stepaniak (1997), a psychrotrophic population of 5.5 log CFU ml^{-1} in raw milk causes ultra-high temperature

(UHT) milk gelation after 20 weeks of storage, while populations between 6.9 and 7.2 logs will cause the same effect between 2 and 10 weeks. The defects of raw milk–derived products can be caused directly by psychrotrophic contamination or indirectly by the activation of their thermoresistant enzymes, proteases (Liu et al., 2007) and lipases (Chen et al., 2003). Many of the produced enzymes retain significant activity after pasteurization (72–75 °C/15–20 s) and even UHT treatment (130–150 °C/2–4 s), and they may subsequently degrade proteins and milk fats present in the processed products. The consequences of these hydrolytic changes on the quality of dairy products are manifested in changes of texture, flavor, reduction in cheese yield, tainting problems, improper coagulation time, and reduction of the quality of the cheese curd, and reduction of shelf-life (Barbano et al., 2006; De Jonghe et al., 2011).

3.3 IMPORTANT SPOILAGE PSYCHROTROPHIC BACTERIA IN MILK

The prolonged refrigerated storage of raw milk resulting from procedures for collecting milk on farms and in management practices at dairies lead to milk plants processing raw milk 2–5 days old, and the cold storage is a selection criterion that enables psychrotrophic bacteria to outgrow other bacteria. It has been well established by independent research groups in different countries that *Pseudomonas* spp. is the most important psychrotroph that dominates the microbiota of raw or pasteurized milk at the time of cold storage. This genus of Gram-negative, aerobic, and rod-shaped bacteria harbors species with a well-established physiological mechanism of adaptation and growth at low temperatures, with the shortest generation times at range of 0–7 °C. *P. fluorescens*, *Pseudomonas fragi*, and *Pseudomonas putida* are the most common species, and they are recognized as producers of extracellular thermostable enzymes, mainly proteolytic and lipolytic (Garcia et al., 1989). The predominance of *Pseudomonas* species was confirmed in various studies by culture-dependent methods worldwide (Decimo et al., 2014; Délbes et al., 2007; Hantsis-Zacharov and Halpern, 2007; Jayarao and Wang, 1999; Machado et al., 2015). After applying culture-independent methods for identifying the spoiler microbiota, other psychrotrophic species belonging to *Pseudomonas* genus have been identified and characterized as contaminants of cold raw milk (Délbes et al., 2007; Ercolini et al., 2009; Rasolofo et al., 2010; Lo et al., 2016; Mallet et al., 2012; Raats et al., 2011).The protease producers such as *Pseudomonas lundensis* were isolated from raw milk samples from Belgium (Marchand et al., 2009a, 2009b); from Germany (von Neubeck et al., 2015) and from Brazil (Machado et al., 2015). Two novel species, *Pseudomonas helleri* and *Pseudomonas weihenstephanensis*, were isolated from cow milk from Germany (von Neubeck et al., 2016). Further studies demonstrated the heat resistance of enzymes produced by *P. weihenstephanensis*, *Pseudomonas proteolytica*, and *Pseudomonas panacis*, which could withstand UHT milk treatment (Baur et al., 2015b; Stoeckel et al., 2016).

Although the predominance of *Pseudomonas* genus is well known, the importance of other psychrotrophic proteolytic bacteria has been discovered to be the *Serratia* genus (Boor and Fromm, 2009; Decimo et al., 2014; Heyndrickx et al., 2010; Machado et al., 2015; Lo et al., 2016). Along with species belonging to *Pseudomonas*, *Serratia* was also detected and characterized as predominant milk spoiler in Italian, Brazilian, and

Australian samples (Decimo et al., 2014; Machado et al., 2015; Lo et al., 2016). *Serratia* was also detected in milk-processing plants and raw milk road tankers (Cleto et al., 2012; Teh et al., 2011). This genus contains species that are recognized to secrete potent extracellular enzymes as proteases and lipases that are thermostable. As well as *Serratia*, other psychrotrophic bacteria belonging to the family of Enterobacteriaceae have been isolated from cold raw milk and have been identified as potential milk spoiler due to heat-resistant enzymes; these are *Klebsiella oxytoca*, *Hafnia alvei*, *Hafnia paralvei*, and *Enterobacter aerogenes*, which are enzymatically active genera (Tondo et al., 2004; Chen et al., 2011; Vithanage et al., 2016). *Acinetobacter*, a genus belonging to Gammaproteobacteria class as well as *Pseudomonas*, is also detected in cold raw milk samples (Munsch-Alatossava and Alatossava, 2006) and, in addition to its psychrotrophic features, may produce enzymes, that could lead to milk spoilage.

Even though some Gram-positive species have been highlighted in raw milk, their numbers are much smaller than those of Gram-negative species. *Bacillus cereus*, *Bacillus licheniformis*, *Bacillus subtilis*, and *Paenibacillus polymyxa* are frequently linked to milk spoilage. Some psychrotrophic *Bacillus* spp. secrete heat-resistant extracellular proteases, lipases, and phospholipases (lecithinases) (Hanamant et al., 2013) that are of comparable heat resistance as those of pseudomonads. *B. cereus*, frequently isolated from milk, has been examined carefully because of its "bitty cream" defect and potential enterotoxin production. Some strains show important spoilage potential in addition to the capacity to survive, possibly as spores, for several weeks at cold temperatures normally used for dairy products storage (De Jonghe et al., 2010; Gopal et al., 2015).

Some *Enterococcus* isolates can grow at 7 °C and have demonstrable proteolytic activity. These bacteria constitute only a minor population of the microbiota in raw milk, but their number may be proportionally higher in pasteurized milk because their resistance to pasteurization temperatures (Wessels et al., 1990).

Most of psychrotrophic bacteria found in raw milk are inactivated by heat treatment commonly used in dairy industries, such as pasteurization and UHT. However, these microorganisms may secrete hydrolytic enzymes, which can be heat resistant. *Pseudomonas*, mainly *P. fluorescens* group, *Serratia*, *Hafnia*, and *Bacillus* have strong proteolytic potential while other species of *Pseudomonas* (mainly nonfluorescent pseudomonads), *Bacillus*, *Enterobacter*, and *Acinetobacter* are strongly lipolytic (Hantsis-Zacharov and Halpern, 2007). *Pseudomonas* is the spoilage-predominant genus that secretes heat-stable hydrolytic enzymes isolated from cold raw milk regardless of sampling location, culture-dependent or culture-independent approaches, and time. Therefore, controlling the proliferation of psychrotrophic bacteria during refrigerated storage of raw milk is crucial to maintaining the shelf lives of the derived dairy products (Griffiths et al., 1987).

3.4 MOLECULAR TOOLS TO CHARACTERIZE PSYCHROTROPHIC BACTERIA

Molecular methods based on the analysis of DNA are usually applied for bacterial identification. These methods include cultural isolation and enrichment with subsequent DNA isolation of grown pure cultures. However, for uncultivable microorganism detection, the DNA must be extracted directly out of the sample.

The molecular identification of psychrotrophic bacteria is often difficult and controversial. The sequence analysis of the 16S rRNA gene, usually employed for the microbial identification, is not always satisfactorily discriminating between the species of *Pseudomonas* (Moor et al., 1996; Anzai et al., 2000; Yamamoto et al., 2000). A study on psychrotrophic spoilage bacteria highlighted that it was difficult to achieve an unequivocal identification of *Pseudomonas* strains at the species level, even though variable regions of the 16S rRNA gene were analyzed (Ercolini et al., 2006). For this reason, several authors have evaluated the use of alternative gene sequences, such as *carA*, *recA*, *gyrB*, *fliC*, *rpoD*, and *rpoB*, for the identification of psychrotrophic bacteria (Mollet et al., 1997; Yamamoto et al., 2000; Bellingham et al., 2001; Hilario et al., 2004). Particularly useful is the *rpoB* gene, encoding the β-subunit of RNA polymerase that was considered a core gene candidate for phylogenetic analyses and identification of bacteria, especially for closely related species (Adékambi et al., 2008). Recently, Sajben et al. (2011) and Decimo et al. (2014) used the *rpoB* gene sequence for the identification of *Pseudomonas*, *Citrobacter*, *Hafnia*, and *Serratia* strains isolated from mushroom caps and milk samples.

In addition to gene sequencing, other methods, like restriction fragment length polymorphism (RFLP) analysis (Rasolofo et al., 2011) and Multiplex-PCR assay, have been used to define the psychrotrophic bacterial species. Porteous et al. (2002) and Bhushan et al. (2013) demonstrate that RFLP analysis, performed with different enzymes (*Alu*I, *Hinf*I, *Rsa*I, *Tru*9I, *Bfa*I, *Dpn*II, *Hae*III), was a powerful tool to identify the fluorescent *Pseudomonas* species (*P. putida*, *Pseudomonas syringae*, *Pseudomonas aeruginosa*, *Pseudomonas stutzeri*, and *P. fluorescens*), while Machado et al. (2013) employed the proteases' gene sequences for a simultaneous identification of *Acinetobacter* spp., *P. fluorescens*, *Aeromonas hydrophila*, and *S. marcescens* in raw milk.

Finally, the detection of psychrotrophic bacteria using PCR to amplify the metalloprotease genes (*aprX* and *ser2*) represents an alternative method for the identification/characterization of proteolytic species, such as *P. aeruginosa*, *P. fluorescens*, *S. marcescens*, and *Serratia liquefaciens* (Martins et al., 2005; Dufour et al., 2008; Decimo et al., 2014; Machado et al., 2016).

One of the main interests of microbiologists is the study of the genetic differences among and within the bacterial species; the recognition of these differences is called *typing*. Traditional typing systems were based on phenotype; however, in the last decades the molecular approach has revolutionized the ability to differentiate strains and bacterial species. Different typing methods were applied to psychrotrophic bacteria.

Pulsed-field gel electrophoresis (PFGE), considered as the gold standard among molecular typing methods, was used successfully for the study of *P. aeruginosa* isolated from clinical and veterinary samples (Kidd et al., 2011; Waters et al., 2012; Scaccabarozzi et al., 2015). Due to the labour intensity, PFGE is not a feasible technique for large-scale typing of isolates, thus other molecular techniques were set up for an effective strain differentiation.

Multilocus sequence typing (MLST) is a universal method for characterizing bacteria based on allelic variation in housekeeping genes. Over the years MLST allele sequences were stored in databases hosted at different sites around the world, but currently the PubMLST.org site (http://pubmlst.org), collects data from all databases and makes them easily accessible (Larsen et al., 2014). Some MLST schemes are available for psychrotrophic bacteria such as: *Citrobacter freundii* (*aspC, clpX, fadD, mdh, arcA, dnaG,* and *lysP*), *Enterobacter cloacae* (*dnaA, fusA, gyrB, leuS, pyrG* and *rplB*), *P. aeruginosa* (*acsA, aroE, guaA, mutL, nuoD, ppsA*, and *trpE*) and *P. fluorescens* (*glnS, gyrB, ileS, nuoD, recA, rpoB*, and *rpoD*). It is interesting to note that some authors (Kidd et al., 2011; Waters et al., 2012), demonstrated a high level of concordance between PFGE and MLST results in *P. aeruginosa* genotyping. Moreover, Andreani et al. (2014) revealed the presence of a connection between the blue pigment produced by *P. fluorescens* and a specific MLST cluster.

Random amplification of polymorphic DNA analysis (RAPD) has been widely reported as a rapid, sensitive and inexpensive method for bacterial typing. In different studies, the use of two RAPD primers in combination (208 and 272 or M13 and OPAA10) allowed observation of a high degree of genotypic variability among *Pseudomonas* and Enterobacteriaceae strains isolated from different sources like bulk tank milk, drinking water, cockroach and human samples (Saitou et al., 2010; Decimo et al., 2014; Wu et al., 2016). In addition, as reported by Waters et al. (2012) the discriminatory power of PFGE, RAPD and MLST, used for *Pseudomonas* typing, was comparable, showing high values of Simpson's index (range: 0.973–0.980).

Other typing methods, such as enterobacterial repetitive intergenic consensus (ERIC) PCR, amplified fragment length polymorphism (AFLP) and PCR-ribotyping have been applied to explore genetic diversity among *Pseudomonas* species isolated from milk, animal, and human samples (Dogan and Boor, 2003; Bedeltavana et al., 2010; Osman et al., 2010; Kidd et al., 2011; Wu et al., 2016).

3.5 INFLUENCE OF PSYCHROTROPHIC CONTAMINATION OF RAW MILK ON DAIRY PRODUCT QUALITY

Modern dairy processing utilizes various preservation treatments that result in an assortment of dairy products having vastly different tastes and textures and a complex spoilage microbiota. However, growth of psychrotrophic bacteria is primarily responsible for limiting the keeping quality of milk and dairy products held at temperatures below 7 °C. The maintenance of raw milk for prolonged cold storage results in simultaneous production of various heat-stable proteolytic and lipolytic enzymes that consequently retain part of their activity after conventional heat treatment applied in dairy industries. Many enzymes resistant to the heat processes applied in the manufacture of processed milk and dairy products, particularly proteases and lipases, are produced by *Pseudomonas* and *Bacillus* species, respectively. Therefore, these enzymes may cause spoilage of the final products during storage and are of great concern to dairy processors.

The proteolysis of milk casein and other milk proteins also occurs due to the activity of the native milk's proteases such as plasmin, a serine protease that enters into milk from blood in the form of plasminogen. Other proteases may be secreted from mammary tissue cells, blood plasma or leucocytes. The action of these enzymes generate peptides, which usually give rise to flavor defects and technological problems as bitter and rancidity flavors as well as browning caused by reactions of the released amino acids upon heating. Besides off-flavor development, proteolytic enzymes strongly contribute to decreased yield during cheese production, milk heat-stability loss, gelation of UHT milk, and reduced shelf life of dairy products (Sørhaug and Stepaniak, 1997; Chen et al., 2003; Datta and Deeth, 2001; Fairbairn and Law, 1996). Proteolytic activity is the main cause of UHT milk spoilage, causing bitterness and gelation problems (Datta and Deeth, 2001; Baglinière et al., 2013). As low levels of proteolytic enzymes are sufficient to cause undesirable amounts of protein degradation in UHT milk during storage at room temperature, sensitive methods for their detection have been sought by the dairy industry. However, no method has been routinely adopted for this purpose.

The number of psychrotrophs required to produce off-flavors varies among species. It is determined not only by the growth rate at the storage temperature but also by the proteolytic and lipolytic activity and heat resistance of their enzymes. Some authors defend the theory that there is a significant correlation between the initial count of psychrotrophs and the storage life of raw milk at refrigeration temperatures. Generally, high levels of psychrotrophic bacteria in raw milk are required to contribute to sufficient quantities of heat-stable proteases and lipases to cause the breakdown of protein and fat after heat treatment. The number of psychrotrophs generally required to initiate spoilage in milk is about 10^6 CFU/ml (Birkeland et al., 1985; Matta et al., 1997). Thus, the detection and enumeration of these bacteria is useful to establish contamination and potential spoilage microbiota.

Lipolysis occurs due to the action of natural or microbial lipolytic enzymes that are able to hydrolyze triglycerides, a milk fat constituent, in the fatty acids such as butyric, caproic, caprylic, and capric acid, which are mainly responsible for off-flavors in milk and for rancidity in cheese (Chen et al., 2003). Free fatty acids with short chain acids (C4–C8) lead mainly to rancid flavors, while the middle-length chains (C10–C12) give rise to most of soapy, unclean or bitter flavors. Microorganisms that produce lipolytic enzymes are of concern in the dairy industry because they make these foods sensorially unacceptable to consumers. Lipolytic enzymes produced by psychrotrophs are more important than proteases in relation to the development of defects of flavor in cheese because proteases are soluble in water and lost in the whey, while lipases are adsorbed in the fatty globules and retained in cheese mass (Fox, 1989).

3.5.1 Bacterial proteases and proteolytic changes in milk

Several terms defining proteolytic enzyme can be found in the literature, such as peptidase, peptide bond hydrolase or protease. Proteolytic enzymes are enzymes able to hydrolyze protein, more particularly peptide bond. According to Barrett (2001), the term *peptidase* is recommended by the International Union of Biochemistry and Molecular Biology (IUBMB). However, the term *protease* is mainly used for proteolytic enzyme from psychrotrophic bacteria.

The majority of spoilage protease found in milk and dairy product are produced by Gram-negative bacteria such as *Pseudomonas, Serratia* or *Hafnia* (Marchand et al., 2009a; Decimo et al., 2014; Machado et al., 2015). Among the Gram-positive organisms, the genus *Staphylococcus*, *Streptococcus*, and particularly, *Bacillus*, have been encountered in milk as proteolytic enzyme producers (Teh et al., 2011). Although *Bacillus* spp. shows important diverse proteolytic activity (Nabrdalik et al., 2010), the genus is not highly detected in cold raw milk as a proteolytic enzyme producer. Considering that major milk spoilage is related to heat-stable protease generally produced by *Pseudomonas* species and other Gram-negative psychrotrophic bacteria, the focus of this section is on proteolytic enzyme produced by *Pseudomonas* and *Serratia*.

The well-studied protease related to milk spoilage is the heat-stable protease AprX, mostly produced by the genus *Pseudomonas* (Liao and McCallus, 1998; Dufour et al., 2008; Marchand et al., 2009b; Baglinière et al., 2013; Martins et al., 2015) but present in other genera such as *Bacillus*, *Acinetobacter*, and *Klebsiella*. Among the genus *Pseudomonas*, the gene of this protease has been detected in the species *P. fluorescens*, *P. fragi*, *P. aeruginosa*, and *Pseudomonas chlororaphis* (Martins et al., 2005; Marchand et al., 2009a; Decimo et al., 2014). The gene *aprX* is located at *aprX-lipA* operon of 14 kb containing eight different genes (McCarthy et al., 2004). The amino acid sequence analysis of the protease AprX showed that this protease is always rich in alanine and glycine residues but poor in cysteine and methionine residues (Dufour et al., 2008). This lack of cysteine residues avoids disulfide bonds on the tertiary structure of the protease, thus increasing its flexibility. AprX is a metalloprotease belonging to the serralysin family, and this enzyme contains the two consensus-binding motives for Ca^{2+} (GGXGXDXUX) and Zn^{2+} (HEIGHTLGLAHP). The requirement of these two metal ions for AprX stability and activity had been confirmed by numerous inhibition studies. While serine protease inhibitors, such as phenylmethane sulfonyl fluoride (PMSF) and leupeptin did not affect AprX activity, the presence of 1–5 mM of typical divalent-ion chelator such as *o*-phenantroline (Zn^{2+} chelator), ethylene glycol-bis (β-aminoethyl ether)-N,N,N′,N′-tetraacetic acid (EGTA, Ca^{2+} chelator), ethylenediamine tetraacetic acid (EDTA, Ca^{2+} and Zn^{2+} chelator), highly inhibited the enzyme activity (Kohlmann et al., 1991a; Schokker and van Boekel, 1997; Liao and McCallus, 1998; Dufour et al., 2008; Matéos et al., 2015).

The protease AprX of *P. fluorescens* has been the most studied among *Pseudomonas* genus. AprX presents the optimum activity between 37 and 47 °C and exhibits activity in a large range of temperatures from 0 to 55 °C. With an optimum activity between pH 7 and 10, AprX is considered an alkaline protease.

The particularity of the protease AprX is the high heat-stability. This protease is resistant to UHT treatment, and the two kinds of pasteurization used in dairy industries, high temperature short time (HTST) and low temperature long time (LTLT). Some characteristics of heat-stable proteases from different *Pseudomonas* strains are presented in the Table 3.1.

Table 3.1 Heat-stability of proteases secreted by *Pseudomonas* genus.

Strain	Heat stability	Reference
P. fluorescens F	Residual activity of 82% after 2 min at 110 °C	Matéos et al., 2015
P. fluorescens CY091	Residual activity of 20% after 10 min at 100 °C	Liao and McCallus, 1998
P. tolaasii	Residual activity of 20% after 1.5 min at 140 °C	Baral et al., 1995
P. fluorescens BJ-10	Residual activity of 39% after 3 min at 130 °C	Zhang et al., 2015
P. fluorescens Rm12	Residual activity of 20% after 20 s at 140 °C	Mu et al., 2009
P. fluorescens	D^a = 124 s, 140 °C	Kroll and Klostermeyer, 1984
P. fluorescens	D = 120 s, 140 °C	Vercet et al., 1977
P. fluorescens AR 11	D = 60 s, 140 °C	Alichanidis and Andrews, 1977

[a] *D* values (time required to reduce 90% of enzyme activity).

The exceptional heat-stability of these proteases could be due to the capacity to fix zinc and calcium and the absence of S-S bridges in the tertiary structure of the enzyme (Adams et al., 1975, 1976).

AprX hydrolyzes various protein substrates as collagen, casein and gelatin, and relating to milk protein, AprX may hydrolyze the four kinds of casein (α_{s1}, α_{s2}, β and κ) composing the micelle with a large activity spectrum. According to Matéos et al. (2015), casein hydrophobic areas could be preferentially hydrolyzed by AprX. Numerous studies using sodium caseinate showed that AprX preferentially hydrolyzed milk caseins in the following order κ->β->α_{S1}-caseins (Fairbairn and Law, 1986; Mu et al., 2009; Pinto et al., 2014). However, according to Baglinière et al. (2012, when the four caseins are in the micelle form, the β-casein is preferentially hydrolyzed by AprX.

The level of proteolytic activity among *Pseudomonas* genus is species and strain dependent. Dufour et al. (2008) and Marchand et al. (2009b) highlighted that, despite the presence of the *aprX* gene, some species of *Pseudomonas* do not show proteolytic activity, and Baglinière et al. (2012) showed that UHT milk destabilization by *P. fluorescens* was strain dependent. The high variability of proteolytic activity among *Pseudomonas* could be a consequence of heterogeneous enzyme expression, secretion, regulated by different factors such as quorum sensing, temperature, oxygen, iron content, and bacterial growth phase.

Generally, *P. fluorescens* produces at least one monomeric extracellular protease of molecular weight varying about 45–56 kDa corresponding to AprX. Nevertheless, secretions of heat-stable proteases with lower molecular weight have been reported by Stuknytė et al. (2016). These authors highlighted that *P. fluorescens* PS19 was able to secrete two heat-stable proteases of about 15 and 25 kDa, with the latter not showing homology to AprX. The presence of several proteolytic bands between 50 and 200 kDa by zymography in the supernatant of three strains of *P. chlororaphis* suggested that this species was able to secrete various proteases with higher molecular mass than AprX (Nicodème et al., 2005). Proteases of molecular masses higher than 56 kDa as yet have not been reported in the literature, and these bands of high molecular mass could correspond to self-aggregation of protease or aggregation with other compounds as described for the protease IV and the elastase B produced by *P. aeruginosa* (Caballero et al., 2001).

As described for *Pseudomonas* genus, the number of proteases secreted by *Serratia* genus is species and strain dependent. Using polyacrylamide gel electrophoresis, Matsumoto et al. (1984) detected four different proteases from *S. marcescens* kums 3958 with molecular weight of 56, 60 and 73 kDa. After isoelectric focusing, it was shown that the band of 73 kDa observed on the polyacrylamide gel was the mix of two proteases. A metalloprotease and a serine protease with molecular mass of 53.5 and 66.5 kDa, respectively, were detected in another strain of *S. marcescens* (Romero et al., 2001).

The proteases secreted by *S. liquefaciens* are encoded by *ser1* and *ser2* genes and were characterized as two serralysin-like metalloproteases with 45 and 52 kDa, respectively (Kaibara et al., 2010, 2012). As the protease AprX, the analysis of the primary structure of these two proteases showed the presence of the consensus motives HEXXHXUGUXH and GGXGXDXUX for the binding of Zn^{2+} and Ca^{2+}, respectively. These two proteases also contain the ABC exporter motif (DXXX), suggesting that they are extracellular enzymes. The main difference between Ser1 and Ser2 proteases seems to be the heat resistance. After heat treatment of the supernatant of *S. liquefaciens* culture at 95 °C for 8.45 min, only the protease Ser2 was able to hydrolyze sodium caseinate (Machado et al., 2016). The protease Ser2 of *S. liquefaciens* L53 presented *D*-values (time required to reduce 90% of enzyme activity) at 75, 85 and 95 °C in semi-skimmed UHT milk higher than the protease AprX of the strain *P. fluorescens* W2a (Machado et al., 2016). This study demonstrated the importance of taking into account the presence of the species *S. liquefaciens* in raw milk because of the secretion of the protease Ser2, an enzyme highly heat-stable and able to hydrolyze casein.

Most technological problems encountered in dairy industries come from the activity of heat-resistant protease secreted in raw milk before processing, resulting in a nondesirable hydrolysis of milk protein. About 80% of total proteins in milk are casein- (α_{s1}-, α_{s2}-, β- and κ-casein) associated in micelle form. Milk destabilization occurs when casein micelles lose their stability, and the protease AprX is one of the main causes of this technological problem. This protease resists UHT treatment and hydrolyzes, during the storage, the four caseins composing the micelle. Casein micelle stability is negatively affected by this hydrolysis, leading to UHT milk destabilization and the presence of aggregates corresponding to casein micelles aggregations. The detection of numerous peptides coming from the κ-casein suggested that this casein was hydrolyzed, affecting the casein micelles charge. When a modification of casein micelle charge occurs, the repulsions between micelles are reduced, resulting in the formation of aggregates and thus milk destabilization.

The proteolysis of casein and other milk proteins may also generate peptides responsible for bitter flavors. The amino acids released can produce browning upon heating and off-flavor of heat-treated dairy products, such as UHT milk. The combination of proteolysis with Maillard reaction may generate new flavor and volatile components in dairy products.

Yield reduction in cheese manufacturing is another technological problem caused by proteolytic activity from psychrotrophic microorganisms, and this phenomenon is frequently encountered in the dairy industry. Generally, psychrotrophic proteases are secreted during the 48–72 h of milk refrigeration, and before heat treatment, these proteases are active towards casein micelles, generating peptides of different sizes. One of the first steps of cheese making is milk acidification at pH 4–5 that precipitates casein. Peptides generated from casein proteolysis cannot precipitate at this pH range, resulting in the loss of nitrogenous matter, leading to cheese yield reduction. Using raw milk stored for 4 days under cold temperatures, Cardoso (2006) showed a reduction of 6.38% of the yield of fresh Minas cheese.

Heat-stable or not, the presence of protease in raw milk may be costly for dairy industries. It seems low levels of these enzymes are sufficient to cause undesirable proteolysis, suggesting that sensitive methods for their detection are necessary in order for the dairy industry to avoid theses technological problems.

3.5.2 Bacterial lipases and phospholipases and their significance in milk

The triglyceride fat fraction of milk, accounted for by more than 96% of total lipid, is located in the core of fat globules, which are arranged from a conjugated ester of long-chained fatty acids with glycerol molecules, enveloped by a thin layer composed of phospholipids, glycolipids, and several proteins (Evers et al., 2008). Obviously, to release the free fatty acids (FFAs) and cause changes and flavor defects in the milk (rancid or soap flavors if, respectively, short-chain or medium- to long-chain fatty acids are released), several consecutive cleavages are necessary, beginning with the breaking of the surrounding globule fat membrane followed by the lysis of the triglycerides into glycerol and fatty acids, which are accumulated in the milk, giving rise to the above mentioned defects. The fat globule is susceptible to the action of many lipolytic agents, such as physiological, physiomechanical and chemical factors (Deeth and Fitz-Gerald, 2006), the latter involving lipases (McPherson and Kitchen, 1983).

Lipases are carboxylesterases that hydrolyze the ester bond of acylglycerols. Those that hydrolyze acylglycerols of less than 10 carbon-chain fatty acids are named *esterases* or carboxylases, whereas those hydrolyzing acylglycerols with more than 10 carbon-chain fatty acids are the *lipases* or *triacylglycerol acylhydrolases* (Jaeger et al., 1999). Lipolytic enzymes can be of endogen origin, in which case most of the lipases are associated with the casein micelle structure (Stepaniak, 2004), or of exogenous origin, secreted as a result of the bacterial metabolic action.

Many lipases have been isolated from plant, animal, and microbial organisms, but microbial lipases are those that find the greatest application. Among the producers of bacterial lipases are species of the genera *Pseudomonas*, *Burkholderia*, *Vibrio*, *Acinetobacter*, *Bacillus*, and *Staphylococcus* (Arpigny and Jaeger, 1999; Snellman et al., 2002). Particular attention is focused on specific classes of enzymes of species *Pseudomonas* because they are among the first studied and used in biotechnological production and because of their involvement in bacterial pathogenesis. Recently, lipases were found in a number of *Pseudomonas* species (Junwal et al., 2003; Kim et al., 2005; Noureddini et al., 2005; Karadzic et al., 2006; Singh and Banerjee, 2007; Wang et al., 2009). On the other hand, microbial lipases represent a problem in the dairy industry due to the hydrolysis of milk triglycerides, leading to "hydrolytic spoilage."

Lipases from raw milk associated–psychrotrophs have indicated as a major problem in the future, highly affecting the level of FFAs in milk. Psychrotrophic lipases, catalyzing the hydrolysis of triacylglycerols, lead to the accumulation of free (nonesterified) fatty acids, partial glycerol esters (monoacylglycerols, diacylglycerols) and even glycerol in some cases in milk (Deeth, 2006). The products of lipolysis are highly detrimental to the formation and stability of milk foams, which are colloidal systems in which air bubbles are stabilized by a matrix composed of milk components (De Jonghe et al., 2011). The depression of milk foaming, caused by lipolysis, is due to the partial glycerides, which are surface active and displace the foam-stabilizing proteins at the air–water interface of the foam bubbles. Besides, although lipolytic degradation of milk is not as predominant as proteolytic degradation, the FFAs released from milk fat hydrolysis, caused by psychrotrophic lipase activity, are the primary cause of changes in product flavor, described as rancid, unclean, butyric, astringent, soapy or bitter, making the product barely acceptable (Ma et al., 2000; Deeth, 2006). The lipolytic flavor defects are particularly pronounced in cream, butter, cheese, and whole UHT milk (Stead, 1986; Champagne et al., 1994). In general, psychrotrophic lipases have molecular masses ranging from 30 to 50 kDa, pH optima between

7 and 9 and a temperature optima range from 22 to 55 °C. From the patterns of hydrolysis of triacylglycerols, bacterial lipases are divided into two major groups, I and II. Group I (nonspecific) lipases catalyze the production of glycerol by releasing fatty acids from all three positions; group II (1,3-specific) lipases catalyze the formation of 1,2- and 2,3-diacylglycerols and 2-monoacylglycerols with the release of fatty acids from positions 1 and 3 of the triglyceride. In the case of group II lipase products, acyl migration leads to the production of 1,3-diacylglycerols and 1-monoacylglycerol, and subsequently, the production of glycerol over an extended period (Macrae, 1983). The ability of psychrotrophic bacteria to produce lipases varies considerably across genera, as well as among species of the same genus.

Pseudomonas spp. are the primary concern with regard to lipolytic degradation of milk fat, and most of the research has been focused on this genus, but there is growing evidence that other genera, belonging to the Enterobacteriaceae family, may have similar relevance by producing strong lipolytic activity. Decimo et al. (2014) reported that the highest number of positive psychrotrophic strains at all incubation temperatures (7, 22, and 30 °C) was observed for lipolytic activity (59/80), followed by proteolytic (31/80) and lecithinase activities (28/80). In the same study, 68 strains displayed a noticeable lipolytic activity after incubation at 22 °C within 3 days; moreover, 49 out of 63 *Pseudomonas* (77.77%) and 9 out of 17 Enterobacteriaceae (52.94%) strains, belonging to *H. alvei* and *S. marcescens*, showed lipase production at 7 °C.

It is generally known that psychrotrophic bacteria produce thermostable lipases active on milk fat and are responsible for spoilage as a result of the production of short- and medium-chain fatty acids. A high FFA accumulation causes the deterioration of technological properties (e.g. cream less able to be whipped) and sensory attributes of milk, which is reflected by a rancid or soap off-flavor that may negatively influence the quality of dairy products. According to Vyletělová et al. (2000), the medium-chain fatty acids (C12–C16) are primarily liberated by the action of *P. fluorescens* lipolytic enzymes, while short-chain fatty acids (C6–C10) are sporadically released or released in very small quantities.

The ability of lipolytic strains to release fatty acids from milk fat varies between and within species related to different capacity of the production of lipolytic enzymes. Triacylglycerol breakdown by psychrotrophs may occur in a greater or lesser extent, but type and amount of released FFAs are not easily predicted (Decimo et al., 2014).

Lecithinases are important groups of phospholipases produced by psychrotrophic bacteria that are able to disrupt the protective membrane structure of fat globules so that the milk fat becomes available to the native milk lipases, resulting in physical degradation of the emulsion in milk and in the development of bitter off-flavors. Little information is available regarding psychrotroph lecithinase activity, and Decimo et al. (2014) observed that the number of *Pseudomonas* producing lecithinases increased with incubation temperatures from 39.68% of positive strains at 7 °C to 55.55% at 30 °C. For Enterobacteriaceae, however, almost half of the (47%) degraded lecithin at 30 and 22 °C, whereas only 10% displayed lecithinase activity at 7 °C. There remain on-going challenges for dairy industry and dairy scientists in elucidating the remaining details of the mechanisms by which lipolysis is promoted and proceeds in milk in order to control a problem that is a constant concern for the dairy industry.

3.6 REGULATION OF EXTRACELLULAR ENZYMES

Despite of the presence of large numbers of studies about psychrotrophic extracellular enzymes in the literature, the understanding of regulation of extracellular hydrolytic enzymes production in milk and dairy products is relatively limited. Regulation of *Pseudomonas* proteases and lipases production was most studied in comparison to the other psychrotrophic genera. For this reason, this section is focuses on the regulation of extracellular proteases and lipases produced by the genus *Pseudomonas*.

Many factors such as temperature, growth phase or oxygen, and calcium content are involved in the protease and lipase production by psychrotrophic bacteria. Because milk is refrigerated and stored under aerobic conditions, the effect of temperature and oxygen content on hydrolytic enzymes production have been extendedly studied. It is well known that *Pseudomonas* species secrete extracellular enzymes at temperatures used for milk refrigeration (4–10 °C) (Sørhaug and Stepaniak, 1997; Baglinière et al., 2012; Vithanage et al., 2016). However, most of *P. fluorescens* strains produce the maximum quantity of extracellular enzymes at temperatures between 17.5 and 25 °C, slightly below the optimum temperature for growth of these bacteria (McKellar, 1982; Fairbairn and Law, 1986; Gugi et al., 1991; Hellio et al., 1993; Nicodème et al., 2005; Anbu, 2014).

Many authors showed that the production of extracellular proteases begins during the exponential phase of growth (Birkeland et al., 1985; Hellio et al., 1993; Schokker and van Boekel, 1997; Rajmohan et al., 2002; Nicodème et al., 2005; Liu et al., 2007; Dufour et al., 2008), while others showed that this production begins at the end of exponential phase and at the stationary phase (McKellar, 1982; Fairbairn and Law, 1986; McKellar and Cholette, 1986; Stead, 1987; Kohlmann et al., 1991b; Stevenson et al., 2003). These contradictory results could be also explained by the different sensitivity of the methods used to detect the proteolytic activity. It is quite possible that in some studies the method used was not sensitive enough to detect proteolysis during the exponential phase of growth. Another factor to be considered is the O_2 concentration in the culture medium, which could be different in these studies. It has been acknowledged that lipases are produced during the late log and early stationary phases of growth (Fox and Stepaniak, 1983; Stuer et al., 1986; Sidhu et al., 1998). As described previously, the beginning of lipase secretion is influenced by the quantity of oxygen present in the culture medium.

Raw milk is normally stored under weak agitation that does not allow good oxygenation. In this condition, cold stored milk is poor in oxygen and could promote the protease production during the exponential growth of psychrotrophic bacteria. *Pseudomonas* species are strictly aerobic and require oxygen to grow. However, the regulation of protease production by O_2 is more complex. Some authors observed in the presence of this gas an inhibition of protease production (Griffiths and Phillips, 1984; Fairbairn and Law, 1986), while others demonstrated that oxygen was essential for its production (Birkeland et al., 1985; Matta et al., 1994). The effect of oxygen on the regulation of protease and lipase production by *Pseudomonas* seems to be concentration dependent. As described by Rowe and Gilmour (1982) and Rowe et al. (1990), the initiation of protease production by *Pseudomonas* is only possible when the concentration of oxygen decreases during the exponential phase of growth. The effect of modified atmosphere by the replacement of air with N_2 and CO_2 in milk inoculated with *Pseudomonas* spp. hindered the growth of *Pseudomonas* and protease and lipase production (Skura et al., 1986; Ma et al., 2003; Munsch-Alatossava et al., 2013).

The effect of calcium is one of the more important factors for the production of active protease by *Pseudomonas*. This mineral is present in high concentration in milk compared to other food products. There is no consensus about the requirement of Ca^{2+} for protease production, and it is quite possible that calcium is necessary for the stability and the activity of the protease but not for its production (Baumann et al., 1993; Liao and McCallus, 1998; Hege and Baumann, 2001). Calcium is also essential for the activity and stability of a few lipases from *Pseudomonas* (Andersson et al., 1979; Martins et al., 2015; Shibata et al., 1998). According to Andersson et al. (1979), heat-stability of *Pseudomonas* lipase can be increased by the combination of polysaccharides and divalent cations, such as Ca^{2+}.

3.7 CONTROL OF PSYCHROTROPHIC BACTERIA AND RELATED ENZYMES

Because of the substantial economic impact of this group of microorganisms upon the global dairy industry, psychrotrophic bacteria have been and continue to be studied extensively, with the main objectives of developing effective control measures and establishing regulations to ensure the quality and safety of milk and dairy products (McPhee and Griffiths, 2011).

The dairy industry has been faced with the control of milk contamination by psychrotrophs, and the implementation of good hygienic practices on dairy farms is a prerequisite for the production of high-quality and microbiologically safe raw milk. Although prevention is technically possible, it is difficult to achieve in practice. Human errors that result in improperly cleaned, sanitized, and handled equipment allow psychrotrophs to enter raw and pasteurized milk. Hence, control must focus on inactivation of psychrotrophs or prevention of their growth in milk. Since proper cleaning and sanitization of dairy equipment are important for production of milk with acceptable microbial quality (AOAC, 2006), control of psychrotrophic should begin at the farm level.

Various methods have been examined for their effectiveness in inhibiting psychrotrophs such as addition of such additives as potassium sorbate, H_2O_2, lysozyme, lactoferrin, lactic acid bacteria, chlorine, CO_2 or ozone injection, and low or high temperatures (Champagne, 1994; Varga and Szigeti, 2016). Applicability of this method of control to dairy products may be limited. The most important is low or high temperatures.

Storage temperature of raw milk will influence the quality of the resulting products, and if the milk was of good quality, that is, less than 10^4 CFU/ml, and was held at 1 °C, after 3 days the total count could be kept under 10^5 CFU/ml. This method is, however, energy intensive and requires large capital outlay for equipment at each stage of collection, transportation and processing. Another alternative is the prompt application of heat treatment after the milking process. Various time/temperature combinations have been applied to eliminate psychrotrophs, and thermization of raw milk at the farm before storage could reduce the total count. This consisted of heating at 60 to 65 °C for 10 s, followed by cooling at 7 °C. Such treatment is presumed to inhibit initial growth through thermal shock, and after the initial 3 days period, growth resumed as normal.

3.8 CONCLUSIONS

Despite the numerous studies on the control of psychrotrophic microorganisms in raw milk, the effective control of processing conditions by adoption of quality systems such as hazard analysis and critical control points (HACCP) or similar procedures is still the best way of ensuring that safe and high-quality products with extended shelf life are produced.

References

Adams DM, Barach JT, Speck ML. 1975. Heat resistant proteases produced in milk by psychrotrophic bacteria of dairy origin. *J Dairy Sci* 58:828–34.

Adams DM, Barach JT, Speck ML. 1976. Effect of psychrotrophic bacteria from raw milk on milk proteins and stability of milk proteins to ultrahigh temperature treatment. *J Dairy Sci* 59:823–27.

Adékambi T, Drancourt M, Raoult D. 2008. The *rpoB* gene as a tool for clinical microbiologists. *Trends Microbiol* 17:38–45. doi:10.1016/j.tim.2008.09.008.

Alichanidis F, Andrews AT. 1977. Some properties of the extracellular proteases produced by psychrotrophic bacterium *Pseudomonas fluorescens* strain AR11. *Biochim Biophys Acta* 485:424–33.

Anbu P. 2014. Characterization of an extracellular lipase by *Pseudomonas koreensis* BK-L07 isolated from soil. *Prep Biochem Biotechnol* 44(3):266–80. doi: 10.1080/10826068.2013.812564.

Andersson RE, Hedlund CB, Jonsson U. 1979. Thermal inactivation of a heat-resistant lipase produced by the psychotrophic bacterium *Pseudomonas fluorescens*. *J Dairy Sci* 62(3):361–67.

Andreani NA, Martino ME, Fasolato L, Carraro L, Montemurro F, Mioni R, Bordin P, Cardazzo B. 2014. Tracking the blue: A MLST approach to characterize the *Pseudomonas fluorescens* group. *Food Microbiol* 39:116–26. doi:10.1016/j.fm.2013.11.012.

Anzai Y, Kim H, Park JY, Wakabayashi H, Oyaizu H. 2000. Phylogenetic affiliation of the pseudomonads based on 16S rRNA sequence. *Int J Syst Evol Microbiol* 50:1563–89. doi: 10.1099/00207713-50-4-1563.

AOAC - Association of Official Analytical Chemists. 2006. *Official methods of analysis*. 18th ed. Washington DC: AOAC.

Arpigny JL, Jaeger KE. 1999. Bacterial lipolytic enzymes: classification and properties. *Biochem J* 1:177–83.

Bade RD, Reinemann DJ, Thompson PD. 2008. Method for assessing teat and udder hygiene. American Society of Agricultural and Biological Engineers (ASABE) annual meeting, Providence, RI, June 29–July 2, 2008. Paper no. 083796.

Baglinière F, Matéos A, Tanguy G, Jardin J, Briard-Bion V, Rousseau F, Robert B, Beaucher E, Gaillard JL, Amiel C, Humbert G, Dary A, Gaucheron F. 2013. Proteolysis of ultra high temperature-treated casein micelles by AprX enzyme from *Pseudomonas fluorescens* F induces their destabilisation. *Int Dairy J* 31(2):55–61. doi: 10.1016/j.idairyj.2013.02.011.

Baglinière F, Tanguy G, Jardin J, Matéos A, Briard V, Rousseau F, Robert B, Beaucher E, Humbert G, Dary A, Gaillard JL, Amiel C, Gaucheron F. 2012. Quantitative and qualitative variability of the caseinolytic potential of different strains of *Pseudomonas fluorescens*: Implications for the stability of casein micelles of UHT milks during their storage. *Food Chem* 135(4): 2593–603. doi: 10.1016/j.foodchem.2012.06.099.

Bai AJ, Rai VR. 2011. Bacterial quorum sensing and food industry. *Compr Rev Food Sci Food Safety* 10 (3):183–93.

Baral A, Fox PF. 1997. Isolation and characterization of an extracellular lipase from *Pseudomonas tolaasii*. *Food Chem* 58:33–38.

Barbano D, Ma Y, Santos M. 2006. Influence of raw milk quality on fluid milk shelf life. *J Dairy Sci* 89:E15-E19. doi: http://dx.doi.org/10.3168/jds.S0022-0302(06)72360-8.

Barrett AJ. 2001. Proteolytic enzymes: nomenclature and classification. In: Beynon R, Bond JS. editors. *Proteolytic enzymes: a practical approach*. 2nd ed. Oxford: Oxford University Press, pp. 1–21.

Baumann U, Wu S, Flaherty KM, McKay DB. 1993. Three-dimensional structure of the alkaline protease of *Pseudomonas aeruginosa*: a two-domain protein with a calcium binding parallel beta roll motif. *EMBO J* 12(9):3357.

Baur C, Krewinkel M, Kranz B, von Neubeck M, Wenning M, Scherer S, Stoeckel M, Hinrichs J, Stressler T, Fischer L. 2015a. Quantification of the proteolytic and lipolytic activity of microorganisms isolated from raw milk. *Int Dairy J* 49:23–29. doi: http://dx.doi.org/10.1016/j.idairyj.2015.04.005.

Baur C, Krewinkel M, Kutzli I, Kranz B, von Neubeck M, Huptas C, Wenning M, Scherer S, Stoeckel M, Hinrichs J, Stressler T, Fischer L. 2015b. Isolation and characterisation of a heat-resistant peptidase from *Pseudomonas panacis* withstanding general UHT processes. *Int Dairy J* 49:46–55. doi: http://dx.doi.org/10.1016/j.idairyj.2015.04.009.

Bava L, Zucali M, Sandrucci A, Brasca M, Vanoni L, Zanini L, Tamburini A. 2011. Effect of cleaning procedure and hygienic condition of milking equipment on bacterial count of bulk tank milk. *J Dairy Res* 78(02):211–19.

Beales N. 2004. Adaptation of microorganisms to cold temperatures, weak acid preservatives, low pH, and osmotic stress: a review. *Compr Rev Food Sci Food Safety* 3(1):1–20.

Bedeltavana A, Haghkhah M, Nazer A. 2010. Phenotypic characterization and PCR-ribotyping of *Pseudomonas fluorescens* isolates, in tracking contamination routes in the production line of pasteurized milk. *Iranian J Vet Res* 11:222–32.

Bellingham NF, Morgan JAW, Saunders JR, Winstanley C. 2001. Flagellin gene sequence variation in the genus *Pseudomonas*. *Syst Appl Microbiol* 24:157–65. doi:10.1078/0723-2020-00031.

Bhunia AK. 2008. *Foodborne microbial pathogens: mechanisms and pathogenesis*. West Lafayette, IN: Springer.

Bhushan A, Joshi J, Shankar P, Kushwah J, Raju SC, Purohit HJ, Kalia VC. 2013. Development of genomic tools for the identification of certain *Pseudomonas* up to species level. *Indian J Microbiol.* 53:253–63. doi 10.1007/s12088-013-0412-1.

Birkeland SE, Stepaniak L, Sørhaug T. 1985. Quantitative studies of heat-stable proteinase from *Pseudomonas fluorescens* P1 by the enzyme-linked immunosorbent assay. *Appl Environ Microbiol* 49:382–87.

Boor K, Fromm H. 2009. Managing microbial spoilage in the dairy industry. In: Blackburn Cde W. editor. *Food spoilage microorganisms*. Cambridge, UK: Woodhead Publishing. pp 171–93.

Caballero AR, Moreau JM, Engel LS, Marquart ME, Hill JM, O'Callaghan RJ. 2001. *Pseudomonas aeruginosa* protease IV enzyme assays and comparison to other Pseudomonas proteases. *Anal Biochem.* 290 330–337.

Cappitelli F, Polo A, Villa F. 2014. Biofilm formation in food processing environments is still poorly understood and controlled. *Food Engin Rev* 6(1-2):29–42.

Cardoso RR. 2006. Influência da microbiota psicrotrófica no rendimento de queijo Minas Frescal elaborado com leite estocado sob refrigeração. *Mestrado em Microbiologia Agrícola* Dissertação, Universidade Federal de Viçosa.

Catanio FS, Inay OM, da Silva AS, Pereira JR, Tamanini R, Beloti V, de Rezende Costa M, de Souza CHB, Aragon-Alegro LC, de Santana EHW. 2012. Refrigerated raw milk quality of a processing plant in the north of Parana after the implementation of changes imposed by NI 62 of 2011. *Semina: Ciências Agrárias* 33(2):3171–80.

Cempírková R. 2002. Psychrotrophic vs. total bacterial counts in bulk milk samples. *Veterinarni Medicina-Praha* 47(8):227–33.

Champagne CP, Laing RR, Roy D, Mafu AA, Griffiths MW. 1994. Psychrotrophs in dairy products: their effects and their control. *Crit Rev Food Sci Nutrition* 34:1–30.

Chen L, Daniel R M, Coolbear T. 2003. Detection and impact of protease and lipase in milk and milk powders. *Int Dairy J* 13:255–75.

Chen TR, Wei QK, Chen YJ. 2011. *Pseudomonas* spp. and *Hafnia alvei* growth in UHT milk at cold storage. *Food Control* 22:697–701.

Cleto S, Matos S, Kluskens L, Vieira MJ. 2012. Characterization of contaminants from a sanitized milk processing plant. *PLoS ONE* 7(6):e40189. doi: 10.1371/journal.pone.0040189.

Constantin OE. 2009. Bacterial biofilms formation at air liquid interfaces. *Innov Romanian Food Biotechnol* 5:18.

Cousin MA. 1982. Presence and activity of psychrotrophic microorganisms in milk and dairy products: a review. *J Food Protec* 45 172–207.

Datta N, Deeth HC. 2001. Age gelation of UHT milk – a review. *Instit Chem Eng* 79:197–210.

Decimo M, Morandi S, Silvetti T, Brasca M. 2014. Characterization of gram-negative psychrotrophic bacteria isolated from Italian bulk tank milk. *J Food Sci* 79(10):M2081–M2090. doi: 10.1111/1750-3841.12645.

Deeth HC, Fitz-Gerald CH. 2006. Lipolytic enzymes and hydrolytic rancidity. In: Fox PF, McSweeney PLH. editors. *Advanced dairy chemistry: lipids*. vol 2. 3rd ed. New York: Springer. pp 1–76.

De Jonghe V, Coorevits A, Van Hoorde K, Messens W, Landshoot A, De Vos P, Heyndrickx M. 2011. Influence of storage conditions on the growth of *Pseudomonas* species in refrigerated raw milk. *Appl Environl Microbiol* 77:460–70. doi: http://dx.doi.org/10.1128/AEM.00521-10.

Delbès C, Ali-Mandjee L, Montel M-C. 2007. Monitoring bacterial communities in raw milk and cheese by culture-dependent and -independent 16S rRNA gene-based analyses. *Appl Environ Microbiol* 73(6):1882–91.

Dogan B, Boor KJ. 2003. Genetic diversity and spoilage potentials among *Pseudomonas* spp. isolated from fluid milk products and dairy processing plants. *Appl Environ Microbiol.* 69:130–38. doi: 10.1128/AEM.69.1.130–138.2003.

Dufour D, Nicodème M, Perrin C, Driou A, Brusseaux E, Humbert G, Gaillard JL, Dary A. 2008. Molecular typing of industrial strains of *Pseudomonas* spp. isolated from milk and genetical and biochemical characterization of an extracellular protease produced by one of them. *Int J Food Microbiol* 125:188–196. doi:10.1016/j.ijfoodmicro.2008.04.004.

Drenkard E, Ausubel FM. 2002. *Pseudomonas* biofilm formation and antibiotic resistance are linked to phenotypic variation. *Nature* 416(6882)–740–43.

Elmoslemany AM, Keefe GP, Dohoo IR, Wichtel JJ, Stryhn H, Dingwell RT. 2010. The association between bulk tank milk analysis for raw milk quality and on-farm management practices. *Prevent Vet Med* 95:32–40.

Ercolini D, Russo F, Ferrocino I, Villani F. 2009. Molecular identification of mesophilic and psychrotrophic bacteria from raw cow's milk. *Food Microbiol* 26:228–31.

Ercolini D, Russo F, Torrieri E, Masi P, Villani F. 2006. Changes in the spoilage-related microbiota of beef during refrigerated storage under different packaging conditions. *Appl Environ Microbiol* 72:4663–71. doi: 10.1128/AEM.00468-06

Evers JM, Haverkamp RG, Holroyd SE, Jameson GB, Mackenzie DDS, McCarthy OJ. 2008. Heterogeneity of milk fat globule membrane structure and composition as observed using fluorescence microscopy techniques. *Int Dairy J* 18:1081–89.

Fairbairn DJ, Law BA. 1986. Proteinases of psycrotrophic bacteria: their production, properties, effects and control. *J Dairy Res* 53:139–77.

Fox PF. 1989. Proteolysis during cheese manufacture and ripening. *J Dairy Sci* 72:1379–1400.

Fox PF, Stepaniak L. 1983. Isolation and some properties of extracellular heat-stable lipases from *Pseudomonas fluorescens* strain AFT 36. *J Dairy Res* 50(1):77–89.

Frank JF, Yousef AE. 2004. Tests for groups of microorganisms. In: Wehr HM, Frank JF. editors. *Standard methods of the examination of dairy products*. 17th ed. Washington, DC: American Public Health Association. pp 227–47.

Garcia ML, Sanz B, Garcia-Collia P, Ordonez JA. 1989. Activity and thermostability of the extracellular lipases and proteinases from pseudomonads isolated from raw milk. *Milchwissenschaft* 44(9):547–50.

Gopal N, Hill C, Ross PR, Beresford TP, Fenelon MA, Cotter PD. 2015. The prevalence and control of In corsivo and related spore-forming bacteria in the dairy industry. *Frontiers Microbiol* 6:1–18.

Griffiths MW, Phillips JD. 1984. Effect of aeration on extracellular enzyme synthesis by psychrotrophs growing in milk during refrigerated storage. *J Food Protection* 47(9):697–702.

Griffiths MW, Phillips JD, Muir DD. 1987. Effect of low-temperature storage on the bacteriological quality of raw milk. *Food Microbiol* 4(4):285–91.

Gugi B, Orange N, Hellio F, Burini JF, Guillou C, Leriche F, Guespin-Michel JF. 1991. Effect of growth temperature on several exported enzyme activities in the psychrotrophic bacterium *Pseudomonas fluorescens*. *J Bacteriol* 173(12):3814–20.

Hanamant PS, Bansilal GM. 2013. Lipolytic psychrotrophic gram positive cocci in milk and fermented milk products. *J Environ Res Develop* 8(2):273.

Hantsis-Zacharov E, Halpern M. 2007. Culturable psychrotrophic bacterial communities in raw milk and their proteolytic and lipolytic traits. *Appl Environ Microbiol* 73(22):7162–68.

Hege T, Baumann U 2001. Protease C of *Erwinia chrysanthemi*: the crystal structure and role of amino acids Y228 and E189. *J Mol Biol* 314(2):187–93. doi: 10.1006/jmbi.2001.5124.

Hellio FC, Orange N, Guespin-Michel JF. 1993. Growth temperature controls the production of a single extracellular protease by *Pseudomonas fluorescens* MF0, in the presence of various inducers. *Res Microbiol* 144(8):617–25.

Heyndrickx M, Marchand S, Jonghe VD, Smet K, Coudijzer K, Block JD. 2010. Understanding and preventing consumer milk microbial spoilage and chemical deterioration. In: Griffiths MW. *Improving the safety and quality of milk — improving quality in milk products*. Vol 2. Cambridge: Woodhead Publishing. pp 96–135.

Hilario E, Buckley TR, Young JM. 2004. Improved resolution of the phylogenetic relationships among *Pseudomonas* by the combined analysis of *atpD, carA, recA* and 16S rDNA. *Antonie Leeuwenhoek* 86:51–64. doi:10.1023/B:ANTO.0000024910.57117.16

Jackson K, Biggerstaff M, Tobin-D'Angelo M, Sweat D, Klos R, Nosari J, Garrison O, Boothe E, Saathoff-Huber L, Hainstock L, Fagan RP. 2011. Multistate outbreak of *Listeria monocytogenes* associated with Mexican-style cheese made from pasteurized milk among pregnant, Hispanic women. *J Food Protection* 74(6):949–53.

Jaeger KE, Dijkstra BW, Reetz MT. 1999. Bacterial biocatalysts: molecular biology, three dimensional structures, and biotechnological applications of lipases. *Annual Rev Microbiol* 53(1):315–51.

Jayarao BM, Wang L. 1999. A study on the prevalence of gram-negative bacteria in bulk tank milk. *J Dairy Sci* 82:2620–24.

Junwal U, Roy U, Chowdhury A, Bhaduri A, Roy P. 2003. Purification and characterization of an alkaline lipase from a newly isolated *Pseudomonas mendoncina* PK-12Cs and chemoselective hydrolysis of fatty acid ester. Bioorg. Medic. Chem. 11 1041–1046.

Kaibara F., Iiyama K., Chieda Y., Man Lee J., Kusakabe T., Yasunaga-Aoki C., Shimizu S., 2012. Construction of serralysin-like metalloprotease-deficient mutants of *Serratia liquefaciens* and their virulence in the silkworm, *Bombyx mori*. J. Insect Biotechnol. Sericology 81(2-3) 55–61. doi: 10.11416/jibs.81.2_3_055.

Kaibara F., Iiyama K., Yasunaga-Aoki C., Shimizu S. 2010. Molecular cloning of serralysin-like melallopro-tease ser1 gene from *Serratia liquefaciens*. *Entomotechnology* 34:25–27.

Karadzic I, Masui A, Zivkovic L, Fujiwara N. 2006. Purification and characterization of an alkaline lipase from *Pseudomonas aeruginosa* isolated from putrid mineral cutting oil as component of metalworking. *Xuid J Biosci Bioeng* 102:82–89.

Kidd TJ, Grimwood H, Ramsay KA, Rainey PB, Bell SC. 2011. Comparison of three molecular techniques for typing *Pseudomonas aeruginosa* isolates in sputum samples from patients with cystic fibrosis. *J Clin Microbiol* 49:263–68. doi:10.1128/JCM.01421-10.

Kim K, Kwon D, Yoon S, Kim W, Kim K. 2005. Purification, refolding, and characterization of recombinant *Pseudomonas fluorescens* lipase. *Protein Expr Purification* 39:24–129.

Kohlmann KL, Nielsen SS, Ladisch MR. 1991a. Purification and characterization of an extracellular protease produced by *Pseudomonas fluorescens* M3/6. *J Dairy Sci* 74(12):4125–36.

Kohlmann KL, Nielsen SS, Steenson LR, Ladisch MR. 1991b. Production of proteases by psychrotrophic microorganisms. *J Dairy Sci* 74(10):3275–83. doi: 10.3168/jds.S0022-0302(91):78513–5.

Kroll S, Klostermeyer H. 1984. Heat inactivation of exogenous proteinases from *Pseudomonas fluorescens*. I. Possibility of inactivation in milk. *Zeitschrift für Lebensmitteluntersuchung und-Forschung A* 179:288–95.

Kumaresan G, Annalvilli R, Sivakumar K. 2007. Psychrotrophic spoilage of raw milk at different tempera-tures of storage. *J Appl Sci Res* 3:1383–87.

Lafarge V, Ogier J-C, Girard V, Maladen V, Leveau J-Y, Gruss A, Delacroix-Buchet A. 2004. Raw cow milk bacterial population shifts attributable to refrigeration. *Appl Environ Microbiol* 70(9):5644–50.

Larsen MV, Cosentino S, Rasmussen S, Friis C, Hasman H, Marvig RL, Jelsbak L, Pontén TS, Ussery DW, Aarestrup FM. Lund O., 2012. Multilocus sequence typing of total genome sequenced bacteria. *J Clinl Microbiol* JCM-06094. doi: 10.1128/JCM.06094-11.

Ledenbach LH, Marshall RT. 2009. Microbiological spoilage of dairy products. In: Sperber WH, Doyle MP. editors. *Compendium of the microbiological spoilage of foods and beverages. Food microbiology and food safety*. New York: Springer. pp 41–67.

LeJeune JT, Rajala-Schultz PJ. 2009. Unpasteurized milk: a continued public health threat. *Clin Infecti Diseases* 48(1):93–100.

Liao C-H, McCallus DE. 1998. Biochemical and genetic characterization of an extracellular protease from *Pseudomonas fluorescens* CY091. *Appl Environ Microbiol* 64(3):914–21.

Liu M, Wang H, Griffiths MW. 2007. Regulation of alkaline metalloprotease promoter by N-acyl homoserine lactone quorum sensing in *Pseudomonas fluorescens*. *J Appl Microbiol* 103(6):2174–84.

Lo R, Turner MS, Weeks M, Bansal N. 2016. Culture-independent bacterial community profiling of carbon dioxide treated raw milk. *Int J Food Microbiol* 233:81–89. doi: 10.1016/j.ijfoodmicro.2016.06.015.

Ma Y, Barbano DM, Santos M. 2003. Effect of CO_2 addition to raw milk on proteolysis and lipolysis at 4 °C. *J Dairy Sci* 86(5):1616–31

Machado SG, Bazzolli DMS, Vanetti MCD. 2013. Development of a PCR method for detecting proteolytic psychrotrophic bacteria in raw milk. *Int Dairy J* 29:8–14. http://dx.doi.org/10.1016/j.idairyj.2012.09.007.

Machado SG, Heyndrickx M, De Block J, Devreese B, Vandenberghe I, Vanetti MCD, Van Coillie E. 2016. Identification and characterization of a heat-resistant protease from *Serratia liquefaciens* isolated from Brazilian cold raw milk. *Int J Food Microbiol* 222:65–71. doi:10.1016/j.ijfoodmicro.2016.01.014.

Machado SG, Silva FL, Bazzolli DMS, Heyndrickx M, Costa PMA, Vanetti MCD. 2015. *Pseudomonas* spp. and *Serratia liquefaciens* as predominant spoilers in Brazilian cold raw milk. *J Food Sci* 80(8): M1842–49.

Macrae AR. 1983. Extracellular microbial lipases. In: Fogarty WM. editor. *Microbial enzymes and biotechnology*. New York: Applied Science Publishers. pp. 225–49.

Malacarne M, Summer A, Franceschi P, Formaggioni P, Pecorari M, Panari G, Vecchia P, Sandri S, Fossa E, Scotti C, Mariani P. 2013. Effects of storage conditions on physico-chemical characteristics, salt equilibria, processing properties and microbial development of raw milk. *Int Dairy J* 29(1):3–41.

Mallet A, Guéguen M, Kauffmann F, Chesneau C, Sesboué A, Desmasures N. 2012. Quantitative and qualitative microbial analysis of raw milk reveals substantial diversity influenced by herd management practices. *Int Dairy J* 27(1-2):13–21.

Mann EE, Wozniak DJ. 2012. *Pseudomonas* biofilm matrix composition and niche biology. *FEMS Microbiol Rev* 36(4):893–916.

Marchand S, De Block J, De Jonghe V, Coorevits A, Heyndrickx M, Herman L. 2012. Biofilm formation in milk production and processing environments; influence on milk quality and safety. *Compr Rev Food Sci Food Safety* 11:133–47. doi: 10.1111/j.1541-4337.2011.00183.x.

Marchand S, Heylen K, Messens W, Coudijzer K, De Vos P, Dewettinck K, Herman L, De Block J, Heyndrickx M. 2009a. Seasonal influence on heat-resistant proteolytic capacity of *Pseudomonas lundensis* and *Pseudomonas fragi*, predominant milk spoilers isolated from Belgian raw milk samples. *Environ Microbiol* 11(2):467–82.

Marchand S, Vandriesche G, Coorevits A, Coudijzer K, De Jonghe V, Dewettinck K, De Vos P, Devreese B, Heyndrickx M, De Block J. 2009b. Heterogeneity of heat-resistant proteases from milk *Pseudomonas* species. *Int J Food Microbiol* 133(1-2):68–77.

Martins ML, de Araujo EF, Mantovani HC, Moraes CA, Vanetti MC. 2005. Detection of the *apr* gene in proteolytic psychrotrophic bacteria isolated from refrigerated raw milk. *Int J Food Microbiol* 102:203–211. doi:10.1016/j.ijfoodmicro.2004.12.016

Martins ML, Pinto UM, Riedel K, Vanetti MCD. 2015. Milk-deteriorating exoenzymes from *Pseudomonas fluorescens* 041 isolated from refrigerated raw milk. *Braz J Microbiol* 46(1): 207–217. doi: 10.1590/s1517-838246120130859.

Matéos A, Guyard-Nicodème M, Baglinière F, Jardin J, Gaucheron F, Dary A, Humbert G, Gaillard J-L, 2015. Proteolysis of milk proteins by AprX, an extracellular protease identified in *Pseudomonas* LBSA1 isolated from bulk raw milk, and implications for the stability of UHT milk. *Int Dairy J* 49:78–88. doi: 10.1016/j.idairyj.2015.04.008.

Matsumoto K, Maeda H, Takata K, Kamata R, Okamura H. 1984. Purification and characterization of four proteases from a clinical isolate of *Serratia marcescens* kums 3958. *J Bacteriol* 157:255–32.

Matta H, Kalra MS, Punj V. 1994. Production of heat stable extracellular protease by *Pseudomonas* sp. AFT-36. *Indian J Dairy Sci* 47(9):785–89.

Matta H, Punj V, Kanwar SS. 1997. An immuno-dot blot assay for detection of thermostable protease from *Pseudomonas* sp. AFT-36 of dairy origin. *Lett Appl Microbiol* 25:300–302.

McCarthy CN, Woods RG, Beacham IR. 2004. Regulation of the aprX-lipA operon of *Pseudomonas fluorescens* B52: differential regulation of the proximal and distal genes, encoding protease and lipase, by ompR-envZ. *FEMS Microbiol Lett.* 241(2):243–48.

McKellar RC. 1982. Factors influencing the production of extracellular proteinase by *Pseudomonas fluorescens*. *J Appl Bacteriol* 53(3):305–16.

McKellar RC, Cholette H. 1986. Possible role of calcium in the formation of active extracellular proteinase by *Pseudomonas fluorescens*. *J Appl Bacteriol* 60(1):37–44.

McPhee JD, Griffiths MW. 2011. Psychrotrophic bacteria *Pseudomonas* spp. In: Fuquay JW, Fox PF, McSweeney PLH. editors. *Encyclopedia of Dairy Sciences*. 2nd ed. San Diego: Academic Press. pp 379–83.

McPherson AV, Kitchen BJ. 1983. Reviews of the progress dairy science: the bovine milk fat globule membrane – its formation, composition, structure and behavior in milk and dairy products. *J Dairy Res* 50:107–33.

Mollet C, Drancourt M, Raoult D. 1997. *rpoB* sequence analysis as a novel basis for bacterial identification. *Mol Microbiol* 26:1005–1011.

Moore ER, Mau M, Arnscheidt A, Böttger EC, Hutson RA, Collins MD, van de Peer Y, De Wachter R, Timmis KN. 1996. The determination and comparison of the 16S rRNA cene sequences of species of the genus *Pseudomonas* (sensu stricto) and estimation of the natural intrageneric relationships. *Syst Appl Microbiol* 19:478–92.

Morandi S, Brasca M, Lodi R. 2011. Technological, phenotypic and genotypic characterisation of wild lactic acid bacteria involved in the production of Bitto PDO Italian cheese. *Dairy Sci Technol* 91(3):341–59.

Mosteller TM, Bishop JR. 1993. Sanitizer efficacy against attached bacteria in a milk biofilm. *J Food Protection* 56:34–41.

Mu Z, Du M, Bai Y. 2009. Purification and properties of a heat-stable enzyme of *Pseudomonas fluorescens* Rm12 from raw milk. *Eur Food Res Technol* 228(5):725–34.

Munsch-Alatossava P, Ghafar A, Alatossava T. 2013. Potential of nitrogen gas (N_2) flushing to extend the shelf life of cold stored pasteurised milk. *Int J Molecular Sci* 14(1):5668–85.

Munsch-Alatossava P, Alatossava T. 2006. Phenotypic characterization of raw milk-associated psychrotrophic bacteria. *Microbiol Res.* 161(4):334–46.

Nabrdalik M, Grata K, Latała A. 2010. Proteolytic activity of *Bacillus cereus* strains. *Proceedings of ECOpole* 4:273–77.

Nicodème M, Grill J-P, Humbert G, Gaillard J-L. 2005. Extracellular protease activity of different *Pseudomonas* strains: dependence of proteolytic activity on culture conditions. *J Appl Microbiol.* 99:641–48.

Nörnberg MF, Friedrich RS, Weiss RD, Tondo EC, Brandelli A. 2010. Proteolytic activity among psychrotrophic bacteria isolated from refrigerated raw milk. *Int J Dairy Technol* 63(1):41–46.

Noureddini H, Gao X, Philkana R. 2005. Immobilized *Pseudomonas cepacia* lipase for biodiesel fuel production from soybean oil. *Bioresource Technol* 96:769–77.

O'Brien B, Guinee TP. 2011. Milk seasonal effects on processing properties of cows. milk. In: Fuquay JW. editor. *Encyclopedia of dairy sciences*. 2nd ed. San Diego: Academic Press. pp 598–606.

Osman KM, Alabady MS, Ata NSSM, Ezzeldin NA, Aly MAK. 2010. Genotypic characterization of *Pseudomonas aeruginosa* isolated from human and animal sources in Egypt. *Zoonoses and Public Health* 57:329–38. doi: 10.1111/j.1863-2378.2008.01212.x.

Pinto CLO, Machado SG, Cardoso RR, Alves RM, Vanetti MCD. 2014. Proteolytic potential of *Pseudomonas fluorescens* isolated from refrigerated raw milk. Revista Brasileira de Agropecuária Sustentável 4:16–25.

Porteous LA, Widmer F, Seidler RJ. 2002. Multiple enzyme restriction fragment length polymorphism analysis for high resolution distinction of *Pseudomonas* (sensu stricto) 16S rRNA genes. *J Microbiol Meth* 51:337–48.

Raats D, Offek M, Minz D, Halpern M. 2011. Molecular analysis of bacterial communities in raw cow milk and the impact of refrigeration on its structure and dynamics. *Food Microbiol* 28:465–71.

Rajmohan S, Dodd CER, Waites WM. 2002. Enzymes from isolates of *Pseudomonas fluorescens* involved in food spoilage. *J Appl Microbiol* 93(2):205–213. doi: 10.1046/j.1365-2672.2002.01674.x.

Rasolofo EA, LaPointe G, Roy D. 2011. Assessment of the bacterial diversity of treated and untreated milk during cold storage by T-RFLP and PCR-DGGE methods. *Dairy Sci Technol* 91(5):573–97.

Rasolofo EA, St-Gelais D, LaPointe G, Roy D. 2010. Molecular analysis of bacterial population structure and dynamics during cold storage of untreated and treated milk. *Int J Food Microbiol* 138(1-2):108–118.

Ray B. 2004. Factors influencing microbial growth in food. *Fundamental food microbiology*. 3rd ed. Boca Raton, FL: CRC Press. pp 74–75.

Romero FJ, García LA, Salas JA, Díaz M, Quirós LM. 2001. Production, purification and partial characterization of two extracellular proteases from *Serratia marcescens* grown in whey. *Process Biochemistry* 36:507–515.

Rowe MT, Johnston DE, Kilpatrck DJ, Dunstall G, Murphy RJ. 1990. Growth and extracellular enzyme production by psychrotrophic bacteria in raw milk stored at low temperature. *Milchwissenschaft* 45(8):597–600.

Rowe MT, Gilmour A. 1982. Growth, enzyme production and changes in oxygen tension occuring during batch cultivation of psychrotrophic *Pseudomonas fluorescens* strains. *Milchwissenschaft* 37(10): 597–600.

Russell NJ. 2002. Bacterial membranes: the effects of chill storage and food processing. An overview. *Int J Food Microbiol* 79(1): 7–34.

Saitou K, Furuhata K, Fukuyama M. 2010. Genotyping of *Pseudomonas aeruginosa* isolated from cockroaches and human urine. *J Infect Chemotherapy* 16:317–321. doi 10.1007/s10156-010-0055-7.

Sajben E, Manczinger L, Nagy A, Kredics L, Vágvölgyi C. 2011. Characterization of pseudomonads isolated from decaying sporocarps of oyster mushroom. *Microbiol Res* 166:255–67. doi:10.1016/j.micres.2010.05.002.

Scaccabarozzi L, Leoni L, Ballarini A, Barberio A, Locatelli C, Casula A, Bronzo V, Pisoni G, Jousson O, Morandi S, Rapetti L, Moroni P. 2015. *Pseudomonas aeruginosa* in dairy goats: genotypic and phenotypic comparison of intramammary and environmental isolates. *PLoS ONE*: e0142973. doi:10.1371/journal.pone.0142973.

Schokker EP, van Boekel MAJS. 1997. Production, purification and partial characterization of the extracellular proteinase from *Pseudomonas fluorescens* 22F. *Int Dairy J* 7(4):265–71.

Shibata H, Kato H, Oda J. 1998. Calcium ion-dependent reactivation of a *Pseudomonas* lipase by its specific modulating protein, LipB. *J Biochem* 123(1):136–41.

Sidhu P, Sharma R, Soni SK, Gupta JK. 1998. Effect of cultural conditions on extracellular lipase production by *Bacillus* sp. RS-12 and its characterization. *Indian J Microbiol* 38:9–12.

Singh S, Banerjee U. 2007. Purification and characterization of trans-3-(4-methoxyphenyl) glycidic acid methyl ester hydrolyzing lipase from *Pseudomonas aeruginosa*. *Process Biochem.* 42:1063–1068.

Skandamis PN, Nychas GJE. 2012. Quorum sensing in the context of food microbiology. *Appl Environ Microbiol* 78(16):5473–82.

Skura BJ, Craig C, McKellar RC, 1986. Effect of disruption of an N2-overlay on growth and proteinase production in milk by *Pseudomonas fluorescens*. *Canadian Inst Food Sci Technol J* 19(3):104–106. doi: 10.1016/s0315-5463(86)71462-4.

Snellman EA, Sullivan ER, Colwell RR. 2002. Purification and properties of the extracellular lipase, LipA, of *Acinetobacter* sp. RAG-1. *Eur J Biochem* 269:5771–79.

Sørhaug T, Stepaniak L. 1997. Psychrotrophs and their enzymes in milk and dairy products: Quality aspects. *Trends Food Sci Technol* 8(2):35–41.

Spiers AJ, Buckling A, Rainey PB. 2000. The causes of *Pseudomonas* diversity. *Microbiol* 146(10): 2345–50.

Stead D. 1986. Lipases: their characteristics, role in food spoilage and industrial uses. *J Dairy Res* 53(3):481–505.

Stead D. 1987. Production of extracellular lipases and proteinases during prolonged growth of strains of psychrotrophic bacteria in whole milk. *J Dairy Res* 54(4):535–43.

Stepaniak L. 2004. Dairy enzymology. *Int J Dairy Technol* 57:153–71.

Stevenson RG, Rowe MT, Wisdom GB, Kilpatrick D. 2003. Growth kinetics and hydrolytic enzyme production of *Pseudomonas* spp. isolated from pasteurized milk. *J Dairy Res* 70(3):293–96.

Stoeckel M, Lidolt M, Achberger V, Glück C, Krewinkel M, Stressler T, von Neubeck M, Wenning M, Scherer S, Fischer L, Hinrichs J. 2016. Growth of *Pseudomonas weihenstephanensis, Pseudomonas proteolytica* and *Pseudomonas* sp. in raw milk: Impact of residual heat-stable enzyme activity on stability of UHT milk during shelf-life. *Int Dairy J* 59:20–28. doi: 10.1016/j.idairyj.2016.02.045.

Stuer W, Jaeger KE, Winkler UK. 1986. Purification of extracellular lipase from *Pseudomonas aeruginosa*. *J Bacteriol* 168(3):1070–1074.

Stuknytė M, Decimo M, Colzani M, Silvetti T, Brasca M, Cattaneo S, Aldini G, De Noni I. 2016. Extracellular thermostable proteolytic activity of the milk spoilage bacterium *Pseudomonas fluorescens* PS19 on bovine caseins. *J Dairy Sci* 99(6):4188–4195. doi:10.3168/jds.2016-10894.

Suhren G. Producer microorganisms. 1989. In: McKellar RC. editor. Enzymes of psychrotrophs in raw foods. Boca Raton, FL: CRC Press. pp 3–34.

Teh KH, Flint S, Palmer J, Lindsay D, Andrewes P, Bremer P. 2011. Thermo-resistant enzyme-producing bacteria isolated from the internal surfaces of raw milk tankers. *Int Dairy J* 21(10):742–47. doi: 10.1016/j.idairyj.2011.04.013.

Tekinson OC, Tothwell JA. 1974. A study on the effect of storage at 5°C on the microbial flora of heat-treated marked cream. *J Soc Dairy Technol* 27:57–59.

Tondo E, Lakus F, Oliveira F, Brandelli A. 2004. Identification of heat stable protease of *Klebsiella oxytoca* isolated from raw milk. *Lett Appl Microbiol.* 38:146–50.

Vacheyrou M, Normand A-C, Guyot P, Cassagne C, Piarroux R, Bouton Y. 2011. Cultivable microbial communities in raw cow milk and potential transfers from stables of sixteen French farms. *Int J Food Microbiol* 146:253–62. doi: 10.1016/j.ijfoodmicro.2011.02.033.

van den Broek D, Chin-A-Woeng TF, Bloemberg GV, Lugtenberg BJ. 2005. Molecular nature of spontaneous modifications in *gacS* which cause colony phase variation in *Pseudomonas* sp. strain PCL1171. *J Bacteriol* 187(2):593–600.

Varga L, Szigeti J. 2016. Use of ozone in the dairy industry: a review. *Int J Dairy Technol* 69:157–68.
Vercet A, Lopez P, Burgos J. 1997. Inactivation of heat-resistant lipase and protease from *Pseudomonas fluorescens* by manothermosonication. *J Dairy Sci* 80:29–36.
Vithanage NR, Dissanayake M, Bolge G, Palombo EA, Yeager TR, Datta N. 2016. Biodiversity of culturable psychrotrophic microbiota in raw milk attributable to refrigeration conditions, seasonality and their spoilage potential. *Int Dairy J* 57:80–90. doi: 10.1016/j.idairyj.2016.02.042.
von Neubeck M, Baur C, Krewinkel M, Stoeckel M, Kranz B, Stressler T, Fischer L, Hinrichs J, Scherer S, Wenning M. 2015. Biodiversity of refrigerated raw milk microbiota and their enzymatic spoilage potential. *Int J Food Microbiol* 211:57–65. doi:10.1016/j.ijfoodmicro.2015.07.001.
von Neubeck M, Huptas C, Glück C, Krewinkel M, Stoeckel M, Stressler T, Fischer L, Hinrichs J, Scherer S, Wenning M. 2016. *Pseudomonas helleri* sp. nov. and *Pseudomonas weihenstephanensis* sp. nov., isolated from raw cow's milk. *Int J Syst Evol Microbiol* 66:1163–73.
Vyletělová M, Ficnar J, Hanuš O. 2000. Effects of lipolytic enzymes *Pseudomonas fluorescens* on liberation of fatty acids from milk fat. *Czech J Food Science* 18:175–82.
Wang S-L, Lin Y-T, Liang T-W, Chio S-H, Ming L-J, Wu P-C. 2009. Purification and characterization of extracellular lipases from *Pseudomonas monteilii* TKU009 by the use of soybeans as the substrate. *J Ind Microbiol Biotechnol.* 36:65–73.
Waters V, Zlosnik JEA, Yau YCW, Speert DP, Aaron SD, Guttman DS. 2012. Comparison of three typing methods for *Pseudomonas aeruginosa* isolates from patients with cystic fibrosis. *Eur J Clin Microbiol Infectious Dis* 31:3341–50. doi 10.1007/s10096-012-1701-z.
Wessels D, Jooste PJ, Mostert JF. 1990 Technologically important characteristics of *Enterococcus* isolates from milk and dairy products. *Int J Food Microbiol.* 10:349–52.
Wu Q, Ye Y, Li F, Zhang J, Guo W. 2016. Prevalence and genetic characterization of *Pseudomonas aeruginosa* in drinking water in Guangdong Province of China. *LWT - Food Sci Technol* 69:24–31. http://dx.doi.org/10.1016/j.lwt.2016.01.014.
Yamamoto S, Kasai H, Arnold DL, Jackson RW, Vivian A, Harayama S. 2000. Phylogeny of the genus *Pseudomonas*: intrageneric structure reconstructed from the nucleotide sequences of *gyrB* and *rpoD* genes. *Microbiol* 146:2385–94.
Zhang S, Li H, Uluko H, Liu L, Pang X, Lv J. 2015. Investigation of protease production by *Pseudomonas fluorescens* BJ-10 and degradation on milk proteins. *J Food Proces Preserv* 39:2466–72.
Zucali M, Bava L, Tamburini A, Brasca M, Vanoni L, Sandrucci A. 2011. Effects of season, milking routine and cow cleanliness on bacterial and somatic cell counts of bulk tank milk. *J Dairy Res* 78(4):436–41.

4 Stabilization of milk quality by heat treatments

Palmiro Poltronieri[1] and Franca Rossi[2]

[1] *Institute of Sciences of food Productions (CNR-ISPA), CNR, National Research Council of Italy, Lecce, Italy*
[2] *Istituto Zooprofilattico Sperimentale 'G. Caporale', Teramo, Italy*

4.1 INTRODUCTION

Heat treatments with different time/temperature combinations allow us to obtain milk, as well as cream, with different degrees of microbiological stability, according to the posttreatment intended use. In this respect, the terms 'thermization', 'pasteurization' and 'sterilization' are applied to products that have been exposed to increasingly intense heating conditions, and these products can be preserved for increasingly longer periods of time. Thermization is a mild heating process aimed at reducing microbial contaminants prior to a process, such as cheese making, in which these are expected to further decrease. Pasteurization is the process in which refrigerated fluid milk or cream is submitted to a heat treatment aimed at short-term microbiological stabilization. It ensures the destruction of all vegetative forms of pathogenic microorganisms and part of the saprophytic microbiota with limited alterations of the physicochemical and organoleptic characteristics. Pasteurized milk can be labelled as 'fresh' if the thermal treatment is carried out within 48 h from milking, and it has a shelf life of 6 days in refrigerated conditions. Enzymatic assays for lactoperoxidase and alkaline phosphatase are used to evaluate the degree of thermal treatment applied to pasteurized milk. One parameter to assess post-pasteurization recontamination is the presence of undesired bacteria and a high microbial content.

The sterilization treatment also kills the bacterial spores and allows storage of milk for several months in hermetically closed containers at room temperature.

4.2 THERMAL TREATMENTS OF MILK

4.2.1 Thermization

Thermization is a process used for industrial purposes only, and it is defined as a heat treatment aimed at reducing the number of microorganisms, thus permitting longer storage of the milk prior to further processing. The heating conditions are 62–65 °C

Microbiology in Dairy Processing: Challenges and Opportunities, First Edition.
Edited by Palmiro Poltronieri.

Published 2018 by John Wiley & Sons Ltd and the Institute of Food Technologists.

for 15–20 seconds (Walstra et al., 1999; Lewis and Jun, 2011; Rukke et al., 2011). In cheese making, thermization is applied to reduce the microbial load and to ensure that *Escherichia coli* and total coliforms in the cheese product will be within the limits of the law (Martin et al., 2016).

4.2.2 Pasteurization

Milk pasteurization is the process in which refrigerated fluid milk or cream is submitted to a heat treatment to achieve short-term microbiological stabilization. It must ensure the destruction of all vegetative forms of pathogenic microorganisms and part of the saprophytic microbiota with limited alterations of the physicochemical and organoleptic characteristics of the product (Lewis, 2003; Hinrichs and Atamar, 2011).

The terms 'pasteurization' and 'pasteurized' imply that every particle of milk or milk product is held continuously at temperatures between 63 °C(145 °F)/30 minutes and 100 °C(212 °F)/0.01 seconds in properly designed and operated equipment. These conditions provide fresh-tasting milk that meets the requirements for consumer safety.

According to the Codex Alimentarius, "Pasteurization is a heat treatment aimed at reducing the number of harmful microorganisms in milk and cream to a level at which they do not constitute a significant health hazard" (CODEX Committee on Milk and Milk Products, 2000). The reduction of the microbial load allows extended shelf life of milk, and in pasteurization only minimal chemical, physical and organoleptic changes occur. Pasteurization conditions were designed to effectively destroy *Mycobacterium tuberculosis*. These conditions are equivalent to 72 °C for 15 seconds, used in continuous flow pasteurization or high-temperature short-time (HTST) pasteurization, or 63 °C for 30 minutes, used in batch pasteurization for milk. For cream, the pasteurization treatments applied must be equivalent to 75 °C for 15 seconds (10–20% fat), 80 °C for 15 seconds (fat content above 20%) and 65 °C for 30 minutes (batch pasteurization). Formulated milks and creams with high sugar content or high viscosity also require pasteurization conditions in excess of the minimum conditions defined for milk.

Other equivalent conditions can be obtained by plotting the line passing through the above-mentioned points on a log time versus temperature graph (CODEX Committee on Milk and Milk products, 2000).

Since 1957, in the United States pasteurization conditions of 145 °F (62.8 °C) for 30 minutes for a batch process or 161 °F (71.7 °C) for 15 seconds for a continuous process have been adopted for the inactivation of *Coxiella burnetii*, the agent of Q fever in humans (Enright et al., 1957).

Heat treatments that give equivalent effects on the reduction of the microbial load are 72 °C (161 °F)/15 seconds; 89 °C (191 °F)/1.0 seconds; 90 °C (194 °F)/0.5 seconds; 94 °C (201 °F)/0.1 seconds; 96 °C (204 °F)/0.05 seconds; 100 °C (212 °F)/0.01 seconds.

Although processing conditions are defined for temperatures above 90 °C, they are rarely used because they can impart an undesirable cooked flavour to milk (Zabbia et al., 2012).

In the continuous process, milk moves in a controlled, continuous flow while subjected to the established pasteurization temperatures for the required time. After pasteurization, the product still contains residual thermoduric microorganisms whose growth is retarded during milk storage by keeping the product refrigerated.

Nations have legal definitions for different categories of pasteurized milk. For instance, in Italy at least five types of pasteurized milk can be distinguished: high-quality fresh pasteurized milk, fresh pasteurized milk, micro-filtered pasteurized milk, pasteurized milk and extended shelf life (ESL) pasteurized milk.

For the 'high-quality fresh pasteurized' milk label, the place or the farm where milking was done must be reported on the label in compliance with the Italian Ministry of Agriculture guidelines on milk traceability (D.M. 14/1/2005). "Fresh" pasteurized milk can be distinguished because it tests positive for lactoperoxidase enzyme, which indicates that milk has not been exposed to heat treatments equivalent to 80 °C for 20 seconds.

Rigorous criteria must be met in order to apply the 'high-quality' labelling. These include the somatic cell count (less than 300,000/ml), the lactic acid concentration (<30 ppm) and fat and protein composition no lower than 3.6% g/g and 32 g/l, respectively (Wehr and Frank, 2004).

Raw milk from which the 'high-quality fresh pasteurized milk' is obtained must come from authorized farms that undergo strict hygiene and composition controls by the official veterinary vigilance authority.

Given the lower initial microbial contamination level, 'high-quality fresh pasteurized milk' is heat treated at 2 °C lower than that used for the 'fresh pasteurized milk' to warrant a soluble serum protein content of at least 15.5% of total proteins.

Micro-filtered pasteurized milk is obtained by first skimming milk by centrifugation and then homogenizing cream, while the skim milk is micro-filtered through ceramic membranes with pores of 1–2.5 μm. Homogenization is a mechanical treatment that results in decrease of size and increase of number and total surface area of fat globules. This reduces cream separation at the surface and on the container walls, enhancing stability of milk.

Homogenized cream, which may or may not be heat treated, is added again to the skim milk to obtain milk with the desired fat content. Finally, the product is pasteurized in conditions milder than those applied in normal pasteurization, given the lower microbial content. This product has a shelf life of up to 10 days but cannot be labelled with the definition 'fresh' (D.L. 24/6/2004, n. 157).

The need for a product with longer shelf life, but with a nutritional and organoleptic quality higher than for sterilized milk, led to the introduction of high-temperature pasteurization, in which milk is heated at temperatures between 80 and 135 °C for 1–4 seconds (ultra-pasteurization or UP).

The product obtained is the ESL pasteurized milk that can be stored in refrigerated conditions (+6 °C) for 25–30 days. The shelf-life duration is fixed by the producer, according to the heat treatment applied, in compliance with the Italian law (D.L. 109/1992), usually for 18 days, according to European laws.

The fat content of milk varies with species (cow, sheep, goat, water buffalo), animal breed, feeding, stage of lactation and other factors, and most milks are standardized in fat content. To this aim milk is processed through centrifugal separators to separate cream, containing usually 40% fat, from the skim milk, which has less than 0.01% fat. The desired fat content of the cream can be varied by changing the separation conditions.

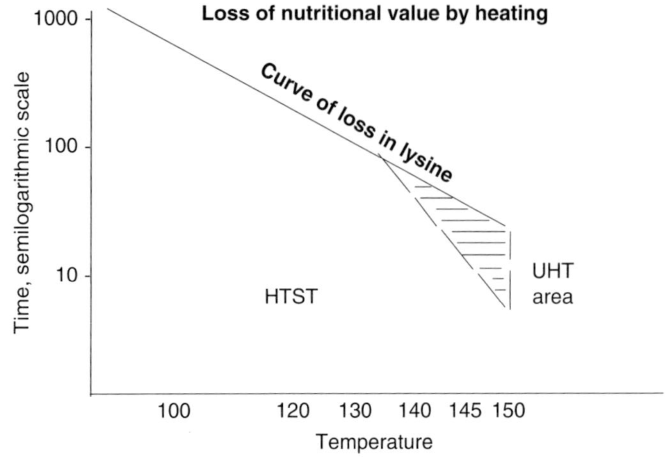

	Thermization (65 °C)	Pasteurization (90 °C)	High pasteurization 120 °C	130 °C
Phosphatase	+	–	–	–
Lactoperoxidase	+	+	–	–

Figure 4.1 Diagram of milk heating vs. time length. Degree of loss in nutritional value in lysine.

The cream is then added back to the skim milk to adjust the fat content for the product. Commercial categories of milk in relation to fat content are whole milk (3.25% fat), partially skimmed milk (1.5– 1.8% fat), and skim milk (less than 0.1% fat).

The native milk alkaline phosphatase (ALP) activity test is an index of the efficiency of the pasteurization process (CODEX Committee on Milk and Milk Products, 2000; Payne and Wilbey, 2009). Complete pasteurization inactivates the enzyme to below levels detectable by conventional methods. Because the heat stability of ALP is greater than that of pathogens that may be present in milk, the enzyme serves as an indicator of product safety. However, the failure to detect ALP activity does not guarantee that the product is pathogen free. Milk after thermization must be phosphatase positive, while pasteurized milk must be phosphatase negative (Figure 4.1). In case an accredited laboratory finds a pasteurized milk sample to be positive for phosphatase, the pasteurization process must be corrected. Instrumental methods for the determination of phosphatase activity are available. The legal limit for phosphatase is 350 mU/l. Values above 100 mU/l are considered to be a sign of possible problems with the pasteurization systems. A pasteurization plant is expected to produce milk with a phosphatase reading of around 30 mU/l (Payne and Wilbey, 2009).

After pasteurization or ultra-pasteurization and aseptic filling of containers, appropriate refrigeration must be applied to milk and milk products to slow the growth of residual microorganisms.

4.2.3 Grade A pasteurized milk

Problems associated with assuring the safety of milk and milk products have become extremely complex because of new products, new processes, new materials and new marketing patterns, which must be evaluated in terms of their public health significance. The Grade A Pasteurized Milk Ordinance (U.S. Dept. of Health and Human Services,

2005), revised in 2009, has translated this knowledge and technology into effective public health practices.

All Grade A raw milk or milk products shall be produced, processed, manufactured and pasteurized, ultra-pasteurized, or aseptically processed to conform to precise chemical, physical, bacteriological and temperature and sanitation parameters.

4.3 MILK STERILIZATION

Milk sterilization is a continuous heat treatment of raw milk at least at 135 °C for not less than 1 second in order to kill microorganism and bacterial spores (Fernandes, 2008). Sterilized milk is packaged in opaque sterile recipients so that physicochemical and organoleptic variations are minimized.

Sterilized milk has a minimum time of storage established by the producer. It is kept at room temperature and does not need refrigeration until opened.

According to the Italian legislation, sterilized milk can be distinguished from long shelf-life sterilized milk that has been heat treated in a closed recipient with a high-temperature/long-time process and ultra high temperature (UHT) treated milk, which is heat treated in a continuous process and is packaged aseptically after heat treatment (Deeth and Datta, 2011). These heat treatments have equivalent bactericidal effects, but the heat-induced chemical effects in milk and cream are greater in the long shelf-life sterilized milk.

UHT involves heating the milk during flow in heat exchangers and filling under aseptic conditions into hermetically sealed packaging. The product is termed 'shelf stable' and does not need refrigeration until opened (see http://www.idfa.org/news-views/media-kits/milk/pasteurization).

Milk sterilization can also be achieved with a UHT treatment consisting of direct addition of saturated steam to the product. In this process, the milk and steam flows must be isolated from pressure fluctuations inside an injection chamber. Preheated milk is introduced at a relatively high pressure of about 5 bar into the injector, where it comes in contact with saturated steam at 150 °C. The latter condenses and transmits heat to milk, which can reach a temperature of 145 °C. The products remains for about 3 seconds at this temperature, then excess water is eliminated by flash-evaporation in a depression bubble, where a pressure of 0.5 bar is created. The evaporation subtracts heat from milk, which is therefore rapidly cooled to 80 °C. Further cooling is achieved by exchange of heat with the raw milk entering the sterilization plant.

4.3.1 Control of proper time/temperature setting for safety of milk and milk products

Milk and milk products that require time/temperature control for safety to limit pathogenic microorganism growth or toxin formation include raw, heat-treated, pasteurized or ultra-pasteurized products (FSSA, 2012).

The phosphatase test, which can be executed by an electronic phosphatase test device, is an index of the efficiency of the pasteurization process. In the event that an accredited laboratory finds that a milk sample is positive for phosphatase, the pasteurization process must be investigated and corrected.

4.4 DISEASES ASSOCIATED WITH UNPASTEURIZED MILK, OR POST-PASTEURIZATION DAIRY-PROCESSING CONTAMINATION

Despite the progress made in the hygiene of milk production and sanitation, occasional milk-borne infection outbreaks still occur (Boor, 1997; Cornell University Milk Quality Improvement Program, 1998; Chambers, 2002; Boor and Murphy, 2002; CDC, 2006).

Brucella abortus and *Brucella melitensis* are bacterial species causing zoonoses (Garin-Bastuji and Verger, 1994), and the brucellosis disease is transmitted to humans by direct animal contact as well as by dairy products, mainly cheese produced using unpasteurized milk. *Listeria monocytogenes*, the causative agent of listeriosis associated with consumption of contaminated foods, can proliferate in yogurts and cheese surfaces when the bacterium survives in milk, even though these products are stored under refrigerated conditions (CDC, 2006).

Tuberculosis is able to be transmitted from cattle via unpasteurized milk products. Crohn's disease has been linked to infection by *Mycobacterium avium* subsp. *paratuberculosis*, found in retail milk in the UK and the United States, similarly to Johne's disease, which affects livestock (Naser and Collins, 2005; Ellingson et al., 2005; Uzoigwe et al., 2007; Hermon-Taylor, 2009).

4.5 CONCLUSIONS

Effective sanitation practices such as microfiltration and thermal treatments ensure a reduction of bacterial contamination of milk and ensure that contaminated milk does not reach the consumer. Differences in compositional quality, degree of freshness, and initial microbial load are reflected in the various labelling attributes applied to the milk product.

References

Boor KJ. 1997. Pathogenic microorganisms of concern to the dairy industry. *Dairy, Food Environ Sanitation* 17:714–17.

Boor KJ, Murphy SC. 2002. The microbiology of market milks. In: Robinson RK. editor. *Dairy microbiology handbook*, 3rd ed. New York: Wiley-Interscience. pp 91–122.

CDC. 2006. Centers for Disease Control and Prevention. Epidemiologic notes and reports. *Morbidity and Mortality Weekly Report* 34:215–16. http://www.cdc.gov/ncidod/ diseases/food/index.htm.

Chambers JV. 2002. The microbiology of raw milk. In: Robinson RK. editor. *Dairy microbiology handbook*, 3rd ed. New York: Wiley-Interscience. pp 39–90.

CODEX Committee on Milk and Milk Products. 2000. Joint FAO/WHO food standards programme. CX/MMP 00/15. Wellington, New Zealand, 28 February – 3 March 2000.

Cornell University Milk Quality Improvement Program. 1998. Sources of microbial contamination as detected by various bacteriological procedures. Cornell University, Dept. Food Science, Ithaca, NY.

Deeth HC, Datta N. 2011. Ultra-high temperature treatment (UHT): heating systems. In: Fuquay J, Fox PF, McSweeney PLH. editors. *Encyclopedia of dairy sciences*, 2nd ed. London: Elsevier Ltd., Academic Press. pp 699–707.

D.L. 109/1992. Decreto Legge "Attuazione delle direttive n. 89/395/CEE e n. 89/396/CEE".

D.L. 24/6/2004. Decreto legge n. 157 "Disposizioni urgenti per l'etichettatura di alcuni prodotti agroalimentari, nonché in materia di agricoltura e pesca". Gazzetta Ufficiale of Italy n. 147, 25 June 2004.

D.M. 14/1/2005. Decreto Ministeriale delle politiche agricole e forestali di concerto con il ministero delle attività produttive. Linee guida per la stesura del manuale aziendale per la rintracciabilità del latte. Gazzetta Ufficiale of Italy n. 30, 7 February 2005.

Ellingson JL, Anderson JL, Koziczkowski JJ, Radcliff RP, Sloan SJ, Allen SE, Sullivan NM. 2005. Detection of viable *Mycobacterium avium* subsp. *paratuberculosis* in retail pasteurized whole milk by two culture methods and PCR. *J Food Prot* 68(5):966–72.

Enright JB, Sadler WW, Thomas RC. 1957. Thermal inactivation of *Coxiella burnetii* and its relation to pasteurization of milk. *Public Health Monogr* 47:1–30.

Fernandes R. 2008. Liquid milk products. In: *Microbiology handbook: dairy products*. Oxford: Leatherhead Publishing, Surrey, and Royal Society of Chemistry. pp 1–15.

FSSA, 2012. *Manual of methods of analysis of foods. Milk and milk products*. Lab manual 1. New Delhi, India: Ministry of Health and Family Welfare. pp 39–41.

Garin-Bastuji B., Verger JM. 1994. *Brucella abortus* and *Brucella melitensis*. In: Monograph on the significance of pathogenic microorganisms in raw milk. Brussels: International Dairy Federation., pp.167–85.

Hermon-Taylor J. 2009. *Mycobacterium avium* subspecies *paratuberculosis*, Crohn's disease and the doomsday scenario. *Gut Pathogens* 1(1):15.

Hinrichs J, Atamar Z. 2011. Sterilization of milk and other products. In: Fuquay J, Fox PF, McSweeney PLH. editors. *Encyclopedia of dairy sciences*, 2nd ed. London: Elsevier Ltd., Academic Press. pp. 761–64.

Lewis M, Jun S. 2011. Thermal processing. In: *Food processing handbook*. Brennan JG, Grandison AS. editors. Weinheim, Germany: Wiley-VCH. pp 60–61.

Lewis MJ. 2003. Improvements in the pasteurisation and sterilisation of milk. In: Smith G. editor. *Dairy processing*. Cambridge, UK: Woodhead Publishing. pp. 79–102.

Martin NH, Trmčić A, Hsieh T-H, Boor KJ, Wiedmann M. 2016. The evolving role of coliforms as indicators of unhygienic processing conditions in dairy foods. *Frontiers Microbiol*. 7:1549.

Naser SA, Collins MT. 2005. Debate on the lack of evidence of *Mycobacterium avium* subsp. *paratuberculosis* in Crohn's disease. *Inflamm Bowel Dis* 11(12):1123.

Payne C, Wilbey RA. 2009. Alkaline phosphatase activity in pasteurized milk: a quantitative comparison of Fluorophos and colorimetric procedures. *Int J Dairy Technol* 62:308–314.

Rukke EO, Sørhaug T, Stephaniak L, 2011. Thermization of milk. In: Fuquay J, Fox PF, McSweeney PLH. editors. *Encyclopedia of dairy sciences*, 2nd ed. London: Elsevier Ltd., Academic Press. pp 693–8.

U.S. Dept. of Health and Human Services. 2005. Grade "A" Pasteurized Milk Ordinance. Public Health Service and Food and Drug Admin. http://www.cfsan.fda.gov/~ear/pmo03toc.html.

Uzoigwe JC, Khaitsa ML, Gibbs PS. 2007. Epidemiological evidence for *Mycobacterium avium* subsp. *paratuberculosis* as a cause of Crohn's disease. *Epidemiol Infect* 135:1057–68.

Walstra P, Geurts TJ, Noomen A, Jellema A, van Boekel MAJS. 1999. *Dairy technology, principles of milk properties and processes*. New York: Marcel Dekker Inc.

Wehr HM, Frank JH. 2004. *Standard methods for the examination of dairy products*. Washington, DC: American Public Health Association.

Zabbia A, Buys EM, De Kock HL. 2012. Undesirable sulphur and carbonyl flavor compounds in UHT milk: a review. *Crit Rev Food Sci Nutr* 52(1):21–30.

5 Genomics of LAB and dairy-associated species

Palmiro Poltronieri[1], Franca Rossi[2], Cesare Cammà[2], Francesco Pomilio[2] and Cinzia Randazzo[3]

[1]Institute of Sciences of food Productions (CNR-ISPA), CNR, National Research Council of Italy, Lecce, Italy
[2]Istituto Zooprofilattico Sperimentale 'G. Caporale', Teramo, Italy
[3]Department of Agriculture, Food and Environment (Di3A), University of Catania, Catania, Italy

5.1 INTRODUCTION

Lactic acid bacteria (LAB) and other relevant dairy microorganisms are able to ferment various carbohydrates and to survive in harsh conditions and low pH, and they have adapted to different environments, including the gastrointestinal tract. This is explained by the ability of such species to exploit the metabolites and substrates available in a microbial community. The ecotypes' biodiversity has been described in depth and have originated from gene loss or gain, by means of insertion or deletion events due to activity of mobile elements, and horizontal gene transfer, leading to bacteria adaptation to the nutrient availability in their environment. The availability of genome sequences of reference strains, and new cheaper sequencing methods, will make it possible to obtain genome sequences from environmental strains for comparative genomics purpose and phylogenomic analyses. This will facilitate the clustering of dairy strains with the ecotypes sharing the same gene pools.

5.2 GENOMICS OF LAB AND DAIRY-ASSOCIATED SPECIES

Genome sequence data can be retrieved from a number of databases, among which the most important, Genomes OnLine Database (GOLD) (https://gold.jgi.doe.gov/index), allows comprehensive access to information regarding genome and metagenome sequencing projects and their associated data around the world. A total of 238 complete LAB genomes deposited in public databases have been recently compared for gene content (Alvarez-Sieiro et al., 2016), while previous studies analysed the sequenced genomes of 213 LAB strains (Sun et al., 2015).

Some other online databases and tools, which can be employed to gain taxonomic data either from entire genomes or single sequences, are Comprehensive Microbial

Microbiology in Dairy Processing: Challenges and Opportunities, First Edition.
Edited by Palmiro Poltronieri.

Published 2018 by John Wiley & Sons Ltd and the Institute of Food Technologists.

Resource (CMR) (http://cmr.jcvi.org/tigr-scripts/CMR/CmrHomePage.cgi), DNA Bank Network (http://www.dnabank-network.org/), Global Genome Biodiversity Network (http://www.ggbn.org/), RibAlign (https://omictools.com/ribalign-tool), SILVA (Comprehensive Ribosomal RNA Databases; https://www.arb-silva.de/), Ribosomal Database Project (http://rdp.cme.msu.edu/).

A sequencing program of the U.S. Department of Energy Joint Genome Institute named Genomic Encyclopedia of Bacteria and Archaea (GEBA) was launched in 2007. In 2016, the database contained 343 sequencing projects of LAB species. GEBA aspires to sequence thousands of bacterial genomes (Kyrpides et al., 2014). Extraction of reliable information requires a standardized description of genomes and analytical procedures to exchange and to integrate genomic data.

It is expected that in the near future species description will include genome sequencing of the type strains, thanks also to the decrease in sequencing costs, and that digital DNA-DNA hybridization (DDH) will replace wet lab DDH (http://ggdc.dsmz.de/) as well as guanine/cytosine (GC) determination (Meier-Kolthoff et al. 2013). It is expected that authorities such as the subcommittee on the taxonomy of *Bifidobacterium, Lactobacillus*, and related organisms will give indications for the integration of genome data in LAB taxonomy (Mattarelli et al., 2014).

5.2.1 Next-generation sequencing of strains, dairy starter genomics and metagenomics

Whole-genome sequencing is a novel and affordable method to identify and select strains based on gene complements and trait characteristics (Kelleher et al., 2015). This approach has been envisaged and is currently pursued for lactococcal and streptococcal starter strains. In the immediate future, genomes of numerous strains belonging to *Lactococcus lactis* and *Streptococcus thermophilus* are going to be sequenced from various niches. A similar approach has been applied to find similarities and differences in probiotic LABs and in defining the clusters of strains mostly related to dairy environments (Douillard et al., 2013a; Sun et al., 2015).

The reads can be *de novo* assembled using CLC Genomics Workbench (Qiagen Bioinformatics). The bacterial genomes can be annotated using the NCBI Prokaryotic Genome Automatic Annotation Pipeline (PGAP) or the Microbial Genome Annotation Pipeline (MiGAP), based on BG7, an open-source tool based on a protein-centred gene calling/annotation, to annotate LAB genomes (Tobes et al., 2015; Tanizawa et al., 2015). MiGAP allows prediction of open reading frames by MetaGeneAnnotator, followed by tRNA prediction, detection of insertion sequences by ISsaga (Varani et al., 2011), identification of phage sequences (Zhou et al., 2011) and integrative and conjugative elements (Wozniak and Waldor, 2010), using ICEberg (Bi et al., 2012), rRNA prediction by RNAmmer (Lagesen et al., 2007), homology search against the RefSeq and TrEMBL, and CRISPRFinder server (Grissa et al., 2007).

Another software program used to annotate bacterial genomes is SEED (http://pubseed.theseed.org/), a web-based tool that performs rapid annotation of microbial genomes using subsystems technology (RAST), hosted on the RAST annotation engine (http://rast.nmpdr.org/) (Overbeek et al., 2014). BLASTN, Mauve, and MUMmer are programs used for genome alignment (Li et al., 2016).

Among the better performing NGS platforms that sequence genomes of 1 to 4 million bases, are platforms by Illumina, Ion Torrent and Pacific Bioscience, which we discuss next.

5.2.2 Pacific Bioscience single-molecule real-time sequencing technology

PacBio single-molecule, real-time (SMRT) sequencing (Chin et al., 2013) of whole bacterial genomes is performed on a PacBio RSII sequencer with P5-C3 chemistry, making use of an array of zero-mode waveguides (ZMW), up to 150,000 ZMW on one chip. When SMRT cells sequence single DNA molecules, the platform yields 350 megabases per SMRT cell. *De novo* assembly can be performed with the hierarchical genome-assembly process (HGAP) method based on the SMRT Analysis package 2.0. PacBio SMRT technology has also been applied to the analysis of methylomes and is able to discriminate a methylated from a nonmethylated base (Murray et al., 2012; Fang et al., 2102; Beaulaurier et al., 2015; O'Callaghan et al., 2015).

5.2.3 Illumina MySeq and HiSeq 2000

The Illumina-Solexa system makes use of sequencing by synthesis and can determine up to 40 million sequences of up to 50 bases in length. In 2008, the early days of NGS, this method was combined with Short Oligonucleotide Analysis Package (SOAP), to perform both ungapped and gapped alignments, using special modules for alignment of pair-ends (Li et al., 2008).

The Illumina MiSeq system, using the 250-bp pair-end, can yield up to 8,600,000 reads, a size well suited to sequence bacterial genomes. Reads containing adaptor sequences can be filtered using the cutadapt software. *De novo* assembly can be performed using Platanus (Kajitani et al., 2014; Tanizawa et al., 2015) or other similar automated genome alignment software.

Illumina HiSeq 2000 combined with Roche 454 pyrosequencing has been used to produce genome sequencing of bacteria, with, on average, 170,000 single-end reads and 557,000,000 pair-end reads, respectively. The reads were extended by an overlap of pair-end reads using the Flash program, followed by filtering for sequence quality and read length. Subsequently, the reads were aligned by Burrows-Wheeler Aligner to the contigs obtained from Newbler assembler, and the aligned sequences were subjected to the Columbus module of Velvet to perform alignment-assisted assembly. The obtained sequences were further improved by scaffolding using Opera (Gao et al., 2011) and by gap-closing using GapFiller (Alkema et al., 2016) to obtain the final draft genome sequences.

5.2.4 Ion Torrent platform

Ion semiconductor sequencing is a method of DNA sequencing based on the detection of hydrogen ions that are released during the polymerization of DNA. This method of sequencing by synthesis, allows a complementary strand to be built based on the sequence of a template strand. A microwell containing a template DNA strand is filled with single species of deoxyribonucleotide triphosphate (dNTP). When the dNTP is complementary to the leading template nucleotide, it is incorporated, causing the release of a hydrogen ion detected by the ion-sensitive field-effect transistor (ISFET) ion sensor. The solution is used as the gate electrode, and a voltage is created between substrate and oxide surfaces. When homopolymer repeats are present in the sequence, multiple dNTP molecules will be incorporated, avoiding any error. There is no need for modified nucleotides or optics. Ion semiconductor sequencing may also be referred to as pH-mediated sequencing, silicon sequencing, or semiconductor sequencing. Each microwell on the semiconductor chip contains many copies of one single-stranded template DNA molecule to be sequenced (Rothberg et al., 2011). This technology, however, requires accurate sample preparation, electrophoretic separation of amplified DNA fragments on agarose gel, and elution from gel of DNA bands with a size of about 350 bp.

5.3 NGS PLATFORM APPLIED TO SEQUENCING OF MICROBIAL COMMUNITIES

Metagenomics has been first applied to produce an overview of the genomes contained in a bacterial community by sequencing of randomly fragmented DNAs, especially for the uncultivable bacteria inhabitants of extreme environments. Many applications in food biotechnology involve complex nutrient environments and mixed cultures. Currently, metagenome-scale models for LAB have been proposed and will be exploited in industrial applications (Branco dos Santos et al., 2013). The availability of a catalogue of reference LAB genomes has opened perspectives for the analysis of food fermentations (Almeida et al., 2014).

Food microbial ecology studies, especially in complex fermentation systems, may gain new knowledge from molecular techniques, such as NGS technology, enabling analysis of microbial communities in fermented foods (Bokulich et al., 2012). The advancement in NGS technologies will allow studies of microbial dynamics within food fermentations and the ecosystems of food-processing built environments through amplicon sequencing, metagenomics and transcriptomics, for better understanding of microbial food communities (Bokulich et al., 2016).

5.3.1 Pangenomics

With the dramatic reductions in the cost and time involved in DNA sequencing, a new approach for the characterisation of bacteria is emerging. It is based on a comparison of complete genome sequences of a number of members of the same species (pangenomics) (Garrigues et al., 2013). Pangenomics opens new opportunities for understanding and improving industrial starter cultures and probiotics (Ljungh and Wadström, 2009). The genomic data can support understanding of texture and flavour formation in dairy products, understanding the functionality of probiotics as well as information exploitable for strain screening, strain improvement, safety assessments and process improvements.

By connecting microbial classes with large numbers of chemical compounds, biological processes, molecular functions and qualities using an array of logical axioms, MicrO software has been developed as a bioinformatic tool to increase the computing power for the automated text mining of prokaryotic taxonomic descriptions, using natural language processing to sum up information on microbial phenotypes and genotypes (Blank et al., 2016).

5.3.2 Omic technologies: transcriptomics, proteomics, functional genomics, systems biology

Oligonucleotide arrays have been extensively used, but a limitation is in the gene selection, often focusing on genes with a clear coding potential and restricted to protein coding genes. To overcome this limitation, tiling arrays covering all the genome sequence have been used to detect either sense or antisense transcripts and to find novel transcripts, detect alternative transcription start sites and define the non-coding RNAome (van der Meulen et al., 2016).

Whole-genome arrays have been made available for LAB species with sequenced genomes: whole-genome arrays for *Lactobacillus acidophilus* NCFM-type strains (Altermann et al., 2004; Klaenhammer et al., 2007) have been used to study the influence of exposure to milk on gene expression and the regulatory networks that promote survival and/or functionality; this technology offers the means to investigate the role of milk and dairy products on gene expression and phenotypic outcomes for bioprocessing dairy LAB. *Lactobacillus plantarum* strains have been object of several studies based on differential gene expression (Stevens, 2008; Stevens et al., 2008; Molenaar et al., 2005; Sturme et al., 2007; Stevens et al., 2010).

Whole-genome transcriptional arrays are providing in-depth views of environmental influences on gene expression and culture behaviour, similarities and differences by comparative genomics, and elucidation of the metabolic and functional roles of these organisms.

RNA sequencing based on NGS technology is a highly performing method to assess whole sets of transcripts and to monitor sense and antisense transcribed genes (Lasa et al., 2001) and non-coding RNAs. It may be applied to sequence whole RNAs from a mixed population or community, obtaining the profile of the metatranscriptomes.

Metatranscriptomics is a technique for extraction of data from microbial communities based on gene expression analysis and species assignment of identified transcripts (Monnet et al., 2016; De Filippis et al., 2016).

All –omic technologies used to analyse a complex environment such as cheese during ripening have been named 'meta-omics' (Dugat-Bony et al., 2015); meta-omics also integrates systems biology studies, especially on probiotic strains (Twardziok et al., 2014).

Transcriptomics allows collection of data that may integrate metagenomic information (Almeida et al., 2014) on microbial communities extracted from NGS sequencing data (Branco dos Santos et al., 2013; Bokulich et al., 2016).

Transcriptome analysis has been one of the tools most exploited in recent years (Douillard et al., 2013a). Transcriptomics has been applied to study the growth response of *Lb. plantarum* (Sieuwerts et al., 2005; Brooijmans et al., 2009) and *Lc. lactis* in various conditions: addition of exogenous heme and resistance to oxidative stress, heat stress, fermentation pH and temperature and salt concentration (Dijkstra et al., 2014). Aeration induced protection against heat stress, whereas higher fermentation temperatures induced enhanced robustness and protection against oxidative stress. The studies allowed setting up the optimum conditions to produce a high biomass of bacteria under oxidative fermentation, exploiting the oxygen consumption by ferrichrome ATP-binding cassette (ABC) transporters and overexpression of ribonucleotide reductases and the ferredoxin (flavin-dependent) reductase (FdR) Yum-like TrxB2 (Cretenet et al., 2014; Chen et al., 2015). The glutathione/glutathione reductase (Gsh-GshR) system can protect *Lc. lactis* from oxidative stress, but in *Lc. lactis* the Gsh-GshR system works in the presence of exogenous glutathione (Chen et al., 2015). The thioredoxin (Trx) system, which is composed of NADPH, thioredoxin reductase (TrxR), and thioredoxin, is a key antioxidant system in defence against oxidative stress through its disulfide reductase activity-regulating protein disulfide homeostasis. The Trx system provides the electrons to thiol-dependent peroxidases to remove reactive oxygen species.

Aerobic growth tolerance has been exploited for biomass growth of *Lc. lactis*, possessing a cytochrome oxidase, by addition of exogenous heme (Gaudu et al., 2002, 2003; Brooijmans et al., 2009; Ianniello et al. 2015) to adapt it to technological stresses. Conditions of oxygenation are relevant at the beginning of fermentation. A recent study (Cretenet et al. 2014) showed the ability of *Lc. lactis* subsp. *cremoris* to respond to oxidative stress, either by the transcriptional or metabolic analyses, in relation to acidification kinetics and growth conditions, especially at an early stage of growth.

Kinetics of growth and acidification in presence of oxygen were good due to an efficient consumption of oxygen within the first 4 hours of culture. In carbon metabolism, the presence of oxygen led to an early shift at the gene level in the pyruvate pathway towards the acetate/2,3-butanediol pathway. A similar positive effect of exposure to oxygen was observed in *Lactobacillus casei* species under aerobic and respiratory growth (Zotta et al., 2014; Reale et al., 2015).

Comparative genomics (Kant et al., 2011, 2014) and phylogenomic studies (Claesson et al., 2008) have provided clues for the evolutionary history of LAB species (Salvetti et al., 2013).

Functional genomics are based on studies performed by molecular methods to find function of transcripts, or the phenotypes in their absence, exploiting gene mutations and inactivation and deletion techniques and engineering strains to analyse the function of each gene to monitor strain adaptation to its environment and behaviour in various culture systems (Licandro-Seraut et al., 2014).

5.4 METABOLOMICS AND PROTEOMICS

Metabolite determination of flavour and volatile profiles can be screened using gas chromatography-mass spectroscopy (GC/MS), mass spectrometry analysis combined with HPLC, nano-HPLC combined with matrix-assisted laser desorption ionization (MALDI) and with a time-of-flight (ToF) mass spectrometer, and ion trap mass analyser (Orbitrap) provided with a compound mass library. Metabolomics approach using headspace solid phase microextraction (SPME) GC/MS and nuclear magnetic resonance (NMR) (Sundekilde et al., 2013) with multivariate statistical analysis are effective methods to determine the metabolite-producing ability of individual strains.

Many food fermentations are performed using mixed cultures of LAB. Interactions between strains are of key importance for the performance of these fermentations.

Yogurt fermentation carried out by a mixed-culture of *S. thermophilus* and *Lactobacillus delbrueckii* subsp. *bulgaricus* has been well described (Sieuwerts et al., 2010). These species are believed to stimulate each other's growth by the exchange of metabolites such as folic acid and carbon dioxide. The interactions between the bacteria were related to purine, and to a long-chain fatty acid metabolism. The results showed that formic acid, folic acid, and fatty acids are provided by *S. thermophilus*. *Lb. bulgaricus* proteolytic activity produced amino acid substrates, while the up-regulation of genes involved in biosynthesis of sulfur and of branched-chain amino acids was necessary for mixed culturing. Several studies have attempted to define the interactions among strains and species, especially in yogurt production, such as interaction between *Lactobacillus plantarum* and *S. thermophilus* (Kort et al., 2015), between *Lb. plantarum* and *Lb. bulgaricus* starters (Settachaimongkon et al., 2016), *Lb. bulgaricus* and *S. thermophilus* (Settachaimongkon et al., 2014), as well as preculturing of probiotics before addition to yogurt (Settachaimongkon et al., 2015) or the contribution of *Lb. bulgaricus* to *Bifidobacterium breve* growth (Ongol et al., 2007).

Proteomic analysis has been often applied to understand the complexity of proteomes and, in particular, secretome, the pool of secreted proteins in LAB species (Zhou et al., 2010), and to understand differences in surface proteins, the so-called exoproteome (Klerebeezem et al., 2010; Johnson et al., 2016), in particular in probiotic or potentially probiotic strains, such as *Lb. acidophilus* (Buck et al., 2005), *Lb. plantarum* (Boekhorst et al., 2006a; Beck et al., 2009; Gross et al., 2010), and *Lactobacillus rhamnosus* (Douillard et al., 2013a, 2013b).

5.4.1 Subcellular localisation (SLC): secretion systems for secreted proteins

Outer proteins are secreted in the extracellular space, mediating interactions with the environment.

Sortase-dependent proteins (SDPs) and S-layer proteins are characterised in lactobacilli. These proteins and respective genes may serve as key markers allowing prediction of probiotic potential of the LAB strains (Van Pijkeren et al. 2006; Zhou et al., 2010; Klerebeezem et al., 2010).

Specific metabolic and physiological characteristics of lactobacilli play a key role in the adaptation to the environment. The cell surface structures of lactobacilli are in direct contact with the environment and can function as both key adaptation and probiotic factors. The LAB cell wall consists of several characteristic structures: a stratified peptidoglycan (PG) decorated with proteins, teichoic acids (the binding to divalent cations, such as Mg^{++}, confers rigidity), and polysaccharides, and in some species, the cell wall is surrounded by an outer semiporous structure of self-assembling proteins packed in a para-crystalline layer (S-layer) (Beck et al., 2009; Kant et al., 2014). *Lb. acidophilus* complex is distinguished in two clades by the presence or absence of S-layers. S-layer–associated proteins (SLAPs) are present in the exoproteomes of various S-layer–forming *Lactobacillus* species that were proteomically identified (Johnson et al., 2016). A recent proteomic study on the S-layer–forming *Propionibacterium freudenreichii*, a probiotic with anti-inflammatory, immunomodulatory capacity, characterised various cell surface proteins, including putative SLAPs (Le Maréchal et al., 2015).

Extracellular localised proteins require a signal peptide and an anchor domain for covalent or non-covalent interaction with the external components. The majority of LAB proteins supported the grouping of secreted proteins by their SLC as anchored to the cytoplasmic membrane via the N- or the C-terminal transmembrane helix, released into the extracellular medium via classic signal peptidase recognition sequence, in the Sec secretion pathway, or through holin or ABC transporters, being part of the competence pseudo-pili (pilin genes associated to pilin-specific sortase gene) (Klerebeezem et al., 2010).

A classification of extracellular proteins has been made based on sortase-dependent signals for anchoring to the cell wall or other domains responsible for anchoring via protein-protein or choline interactions (Klerebeezem et al., 2010). The analysis of LAB genomes contributed to the characterisation of genes encoding the Sec, the fimbrillin-protein exporter (FPE), peptide-efflux ABC and holin systems.

Cell surface proteins can be anchored to the cell wall or to the membrane by different mechanisms. The proteins can be attached to the membrane layer by means of an N- or C-terminal transmembrane (hydrophobic) anchor or a lipoprotein-anchor (lipobox): the lipoproteins are anchored to the membrane lipids.

LPxTG-type anchors (e.g. LPxTG proteins) are attached to the cell wall covalently. These surface proteins are anchored by sortase enzymes that recognise a highly conserved LPxTG sequence motif. The studies by Sun (Sun et al., 2015) identified 1628 predicted LPxTG-containing proteins and 357 sortase enzymes in 213 LAB genomes. Sortase-dependent pilus gene clusters are common in Gram-positive pathogens; these pili are also produced by probiotic species. *Lactobacillus rhamnosus* strain GG pilus gene complex is formed by a cluster of three pilin genes and one pilin-specific sortase gene (Douillard et al., 2013a). SpaCBA pili have been shown to contribute to probiotic properties by mucin binding and cellular signalling (Klerebezem et al., 2010). It is interesting to note that some *Lb. casei* strains such as LOCK 0919 (Koryszewska-Bagińska et al., 2013) harbours a spaCBA-sortase gene cluster on the plasmid pLOCK0919, conferring adhesion potential, and this cluster is expressed (Aleksandrzak-Piekarczyk et al., 2015).

Proteins can be attached to the cell wall noncovalently, through cell wall–binding (repeated) domains, such as LysM, SLH or WxL domains or glycine-tryptophan dipeptide motifs (Van Pijkeren et al., 2006).

These studies showed that several groups of extracellular proteins, devoid of signal peptides, are translocated into the outer space through additional secretory mechanisms. Some secreted proteins have been shown to associate to the cell wall by electrostatic interactions (Bath et al., 2005).

The analysis of the exoproteome in *Lactobacillus* species (Zhou et al., 2010; Klerebeezem et al., 2010) classified the extracellular space proteins in several secretome groups, according to pathways of transport and secretion (http://www.cmbi.ru.nl/lab_exoproteome/).

5.4.2 Interactome for cell adhesion and pathogen exclusion

Surface proteins of lactobacilli include key interaction receptors for probiotics and enzymes for growth in milk. The extracellular transglycosylases, also referred to as aggregation-promoting factors (APFs), mannose-binding proteins (MBPs) (Boekhorst et al., 2006b; Gross et al., 2010; MacKenzie et al., 2010) and other groups of proteins are involved in adhesion to glucans and to cellular receptors. Adhesins (MBPs) are large proteins, consisting of repeating modules or domains. Many extracellular proteins contain domains for binding to macromolecular substrates, often duplicated and arranged in several repeats. In addition to domains for binding to the cell wall, other domains are related to binding to host macromolecules.

5.4.3 LAB peptidome

Furthermore, adding complexity to our understanding of the effectors produced by individual strains is the progress in identification of LAB peptidome (Boekhorst et al., 2011). Putative small proteins encoded in not well-identified short gene sequences, the ORFeome (the ensemble of short open reading frames), are widespread in bacterial genomes, and an implementation in the annotation of the *Lb. plantarum* WCFS1 genome allowed inclusion of some of the ORFs not previously seen. The identification of small peptides originated from an ORF or by proteolysis of a larger protein is difficult, since peptides elute fast from polyacrylamide gels and require dedicated analytical methods focusing on small compounds. In fact, several ongoing studies using metabolomic tools applied to study the LAB supernatants could help in identifying the peptidome world.

5.5 COMPARATIVE GENOMICS OF DAIRY-ASSOCIATED BACTERIA: THE *LACTOBACILLUS* GENUS COMPLEX, *STREPTOCOCCI/LACTOCOCCI, ENTEROCOCCI, PROPIONIBACTERIA* AND *BIFIDOBACTERIA*

The *Lactobacillus* group complex includes, in addition to *Lactobacillus* genus, five other genera: *Pediococcus, Weissella, Leuconostoc, Oenococcus* and *Fructobacillus* (Salvetti et al., 2013; Holzapfel and Wood, 2014; Sun et al., 2015). *Lactobacillus* species are important as starter or adjunct non-starter strains in dairy productions (Mayo et al., 2008). Among dairy lactobacilli, *Lactobacillus helveticus* (Cremonesi et al., 2013), *Lb. casei* (Maze et al., 2010; Douillard et al., 2013a; Smokvina et al., 2013; Sun et al., 2015), *Lactobacillus delbrueckii* subsp. *bulgaricus* and subsp. *lactis*, *Lb. plantarum* (van den Nieuwboer et al., 2016) and *Lb. rhamnosus* (Douillard et al., 2013a) have been extensively analysed.

The family Lactobacillaceae includes the highest number of Generally Recognised as Safe species (GRAS), and many strains are among the most important bacteria in food microbiology and human nutrition due to their contribution to fermented food production or their use as probiotics. The description of novel species in the past years has led to a continue revision of the genus with the recognition of a growing number of phylogenetic groups (Dellaglio and Felis, 2005; Salvetti et al., 2013; Zheng et al., 2015).

More than 200 *Lactobacillus* spp. genomes have been sequenced (Sun et al., 2015). Other sequencing projects on LAB strains are ongoing or have been completed by the U.S. Department of Energy, Joint Genome Institute (JGI, http://www.jgi.doe.gov) in collaboration with the LAB Genome Consortium and placed in the National Center for Biotechnology Information.

At taxonomic level, considering the family of Lactobacillaceae, differences in GC content and organization of genomes vary considerably. The genomes of the lactobacilli range in size from 1.23 Mb in *Lactobacillus sanfranciscensis* to 4.91 Mb in *Lactobacillus parakefiri*. The GC content varies from 31.93 to 57.02%.

Sun and colleagues (Sun et al., 2015) constructed a phylogenetic tree with the lactobacilli and representative genomes of 452 selected genera from 26 phyla using 16 proteins common to all taxa and 73 core proteins shared by the 213 genomes of the lactobacilli. The tree reinforced the definition of the *Lactobacillus* group complex as by previous phylogenetic analyses (Salvetti et al., 2013). A high genetic diversity is highly characteristic for the LAB, especially for the species within the *Lb. acidophilus* group (Zhou et al., 2010), *Lb. plantarum* (Siezen and van Hylckama Vlieg 2011) and *Lb. casei* groups (Douillard et al., 2013b; Smokvina et al., 2013).

Glycolysis occurs in obligate homofermentative and facultative heterofermentative LAB species and is the effect of the presence of the *FdpA* gene coding for fructose 1,6-bisphosphate aldolase. Glycolysis may lead to heterolactic fermentation under certain conditions in some LAB species, considered homofermentative, that exploit the pentose phosphate pathway when using certain substrates.

The absence of pyruvate dehydrogenase, coded in the *PDHc* operon, and the presence of phosphofructokinase (*pfkA* gene) are related to exploitation of glycolytic and alternative pathways.

The *Lb. delbrueckii* group is lacking the pyruvate dehydrogenase complex (PDHc), explaining their phenotype: the species are homofermentative and glucose is converted exclusively into lactic acid via lactate dehydrogenase (Ldh). Obligate heterofermentative lactobacilli use the 6-phosphogluconate/phosphoketolase (6-PG/PK) pathway as their central metabolic pathway and are unable to grow on fructose as a unique carbon source. Heterofermentative LAB (*Leuconostoc* and some *Lactobacillus* species) use pentoses in the phosphoketolase pathway. The extra NAD(P)H, which is produced during growth on hexoses, is transferred to acetyl-CoA, yielding ethanol. The facultative heterofermenting LAB species use both the Embden-Meyerhof-Parnas (EMP) pathway and the phosphoketolase pathway. In *Lactobacillus brevis*, fructose-specific phosphotransferase system and glycolytic enzymes allow fructose to be metabolised via the EMP pathway.

The *gap* operon codes for genes codifying enzymes converting triose phosphates into phosphoenolpyruvate (PEP), important in homolactic and facultative heterolactic fermentative LAB.

The *Lactobacillus, Leuconostoc, Weissella, Fructobacillus* and *Oenococcus* species that lack phosphofructokinase (*Pfk*$^-$) form a distinct monophyletic group that includes *Lactobacillus reuteri, Lactobacillus brevis, Lactobacillus buchneri, Lactobacillus collinoides, Lactobacillus vaccinostercus* and *Lactobacillus fructivorans* groups. The absence of *PfkA* gene correlates with the metabolism, as glucose fermentation is driven through the PK pathway, leading to the production of CO_2 in obligate heterofermenters (Salvetti et al., 2013; O'Connell Motherway et al., 2013). Most species (87%) within the *PfkA*$^-$ group are classified as obligate heterofermentative, with the rest being facultatively heterofermentative, due to the ability to produce dihydroxyacetone phosphate/glyceraldehyde phosphate (DHAP+GAP) and exploit the EMP pathway to produce an additional NADH. The obligate heterofermentative LAB rely on fast assimilation of pentoses, while hexoses are slowly assimilated. Most species (75%) within this *PfkA*⁻ clade also lack fructose 1,6 diphosphate aldoalase (FdpA). *PfkA* gene is present in the *Weissella* clade as well as in some *Leuconostoc* and in some species of the *Lb. reuteri* and *Lb. fructivorans* groups.

As for species clustering based on phosphoketolase pathway (PKP) genes distribution, *Lb. delbrueckii* species cluster independently. *Leuconostocaceae* with few lactobacilli and the remaining *Lactobacillus* species, together with *Enterococcus*, *Bacillus* and members of *Lc. lactis,* cluster separately.

The *Leuconostoc mesenteroides* cluster includes, among others, *Ln. mesenteroides* species metabolising citrate, frequently involved in dairy productions.

The *Pediococcus* genus includes *Pediococcus acidilactici* or *Pediococcus pentosaceous* as adventitious (nonpredominant, non-starter LAB) strains in cheese (such as in cheddar), that exploit the availability of pyruvate and fatty acids, with scarce ability to ferment lactose (Caldwell et al., 1996), in some cases used as probiotics (Casey et al., 2007) and for bacteriocin production.

S. thermophilus belongs to the family Streptococcaceae. The thermophilic *S. thermophilus* is the main GRAS species, utilised in many dairy productions (Dellaglio and Felis, 2005). *Streptococcus macedonicus* is associated with certain fresh dairy products (Georgalaki et al., 2000; De Vuyst and Tsakalidou, 2008), with acidifying and lipolytic properties, synthesising exopolysaccharides and anti-clostridial bacteriocins; this species belongs to a group of species associated with endocarditis cases, but has lost the pathogenicity-related genes and virulence factors, except one pilus gene for cell adhesion (Papadimitriou et al., 2014; Papadimitriou et al., 2016).

The *Lactococcus* genus includes three main mesophilic species producing lactic acid. *Lc. lactis* is a widely used LAB species. The genomes of *Lc. lactis* subsp. *lactis* bv. *diacetylactis* CRL264 (Zuljan et al., 2016) and *Lc. lactis* subsp. *cremoris* HPT, a well-defined dairy starter (Lambie et al., 2014), have been recently made available. Kelly proposed PFGE profiles (pulsotypes) as a method to differentiate environmental strains based on chromosome composition, to distinguish them from domesticated strains (Kelly et al., 2010; Passerini et al., 2010). Bayjanov identified relations between gene presence and a collection of 207 phenotypic characters across 38 *Lc. lactis* strains of dairy and plant origin (Bayjanov et al., 2013). In *Lc. lactis,* RNA sequencing has provided new in-depth knowledge regarding transcribed portions of the genome, and novel small RNAs have been identified and involved in carbon uptake and metabolism (van der Meulen et al., 2016). On the other side, *Lactococcus garvieae,* a species involved in cheese products (Flórez and Mayo, 2015) has been included in a list of emerging pathogens also in humans (Ferrario et al., 2013).

Enterococcus is a large genus of LAB in the order Lactobacillales (Klein, 2003). For their salt tolerance, aerotolerance and thermotolerance, *Enterococcus faecium, Enterococcus durans* and *Enterococcus faecalis* strains are found in microbiota of local cheese productions (Morales et al., 2003; Abeijón et al., 2006; Chaves-López et al., 2006; Cabral et al., 2007), with strains able to produce bacteriocins (Leroy et al., 2003; Henning et al., 2015; Maki et al., 2015). Some strains are used as probiotics for human and animal health (Franz et al., 2011), such as *E. faecium* NCIMB 10415 (Kreuzer-Redmer et al., 2016), *E. durans* TN3 (Kanda et al., 2016), *Enterococcus mundtii* 50H (Haghshenas et al., 2016), and *Enterococcus lactis* IW5 (Nami et al., 2015). Enterocin AS-48–producing *E. faecalis* (Maky et al., 2015) was used as starter or co-culture with a commercial lactic starter in the manufacture of raw milk Manchego cheese (Arqués et al., 2015). In *Enterococcus* spp., with interest in pathogenic strains, small RNAs have been described with a great complexity of RNA transcripts (Fouquier et al., 2011): the application of intergenic tiling microarrays (with oligonucletides reflecting the genome complexity) and tagged RNA sequencing allowed the identification of hundreds of sRNA under different stress conditions in *E. faecalis* (Shioya et al., 2011; Innocenti et al., 2015).

The dairy propionic bacteria *P. freundenreichii* (PF), *Propionibacterium jensenii, Propionibacterium acidipropionici, Propionibacterium ciclohexanicum,* and *Propionibacterium microaerophilum* belong to the class Actinobacteria, having lipolytic activity and producing short-chain fatty acids (SCFA), are at the basis of Emmental and Swiss type cheese productions; selected strains are used as probiotics for their immunomodulatory action and bifidogenic effect in human gut (Cousin et al., 2010; Thierry et al., 2011; Kwon et al., 2016). PF is used in cheese technology and for its probiotic properties. PF easily adapts to ecological niches, being able to grow on different carbon and nitrogen sources. Twenty-one strains of PF were sequenced, and ability to use 50 different sugars was studied (Koskinen et al., 2015).

The genus *Bifidobacterium* belongs to the phylum and class Actinobacteria, family Bifidobacteriaceae (Dellaglio and Felis, 2005). Bifidobacteria are Gram-positive bacteria, anaerobic or microaerophilic, depending on strain, and are shaped in polymorphic rods. They are non-spore forming, non-motile and non-filamentous, acidify through oligosaccharides and fructose fermentation, and produce acetate. *Bifidobacteria* are one of the dominant microbial groups that occur in the gut of various animals, being particularly prevalent during the suckling period of humans and other mammals. Their ability to compete and prevail on other gut bacteria and pathogenic species is attributed to the production of acetate (Fukuda et al., 2012).

Bifidobacteria are present in traditional yogurt and fermented milk production or added as probiotic strains to dairy products. *Bifidobacterium longum* are a prevalent species in the intestinal microbiota, especially in infants, and show intra-species genomic diversity reflecting ecological specialisation (Chaplin et al., 2015).

The *B. longum* reference strain has been set to be *B. longum* ATCC 15697, and in the race to individuate similarities and differences, new potential probiotic strains have been sequenced (Wei et al., 2010; Zakharevich et al., 2015; O'Callaghan et al., 2015; Chaplin et al., 2015), determining gene differences between the three known subspecies (*B. longum, Bifidobacterium infantis* and *Bifidobacterium suis*). Other studies have focused on species-specific differences (Milani et al., 2015), characterising the *Bifidobacterium adolescentis* genome (Duranti et al., 2016), *Bifidobacterium animalis* sbsp. *lactis* (Garrigues et al., 2010; Jungersen et al., 2014) and *B. breve* (Shanahan et al., 2011). Recently, studies have determined also metabolism in milk (Turroni et al., 2011), nutrient needs (Ferrario et al., 2015), probiotic determinants and functional genomics analysis in *B. breve* (O'Connell Motherway et al., 2011; Ferrario et al., 2015, 2016).

5.5.1 Comparative genomics of *Lb. rhamnosus* and *Lb. casei*

In *Lb. rhamnosus* and *Lb. casei*, phylogenetic distribution and organization of genes have been described (Maze et al., 2010; Douillard et al., 2013a, 2013b; Kant et al., 2014) and genomes compared (Cai et al., 2009; Sun et al. 2015).

Lb. rhamnosus has been isolated from a large variety of ecological niches, for example, human intestinal tract, vaginal cavity, oral cavity and cheese, exemplifying its remarkable ecological adaptability.

To cover the phenotypic and genomic diversity of the *Lb. rhamnosus* species, 100 *Lb. rhamnosus* strains were isolated from different ecological niches, among which 23 strains were of dairy origins, including artisanal cheeses and products marketed as probiotics. The genomes of all strains were sequenced and mapped onto the *Lb. rhamnosus* GG chromosome. Gene duplication events or horizontal gene transfer (HGT) of genes involved in sugar metabolism are at the basis of genetic diversity. The number of genes shared between 100 *Lb. rhamnosus* isolates and *Lb. rhamnosus* GG ranged from 86.9 to 100% of genes. As for the gene content, dairy isolates showed a highest diversity from *Lb. rhamnosus* GG, suggesting that the dairy isolates are genetically distant. Based on comparative genomic analysis, the hierarchical clustering of the *Lb. rhamnosus* species resulted in distinct clusters (Douillard et al., 2013a). The majority of dairy strains were found to cluster together and showed great differences with the non-dairy strains. In contrast, intestinal isolates, including *Lb. rhamnosus* strains marketed as probiotics, shared similarities with other human isolates. This supports the hypothesis that the genomes of probiotic-marketed strains reflect their adaptation to the human intestinal tract.

Although no major differences are present in the Clusters of Orthologous Groups (COG) of proteins distribution between the different subgroups, it was noted that 87 *Lb. rhamnosus* GG genes assigned to the COG carbohydrate transport and metabolism are not in the estimated core genome and are predicted to encode mostly phosphotransferase system (PTS) and other sugar transport systems, possibly essential for the adaptation to different environments. These genes were located in highly variable regions of the *Lb. rhamnosus* genome, reflecting the metabolic diversity of this species. The most variable chromosomal regions include genomic islands rich in transposases and other mobile genetic elements. The presence of genomic islands varies among strains of the species *Lb. rhamnosus*, as observed previously for the strain LC705. This suggests that HGT events have contributed significantly to the diversity of the *Lb. rhamnosus* species.

Comparative genomic analysis of the 100 strains also revealed metabolic capability to utilise different carbon sources.

Carbohydrate utilisation profiling showed that most *Lb. rhamnosus* strains use a large range of simple and complex carbohydrates. However, some differences may reflect their genomic diversity, with the acquisition or loss of metabolism-associated genes. The ability to utilise carbohydrates mostly relies on the presence of functional transporter machinery and intact metabolic pathways.

Two prevailing *Lb. rhamnosus* geno-phenotypes, A and B, were found in the *Lb. rhamnosus* species. The strains belonging to the geno-phenotype A are characterised by a lack of SpaCBA pili, a different carbohydrate metabolism (D-lactose, D-maltose and L-rhamnose) and a distinct clustered regularly interspaced short palindromic repeats

(CRISPR) system profile, indicative of a continuous evolutionary race with their phages in dairy environments. Of 23 strains of dairy origins, most dairy strains showed marked differences with non-dairy strains. These strains present a metabolic profile similar to that of industrial dairy strain *Lb. rhamnosus* LC705. Most dairy strains utilise D-lactose: *lacT* and *lacG* genes have remained functional in these strains. Almost 74% of all isolates analysed can partially or fully utilise L-rhamnose, a carbohydrate from which the species name derives (Douillard et al., 2013a).

5.5.2 *Lb. casei* core genome and ecotype differences in dairy adapted strains

The PTS and metabolic-related genes nonessential in dairy products were lost or decayed, such as loss of L-fucose utilisation. Several factors previously associated with host–microbe interactions such as pili, cell-envelope proteinase, hydrolases p40 and p75, or the capacity to produce short branched-chain fatty acids (*bkd* operon) are part of the *Lb. casei* core genome. The variome consists mainly of hypothetical proteins, phages, plasmids, transposon/conjugative elements and other functions (Smokvina et al., 2013).

By means of duplication or HGT of genes involved in sugar metabolism, a variety of genes for utilisation of sugars and as sugar metabolism have been created. A large variety of sugar utilisation gene cassettes were identified, with strains harbouring between 25 and 53 cassettes, reflecting the high adaptability of *Lb. casei* to different niches (Smokvina et al., 2013).

Similar mechanisms are at the basis of heterogeneity of cell surface proteins, transporters and EPS biosynthesis proteins.

The *lac* operon is generally plasmid associated and may be transferred horizontally: it contains genes for the tagatose-6-phosphate pathway (*lacABCD*), a lactose PTS (*lacFEG*), with *lacG* coding for phospho-β-galactosidase, and the antiterminator *lacT*.

5.6 CLUSTERED REGULARLY INTERSPACED SHORT PALINDROMIC REPEATS (CRISPR) IN ADAPTIVE IMMUNITY

The CRISPR/Cas endonuclease system, an adaptive bacterial immunity system based on target sequence–guided DNA cleavage, resembles the RNA interference system based on small interfering RNAs (siRNAs) present in the other kingdoms of life. The Nobel laureates Charpentier and Doudna found the mechanism in bacteria allowing to fight viral and bacteriophage infections. CRISPR-encoded RNAs form duplex structures with the target sequence that guides the CRISPR-associated nuclease Cas9 to degrade invading DNA molecules in a sequence-specific manner. In Archaea, the CRISPR system is constitutive, while in bacteria it is inducible. CRISPR-Cas systems provide a mechanism for prokaryotes to incorporate short stretches of foreign DNA (spacers) into their genomes to archive sequence information on "non-self" DNA, such as viruses or plasmids. This is the adaptation stage of the immune process. Once integrated, these spacers serve as templates for the synthesis of RNA that then directs Cas nucleases to specific foreign nucleic acids in order to degrade them (Westra et al., 2012). Several different types of CRISPR systems have been identified, and each is associated with a distinct set of Cas proteins (Makarova et al., 2013). Only two proteins, Cas1 and Cas2, appear to be strictly conserved among the various CRISPR systems, and they are both metal-dependent nucleases. Researchers have found a diversity of CRISPR systems in *Bifidobacteria* (Briner et al., 2015), in *Lc. lactis* (Millen et al., 2012), *Lb. buchneri* (Briner and Barrangou, 2014), *Lb. rhamnosus* (Douillard et al., 2013a), *S. thermophilus*, and in *Lb. paracasei* (Smokvina et al., 2013), as well as in other LAB species (Sun et al., 2015). A large number of CRISPR spacers in bifidobacteria show perfect homology to prophage sequences harboured in the chromosomes of *Bifidobacterium* species: spacer sequences can be considered as a signature of past exposure to exogenous DNA, including some spacers that self-target the *Bifidobacterium* chromosome (Briner et al., 2015). Comparative studies on CRISPR systems content and diversity in LAB species have been performed (Horvath et al., 2009), providing insights on this immunity system to LAB adaptation to viruses and phages (Desiere et al., 2002), with a selective pressure for strains inhabiting dairy environments (Douillard et al., 2013a).

The CRISPR system provides a protection/immunization system against plasmid conjugation and phages determining that CRISPR sequences and the presence of the *cas* gene could be used as an indicator of a specific niches. Douillard and colleagues produced a CRISPR profile based on spacer oligotyping for *Lb. rhamnosus* strains. In addition to the high degree of diversity among various strains, the strains from dairy cluster shared a similar CRISPR spacer set due to the similar exposure to phages and other mobile genetic elements in that habitat.

In the most extensive analysis on LAB genomes, performed on 213 strains (Sun et al., 2015), CRISPR Type II systems were detected in 36% of the *Lactobacillus* genus complex and associated genera, although they occur in only 5% of all bacterial genomes analysed, and also detected a huge variability in locus size and spacer content, ranging from 2 to 135 CRISPR spacers.

In other species, different groups of scientists performed strain identification based on spacer diversity. CRISPR-based genotyping methods may give insights into genome microevolution, and several studies have shown potential for CRISPR-based typing of industrial starter cultures, probiotic strains and problematic pathogens (Sola et al., 2015; Barrangou and Dudley, 2016). These researches have led to the development of molecular tools for spacer oligotyping analysis and strain differentiation.

5.7 REGULATION IN CARBON METABOLISM

The utilisation of carbohydrates relies on several mechanisms of control, from substrate competition, regulation of catabolite control protein (ccpA), activation of repression of catabolite-responsive elements in gene promoters, to transcriptional and posttranscriptional regulation, and finally to riboregulation by means of non-coding RNAs.

5.7.1 Transcriptional and posttranscriptional regulation in carbon metabolism

LAB strains have been tested for industrial applications by modifying the carbon catabolite control (Titgemeyer and Hillen, 2002). LAB impose regulatory mechanisms on metabolic processes to ensure that the needs of the cell are met and not exceeded. These mechanisms interfere with the desired exhaustive industrial production of compounds by bacteria and their use in food industry. The most important regulatory mechanism is catabolite repression (CR), which controls carbon and nitrogen metabolism and carbon fluxes under conditions in which rapidly metabolising carbon sources are abundant. Several groups researched methods to improve strains insensitive to CR and deregulated at the enzyme level (Starlab biotech project, http://www.amalnet.k12.il/micro/contractors.html#proteolysis). To improve the storage properties of fermented milk, optimize fermentation of pentoses and improve nutritive value of feed, genetic modification of *Lb. casei* strains has been attempted in which carbon fluxes are controlled, which is important for milk fermentation and flavour production.

5.7.2 Two-component systems and phosphorylation in sugar substrate regulation

Carbon CR is mediated via several mechanisms. The metabolic state of the cell is sensed by the metabolite-activated HPr kinase/phosphorylase (HPrK/P). The enzyme catalyses the ATP as well as the pyrophosphate-dependent phosphorylation of Ser-46 in HPr, a phosphocarrier protein of a sugar transport and phosphorylation system. HprK/P also catalyses the pyrophosphate-producing, inorganic phosphate-dependent dephosphorylation (phosphorolysis) of seryl-phosphorylated HPr (P-Ser-HPr).

PEP acts as phosphoryl donor: it phosphorylates enzyme I, and then the phosphoryl group is sequentially transferred to HPr, enzyme IIA, enzyme IIB, and finally to an incoming sugar in a reaction dependent on the transmembrane translocator, enzyme IIC (Lorca et al., 2010). P-Ser-HPr functions as catabolite co-repressor by interacting with the LacI-type repressor, catabolite control protein A (CcpA). The IIA domain of the lactose/H+ symport protein LacS (IIALacS) possesses a histidine residue that can be phosphorylated by HPr(His-P), a component of the PTS. HPr-ser-P prevents the entry of less-preferred carbon sources but may also regulate the sugar-specific permease.

Catabolite response elements (*cre*) preceding numerous genes and operons are the target sites for the P-Ser-HPr/CcpA complex. In Gram-positive bacteria with low GC content, carbon regulation is mediated by a shared molecular mechanism of transcriptional control by CcpA/HPr-ser-P complex binding to *cre*. Depending on the location of *cre* with respect to the promoter, this can lead to gene repression or activation.

This global response depends on the metabolic state of the cell, sensed by the metabolite-activated HPrK/P that adjusts the level of HPr-ser-P. The process assures economic utilisation of a preferred carbon source by concomitant regulation of target systems: (i) Catabolic operons are repressed, (ii) genes of central pathways such as glycolysis and genes involved in secretion of metabolic end products are activated, (iii) when required, transport of the preferred carbon source may be up- or down-regulated, and (iv) expression of *ccpA* itself may be autoregulated.

For the regulatory process, the signal is provided by the phosphorylation state of the PTS components, which varies according to the availability of PTS substrates and the metabolic state of the cell.

PTS regulation domain-containing antiterminators and transcription activators are present in *L. casei* strains. Their activity is controlled by PTS-mediated phosphorylation reactions: P ~ EIIB-dependent phosphorylation regulates induction of the corresponding genes, and P ~ His-HPr-mediated phosphorylation plays a role in carbon CR.

Any mutation in the genes encoding a PTS component, *CcpA* or *HprK/P*, changes sugar fermentation ability via the regulatory network of carbon control. Enzyme I (E I) mutants are pleiotropically deficient in PTS-dependent sugar uptake, and HPr mutants are further deficient in CcpA-mediated regulation and in inducer exclusion. Loss of *CcpA* or *HPrK/P* renders the cell insensitive to global carbon control. Accordingly, a *ccpA* mutant of *Lc. lactis* exhibited mixed acid fermentation characteristics, while the wild type was homolactic (Luesink et al., 1998).

The transcriptional control is reinforced by protein activity control via inducer exclusion. This is mediated by HPr-ser-P, which prevents the entry of less-preferred carbon sources but may also regulate the permease that imports the preferred carbon source, as was shown for *LacS* of *S. thermophilus*.

Carbohydrate transporters belong mainly to the phosphoenolpyruvate/sugar PTS, but other kinds of non-PTS transporters are present (Lathinen et al., 2012). In *Lb. casei*, regulation of carbohydrate transport and carbon metabolism is achieved by PTS proteins. Carbohydrate transport of *Lb. casei* is also regulated via inducer exclusion and inducer expulsion. The presence of glucose, fructose, and so on, leads to inhibition of the transport or metabolism of less-favourable carbon sources (inducer exclusion) or to the export of accumulated non-metabolised carbon sources (inducer expulsion). Recent evidence suggests that the PTS of *Lb. casei* also plays a role in cold shock response (Monedero et al., 2007).

5.7.3 Regulatory RNAs and alternative sigma factors in gene expression

Sigma factors direct specific binding of the bacterial RNA polymerase to the promoter. In *Lb. plantarum,* a sequence-based prediction of sigma(54)-dependent promoters revealed an operon encoding a mannose PTS as candidate for sigma(54)-mediated control (Stevens et al., 2010) in concert with the sigma(54)-activator ManR. Genome-wide transcription comparison of the wild-type and the rpoN-deletion strain revealed nine upregulated genes in the wild-type, and 21 upregulated genes in the rpoN mutant. The sigma(54)-controlled mannose PTS was shown to transport glucose in *Lb. plantarum*. The mannose PTS drained the phosphoenolpyruvate (PEP) pool in resting cells. The sigma 54 in *Lb. plantarum* controls the expression of glucose transporter and mannose PTS, contributing to glucose-mediated catabolite control via modulation of the PEP pool. The activator protein binds upstream of the σ 54 promoter site, and DNA looping is required for the activator to contact the RNA polymerase. Therefore, the σ 54 activators are called bacterial enhancer-binding proteins. In *Lb. casei* the expression of the operon encoding the mannose phosphotransferase system (mannose PTS) appears to be σ 54–dependent (Yebra et al., 2004).

Regulatory RNAs have been studied in depth in enterobacteria, in *Bacillus, Clostridia, Listeria* spp. and *S. aureus*. They exert a fine-tuning regulation of transcripts protecting them from endonuclease activity or intervening on transcription initiation start sites. Considering LAB species, RNA sequencing in *Lc. lactis* has provided new in-depth knowledge of transcribed portions of genome, and novel small RNAs have been identified and involved in carbon uptake and metabolism (van der Meulen et al., 2016).

5.8 CONCLUSIONS

Knowledge of the genetic basis of metabolism at the species and strain level and nitrogen and sugar degradation in *Lactobacillaceae*, *Lactococcus* spp, *S. thermophilus*, *Leuconostoc* spp., *P. freudenreichii* and other dairy-associated species are opening the pathway to new strategies for the screening of LAB and dairy-associated strain collections to enable optimized selection of cheese starters and dairy products applications.

References

Abeijón MC, Medina RB, Katz MB, González SN. 2006. Technological properties of *Enterococcus faecium* isolated from ewe's milk and cheese with importance for flavour development. *Can J Microbiol* 52(3):237–45.

Aleksandrzak-Piekarczyk T, Koryszewska-Bagińska A, Grynberg M, Nowak A, Cukrowska B, Kozakova H, Bardowski J. 2016. Genomic and functional characterization of the unusual pLOCK 0919 plasmid harboring the spaCBA pili cluster in *Lactobacillus casei* LOCK 0919. *Genome Biol Evol* 8(1):202–217.

Alkema W, Boekhorst J, Wels M, van Hijum SAFT. 2016. Microbial bioinformatics for food safety and production. *Brief Bioinform* 17(2):283–92.

Almeida M, Hébert A, Abraham AL, Rasmussen S, Monnet C, Pons N, Delbès C, Loux V, Batto JM, Leonard P, Kennedy S, Ehrlich SD, Pop M, Montel MC, Irlinger F, Renault P. 2014. Construction of a dairy microbial genome catalog opens new perspectives for the metagenomic analysis of dairy fermented products. *BMC Genomics* 15:1101.

Altermann E, Russell WM, Azcarate-Peril MA, Barrangou R, Buck BL, McAuliffe O, Souther N, Dobson A, Duong T, Callanan M, Lick S, Hamrick A, Cano R, Klaenhammer TR. 2004. Complete genome sequence of the probiotic lactic acid bacterium *Lactobacillus acidophilus* NCFM. *Proc Natl Acad Sci USA* 102:3906–912.

Alvarez-Sieiro P, Montalbán-López M, Mu D, Kuipers OP. 2016. Bacteriocins of lactic acid bacteria: extending the family. *Appl Microbiol Biotechnol* 100:2939–51.

Arqués JL, Rodríguez E, Langa S, Landete, JM, Medina M. 2015. Antimicrobial activity of lactic acid bacteria in dairy products and gut: effect on pathogens. *Biomed Res Int* 584183.

Barrangou R, Dudley EG. 2016. CRISPR-based typing and next-generation tracking technologies. *Ann Rev Food Sci Technol* 7:395–411.

Bath K, Roos S, Wall T, Jonsson H. 2005. The cell surface of *Lactobacillus reuteri* ATCC 55730 highlighted by identification of 126 extracellular proteins from the genome sequence. *FEMS Microbiology Lett* 253:75–82.

Bayjanov JR, Starrenburg MJC, van der Sijde MR, Siezen RJ, van Hijum SAFT. 2013. Genotype-phenotype matching analysis of 38 *Lactococcus lactis* strains using random forest methods. *BMC Microbiology* 13:68.

Beaulaurier J, Zhang XS, Zhu S, Sebra R, Rosenbluh C, Deikus G, Shen N, Munera D, Waldor MK, Chess A, Blaser MJ, Schadt EE, Fang G. 2015. Single molecule-level detection and long read-based phasing of epigenetic variations in bacterial methylomes. *Nat Commun* 6:7438.

Beck HK, Madsen SM, Petersen J, Israelsen H, Norrelykke MR, Antonsson M, Hansen AM. 2009. Proteomic analysis of cell surface-associated proteins from probiotic *Lactobacillus plantarum*. *FEMS Microbiology Lett* 297:61–66.

Bi D, Xu Z, Harrison EM, Tai C, Wei Y, He X, Jia S, Deng Z, Rajakumar K, Ou HY. 2012. ICEberg: a web-based resource for integrative and conjugative elements found in Bacteria. *Nucleic Acids Res* 40:D621–626.

Blank CE, Cui H, Moore LR, Walls RL. 2016. MicrO: an ontology of phenotypic and metabolic characters, assays, and culture media found in prokaryotic taxonomic descriptions. *J Biomed Seman* 7:18.

Boekhorst J, Helmer Q, Kleerebezem M, Siezen RJ. 2006b. Comparative analysis of proteins with a mucus binding domain found exclusively in lactic acid bacteria. *Microbiology* 152:273–80.

Boekhorst J, Wels M, Kleerebezem M, Siezen RJ. 2006a. The predicted secretome of *Lactobacillus plantarum* WCFS1 sheds light on interactions with its environment. *Microbiology* 152:3175–83.

Boekhorst J, Wilson G, Siezen RJ. 2011. Searching in microbial genomes for encoded small proteins. *Microbial Biotechnol* 4:308–13.

Bokulich NA, Lewis ZT, Boundy-Mills K, Mills DA. 2016. A new perspective on microbial landscapes within food production. *Curr Opin Biotechnol.* 37, 182–9.

Bokulich NA, Mills DA. 2012. Next-generation approaches to the microbial ecology of food fermentations. *BMB Rep* 45(7):377–89.

Branco dos Santos F, de Vos WM, Teusink B. 2013. Towards metagenome-scale models for industrial applications--the case of Lactic Acid Bacteria. *Curr Opin Biotechnol* 24(2):200–6.

Briner AE, Barrangou R. 2014. *Lactobacillus buchneri* genotyping on the basis of clustered regularly interspaced short palindromic repeat (CRISPR) locus diversity. *Appl Environ Microbiol* 80(3): 994–1001.

Briner AE, Lugli GA, Milani C, Duranti S, Turroni F, Gueimonde M, Margolles A, van Sinderen D, Ventura M, Barrangou R. 2015. Occurrence and diversity of CRISPR-Cas systems in the genus Bifidobacterium. *PLoS One* 10(7):e0133661.

Brooijmans RJ, de Vos WM, Hugenholtz J. 2009b. *Lactobacillus plantarum* WCFS1 electron transport chains. *Appl Environ Microbiol* 75:3580–85.

Brooijmans RJW, de Vos WH, Hugenholtz J. 2009a. The electron transport chains of *Lactobacillus plantarum* WCFS1. *Appl Environ Microbiol* 75:3580–85.

Buck BL, Altermann E, Svingerud T, Klaenhammer TR. 2005. Functional analysis of putative adhesion factors in *Lactobacillus acidophilus* NCFM. *Appl Environ Microbiol* 71(12):8344–51.

Cabral ME, Abeijón Mukdsi MC, Medina de Figueroa RB, González SN. 2007. Citrate metabolism by *Enterococcus faecium* and *Enterococcus durans* isolated from goat's and ewe's milk: influence of glucose and lactose. *Can J Microbiol* 53(5):607–615.

Cai H, Thompson R, Budinich MF, Broadbent JR, Steele JL. 2009. Genome sequence and comparative genome analysis of *Lactobacillus casei*: insights into their niche-associated evolution. *Genome Biol Evol* 1:239–57.

Caldwell SL, McMahon DJ, Oberg CJ, Broadbent JR. 1996. Development and characterization of lactose-positive *Pediococcus* species for milk fermentation. *Appl Environ Microbiol* 62:963–41.

Casey PG, Gardiner GE, Casey G, Bradshaw B, Lawlor PG, Lynch PB, Leonard FC, Stanton C, Ross RP, Fitzgerald GF, Hill C. 2007. A five-strain probiotic combination reduces pathogen shedding and alleviates disease signs in pigs challenged with *Salmonella enterica* Serovar Typhimurium. *Appl Environ Microbiol* 73(6):1858–63.

Chaplin AV, Efimov BA, Smeianov VV, Kafarskaia LI, Pikina AP, Shkoporov AN. 2015. Intraspecies genomic diversity and long-term persistence of *Bifidobacterium longum*. *PLoS One* 10(8):e0135658.

Chaves-López C, De Angelis M, Martuscelli M. Serio A, Paparella A, Suzzi G. 2006. Characterization of the *Enterobacteriaceae* isolated from an artisanal Italian ewe's cheese (Pecorino Abruzzese). *J Appl Microbiol* 101(2):353–60.

Chen J, Shen J, Solem C, Jensen PR. 2015. A new type of YumC-like ferredoxin (flavodoxin) reductase is involved in ribonucleotide reduction. *MBio* 6(6):e01132–15.

Chin C-S, Alexander DH, Marks P, Klammer AA, Drake J, Heiner C, Clum A, Copeland A, Huddleston J, Eichler EE, Turner SW, Korlach J. 2013. Nonhybrid, finished microbial genome assemblies from long-read SMRT sequencing data. *Nat Meth* 10:563–69.

Claesson MJ, van Sinderen D, O'Toole PW. 2008. Lactobacillus phylogenomics – towards a reclassification of the genus. *Int J Syst Evol Microbiol* 58:2945–54.

Cousin FJ, Mater DDG, Foligné B, Jan G. 2010. Dairy propionibacteria as human probiotics: a review of recent evidence. *Dairy Sc Technol* 91:1–26.

Cremonesi P, Chessa S, Castiglioni B. 2013. Genome sequence and analysis of *Lactobacillus helveticus*. *Front Microbiol* 3:435.

Cretenet M, Le Gall G, Wegmann U, Even S, Shearman C, Stentz R, Jeanson S. 2014. Early adaptation to oxygen is key to the industrially important traits of *Lactococcus lactis* ssp. *cremoris* during milk fermentation. *BMC Genomics* 15:1054.

De Filippis F, Genovese A, Ferranti P, Gilbert JA, Ercolini D. 2016. Metatranscriptomics reveals temperature-driven functional changes in microbiome impacting cheese maturation rate. *Sci Rep* 6:21871.

De Vuyst L, Tsakalidou E. 2008. *Streptococcus macedonicus*, a multi-functional and promising species for dairy fermentations. *Int Dairy J* 18(5):476–85.

Dellaglio F, Felis GE. 2005. Taxonomy of lactobacilli and bifidobacteria. In: Tannock GW. editor. *Probiotics and prebiotics: scientific aspects*. Norfolk: Caister Academic Press. pp 25–50.

Desiere F, Lucchini S, Canchaya C, Ventura M, Brüssow H. 2002. Comparative genomics of phages and prophages in lactic acid bacteria. *Antonie Van Leeuwenhoek* 82(1–4):73–91.

Douillard FP, Ribbera A, Järvinen HM, Kant R, Pietilä TE, Randazzo C, Paulin L, Laine PK, Caggia C, von Ossowski I, Reunanen J, Satokari R, Salminen S, Palva A, de Vos WM. 2013b. Comparative genomic and functional analysis of *Lactobacillus casei* and *Lactobacillus rhamnosus* strains marketed as probiotics. *Appl Environ Microbiol* 79(6):1923–33.

Douillard FP, Ribbera A, Kant R, Pietilä TE, Järvinen HM, Messing M, Randazzo C, Paulin L, Laine P, Ritari, J, Caggia C, Lähteinen T, Brouns SJ, Satokari R, von Ossowski I, Reunanen J, Palva A, de Vos WM. 2013a. Comparative genomic and functional analysis of 100 *Lactobacillus rhamnosus* strains and their comparison with strain GG. *PLoS Genet* 9(8):e1003683.

Dugat-Bony E, Straub C, Teissandier A, Onésime D, Loux V, Monnet C, Irlinger F, Landaud S, Leclercq-Perlat MN, Bento P, Fraud S, Gibrat JF, Aubert J, Fer F, Guédon E, Pons N, Kennedy S, Beckerich JM, Swennen D, Bonnarme P. 2015. Overview of a surface-ripened cheese community functioning by meta-omics analyses. *PLoS One* 10(4):e0124360.

Duranti S, Milani C, Lugli GA, Mancabelli L, Turroni F, Ferrario C, Mangifesta M, Viappiani A, Sánchez B, Margolles A, van Sinderen D, Ventura M. 2016. Evaluation of genetic diversity among strains of the human gut commensal *Bifidobacterium adolescentis*. *Sci Rep* 6:23971.

Fang G, Munera D, Friedman DI, Mandlik A, Chao MC, Banerjee O, Feng Z, Losic B, Mahajan MC, Jabado OJ, Deikus G, Clark TA, Luong K, Murray IA, Davis BM, Keren-Paz A, Chess A, Roberts RJ, Korlach J, Turner SW, Kumar V, Waldor MK, Schadt EE. 2012. Genome-wide mapping of methylated adenine residues in pathogenic *Escherichia coli* using single-molecule real-time sequencing. *Nature Biotechnol* 30:1232–1239.

Ferrario C, Duranti S, Milani C, Mancabelli L, Lugli GA, Turroni F, Mangifesta M, Viappiani A, Ossiprandi MC, van Sinderen D, Ventura M. 2015. Exploring amino acid auxotrophy in *Bifidobacterium bifidum* PRL2010. *Front Microbiol* 6:1331.

Ferrario C, Milani C, Mancabelli L, Lugli GA, Duranti S, Mangifesta M, Viappiani A, Turroni F, Margolles A, Ruas-Madiedo P, van Sinderen D, Ventura M. 2016. Modulation of the eps-ome transcription of bifidobacteria through simulation of human intestinal environment. *FEMS Microbiol Ecol* 92(4): pii:fiw056.

Ferrario C, Ricci G, Milani C, Lugli GA, Ventura M, Eraclio G, Borgo F, Fortina MG. 2013. *Lactococcus garvieae*: where is it from? A first approach to explore the evolutionary history of this emerging pathogen. *PLoS One* 8(12):e84796.

Flórez AB, Mayo B. 2015. The plasmid complement of the cheese isolate *Lactococcus garvieae* IPLA 31405 revealed adaptation to the dairy environment. *PLoS One* 10:e0126101.

Fouquier d'Hérouel A, Wessner F, Halpern D, Ly-Vu J, Kennedy, SP, Serror P, Aurell E, Repoila F. 2011. A simple and efficient method to search for selected primary transcripts: non-coding and antisense RNAs in the human pathogen *Enterococcus faecalis*. *Nucleic Acids Res* 39(7):e46.

Franz CM, Stiles ME, Schleifer KH, Holzapfel WH. 2003. Enterococci in foods–a conundrum for food safety. *Int J Food Microbiol* 88(2-3):105–22.

Fukuda S, Toh H, Taylor TD, Ohno H, Hattori M. 2012. Acetate-producing bifidobacteria protect the host from enteropathogenic infection via carbohydrate transporters. *Gut Microbes* 3(5):449–54.

Gao S, Sung W-K, Nagarajan N. 2011. Opera: reconstructing optimal genomic scaffolds with high-throughput paired-end sequences. *J Comput Biol* 18:1681–91.

Garrigues C, Johansen E, Crittenden R. 2013. Pangenomics–an avenue to improved industrial starter cultures and probiotics. *Curr Opin Biotechnol* 24(2):187–91.

Garrigues C, Johansen E, Pedersen MB. 2010. Complete genome sequence of *Bifidobacterium animalis* subsp. *lactis* BB-12, a widely consumed probiotic strain. *J Bacteriol* 192:2467–68.

Gaudu P, Lamberet G, Poncet S, Gruss A. 2003. CcpA regulation of aerobic and respiration growth in *Lactococcus lactis*. *Mol Microbiol* 50:183–92.

Gaudu P, Vido K, Cesselin B, Kulakauskas S, Tremblay J, Rezaïki L, Lamberret G, Sourice S, Duwat P, Gruss A. 2002. Respiration capacity and consequences in *Lactococcus lactis*. *Antonie Van Leeuwenhoek* 82(1-4):263–9.

Georgalaki M, Papadelli M, Chassioti E, Anastasiou R, Aktypis A, De Vuyst L, Van Driessche, G, Devreese B, Tsakalidou E. 2010. Milk protein fragments induce the biosynthesis of macedocin, the lantibiotic produced by *Streptococcus macedonicus* ACA-DC 198. *Appl Environ Microbiol* 76(4):1143–51.

Grissa I, Vergnaud G, Pourcel C. 2007. CRISPRFinder: a web tool to identify clustered regularly interspaced short palindromic repeats. *Nucleic Acids Res* 35:W52–W57.

Gross G, Snel J, Boekhorst J, Smits MA, Kleerebezem M. 2010. Biodiversity of mannose-specific adhesion in *Lactobacillus plantarum* revisited: strain-specific domain composition of the mannose-adhesin. *Beneficial Microbes* 1:61–66.

Haghshenas B, Haghshenas M, Nami Y, Yari Khosroushahi A, Abdullah N, Barzegari A, Rosli R, Hejazi MS. 2016. Probiotic assessment of *Lactobacillus plantarum* 15HN and *Enterococcus mundtii* 50H isolated from traditional dairies microbiota. *Adv Pharm Bull* 6(1):37–47.

Henning C, Gautam D, Muriana P. 2015. Identification of multiple bacteriocins in *Enterococcus* spp. using an *Enterococcus*-specific bacteriocin PCR array. *Microorganisms* 3(1):1–16.

Holzapfel W, Wood B. 2014. *Lactic acid bacteria. Biodiversity and taxonomy*. Weinheim, Germany: Wiley Blackwell.

Horvath P, Couté-Monvoisin A-C, Romero DA, Boyaval P, Fremaux C, Barrangou R. 2009. Comparative analysis of CRISPR loci in lactic acid bacteria genomes. *Int J Food Microbiol* 131:62–70.

anniello RG, Ricciardi A, Parente E, Tramutola A, Reale A, Zotta T. 2015. Aeration and supplementation with heme and menaquinone affect survival to stresses and antioxidant capability of *Lactobacillus casei* strains. *LWT Food Sci Technol* 60:817–24.

Innocenti N, Golumbeanu M, Fouquier d'Hérouël A, Lacoux C, Bonnin RA, Kennedy SP, Wessner F, Serror P, Bouloc P, Repoila F, Aurell E. 2015. Whole-genome mapping of 5' RNA ends in bacteria by tagged sequencing: a comprehensive view in *Enterococcus faecalis*. *RNA* 21(5):1018–30.

Johnson BR, Hymes J, Sanozky-Dawes R, DeCrescenzo Henriksen E, Barrangou R, Klaenhammer TR. 2016. Conserved S-layer-associated proteins revealed by exoproteomic survey of S-layer-forming lactobacilli. *Appl Environ Microbiol* 82:134–45.

Jungersen M, Wind A, Johansen E, Christensen JE, Stuer-Lauridsen B, Eskesen D. 2014. The science behind the probiotic strain *Bifidobacterium animalis* subsp. *lactis* BB-12®. *Microorganisms* 2:92–110.

Kajitani R, Toshimoto K, Noguchi H, Toyoda A, Ogura Y, Okuno M, Yabana M, Harada M, Nagayasu E, Maruyama H, Kohara Y, Fujiyama A, Hayashi T, Itoh T. 2014. Efficient de novo assembly of highly heterozygous genomes from whole-genome shotgun short reads. *Genome Res* 24:1384–95.

Kanda T, Nishida A, Ohno M, Imaeda H, Shimada T, Inatomi O, Bamba S, Sugimoto M, Andoh A. 2016. *Enterococcus durans* TN–3 induces regulatory T cells and suppresses the development of dextran sulfate sodium (DSS)-induced experimental colitis. *PLoS One* 11(7):e0159705.

Kant R, Blom J, Palva A, Siezen RJ, de Vos WM. 2011. Comparative genomics of Lactobacillus. *Microb Biotechnol* 4(3):323–32.

Kant R, Rintahaka J, Yu X, Sigvart-Mattila P, Paulin L, Mecklin JP, Saarela M, Palva A, von Ossowski I. 2014. A comparative pan-genome perspective of niche-adaptable cell surface protein phenotypes in *Lactobacillus rhamnosus*. *PLoS One* 9(7):e102762.

Kelleher P, Murphy J, Mahony J, van Sinderen D. 2015. Next-generation sequencing as an approach to dairy starter selection. *Dairy Sci Technol* 95:545–68.

Kelly WJ, Ward LJ, Leahy SC. 2010. Chromosomal diversity in *Lactococcus lactis* and the origin of dairy starter cultures. *Genome Biol Evol* 2:729–44.

Klaenhammer TR, Azcarate-Peril MA, Altermann E, Barrangou R. 2007. Influence of the dairy environment on gene expression and substrate utilization in lactic acid bacteria. *J Nutr* 137(3 Suppl 2):748S–50S.

Kleerebezem M, Hols P, Bernard E, Rolain T, Zhou M, Siezen RJ, Bron PA. 2010. The extracellular biology of the lactobacilli. *FEMS Microbiol Rev* 34:199–230.

Klein G. 2003. Taxonomy, ecology and antibiotic resistance of enterococci from food and the gastro-intestinal tract. *Int J Food Microbiol* 88(2-3):123–31.

Kort R, Westerik N, Serrano LM, Douillard FP, Gottstein W, Mukisa IM, Tuijn CJ, Basten L, Hafkamp B, Meijer WC, Teusink B, de Vos WM, Reid G, Sybesma W. 2015. A novel consortium of *Lactobacillus rhamnosus* and *Streptococcus thermophilus* for increased access to functional fermented foods. *Microbial Cell Factories* 14:195.

Koryszewska-Baginska A, Aleksandrzak-Piekarczyk T, Bardowski J. 2013. Complete genome sequence of the probiotic strain *Lactobacillus casei* (formerly *Lactobacillus paracasei*) LOCK919. *Genome Announc* 1(5):pii:e00758-13.

Koskinen P, Deptula P, Smolander OP, Tamene F, Kammonen J, Savijoki K, Paulin L., Piironen V., Auvinen P., Varmanen P. 2015. Complete genome sequence of *Propionibacterium freudenreichii* DSM 20271(T). *Stand Genomic Sci* 10:83.

Kreuzer-Redmer S, Bekurtz JC, Arends D, Bortfeldt R, Kutz-Lohroff B, Sharbati S, Einspanier R, Brockmann GA. 2016. Feeding of *Enterococcus faecium* NCIMB 10415 leads to intestinal miRNA-423-5p-induced regulation of immune-relevant genes. *Appl Environ Microbiol* 82(8):2263–69.

Kwon G, Lee J, Lim YH. 2016. Dairy Propionibacterium extends the mean lifespan of *Caenorhabditis elegans* via activation of the innate immune system. *Sci Rep* 6:31713.

Kyrpides NC, Woyke T, Eisen JA, Garrity G, Lilburn TG, Beck BJ, Whitman WB, Hugenholtz P, Klenk HP. 2014. Genomic encyclopedia of type strains, phase I: The one thousand microbial genomes (KMG-I) project. *Stand Genomic Sci* 9:1278–84.

Lagesen K, Hallin P, Rødland EA, Staerfeldt H-H, Rognes T, Ussery DW. 2007. RNAmmer: consistent and rapid annotation of ribosomal RNA genes. *Nucleic Acids Res* 35:3100–108.

Lambie SC, Altermann E, Leahy SC, Kelly WJ. 2014. Draft genome sequence of *Lactococcus lactis* subsp. *cremoris* HPT, the first defined-strain dairy starter culture bacterium. *Genome Announc* 2(2):pii:e00107–14.

Lasa I, Toledo-Arana A, Dobin A, Villanueva M, Ruiz de losMozos I, Vergara-Irigaray M, Segura V, Fagegaltier D, Penades JR, Valle J, Solano C, Gingeras TR. 2011. Genome-wide antisense transcription drives mRNA processing in bacteria. *Proc Natl Acad Sci USA* 108:20172–77.

Lathinen S, Ouwehand AC, Salminen S, von Wright A. 2012. *Lactic acid bacteria: microbiological and functional aspects*. Boca Raton, FL: CRC Press.

Le Maréchal C, Peton V, Plé C, Vroland C, Jardin J, Briard-Bion V, Durant G, Chuat V, Loux V, Foligné B, Deutsch SM, Falentin H, Jan G. 2015. Surface proteins of *Propionibacterium freudenreichii* are involved in its anti-inflammatory properties. *J Proteomics* 113:447–61.

Leroy F, Foulquié Moreno MR, De Vuyst L. 2003. *Enterococcus faecium* RZS C5, an interesting bacteriocin producer to be used as a co-culture in food fermentation. *Int J Food Microbiol* 88(2-3):235–40.

Li P, Li X, Gu Q, Lou XY, Zhang XM, Song DF, Zhang C. 2016. Comparative genomic analysis of *Lactobacillus plantarum* ZJ316 reveals its genetic adaptation and potential probiotic profiles. *J Zhejiang Univ Sci B* 17(8):569–79.

Li R, Li Y, Kristiansen K, Wang J. 2008. SOAP: short oligonucleotide alignment program. *Bioinformatics* 24(5):713–14.

Licandro-Seraut H, Scornec H, Pédron T, Cavin JF, Sansonetti PJ. 2014. Functional genomics of *Lactobacillus casei* establishment in the gut. *Proc Natl Acad Sci USA* 111(30):E3101–9.

Ljungh Å, Wadström T. 2009. editors. *Lactobacillus molecular biology: from genomics to probiotics*. Norfolk: Caister Academic Press.

Lorca G, Reddy L, Nguyen A, Sun EI, Tseng J, Yen M-R, Saier MH Jr. 2010. Lactic acid bacteria: comparative genomic analyses of transport systems. In: Mozzi F, Raya RR, Vignolo GM. editors. *Biotechnology of lactic acid bacteria. Novel applications*. Ames, IA: WileyBlackwell. pp 73–87.

Luesink E, Van Herpen R, Grossiord B, Kuipers O, De Vos W. 1998. Transcriptional activation of the glycolytic las operon and catabolite repression of the gal operon in *Lactococcus lactis* are mediated by the catabolite control protein *CcpA Mol Microbiol* 30:789–98.

MacKenzie DA, Jeffers F, Parker ML, Vibert-Vallet A, Bongaerts RJ, Roos S, Walter J, Juge N. 2010. Strain-specific diversity of mucus-binding proteins in the adhesion and aggregation properties of *Lactobacillus reuteri*. *Microbiology* 156:3368–78.

Makarova KS, Wolf YI, Koonin EV. 2013. The basic building blocks and evolution of CRISPR-Cas systems. *Biochem Soc Trans* 41:1392–1400.

Maky MA, Ishibashi N, Zendo T, Perez RH, Doud JR, Karmi M, Sonomoto K. 2015. Enterocin F4-9, a novel O-linked glycosylated bacteriocin. *Appl Environ Microbiol* 81(14):4819–26.

Mattarelli P, Holzapfel W, Franz CM, Endo A, Felis GE, Hammes W, Pot B, Dicks L, Dellaglio F. 2014. Recommended minimal standards for description of new taxa of the genera Bifidobacterium, Lactobacillus and related genera. *Int J Syst Evol Microbiol* 64(Pt 4):1434–51.

Mayo B, van Sinderen D, Ventura M. 2008. Genome analysis of food grade lactic acid-producing bacteria: from basics to applications. *Curr Genomics* 9(3):169–83.

Maze A, Boel G, Zuniga M, Bourand A, Loux V, Yebra MJ, Monedero V, Correia K, Jacques N, Beaufils S, Poncet S, Joyet P, Milohanic E, Casaregola S, Auffray Y, Perez-Martinez G, Gibrat JF, Zagorec M, Francke C, Hartke A, Deutscher J. 2010. Complete genome sequence of the probiotic *Lactobacillus casei* strain BL23. *J Bacteriol* 192:2647–48.

Meier-Kolthoff JP, Auch AF, Klenk HP, Göker M. 2013 Genome sequence-based species delimitation with confidence intervals and improved distance functions *BMC Bioinformatics* 14:60

Milani C, Turroni F, Duranti S, Lugli GA, Mancabelli L, Ferrario C, van Sinderen D, Ventura M. 2015. Genomics of the genus Bifidobacterium reveals species-specific adaptation to the glycan-rich gut environment. *Appl Environ Microbiol* 82(4):980–91.

Millen AM, Horvath P, Boyaval P, Romero DA. 2012. Mobile CRISPR/Cas-mediated bacteriophage resistance in *Lactococcus lactis*. *PLoS One* 7(12):e51663.

Molenaar D, Bringel F, Schuren FH, de Vos WM, Siezen RJ, Kleerebezem M. 2005. Exploring *Lactobacillus plantarum* genome diversity by using microarrays. *J Bacteriol* 187:6119–127.

Monedero V, Mazé A, Boël G, Zúñiga M, Beaufils S, Hartke A, Deutscher J. 2007. The phosphotransferase system of *Lactobacillus casei*: regulation of carbon metabolism and connection to cold shock response. *J Mol Microbiol Biotechnol* 12(1-2):20–32.

Monnet C, Dugat-Bony E, Swennen D, Beckerich JM, Irlinger F, Fraud S, Bonnarme P. 2016. Investigation of the activity of the microorganisms in a reblochon-style cheese by metatranscriptomic analysis. *Front Microbiol* 7:536.

Morales P, Fernández-García E, Nuñez M. 2003. Caseinolysis in cheese by Enterobacteriaceae strains of dairy origin. *Lett Appl Microbiol* 37(5):410–14.

Murray IA, Clark TA, Morgan RD, Boitano M, Anton BP, Luong K, Fomenkov A, Turner SW, Korlach J, Roberts RJ. 2012. The methylomes of six bacteria. *Nucleic Acids Research* 40(22):11450–62.

Nami Y, Haghshenas B, Haghshenas M, Abdullah N, Yari Khosroushahi A. 2015. The prophylactic effect of probiotic *Enterococcus lactis* IW5 against different human cancer cells. *Front Microbiol* 6:1317.

O'Callaghan A, Bottacini F, O'Connell Motherway M, van Sinderen D. 2015. Pangenome analysis of *Bifidobacterium longum* and site-directed mutagenesis through by-pass of restriction–modification systems. *BMC Genomics* 16:832.

'Connell Motherway M, Zomer A, Leahy SC, Reunanen J, Bottacini F, Claesson MJ, O'Brien F, Flynn K, Casey PG, Moreno Munoz JA, Kearney B, Houston AM, O'Mahony C, Higgins DG, Salvetti E, Fondi M, Fani R, Torriani S, Felis GE 2013. Evolution of lactic acid bacteria in the order *Lactobacillales* as depicted by analysis of glycolysis and pentose phosphate pathways. *Syst Appl Microbiol* 36:291–305.

O'Connell Motherway M, Zomer A, Leahy SC, Reunanen J, Bottacini F, Claesson MJ, O'Brien F, Flynn K, Casey PG, Moreno Munoz JA, Kearney B, Houston AM, O'Mahony C, Higgins DG, Shanahan F, Palva A, de Vos WM, Fitzgerald GF, Ventura M, O'Toole PW, van Sinderen D. 2011. Functional genome analysis of *Bifidobacterium breve* UCC2003 reveals type IVb tight adherence (Tad) pili as an essential and conserved host-colonization factor. *Proc Natl Acad Sci USA* 108:11217–22.

Ongol MP, Sawatari Y, Ebina Y, Sone T, Tanaka M, Tomita F, Yokota A, Asano K. 2007. Yoghurt fermented by *Lactobacillus delbrueckii* subsp. *bulgaricus* H+-ATPase-defective mutants exhibits enhanced viability of *Bifidobacterium breve* during storage. *Int J Food Microbiol* 116(3):358–66.

Overbeek R, Olson R, Pusch GD, Olsen GJ, Davis JJ, Disz T, Edwards RA, Gerdes S, Parrello B, Shukla M, Vonstein V, Wattam AR, Xia F, Stevens R. 2014. The SEED and the Rapid annotation of microbial genomes using subsystems technology (RAST). *Nucleic Acids Res* 42(Database issue):D206–14.

Papadimitriou K, Mavrogonatou E, Bolotin A, Tsakalidou E, Renault P. 2016. Whole-genome sequence of the cheese isolate *Streptococcus macedonicus* 679. *Genome Announc* 4(5):pii:e01025–16.

Papadimitriou K, Anastasiou R, Mavrogonatou E, Blom J, Papandreou NC, Hamodrakas SJ, Ferreira S, Renault P, Supply P, Pot B, Tsakalidou E. 2014. Comparative genomics of the dairy isolate *Streptococcus macedonicus* ACA-DC 198 against related members of the *Streptococcus bovis/Streptococcus equinus* complex. *BMC Genomics* 15:272.

Passerini D, Beltramo C, Coddeville M, Quentin Y, Ritzenthaler P, Daveran-Mingot ML, Le Bourgeois P. 2010. Genes but not genomes reveal bacterial domestication of *Lactococcus lactis*. *PLoS One* 5(12):e15306.

Reale A, Di Renzo T, Rossi F, Zotta T, Iacumin L, Preziuso M, Parente E, Sorrentino E, Coppola R. 2015. Tolerance of *Lactobacillus casei, Lactobacillus paracasei* and *Lactobacillus rhamnosus* strains to stress factors encountered in food processing and in the gastro-intestinal tract. *LWT - Food Sci Technol* 60:721–28.

othberg JM, Hinz W, Rearick TM, Schultz J, Mileski W, Davey M, Leamon JH, Johnson K, Milgrew MJ, Edwards M, Hoon J, Simons JF, Marran D, Myers JW, Davidson JF, Branting A, Nobile JR, Puc BP, Light D. Clark TA, Huber M, Branciforte JT, Stoner IB, Cawley SE. Lyons M, Fu Y, Homer N, Sedova M, Miao X, Reed B, Sabina J, Feierstein E, Schorn M, Alanjary M, Dimalanta E, Dressman D. Kasinskas R, Sokolsky T, Fidanza JA, Namsaraev E, McKernan KJ, Williams A, Roth GT, Bustillo J. 2011. An integrated semiconductor device enabling non-optical genome sequencing. *Nature* 475(7356):348–52.

Salvetti E, Fondi M, Fani R, Torriani S, Felis GE. 2013. Evolution of lactic acid bacteria in the order *Lactobacillales* as depicted by analysis of glycolysis and pentose phosphate pathways. *Syst Appl Microbiol* 36:291–305.

Settachaimongkon S, Nout MJ, Antunes Fernandes EC, Hettinga KA, Vervoort JM, van Hooijdonk TC, Zwietering MH, Smid EJ, van Valenberg HJ. 2014. Influence of different proteolytic strains of *Streptococcus thermophilus* in co-culture with *Lactobacillus delbrueckii* subsp. *bulgaricus* on the metabolite profile of set-yoghurt. *Int J Food Microbiol* 177:29–36.

Settachaimongkon S, van Valenberg HJ, Gazi I, Nout MJ, van Hooijdonk TC, Zwietering MH, Smid EJ. 2016. Influence of *Lactobacillus plantarum* WCFS1 on post-acidification, metabolite formation and survival of starter bacteria in set-yoghurt. *Food Microbiol* 59:14–22.

Settachaimongkon S, van Valenberg HJ, Winata V, Wang X, Nout MJ, van Hooijdonk TC, Zwietering MH, Smid EJ. 2015. Effect of sublethal preculturing on the survival of probiotics and metabolite formation in set-yoghurt. *Food Microbiol* 49:104–15.

Shanahan F, Palva A, de Vos WM, Fitzgerald GF, Ventura M, O'Toole PW, van Sinderen D. 2011. Functional genome analysis of *Bifidobacterium breve* UCC2003 reveals type IVb tight adherence (Tad) pili as an essential and conserved host-colonization factor. *Proc Natl Acad Sci USA* 108:11217–22.

Shioya K, Michaux C, Kuenne C, Hain T, Verneuil N, Budin-Verneuil A, Hartsch T, Hartke A, Giard JC. 2011. Genome-wide identification of small RNAs in the opportunistic pathogen *E. faecalis* V583. *PLoS One* 6(9):e23948.

Sieuwerts S, Molenaar D, Van Hijum SAFT, Beerthuyzen M, Stevens MJA, Janssen PWM, Ingham CJ, De Bok FAM, De Vos WM, Van Hylckama Vlieg JET. 2010. Mixed-culture transcriptome analysis reveals the molecular basis of mixed-culture growth in *Streptococcus thermophilus* and *Lactobacillus bulgaricus*. *Appl Environ Microbiol* 76:7775–84.

Siezen RJ, Van Hylckama Vlieg, JET. 2011. Genomic diversity and versatility of *Lactobacillus plantarum*, a natural metabolic engineer. *Microbial Cell Factories* 10(Suppl.1):S3.

Smokvina T, Wels M, Polka J, Chervaux C, Brisse S, Boekhorst J, van Hylckama Vlieg JE, Siezen RJ. 2013. *Lactobacillus paracasei* comparative genomics: towards species pan-genome definition and exploitation of diversity. *PLoS One* 8(7):e68731.

Sola C, Abadia E, Le Hello S, Weill FX. 2015. High-throughput CRISPR typing of *Mycobacterium tuberculosis* complex and *Salmonella enterica* serotype Typhimurium. *Methods Mol Biol* 1311:91–109.

Stevens M. 2008. *Transcriptome response of* Lactobacillus plantarum *to global regulator deficiency, stress and other environmental conditions*. PhD thesis, Wageningen University, Wageningen, The Netherlands.

Stevens MJ, Molenaar D, de Jong A, De Vos WM, Kleerebezem M. 2010. Sigma54-mediated control of the mannose phosphotransferase system in *Lactobacillus plantarum* impacts on carbohydrate metabolism. *Microbiology* 156:695–707.

Stevens MJ, Wiersma A, de Vos WM, Kuipers OP, Smid EJ, Molenaar D, Kleerebezem M. 2008. Improvement of *Lactobacillus plantarum* aerobic growth as directed by comprehensive transcriptome analysis. *Appl Environ Microbiol* 74:4776–78.

Sturme MH, Francke C, Siezen RJ, de Vos WM, Kleerebezem M. 2007. Making sense of quorum sensing in lactobacilli: a special focus on *Lactobacillus plantarum* WCFS1. *Microbiology* 153(Pt 12):3939–47.

Sun Z, Harris HM, McCann A, Guo C, Argimón S, Zhang W, Yang X, Jeffery IB, Cooney JC, Kagawa TF, Liu W, Song Y, Salvetti E, Wrobel A, Rasinkangas P, Parkhill J, Rea MC, O'Sullivan O, Ritari J, Douillard FP, Paul Ross R, Yang R, Briner AE, Felis GE, de Vos WM, Barrangou R, Klaenhammer TR, Caufield PW, Cui Y, Zhang H, O'Toole PW. 2015. Expanding the biotechnology potential of lactobacilli through comparative genomics of 213 strains and associated genera. *Nat Commun* 6:8322.

Sundekilde UK, Larsen LB, Bertram HC. 2013. NMR-based milk metabolomics. *Metabolites* 3(2):204–222.

Tanizawa Y, Tohno M, Kaminuma E, Nakamura Y, Arita M. 2015. Complete genome sequence and analysis of *Lactobacillus hokkaidonensis* LOOC260T, a psychrotrophic lactic acid bacterium isolated from silage. *BMC Genomics* 16:240.

Thierry A, Deutsch SM, Falentin H, Dalmasso M, Cousin FJ, Jan G. 2011. New insights into physiology and metabolism of *Propionibacterium freudenreichii*. *Int J Food Microbiol* 149(1):19–27.

Titgemeyer F, Hillen W. 2002. Global control of sugar metabolism: a Gram-positive solution. *Antonie van Leeuwenhoek* 82:59–71.

Tobes R, Pareja-Tobes P, Manrique M, Pareja-Tobes E, Kovach E, Alekhin A, Pareja E. 2015. Gene calling and bacterial genome annotation with BG7. *Methods Mol Biol* 1231:177–89.

Turroni F, Foroni E, Serafini F, Viappiani A, Montanini B, Bottacini F, Ferrarini A, Bacchini PL, Rota C, Delledonne M, Ottonello S, van Sinderen D, Ventura M. 2011. Ability of *Bifidobacterium breve* to grow on different types of milk: exploring the metabolism of milk through genome analysis. *Appl Environ Microbiol* 77(20):7408–7417.

Twardziok SO, Pieper R, Aschenbach JR, Bednorz C, Brockmann GA, Fromm M, Klingspor S, Kreuzer S, Lodemann U, Martens H, Martin L, Richter JF, Scharek-Tedin L, Siepert BF, Starke IC, Tedin K, Vahjen W, Wieler LH, Zakrzewski SS, Zentek J, Wrede P. 2014. Cross-talk between host, microbiome and probiotics: a systems biology approach for analyzing the effects of probiotic *Enterococcus faecium* NCIMB 10415 in piglets. *Mol Inform* 33(3):171–82.

van den Nieuwboer M, van Hemert S, Claassen E, de Vos WM., 2016. *Lactobacillus plantarum* WCFS1 and its host interaction: a dozen years after the genome. *Microb Biotechnol* 9(4):452–465.

van der Meulen SB, de Jong A, Kok J. 2016. Transcriptome landscape of Lactococcus lactis reveals many novel RNAs including a small regulatory RNA involved in carbon uptake and metabolism. *RNA Biol* 13(3):353–66.

Van Pijkeren JP, Canchaya C, Ryan KA, Li Y, Claesson MJ, Sheil B, Steidler L, O'Mahony L, Fitzgerald GF, Van Sinderen D, O'Toole PW. 2006. Comparative and functional analysis of sortase-dependent proteins in the predicted secretome *of Lactobacillus salivarius* UCC118. *Appl Environ Microbiol* 72:4143–53.

Varani AM, Siguier P, Gourbeyre E, Charneau V, Chandler M. 2011. ISsaga is an ensemble of web-based methods for high throughput identification and semi-automatic annotation of insertion sequences in prokaryotic genomes. *Genome Biol* 12:R30.

Wei Y-X, Zhang Z-Y, Liu C, Zhu Y-Z, Zhu Y-Q, Zheng H, Zhao G-P, Wang S, Guo X-K. 2010. Complete genome sequence of *Bifidobacterium longum* JDM301. *J Bacteriol* 192:4076–4077.

Westra ER, Swarts DC, Staals RHJ, Jore MM, Brouns SJJ, van der Oost J. 2012. The CRISPRs, they are a-changin?: how prokaryotes generate adaptive immunity. *Annu Rev Genet* 46:311–339.

Wozniak RAF, Waldor MK. 2010. Integrative and conjugative elements: mosaic mobile genetic elements enabling dynamic lateral gene flow. *Nat Rev Microbiol* 8:552–63.

Yebra MJ, Viana R, Monedero V, Deutscher J, Perez-Martinez G. 2004. An esterase gene from *Lactobacillus casei* cotranscribed with genes encoding a phosphoenolpyruvate:sugar phosphotransferase system and regulated by a LevR-like activator and σ54 factor. *J Mol Microbiol Biotechnol* 8:117–128.

Zakharevich NV, Averina OV, Klimina KM, Kudryavtseva AV, Kasianov AS, Makeev VJ, Danilenko VN. 2015. Complete genome sequence of *Bifidobacterium longum* GT15: identification and characterization of unique and global regulatory genes. *Microb Ecol* 70(3):819–34.

Zheng J, Ruan L, Sun M, Gänzle M. 2015. A genomic view of Lactobacilli and Pediococci demonstrates that phylogeny matches ecology and physiology. *Appl Environ Microbiol* 81(20):7233–43.

Zhou M, Theunissen D, Wels M, Siezen RJ. 2010. LAB-Secretome: a genome-scale comparative analysis of the predicted extracellular and surface-associated proteins of lactic acid bacteria. *BMC Genomics* 11:651.

Zhou Y, Liang Y, Lynch KH, Dennis JJ, Wishart DS. 2011. PHAST: a fast phage search tool. *Nucleic Acids Res* 39:W347–52.

Zotta T, Ricciardi A, Ianniello RG, Parente E, Reale A, Rossi F, Iacumin L, Comi G, Coppola R. 2014. Assessment of aerobic and respiratory growth in the *Lactobacillus casei* group. *PLoS One* 9(6):e99189.

Zuljan F, Espariz M, Blancato VS, Esteban L, Alarcón S, Magni C. 2016. Draft genome sequence of *Lactococcus lactis* subsp. *lactis* bv. *diacetylactis* CRL264, a citrate-fermenting strain. *Genome Announc* 4(1):pii:e01575–15.

6 Metabolism and biochemistry of LAB and dairy-associated species

Palmiro Poltronieri[1], Giovanna Battelli[2] and Nicoletta Pasqualina Mangia[3]

[1]Institute of Sciences of food Productions (CNR-ISPA), CNR, National Research Council of Italy, Lecce, Italy
[2]ISPA-CNR, Institute of Sciences of Food Productions, Milano, Italy
[3]Department of Agriculture, University of Sassari, Italy

6.1 INTRODUCTION

Lactic acid bacteria (LAB) are widely used for the industrial production of fermented dairy products and form a group of related low-GC-content Gram-positive bacteria. The LAB have the ability to produce lactic acid during homo- or heterofermentative metabolism. The acidification and enzymatic processes accompanying the growth of LAB impact the flavour, texture and preservative qualities. Industrial dairy applications of LAB species rely on *Lactococcus lactis* (*Lc. lactis*), *Lactobacillus* (*Lb.*) spp., *Leuconostoc* (*Ln.*) spp., *Pediococcus* (*Pd.*) spp., and *Streptococcus* (*S.) thermophilus*. Technologically relevant LAB strains are applied as starter cultures or found as main nonstarter species, or they are added as probiotics delivered through dairy products. The benefits include a positive influence on the normal microbiota, competitive exclusion of pathogens, and modulation of mucosal immunity.

Cell wall composition differs slightly in various LAB species. Peptidoglycan differences are found in the amino acids used in the peptide sequences linked through inter-chain peptidic bonds and in teicoic acids, giving strength and rigidity through the binding of magnesium ions. The presence of *meso*-diaminopimelic acid in the cell wall of *Lactobacillus plantarum* is one of the key differences from the other species. The Lys-D-Asp type is predominant in the genus *Lactobacillus* and is useful to differentiate, at species level, *Weissella* from other LAB and *Lactobacillus reuteri* from *Lactobacillus fermentum* (Orn-D-Asp) (Dellaglio and Felis, 2005).

This chapter reviews the biochemistry of LAB species, the carbohydrate use for energy production, and the proteolysis of LAB. Flavour production and origin of specific compounds are presented.

Microbiology in Dairy Processing: Challenges and Opportunities, First Edition.
Edited by Palmiro Poltronieri.

Published 2018 by John Wiley & Sons Ltd and the Institute of Food Technologists.

6.2 CARBOHYDRATE SUBSTRATES, GLYCOLYSIS AND ENERGY PRODUCTION

LAB species can utilize an extensive set of carbon sources, an ability that is likely to contribute to their adaptive ability. Adaptation to environmental niches has influenced the genome of LAB species, with loss of ability to use sugars not available in dairy environment and exploit either the Embden-Meyerhof-Parnas (EMP) pathway as well as the pentose phosphate pathway (PPP).

LAB species metabolize glucose to phosphoenolpyruvate (PEP) via glycolysis. PEP may serve as the branch point in the formate-, acetate-, ethanol-producing pathway (C3 pathway).

In *Lactobacillus brevis*, *Lb. reuteri* and other species relying on facultative heterofermentation (Table 6.1), the phosphotransferase systems (PTS) and glycolytic enzymes allow sugars to be metabolised via the EMP pathway (Saier et al., 1996). Fructose is used as external electron acceptor for reoxidation of NAD(P)H produced in pentose fermentation (Zaunmüller et al., 2006; Kang et al., 2013a, 2013b): presence of fructose-1,6-bisphosphate aldolase, phosphofructokinase and phosphoglucoisomerase ensure the exploitation of the glycolytic pathway (Arsköld et al., 2008).

Carbohydrate uptake can occur in LAB through PTS, ATP-binding cassette (ABC) pumps or glycoside transporters (Lorca et al., 2010). In Bifidobacteria, it was shown that probiotic strains adapted to the GI-tract possess specific ABC-type carbohydrate transporters for oligosaccharides, which are metabolized to acetate (Fukuda et al., 2012).

Glycolysis is characterised by the formation of fructose-1,6-diphosphate (FDP), which is split by the FDP aldolase into dihydroxyacetone-phosphate (DHAP) and

Table 6.1 LAB metabolism and type of fermentation.

A	Homofermentative	Heterofermentative
Glucose fermented	≥85%	50%
Produced: CO_2, Acetate, ethanol	–	+
Glucose-CO^2	–	+
Fructose diphosphate Aldolase	+	–

B	Homofermentative	Facultative heterofermentative	[a]Obligate heterofermentative
Hexose: lactate	+	+	–
Hexose: lactate, acetate, ethanol, CO_2	–	–	+
Lactate, acetate, formate, ethanol, under glucose limitation		Strain dependent	+
Pentose phosphoketolase	–	+	+
Gluconate fermented	–	+	+

[a] Source: Hammes and Hertel (2003).

glyceraldehyde-3-phosphate (GAP). GAP is converted to pyruvate in a metabolic sequence including substrate level phosphorylation. One mole of glucose results in 2 moles of lactic acid and a net gain of two ATP.

It needs to be noted that glycolysis may lead to heterolactic fermentation under certain conditions, and some LAB, regarded as homofermentative, use the PPP when metabolising certain substrates (Axsellson and Ahrné, 2000).

The EMP pathway (glycolysis) is not used by obligate heterofermentative lactobacilli, *Leuconostoc, Weissella* spp. and few other LAB species.

Lactose is taken up via the phosphoenolpyruvate-dependent PTS and enters the cytoplasm as lactose phosphate, which is cleaved by phospho-β-D-galactosidase to yield glucose and galactose-6-phosphate. The lactose operon may be located on plasmids. The instability of several phenotypic properties of LAB is linked to the presence of plasmids. The enzyme systems of lactose-PTS and P-β-gal are generally inducible and repressed by glucose.

Glucose is phosphorylated by glucokinase and metabolised through the glycolytic pathway or the PPP. Galactose-6-phosphate is metabolised through the tagatose-6-phosphate pathway, while the Leloir pathway is used by galactose-fermenting LAB, which transport galactose with a permease and are devoid of the galactose-PTS (Grossiord et al., 2003).

Some of thermophilic lactobacilli, such as *Lactobacillus delbrueckii* subsp. *bulgaricus, Lb. delbrueckii* subsp. *lactis* and *Lactobacillus acidophilus* only metabolise the glucose moiety after transport of lactose and β-gal activity, while galactose is excreted into the medium.

6.2.1 Pentose phosphate pathway

In the PPP, the key step is the phosphoketolase (PK) split of xylulose-5-phosphate to GAP and acetylphosphate. GAP is converted to lactate and acetylphosphate to acetate and ethanol. Heterolactic fermentation gives 1 mole each of lactic acid, ethanol and CO_2 and 1 ATP per mole of glucose.

In general, facultative and obligate heterofermenting lactobacilli use pentoses for energy production. Specific permeases transport sugars intracellularly, where pentoses are phosphorylated and converted by epimerases and isomerases to ribulose-5-phosphate or xylulose-5-phosphate, which can be metabolised by the lower half of the PPP.

6.2.2 Citrate fermentation

Citrate serves as an energy source for LAB. LAB species possess a plasmid containing genes coding for citrate transporter (Bandell et al., 1998). Citrate is converted by citrate lyase to acetate and oxaloacetate, than decarboxylated to pyruvate (Wagner et al., 2005). This may be converted to acetyl-CoA (via pyruvate dehydrogenase), which leads to acetate (via acetate kinase), and acetaldehyde/ethanol (via alcohol dehydrogenase), or to formate (via pyruvate formate lyase), or to α-acetolactate (via acetolactate synthase), which leads to acetoin (acetolactate decarboxylase), to diacetyl and 2,3 butanediol (via diacetyl/acetoin reductase), and to lactate (via lactate dehydrogenase).

Energy is produced by conversion of acetyl-CoA to acetate: citrate acts as electron acceptor, resulting in high production of acetate and ATP via acetate kinase pathway.

Energy is produced during initial breakdown of citrate into pyruvate. In *Lc. lactis* subsp *lactis* biovar *diacetylactis* the uptake of citrate is coupled to generation of a proton motive force strong enough to drive additional ATP synthesis. Diacetyl, acetoin and acetaldehyde have distinct aroma properties. Diacetyl gives aroma to fresh cheese, fermented milk, butter and cream. In addition, CO_2 is produced, adding to texture of fermented dairy products (Kimoto et al., 1999).

The production of succinate from citrate is due to presence of citrate lyase, malate dehydrogenase, fumarase and succinate dehydrogenase. A putative citric acid cycle (PCAC) in *Lb. casei* was studied (Diaz- Muñiz et al., 2006). This may lead two ATPs per molecule of citrate. By analysing end products during growth in defined medium and cheddar cheese, these authors identified several organic acids. Additionally, *Lb. casei* produced D-lactic acid, acetoin, formic acid, ethanol, and diacetyl, but not butanediol or malic acid.

6.3 PROTEOLYSIS, PROTEIN SUBSTRATES AND AMINO ACID AVAILABILITY INFLUENCING GENE EXPRESSION

Proteolysis is considered one of the most important biochemical processes involved in manufacturing of many fermented dairy products (Fox, 1989) because LAB strains have evolved to adapt to grow on milk proteins (caseins, whey proteins).

In cheese manufacture, the proteolysis of casein plays an essential role because amino acids resulting from proteolysis are major precursors of specific flavour compounds due to secondary catabolic changes (e.g. deamination, decarboxylation, transamination, desulfuration, catabolism of aromatic compounds); changes to the cheese matrix, changes of the cheese curd texture, decrease in water activity (a_w) through water binding by carboxyl and amino groups, and increase in pH through amine release and NH_3 (McSweeney and Sousa, 2000).

In most cheese varieties, the initial hydrolysis of caseins is caused by the coagulant and to a lesser extent by plasmin. Coagulants are responsible of primary proteolysis because destabilize the colloidal structure of casein by cutting the peptide bond between Phe_{105} and Met_{106} in κ-casein fraction while secondary proteolysis results from the activity of proteases and peptidases of starter or nonstarter bacteria that convert casein to lower molecular weight peptides and free amino acids (FAAs) (Ozer and Akdemir-Evrendilek, 2015).

The main role of LAB is the acidifying activity; LAB provide the proteolytic enzymes, including oligoendopeptidase, di- and tri-peptidases, aminopeptidase and a number of proline-specific peptidases that contribute to the cheese ripening with the secondary proteolysis in cheese.

Proteases and peptidases are otherwise released by microorganisms upon their lysis, and the activities thereof depend on the environmental conditions prevailing in the curd (i.e. water activity, pH, mineral content and redox potential), as well as on the conditions provided throughout ripening (i.e. temperature, type of salt addition, nature of the secondary microbiota added or naturally developing) (Law and Wigmore, 1983; Christensen et al., 1999; Pereira et al. 2010; Hayaloglu and McSweeney, 2015).

Many authors reported the influence of proteolytic LAB in cheese manufacturing (Oommen et al., 2002; Madrau et al., 2006; Di Cagno et al., 2010; Chaves-López et al., 2014; Poveda et al., 2015; Pedersen et al., 2016), highlighting the importance of the selection of proteolytic strains to manufacture cheeses with defined functionality (Oommen et al., 2002).

Caseins are classified as αs1-, αs2-, β-, and κ-caseins; each type contains a large number of proline residues that prevent the formation of either α-helices and β-sheets and promote the formation of random coils. Together, the secondary structure characteristics lead to an unstructured, open molecule susceptible to action of cell-envelope proteinases (CEPs) (Fox, 1989, 1993).

CEPs contribute to the formation of small peptides in cheese by hydrolysing larger peptides produced from αs1-casein by chymosin or from β-casein by plasmin.

6.3.1 Cell-envelope proteinases: the Prt system

The complex proteolytic system of LAB comprises a cell-envelope proteinase (CEP), which has a strong preference for hydrophobic caseins (Savijoki et al., 2006; Strahinic et al., 2010; Sadat-Mekmene et al., 2011), producing oligopeptides; specific peptide transport systems, which effect the intake of peptides into the cells; various intracellular peptidases; and different enzymes that convert amino acids into flavour compounds.

CEPs are classified on the basis of the signal for export to the extracellular space, depending on the anchoring mechanism. For instance, *Lactobacillus buchneri CEP* gene is lacking the anchoring signal, so the protease is released in the extracellular space (Sun et al., 2015).

In *Lc. lactis*, two types of proteinase (P_I and P_{III}) are anchored to the cell wall, being released by lysozyme treatment or Ca^{2+} chelating agents: P_I degrades only β-casein, while P_{III} is active mainly on αs1-casein. β-casein is cleaved into a hundred different oligopeptides ranging from 4 to 30 amino acid residues, and κ-casein is cleaved to a lesser extent by the P_I-type enzyme, whereas the P_{III}-type is able to cleave αS1-, β-, and κ-caseins equally well (Pritchard and Coolbear, 1993). PrtPs are further classified into seven groups (a, b, c, d, e, f and g) by their specificity toward the α_{S1}-casein fragment from positions 1 to 23 (f1–23) (Kunji et al., 1996). In *Lb. casei* the extracellular proteinase is a serine protease, sensitive to phenyl methyl sulfonyl fluoride. Several proteinases have been described in *Lb. delbrueckii* subsp. *bulgaricus*. One of these is a metallo-enzyme, inhibited by EDTA, and a second is a cysteine protease. *S. thermophilus* is considered less proteolytically active than *Lactobacilli*. Caseinolytic activity was studied in *Bifidobacteria*, in *Enterococcus faecalis* and, using cell-free extracts, in *Leuconostoc mesenteroides* and in dairy *Pediococci*.

While many LAB strains contain CEP, several strains, that is, nonstarter LAB, do not. Thus, they rely on starter LAB for the production of peptides and amino acids; peptides are then taken up by the cells via specific peptide transport systems.

6.3.2 Oligopeptide permeases and other transporters for peptides and amino acids

The oligopeptides are then transported in the inside by the oligopeptide permease (OPP). When analysing the kinetics of OPP toward different peptides (Detmers et al., 1998,

2001), a second oligopeptide ABC transporter was found, the so-called Opt system, present in wild-type strains such as *Lc. lactis* SK11 and Wg2, whereas the plasmid-free strains MG1363 and IL-1403 each synthesise only one system, Opp and Opt, respectively (Charbonnel et al., 2003, Lamarque et al., 2004).

Peptide uptake occurs via oligopeptide transport systems and di-/tri-peptide transporters. In addition, various amino acid transport systems have been identified with a high specificity for structurally similar amino acids (Schweikhard and Ziegler, 2012; Trip et al., 2013).

In *Lc. lactis*, the Opp_{Ll} system possesses only one binding protein (OppA), whereas the transport systems of *S. thermophilus* and *Lb. bulgaricus* are composed of three and two oligopeptide-binding proteins (OBPs), respectively (Peltoniemi et al., 2002).

OppA is able to bind peptides with up to 35 amino acid (aa) residues, of which only the six N-terminal residues of the peptides are entrapped in the binding pocket of OppA, whereas the remainder of the peptide interacts with the surface of the binding protein in a nonopportunistic manner. Nevertheless, a recent study has demonstrated that the binding protein OppA is not the sole determinant of the specificity of peptide transport in *Lc. lactis* (Doeven et al., 2005).

A second oligopeptide transport system called the 'Opt system' differs from the well-known Opp_L system in its genetic organization and in the number of OBPs synthesised.

Di- and tripeptides transporters carry smaller peptides in the inside through active mechanisms, the ABC transporter Dpp, and proton motive force–driven transporter DtpT.

GlnPQ is an ABC transporter for glutamine, asparagine and glutamic acid, with energy provided by hydrolysis of ATP. The system is essential in various Gram-positive bacteria, including *Lc. lactis* (Fulyani et al., 2015).

6.3.3 Peptidolysis and free amino acids

Intracellularly, amino- and carboxypeptidases, endopeptidases, tripeptideses and dipeptidases produce free amino acids that are substrates for production of flavour compounds (Smit et al., 2005b). Bacterial proteolytic enzymes have a particularly intense involvement during the first weeks of dairy productions, as well as in the second half of the cheese-ripening period (30–60 d).

Peptidases capable of acting on oligopeptides are described in Table 6.2 (Kunji et al., 1996; Christensen et al., 2003). To date, no enzyme with carboxypeptidase activity was reported for any LAB. However, degradation products originated from carboxypeptidase activity have been described, while phosphorylated peptides remained weakly hydrolysed, indicating high intrinsic resistance to peptidase activity (Deutsch et al., 2000).

β-casein is degraded to a higher extent by *Lc. lactis* (Kunji et al., 1996) and *Lactobacillus helveticus* (Christensen et al.,1999), while *Lb. delbrueckii* subsp. *lactis* and *S. thermophilus* are less proteolytic, with this second unable to freeing proline from peptides.

A specificity on αs1-casein f1–23 was shown for PepO from *Lb. rhamnosus* HN001, and on post-proline residues of β-casein f203–209 was demonstrated for PepO2 from

Table 6.2 LAB Peptidases.

Peptidases	Type of enzyme activity	
Aminopeptidases		
PepA1, in *L. lactis*	Metallopeptidase, with a cysteine sensitive to DTT possibly involved in folding	Glutamyl/aspartyl specific aminopeptidase, freeing N-terminal acidic residues from three- to nine-residue peptides
PepA2 in *L. delbrueckii*	Metallopeptidase	Not inhibited by dithiothreitol (DTT), neither by its product glutamate
PepC	Cysteine peptidase	
PepL	Serine peptidase	
MetAP methionine aminopeptidase	Metallopeptidase, with preference for Co, Mn divalent cations	Cleaves N-terminal Met in peptides with Gly, Ala, Ser, Val, Pro, or Thr penultimate amino acid. The enzyme is active on tetrapeptides (in *L. lactis, S. thermophilus*)
PepN general aminopeptidase	Metallopeptidase	
PepP general aminopeptidase	Metallopeptidase	
PepS	Metall peptidase	Shows preference for peptides containing two to five residues with Arg or aromatic amino acid residues in the N-terminal position, debittering activity on bitter peptides
PepX	Serine peptidase	X-prolyl dipeptidyl peptidase
Endopeptidases		
PepD	Cysteine peptidase	Dipeptidase
PepE (E1, E2) endopeptidase	Metallopeptidase	
PepF (F1, F2)	Metallopeptidase	Cleaves oligopeptides in the range 7- to 17-residue-long
PepG	Cysteine peptidase	
PepIP	Serine peptidase	Proline iminopeptidase, acts on sequences containing proline in the penultimate position
PepO (O1, O2, O3)	Metallopeptidase	PepO2 in *L. helveticus* has specificity for peptide bonds C terminal to Pro residues, degrading αS1-casein(f1-9) and β-casein(f193-209)
PepQ, prolidase	Metallopeptidase	Cleaving dipeptides carrying proline in the second position
PepR	Serinepeptidase	Prolinase, acts on dipeptides containing proline in the penultimate position
PepT tripeptidase	Metallopeptidase	Prefer peptides containing hydrophobic amino acids including leucine, methionine, phenylalanine, or glycine
PepV dipeptidase	Metallopeptidase	

Lb. helveticus CNRZ32 (Christensen et al., 2003). Besides cleaving oligopeptides from 7- to 17-residue-long, PepF is also important for protein turnover under conditions of nitrogen starvation in *Lc. lactis* (Monnet et al., 1994). To identify the substrates of Opp present in the casein hydrolysate, the PrtP-generated peptide pool was made available to mutants lacking the proteinase, but containing Opp, and the disappearance of peptides from the culture medium as well as the intracellular accumulation of amino acids and peptides was monitored in peptidase-positive and peptidase-negative genetic backgrounds. The results showed that (i) the carboxyl-terminal end of β-casein is degraded preferentially despite the broad specificity of the proteinase; (ii) peptides smaller than five residues are not formed in vivo; (iii) oligopeptides of 5–10 residues are utilized after uptake via Opp; (iv) of the peptides generated by PrtP, only a limited number are used.

Savijoki et al. (2006) utilized proteomic tools to characterise the Emmental cheese-ripening process. This approach provided information on the peptidases released into the cheese by the starter bacteria *Lb. helveticus* and *S. thermophilus* during ripening. Different peptidases were originating from *Lb. helveticus* (PepN, PepO, PepE and PepQ) and *S. thermophilus* (PepN, PepX and PepS). The different composition in released nitrogen fractions, such as small peptides and free amino acids, were related to the proteolytic activity of *Lb. helveticus* and *Lb. bulgaricus* (PepA) (Stressler et al., 2010). A stronger peptidase activity of lactobacilli in comparison with streptococci and an acceleration of ripening of ewe's pasteurized cheese were found, with release of *Lb. helveticus* and *Lb. bulgaricus* proteolytic enzymes in the matrix (Candioti et al., 2010). The high activity of transformation of pH 4.6-soluble nitrogen into trichloro-acetic acid (TCA)-soluble nitrogen was directly proportional to proteolytic activity of the *Lb. casei* RP5 strain. In hot brined cheese 80% of the bacterial proteases maintained their activity after hot brining and actively participated in cheese ripening. The concentrations of free amino acids released by *Lb. bulgaricus* 2–11 were similar to that of *Lactobacillus casei* RP5. It was found that this strain possesses aminopeptidases able to free glutamic acid. Studies on the effect of starters and nonstarter strains in the ripening of various types of cheese have been performed, such as Parmigiano Reggiano (Qian and Reineccius, 2002), gorgonzola (Moio et al., 2000), mozzarella (Moio et al., 1993), and in laboratory cheese models (Randazzo et al., 2007; Ruggirello et al., 2014).

Mesophilic and thermophilic LAB enzymes have a particularly intense involvement during cheese proteolysis, especially in the second half of the cheese-ripening period. In different types of cheese, LAB ability to transform soluble nitrogen into low molecular and nitrogen compounds suggests high peptidase and aminopeptidase activity of the mesophilic and thermophilic strains (Simov and Ivanov, 2005; Madrau et al., 2006; Candioti et al., 2010; Tabet et al., 2016).

In particular in cheese manufacturing, mesophilic strains, such as *Lc. lactis* ssp. *lactis* (LMG S 19870), *Lb. brevis* (LMG 6906) and *Lb. plantarum* (LMG S 19557), were able to hydrolyse peptides into FAA (Pereira et al., 2010).

Distinctive features of ripened cheese were determined by concentrations of individual free amino acid. Employment of LAB with specific proline-specific peptidase have a crucial importance as the caseins are rich in this amino acid residue (Ozer and Akdemir-Evrendilek, 2015).

Swiss cheeses are rich in free proline, which is associated with the sweet flavour. In Emmental cheese, *S. thermophilus* participates in peptide degradation, with the production of general aminopeptidase PepN, PepX and aminopeptidase PepS (Gagnaire et al., 2004). Since *S. thermophilus* cannot release it, the amount of free proline directly depends on the activity from the peptidases (PepIP and PepQ) of thermophilic lactobacilli and on their availability in the cheese matrix after bacterial lysis (Deutsch et al., 2000).

6.3.4 Peptidolysis and catabolite repression

The transcriptional repressor CodY senses the internal pool of branched-chain amino acids (isoleucine, leucine and valine) (Guédon et al., 2001); using these residues as cofactors, CodY represses the expression of *Opp* and of the proteolytic system genes in *Lc. lactis*. A branched-chain amino acid (BCAA) responsive transcription regulator, BCARR, repressing the expression of the genes in the proteolyitic system, was described in various LAB species (Klaenhammer et al., 2007) and in *Lb. helveticus* (Wakai and Yamamoto, 2013). The availability of BCAA increases the inhibition of expression of the proteolytic system genes. In *S. thermophilus*, CodYSt repressed the transformation of the central metabolic pathway to amino acid metabolism and improved lactose utilization (Lu et al., 2015).

Although peptides and amino acids do have specific flavour characteristics, like sweet, bitter, or malty, it is believed that the basic taste of cheese is already defined. Stimulation or overexpressing of several proteolytic enzymes and also the addition of amino acids to cheese did not influence the positive flavour perception of cheese. However, unbalanced proteolysis might lead to excess of bitter peptides (Lemieux and Simard, 1992), which can lead to decreased cheese flavour perception. Specific cultures have been selected with ability to degrade bitter-tasting peptides. These cultures are used in cheese and curd processing. Free amino acids are then substrates for successive flavour-forming reactions. Thus, proteolysis and peptidolysis contribute to flavour formation but are not rate controlling in flavour formation. However, ripening process may be accelerated by addition of exogenous enzymes, use of adjunct strains, or starters with high proteolytic activities, attenuated (heat or freeze-shocked) or inactivated starters, and cheese slurries, or by addition of free amino acids (Fox et al., 1996).

6.3.5 Amino acid biosynthesis and auxotrophy

A set of amino acids (arginine, cysteine, glutamic acid, lysine) are essential in certain LAB strains that are auxotrophic for them, and need to release free amino acids from peptides. Most lactobacilli need some free amino acid (arginine, cysteine) to remain alive during harsh conditions, such as storage under low temperature, or acid and oxygen stress. Cysteine uptake system was shown essential in oxygen tolerance in *Lb. fermentum* (Turner et al., 1999). L-cysteine might improve the anaerobic growth of lactobacilli, provide amino nitrogen as a growth-favouring factor and behave as an anti-oxidative stress factor (Sutula et al., 2012). Lactococci used in dairy fermentations have a complex nutritional requirement and require at least glutamate, valine, methionine, histidine, serine, leucine and isoleucine for growth, and the number of essential amino acids is strain-dependent. Industrial *Lc. lactis* subsp. *cremoris* strains require different amino acids for growth. Wild *Lc. lactis* subsp. *cremoris* strains generally require two to three amino acids, while some *Lc. lactis* subsp. *lactis* strains only need one or two amino acids.

It is known that various LAB strains differ in amino acid–converting abilities and that these activities are linked to the ability to synthesise amino acids (Dudley and Steele, 2001; Smit et al., 2005a). The strains with the highest activity are *Lc. lactis* strains isolated from natural sources and non-dairy environments, the so-called 'wild lactococci'

(Ayad et al., 2000). For instance, many strains do not degrade caseins, produce antimicrobial compounds and/or have low acidifying activity. However, when the dependency of these strains for amino acids in the growth medium was determined using the single omission technique, it was found that these strains had a much larger potential to synthesise their own amino acids as compared to industrial strains. The absence of some amino acid biosynthetic pathways in dairy lactococci might be a consequence of their adaptation to dairy products, since in milk, the amino acids are readily available from the proteolytic degradation of caseins. Wild strains are not naturally associated with a rich environment such as milk, which makes them more dependent on their own biosynthesis of amino acids compared to industrial strains.

Casein proteolysis and the derived peptides, urea and non-protein nitrogen (NPN) fractions released by thermal treatment of milk are able to induce the expression of genes such as biosynthetic pathways for cysteine and methionine in *Bifidobacterium bifidum* (Ferrario et al., 2015). Transcriptional analysis involving *B. bifidum* PRL2010 cultivated on whey proteins or casein hydrolysate revealed that the biosynthetic pathways for cysteine and methionine are modulated by the presence of casein hydrolysate (Ferrario et al., 2015).

Amino acid auxotrophy or prototrophy in *Bifidobacteria* and *Lactobacilli* is dependent on species (Deguchi and Morishita, 1992) and linked to the presence in the genome of genes coding for transaminases and amino acid metabolic pathways (Ferrario et al., 2015). A branched-chain aminotransferase (ilvE) from *Lc. lactis* LM0230 (Atiles et al., 2000; Ganesan and Weimer, 2004) was shown to synthesise Ile and Val from precursor keto acids (Santiago et al., 2012).

Methylation of homocysteine, the final step of methionine biosynthesis, can be catalysed by two non-homologous enzymes, that is, a cobalamin-dependent homocysteine methyltransferase (MetH) or a cobalamin-independent methionine synthase (MetE).

Methionine biosynthesis from homoserine can be initiated by homoserine kinase (HSK), homoserine trans-succinylase (HSST) and homoserine *O*-acetyltransferase (HSAT). In the sequenced LAB genomes, HSK and HSST are well distributed. HSST and HSAT share little sequence similarity but have similar substrate specificity (Liu et al., 2008). The orthologous gene in *Lb. plantarum* is located in the same operon as the genes encoding *O*-acetylhomoserine sulfhydrylase (AHSH) and homoserine dehydrogenase (HSDH), with functional correlation between HSDH, HSST and AHSH. Because *Lb. plantarum* is a species lacking a tricarboxylic acid cycle to make succinyl-CoA, it needs to use acetyl-CoA to synthesise methionine (Liu et al., 2008).

6.4 LIPOLYSIS, LIPASES, ESTERASES

In addition to renneting enzyme, lipases from lamb or pig stomach are used in the transformation of ewe's milk, contributing to the release of flavours in ewe's and goat cheeses. The lipases and esterases influence these types of cheese products.

Lipases and esterases of LABs are intracellular, therefore enzymes may be released in the cheese matrix after cell lysis. Adjunct starters and nonstarter LAB (NSLAB) are considered responsible for the production of free fatty acids (FFAs) and short-chain FFA (Dherbécourt et al, 2010; Wolf et al., 2016), which have volatile properties and provide flavouring.

Lipolysis results in the formation of FFAs, which can be precursors of flavour compounds such as methyl ketones, secondary alcohols, esters and lactones. LAB contribute relatively little to lipolysis, but additional cultures, as in the case of surface-ripened cheeses, often have high activities in fat conversion. Flavours derived from the conversion of fat are particularly important in soft cheeses like Camembert and Roquefort.

Penicillium roqueforti blue mould and other fungal lipases release FFA and convert them into alk-2-ones in blue cheese (McSweeney, 2004). Various flavours have been described as dependent on the application of fungal species in dairy products: *Penicillium casicolum* Beiner, *Penicillium candidum, Penicillium camemberti, Penicillium album, Penicillium nalgiovense, Penicillium glaucum, Penicillium cyclopium* (Saint Nectaire), *Geotrichum candidum*. Their activity favours the growth of yeasts, such as *Torulopsis, Candida, Kluyveromyces* and *Debariomyces* spp. A complex interaction with surface microorganisms, such as Micrococci, involves the use of lactate as substrate, growth at temperature of cellar, and requirement of salt concentration under 1%. After the lipolysis, FFAs are converted into β-ketoacids and dimethyl ketones.

In studies on acceleration of ripening (Wallace and Fox, 1997), *Mucor miehei* lipase, *P. roqueforti* and *P. candidum* lipases were tested with good results (Fox et al., 1996). FlavorAge product contains a unique lipase from a strain of *Aspergillus oryzae* with high specificity for C6-C8 fatty acids, and applied to cheese, produced FFA profiles similar to those found in control cheeses. Feta cheese produced from cow's milk using a blend of *Lb. casei* and *Lc. lactis* as starters, and a blend of kid and lamb pre-gastric esterase (PGE), developed a body, flavour and texture similar to authentic feta cheese.

Esters, such as ethylbutyrate, contribute to cheddar and Gouda flavour, although an excess of esters in proportion to other flavour compounds could be responsible for fruity defects. In Camembert, phenylacetaldehyde, 2-phenylethanol and the ester phenylethyl acetate, which all result from phenylalanine degradation, are identified in fractions with floral rose-like odour, producing the floral note.

Esters are formed in a reaction between an alcohol and an organic acid, which also might be activated by coupling to CoA. Besides amino acid metabolism, also sugar and fat metabolism provide substrates for ester formation. Although ester formation is generally considered to be an enzymatic catalysed reaction, the reaction between acetyl-CoA and methanethiol is spontaneous. Esterases are serine hydrolases capable of synthesising or hydrolysing esters, depending on the environmental conditions, while alcohol acetyltransferases only catalyse ester synthesis. By knocking out the *Lc. lactis esterase* gene (*estA*), it was shown to be responsible for ester hydrolysing activity and for the formation of short-chain fatty acid esters in vitro.

Liu et al. (2008) described an alternative reaction for formation of esters by dairy LAB. The process was named alcohol lysis, in which an acyl-transferase reacts with fatty acyl groups and acyl groups from acyl-CoA derivatives that are directly transferred to alcohols.

6.5 AROMA AND FLAVOUR PRODUCTS OF METABOLISM

An intriguing issue of LAB is that they are a flavour factory, and this aspect has a dramatic importance in food processing and quality assessment. Flavour characterises a cheese, defines its sensorial specificity and, in some cases, a defect. In fact, the final step of catabolism of the three main constituents of milk (carbohydrates, proteins and lipids) is the formation of many molecules of low molecular mass (44–200 Da), most of which are odour active and impact strongly the flavour, along with the molecules derived directly by feeding and/or environment (i.e. terpenes). When ripening of a particular cheese follows its 'normal' development, flavour compounds express its typicality, while an abnormal production of some compounds due to the growth of undesirable microbiota, causes off-flavour and possibly texture defects (blowing).

Dairy flavour involves smell and taste in a ratio of approximately 75:25. This means that taste, although in a minor extent, also has an important role in flavour. Dealing with taste, the compounds involved are mostly amino acids, organic acids, biogenic amines, peptides with low molecular weight and mineral salts that are water-soluble. Amino acids, along with small peptides (300–400 Da) are typically responsible for a bitter taste. Aromatic compounds involving nose receptors are mainly lipophilic and are derived not only from the lipid breakdown but also from the catabolism of amino acids and carbohydrates. These molecules are characterised by different kinds of scents that in diverse combinations form a bouquet that can be simple, as in fermented milk, yogurt and quark cheese with a very short ripening period (few days), or very complex, as in mould-ripened cheese and cheese ripened a long time, like Parmigiano Reggiano (>2 years). Moreover, the olfactory impact of a volatile compound is related to odour activity and matrix effect rather than its concentration.

For cheeses ripened a short time, the most important flavour compounds derive by the metabolism of carbohydrates (lactose and citrate) by mesophilic LAB giving acetate, diacetyl, acetaldehyde, acetoin, ethanol, 2,3-butanediol, which are strongly odour active. For instance, diacetyl and acetoin are derived from co-metabolism of citrate and carbohydrates in *Ln. mesenteroides* (Schmitt et al., 1997).

For cheeses of a longer ripening period, lipid and protein breakdown also occurs, resulting in the production of a numerous compounds, which in turn can react producing other volatiles giving cheese a more complex flavour. Lipolysis is responsible for the production of short-chain fatty acids that are strongly odour active and give a piquant taste. Saturated fatty acids can be oxidized to β-ketoacids, then decarboxylated to methyl ketones with one less carbon, which can be reduced to secondary alcohols. Both methyl ketones and secondary alcohols possess a very low odour threshold and are typical of mould-ripened cheeses.

FFAs can react with alcohols, giving esters, characterised by a mild, floral odour. Free hydroxylated fatty acids can undergo intramolecular transesterification between the acid and hydroxyl groups, forming lactones.

Proteolysis and particularly amino acid catabolism is a more complex issue during ripening and is responsible for many important flavour compounds. BCAAs (leucine, isoleucine and valine) can undergo deamination producing iso-fatty acids, aldehydes and secondary alcohols that can be esterified by fatty acids. Aromatic amino acids (phenylalanine, tyrosine, tryptophan and histidine) can be transaminated and reduced to aldehydes (benzylaldehyde, phenylacetaldehyde), acids (benzoic acid, phenylacetic acid) and alcohols (phenylethanol) that can be esterified with acids (phenyl ethyl acetate, ethyl benzoate). Degradation of aromatic amino acids is also responsible for faecal, burnt and mothball off-flavours (indole, skatole). The sulfur amino acid methionine can give methanethiol that can be oxidized to dimethyl sulfide (DMS), dimethyl disulfide

(DMDS) and dimethyl trisulfide (DMTS) (Fernandez et al., 2002). These sulfur compounds are strong odour active and are typical of blooming rind cheeses like Brie and Camembert. Another sulfur amino acid, cysteine, gives hydrogen sulfide, found in cheddar. Flavours originating from amino acids are described in Tables 6.3, 6.4 and 6.5. In Table 6.4 volatile, oil- or water-soluble compounds are reported, along with their perception thresholds.

Table 6.3 Major volatile odor-active compounds in dairy products.

Volatile compounds	Main precursors	Odour descriptor	Dairy product
Acids			
Acetic	Carbohydrates	Vinegar, peppers, green, sour	Cheddar, Parmesan, Camembert
Propanoic	Carbohydrates	Pungent, rancid	Swiss
2-methylpropanoic	Proteins	Rancid	
Butanoic	Carbohydrates/ lipids	Sweat, fecal, rancid, dirty sock	Cheddar, Parmesan, Blue
3-Methylbutanoic	Proteins	Cheese, sweat, rancid, rotten fruit	Gouda
Hexanoic	Lipids	Sweat, goat, bad breath	Cheddar, Gouda, Parmesan
Octanoic	Lipids	Cheese, rancid, sweat	Gouda, Parmesan
Decanoic	Lipids	Stale, butter, grassy, fatty, cheese,	Cheddar, Gouda
Phenylacetic	Proteins	Honey, flower	
Alcohols			
Ethanol	Carbohydrates	Dry, dust, alcohol	Swiss
2-Propanol	Carbohydrates	Fruity, ethereal, wine-like	Swiss
1-Butanol	Carbohydrates	Banana-like, wine-like, fuel oil	Parmesan
2-Butanol	Proteins	Sweet, fruity, fuel oil, wine-like	Swiss
2-Methy-1-lbutanol	Proteins	Malty, wine, onion	Swiss
3-Methyl-1-butanol	Proteins	Fresh cheese, alcoholic, fruity, solvent-like	Mozzarella, Gouda
2-Pentanol	Proteins	Green, alcoholic, fruity, fresh	Parmesan
2-Heptanol	Proteins	Fruity, earthy, green, sweetish, dry, dusty carpet	Camembert
1-Octen-3-ol	Proteins	Mushroom-like, mouldy, earthy	Camembert
2-Nonanol	Proteins	Fatty green	Camembert
Phenyl ethanol	Proteins	Unclean, rose, violet-like, honey, floral	Mozzarella, Camembert, blue
Esters			
Ethyl acetate	Carbohydrates	Solvent, fruity, pineapple	Parmesan, Swiss
Ethyl butyrate	Lipids	Fruity, green, apple, banana	Cheddar, Gouda, Parmesan, Swiss
Ethyl caproate	Lipids	Pineapple, sweet, fruity, banana	Cheddar, Parmesan, blue
Ethyl octanoate	Lipids	Pear, apricot, sweet, fruity, banana, pineapple	Parmesan
Ethyl phenyl acetate	Proteins	Floral, rose, lily-jasmine, honey	Camembert
3-Methyl-1-butanol acetate	Proteins		

(Continued)

Table 6.3 (Continued)

Volatile compounds	Main precursors	Odour descriptor	Dairy product
Carbonyl compounds			
Acetone	Carbohydrates	Sweet, fruity, ethereal, wood pulp, hay	Swiss
Acetaldehyde	Carbohydrates	Ethereal, fresh, green, pungent	Yogurt
Acetoin	Carbohydrates	Buttery, sour milk	Butter, Cheddar, Gouda
Diacetyl	Carbohydrates	Strong buttery, caramel, toffee	Butter
2-Butanone	Lipids	Sweet, ethereal, slightly nauseating	Swiss
2-Pentanone	Lipids	Orange peel, sweet, fruity	Gouda, Parmesan, Camembert, Swiss
2-Heptanone	Lipids	Blue cheese, spicy, Roquefort, fruity	Cheddar, Camembert Swiss
1-Octan-3-one	Lipids	Mushroom-like	Cheddar, Camembert
2-Nonanone	Lipids	Malty, rotten fruit, hot milk, green, earthy	Gouda, Parmesan, Camembert, Swiss, blue
2-Undecanone	Lipids	Floral, fruity, green, musty, tallow	Gouda, Camembert, blue
Phenylacetaldehyde	Proteins	Honey-like, rosey, violet-like, styrene	Cheddar
3-Methyl butanal	Proteins	Malty, powerful, cheese, green, dark chocolate	Cheddar, Gouda, Parmesan, Camembert, Swiss, blue
Others			
Hydrogen sulfide	Proteins	Rotten egg	Cheddar
Methional	Proteins	Baked potato	Cheddar, Parmesan, Camembert
Methanethiol	Proteins	Rotting cabbage, cheese, vegetative, sulfur	Camembert
Di-methyl sulfide	Proteins	Sulfur, cabbage-like, pomegranate	Camembert
Di-methyl-disulfide	Proteins	Green, sour, onion	Camembert
Di-methyl-trisulfide	Proteins	Sulfurous, garlic, putrid, cabbage-like	Cheddar, Gouda, Parmesan, Camembert
Tetra-methyl pyrazine	Proteins	Baked, beans, raw potato	Parmesan
δ-Decalactone	Lipids	Coconut-like, peachy, creamy, milk fat	Cheddar, Gouda, Camembert
δ-Dodecalactone	Lipids	Cheesy, coconut, sweet, soapy, buttery, peach,	Gouda
γ-Dodecalactone	Lipids	Peach, almonds, herbs, lilacs, fruit, toffee	Gouda

6.5.1 Aldehydes, alcohols and carboxylic acids

Alcohol dehydrogenase (AlcDH) and aldehyde dehydrogenase (AldDH) catalyse the conversion of aldehydes to alcohols and carboxylic acids. α-ketoacid intermediates can be converted to aldehyde carboxylic acids and CoA-esters via oxidative decarboxylation, or they can be reduced to hydroxyacids by hydroxyacid dehydrogenases (HADH). In LAB, dehydrogenases play an important role in maintaining the intracellular redox balance, and are required to restore NAD from NADH in excess. The hydroxyacids

Table 6.4 Descriptions of some important flavour components and their thresholds.

Flavour compound	Description	Odour threshold in ppb (μM) In water	In oil
2-Methylpropanal	Banana, malty, chocolate		0.1–2.3
3-Methylbutanal	Malty		13 (0.15)
3-Methylbutanal	Fresh cheese, breathtaking, alcoholic	250 (2.8)	–
3-Methylbutyric acid	Sweaty, cheese	120–700 (1.2–6.9)	–
Butyric acid	Butter, cheese, strong, acid	240 (2.7)	–
Propionic acid	Pungent, sour milk, cheese	2×10^4 (270)	–
Ethylbutyrate	Fruity, buttery, ripe fruit	1 (0.008)	–
Diacetyl	Buttery, strong	2.3 (0.026)	–
Acetaldehyde	Yoghurt, green, nutty, pungent	15 (0.34)	–
Methional	Cooked potato, meat like, sulfur	0.05–10 1×10^{-4}	–
Methanethiol	"Rotting" cabbage, cheese, sulfur	0.02–2 (3×10^{-4})	–
Benzaldehyde	sweet cherry	350 (3.3)	–
Phenyl acetate	Rough, lily-jasmine with metallic note	25 (0.5)	–

Data according to Table 2 in: Smit, G., Smit B.A., Engels, W.J.M., 2005. Flavour formation by lactic acid bacteria and biochemical flavour profiling of cheese products. *FEMS Microbiol. Rev.* 29, 591–610. Copyright by Oxford University Press. Licence 4101330821266, Invoice RLNK502301863, Copyright Clearance Center's RightsLink® service.

Table 6.5 Flavour compounds from amino acids by transamination and α-keto acid formation.

Amino acid	Keto acid	Aldehyde	Organic acid	Alcohol or thiol	Esters
Ile	α-Keto-3-methyl-pentanoic acid	2-Methylbutanal	2-Methylbutyric acid	2-Methylbutanol	
Leu	α-Ketoisocaproic acid	3-Methylbutanal	3-Methylbutyric acid	3-Methylbutanol	Ethyl-3-methylbutanoate
Leu		2-Methylpropanal	2-Methyl propanoic acid	2-Methylpropanol	
Val	α-Ketoisovaleric acid	2-Methylpropanal	2-Methyl propanoic acid	2-Methylpropanol	Ethyl isobutanoate
Phe	Phenyl pyruvate	Benzaldehyde	Benzoic acid	Phenylmethanol	Ethyl benzoate
Phe		Phenylacetaldehyde	Phenylacetic acid	Phenylethanol	Phenylethyl acetate
Trp	Indole-3-pyruvate	Indole-3-acetaldehyde	Indole-3-acetic acid		
Met	α-Keto methylthio butyrate	Methional	Methylthiobutyric acid	Methionol	Ethyl-3-methylthio propionate
Met		Methylthio-acetaldehyde		Methanethiol	Methylthioacetate

derived from BCAAs, ArAAs and methionine, observed in many dairy fermentations, can lead to low α-ketoacid concentrations, negatively affecting the flux toward flavour compounds such as aldehydes. The 2-hydroxyacid dehydrogenases better studied are lactate dehydrogenase (LDH) and hydroxy isocaproate dehydrogenase (HicDH), characterised in several LAB (Broadbent et al., 2004). Several enzymes are named hydroxy isocaproate dehydrogenase (HicDH), due to their preference for α-ketoisocaproate. The enzymes have a broad substrate specificity and catalyse the stereospecific hydrogenation of α-ketoacids, using NADH as hydrogen donor.

6.5.2 Amino acids as precursor flavour compounds

Various enzymes are involved in the conversion of amino acids to flavour compounds. The genome sequence analysis of LAB species provided insights into the metabolic pathways involving amino acids (Engels et al., 2000; van Kranenburg et al., 2002; Do Carmo et al., 2011, 2012). The enzymes catalyse reactions of transamination, deamination, decarboxylation, and cleavage of amino acid side chains.

The enzymes aminotransferases (AT); branched-chain amino acid transferase (BCAT), aromatic AT (ArAT), aspartate AT (AspAT), glutamate dehydrogenase (GDH), hydroxyacid dehydrogenase (HADH), alcohol DH, aldehyde DH, branched-chain α-ketoacid decarboxylase KdcA, lysine/ornithine decarboxylase and esterase A (EstA) are involved in flavour formation (Nardi et al., 2002). Various enzymes belong to large protein families. For instance, aminotransferases ArAT, AspAT, and cystathionine lyases CBL/CGL fall into the same aminotransferase family I (Liu et al., 2008). Some of the enzymes exhibited overlapping substrate specificities, for example, ArAT, AspAT and BCAT (Yvon et al., 2000).

α-Ketoglutarate is generally the preferred amino group acceptor (co-substrate) for transamination reactions (Yvon et al., 1998; Banks et al., 2001). The uptake of α-ketoglutarate by citrate transporters has been shown important to sustain the intracellular need (Pudlik and Lolkema, 2012, 2013). In LAB, the transaminase reaction linked to the deamination of glutamate to α-ketoglutarate is catalysed by glutamate dehydrogenase (GDH). The lack of GDH activity is common in strains used for cheese manufacture (Tanous et al., 2002). Glutamate dehydrogenase activity is a criterion for the selection of flavour-producing LAB strains (Tanous et al., 2002). α-Ketoglutarate serves as amino acceptor in the transamination reaction of leucine, phenylalanine and other amino acids, and the addition of α-ketoglutarate strongly increases amino acid conversion of *Lactobacillus sakei* and *Lb. plantarum* (Larrouture et al., 2000). LAB GDH activity is exerted in a strain-specific manner. The enzyme catalyses the NAD(P)H-dependent recycling of glutamate to α-ketoglutarate.

BCAAs can be transaminated to ketoacids, which undergo spontaneous degradation or are enzymatically converted to corresponding aldehydes or carboxylic acids (Smit et al., 2004b).

Methanethiol and other key sulfuric aroma compounds such as dimethyl sulfide (DMS), dimethyl disulfide (DMDS), and dimethyl trisulfide (DMTS) have a significant impact on cheese sensory profiles (Weimer et al., 1999). Key sulfur-containing flavour compounds are 4-methylthio-2-oxobutanoate (KMBA) and 2-hydroxy-4-methylsulfanylbutyric acid (HMBA).

The *cbs* genes from LAB code for cystathionine β-synthase. Other enzymes were predicted to catalyse different types of reactions; for instance, proteins from the CBL/CGL subfamily may carry out both α,β-elimination (cystathionine β-lyase, CBL) and α,γ-elimination reactions (cystathionine γ-lyase, CGL) (Liu et al., 2008). In the last step in metabolism of methionine and cysteine, CBL converts cystathionine to homocysteine (Weimer et al., 1999).

Proteins from cystathionine γ-synthase (CGS) and *O*-acetyl homoserine sulfhydrylase (AHSH) subfamiles might display both CGS and AHSH activity (Liu et al. 2008).

Among enzymes involved in the flavour synthesis, cysteine serine acetyltransferases (CysE, CysK) were identified. Degradation of methionine involves the transamination pathway that is initiated by aminotransferases and the elimination pathway initiated by C-S lyases.

Threonine aldolase, belonging to another class of lyases, is able to convert threonine directly to acetaldehyde.

The oxidative decarboxylation of α-ketoacid dehydrogenase (DH) complex in *Lactococcus* spp. and in related species is relevant for flavour formation in cheese. The oxidative decarboxylation of α-ketoacids results in the formation of organic acids. Carboxylic acids, such as isovaleric acid, are important flavour compounds. Furthermore, carboxylic acids are precursors for other aroma compounds, such as esters, thioesters, cresol and skatole. Cresol and skatole can chemically as well as enzymatically be formed from tyrosine and tryptophan.

In particular, the α-ketoacid decarboxylase enzyme in the pathway leading to 3-methylbutanal was deeply studied (Smit et al., 2004b, 2004c).

6.6 NONENZYMATIC PRODUCTION OF FLAVOURS

Nonenzymatic conversions to produce flavour-active compounds have been found to have an important role in cheese. Some chemical, nonenzymatic conversion steps have been described, such as the formation of benzaldehyde from phenylpyruvic acid. The aldehydes formed can be dehydrogenated or hydrogenated to their corresponding alcohols or organic acids, which are in turn substrates for esterases and acyltransferases, leading to (thio)esters. One of the biological roles of these amino acid–degrading pathways is the generation of precursors, which are needed, for example, in the sterol and branched-chain fatty acids synthesis. On the other hand, the hydrogenation of the α-ketoacids may act as a sink for excessive redox potential (NADH). The conversion of amino acids to α-ketoacids and to alcohols to produce short-branched-chain alcohols, was discovered in yeast and named 'Ehrlich's pathway'.

Chemical conversion reactions are a category of flavour-forming reactions. A variety of chemical reactions proceed at the mild conditions of the cheese-ripening process and thereby contribute to flavour formation. LAB might influence the chemical reactions by influencing environmental characteristics such as pH, cofactor and oxygen availability.

The α-ketoacids of phenylalanine (phenylpyruvic acid) and methionine can non-enzymatically be converted into flavour compounds such as benzaldehyde and methylthioacetaldehyde. The existence of the chemical conversion of phenylpyruvic acid was shown (Nierop Groot et al., 1998; Smit et al., 2004c), as well as the conversion of indole-3-pyruvate and α-ketoisocaproic acid and the conversion to methylthioacetaldehyde. All these conversions seem to proceed via a similar reaction mechanism. The spontaneous degradation of hydroxyphenylpyruvate to hydroxybenzaldehyde also occurs, and both benzaldehyde and hydroxybenzaldehyde were found in significant amounts in semi-hard cheeses. The reaction is catalysed by divalent metal ions. The conversion products of indole-3-pyruvate (indole acetic acid, indol-3-acetaldehyde and skatole) have been identified as off-flavours in some type, of cheese.

This chemical conversion of α-ketoacids, especially the α-ketoacid of leucine, was studied in detail (Smit et al, 2004c). 2-Methylpropanal thus can be formed by two mechanisms: an enzymatic conversion originating from valine and a nonenzymatic chemical conversion with leucine as the substrate. The reaction is strongly influenced by the availability of substrates (α-ketoacid and oxygen), manganese, and by pH. In the presence of 10 mM $MnCl_2$, the chemical conversion occurred. The pH in ripening cheese seems to be optimal. Oxygen is consumed quickly in the fermentation process in semi-hard cheeses, thereafter being almost absent. On the other hand, on the surface of smear-cheese and fungal-ripened cheese, the situation favours such conversions.

An important chemical reaction from amino acids leading to flavour compounds is the Strecker degradation (Dunn and Lindsay, 1985; Urbach, 1993). The Strecker degradation is described as the reaction of the amino group of an amino acid with an α-dicarbonyl like a reducing sugar, and it is an important step in the Maillard reaction. However, at high temperatures, direct oxidative decarboxylation of amino acids can also lead to the same aldehydes. These reactions are especially intense at high temperatures, and they contribute to the flavour of baked products. However, even at lower temperatures, the reactions proceed, as in the case of cheese and beer production. Strecker degradation of leucine can result in 3-methylbutanal, but also the conversion of valine results in this aldehyde.

6.7 METHODS OF ANALYSIS OF FLAVOURS IN DAIRY PRODUCTS: HPLC, GAS CHROMATOGRAPHY/MASS ANALYSIS (GC/MS)

The techniques of detection of flavour-producing strains are achieved either by high-throughput assays and sensory profiles (Smit et al., 2004a; Licitra and Carpino, 2014), genotyping, by enzymatic analysis, or by physico-chemical analysis such as liquid chromatography (HPLC) or ultra-performance liquid chromatography nanoflow electrospray ionization tandem mass spectrometry (UPLC-nano-ESI-MS), by GC/MS (Bosset et al., 1993; Rychlik and Bosset, 2001) or metabolomic platforms and electronic nose (Marilley and Casey, 2004).

The most potent flavour compounds are the aldehydes, alcohols, carboxylic acids and esters. Especially important are the aldehydes, alcohols, carboxylic acids and esters derived from the amino acids methionine, phenylalanine, threonine and the BCAAs. The importance of these amino acids for the cheese flavour is a combination of their abundance, their conversion rates, and the odour threshold of the compounds derived from them.

6.8 NATURAL BIODIVERSITY OF STRAINS IN DAIRY PRODUCTIONS

The robustness of the starter culture during cheese fermentation is enhanced by the presence of a rich consortium of microbes. Natural starters are consortia of microbes undoubtedly richer than selected starters. Among natural starters, natural whey starters (NWSs) are the most common cultures currently used to produce different varieties of cheeses. Undefined NWS are typically used for Italian cooked, long-ripened, extra-hard, raw milk cheeses, such as Parmigiano Reggiano and Grana Padano. Together with raw milk microbiota, NWSs are responsible for most cheese characteristics (Smid et al., 2014). The microbial ecology of the two cheese varieties is based on a complex interaction among starter LAB (SLAB) and nonstarter LAB (NSLAB), which are characterised by their different abilities to grow in a changing substrate. The dynamics of SLAB, which mainly arise from NWS, and NSLAB, which mainly arise from raw milk, determine the characteristics of Parmigiano Reggiano and Grana Padano cheeses. The microbiological and chemical properties of the ripened cheese are dependent on cheese-making process variables, as these variables may affect microbial growth. The microbiota of raw milk is affected by specific milk treatments, from milking to the filling of the cheese milk vat. The microbiota of NWS and the microbial dynamics from curd to ripened cheese contribute to microbial functionality in cheese.

When used as adjunct starters, NSLAB contribute positively with a more intense and rapid secondary proteolysis through the release of free amino acids and accelerate cheese ripening (Lane and Fox, 1996; Lynch et al., 1996).

Some NSLAB have also been reported to possess key enzymes for cheese flavour formation through the catabolism of free amino acids. Other authors observed that NSLAB used as adjunct starters in cheddar cheese manufacture are involved in the release of amino acids and peptides (McSweeney et al.,1994).

Multiple interactions between *S. thermophilus*, *Lb. helveticus* and *Lb. delbrueckii* have been described during cheese making in a hard cooked cheese (Charlet et al., 2009). There was antagonism between *Lb. helveticus* and *Lb. delbrueckii*. The lactobacilli had a positive effect on *S. thermophilus*, which was reciprocal for *Lb. helveticus*. *Lb. helveticus* had a negative effect on *S. thermophilus* cultivability. And the combination of *S. thermophilus* inoculated in large quantities and *Lb. helveticus* strain H2 had a negative effect on the growth of the *Lb. delbrueckii* strain D2. While the positive effect of *Lb. delbrueckii* on *S. thermophilus* interactions in milk has been already described (Rajagopal and Sandine, 1990; Courtin et al., 2002; Herve-Jimenez et al., 2009; Sieuwerts et al., 2010; Thevenard et al., 2011; Sasaki et al., 2014), other interactions have been only recently made known. These interactions are of major importance for the growth kinetics of streptococci and thermophilic lactobacilli during cheese-making (Charlet et al., 2009).

Sanna et al. (2005) reported a negative interaction between *S. thermophilus* and thermophilic lactobacilli about folate production; in particular, the low amount of folate produced by mixed culture (compared to single culture) in fermented goat milk could be due to some inhibitory effect of the lactobacilli toward *S. thermophilus*.

In Pecorino Sardo (Madrau et al., 2006) and Pecorino Romano PDO (Mangia et al., 2011) cheeses, traditionally made from natural starter cultures (whey and 'scotta'), when produced with autochthonous and selected strains, the cheeses resulted in better technological and nutritional characteristics.

The evolution of different species during ripening of Fiore Sardo showed that, when present, *Lb. paracasei/casei* persists and dominates the microbiota of the cheese in the last period of ripening, suggesting that this species, more resistant to the constraints of the mature cheese, could be involved in proteolysis and in other enzymatic processes occurring during cheese ripening, according to Mangia et al. (2008). The same authors (Mangia et al., 2016) reported the persistence of *Lb. paracasei*, as well as of other LAB, from milk to cheese in Casizolu pasta filata cheese qualifying the raw milk as a reservoir of microbial biodiversity. In contrast, the stretching step typical of 'pasta filata' cheese, such as Ragusano, induced a simplification of the raw milk profiles, allowing the persistence of only some predominant species, such as *S. thermophilus, Lb. delbrueckii* subsp. *lactis, Lc. lactis*, and *S. macedonicus*.

In Roncal (Ortigosa et al., 2006) and Fiore Sardo (Mangia et al., 2008) cheeses, both made from ewe's milk, a possible interaction between mesophilic lactococci and lactobacilli have been reported. In Fiore Sardo, in particular, the synergistic effect appeared particularly evident for the growth rates and seemed to be more relevant for rods and for *Lb. plantarum* in particular.

It was shown that the viable numbers of *Lc. lactis* ssp. *lactis* (either inoculated alone or in co-culture with a *Lactobacillus* strain) decreased gradually until the end of ripening. This implicates autolysis, resulting in the release of intracellular enzymes contributing to free AA release (Kieronczyk et al. 2003, Gatti et al., 2014).

Different strains could influence each other in formation of flavour components (Ayad et al., 2000). The combination of *Lc. lactis* B1157 and SK110 strains resulted in the formation of high levels of 3-methylbutanal. In SK110, a highly proteolytic strain from industrial origin, the complete pathway from casein via leucine to 3-methylbutanal cannot proceed due to the lack of an enzyme. *Lc. lactis* strain B1157, on the other hand, is a non-proteolytic wild strain and thus unable to produce amino acids as substrate for the transamination and decarboxylation reactions. When cultivated together, the B1157 and SK110 strains complemented each other with regard to their enzyme activities, resulting in a high production levels of 3-methylbutanal. This proto-cooperation between strains offers new possibilities for the construction of tailor-made starter cultures because not all the desired enzyme activities in a certain flavour pathway leading to flavour needs to be present in one strain.

6.9 CONCLUSIONS

LAB isolated from natural niches and dairy products have a larger potential for exploitation, thanks to the variability of phenotypes, differences in amino acid requirements, and arrays of enzymes coded in their genomes. Their application can allow production of a wide range of flavour components and flavour profiles. This natural biodiversity could offer further potential to produce new types of cheese and dairy products.

References

Arsköld E, Lohmeier-Vogel E, Cao R, Roos S, Rådström P, van Niel EW. 2008. Phosphoketolase pathway dominates in *Lactobacillus reuteri* ATCC 55730 containing dual pathways for glycolysis. *J Bacteriol* 190(1):206–212.

Atiles MW, Dudley EG, Steele JL. 2000. Gene cloning, sequencing, and inactivation of the branched-chain aminotransferase of *Lactococcus lactis* LM0230. *Appl Environ Microbiol* 66(6):2325–29.

Axelsson LT, Ahrné S. 2000. Lactic acid bacteria. In: Priest F, Goodfellow M (Eds). *Applied microbial systematics*. New York: Springer, pp 367–88.

Ayad EHE, Verheul A, Wouters JTM, Smit G. 2000. Application of wild starter cultures for flavour development in pilot plant cheese making. *Int Dairy J* 10:169–79.

Bandell M, Lhotte ME, Marty-Teysset C, Veyrat A, Prevost H, Dartois V, Divies C, Konings WN, Lolkema JS. 1998. Mechanism of the citrate transporters in carbohydrate and citrate cometabolism in Lactococcus and Leuconostoc species. *Appl Environ Microbiol* 64:1594–1600.

Banks JM, Yvon M, Gripon JC, de la Fuente MA, Brechany EY, Williams AG, Muir DD. 2001. Enhancement of amino acid catabolism in Cheddar cheese using alpha-ketoglutarate: Amino acid degradation in relation to volatile compounds and aroma character. *Int Dairy J* 11:235–43.

Bosset JO, Gauch R. 1993. Comparison of the volatile flavour compounds of six European 'AOC' cheeses by using a new dynamic headspace GC–MS method. *Int Dairy J* 3:359–77.

Broadbent JR, Gummalla S, Hughes JE, Johnson ME, Rankin SA, Drake MA. 2004. Overexpression of *Lactobacillus casei* d-hydroxyisocaproic acid dehydrogenase in cheddar cheese. *Appl Environ Microbiol* 70:4814–20.

Candioti, M, Bergamini C, Palma S, Brusetti M, Mainardi C, Salazar C. 2010. Characterization of proteolysis profile of Argentinean sheep cheese made by two different production methods. *J Sci Food Agric* 90:36–42.

Charbonnel P, Lamarque M, Piard J, Gilbert C, Juillard V, Atlan D. 2003. Diversity of oligopeptide transport specificity in *Lactococcus lactis* species – A tool to unravel the role of OppA in uptake specificity. *J Biol Chem* 278:14832–40.

Charlet M, Duboz G, Faurie F, Le Quéré J-L, Berthier F. 2009. Multiple interactions between *Streptococcus thermophilus*, *Lactobacillus helveticus* and *Lactobacillus delbrueckii* strongly affect their growth kinetics during the making of hard cooked cheeses. *Int J Food Microbiol* 131:10–19.

Chaves-López C, Serio A, Paparella A, Martuscelli M, Corsetti A, Tofalo R, Suzzi G. 2014. Impact of microbial cultures on proteolysis and release of bioactive peptides in fermented milk. *Food Microbiol* 42:117–21.

Christensen JE, Broadbent JR, Steele JL. 2003. Hydrolysis of casein-derived peptides alpha(S1)-casein(f1-9) and beta-casein(f193-209) by *Lactobacillus helveticus* peptidase deletion mutants indicates the presence of a previously undetected endopeptidase. *Appl Environ Microbiol* 69(2):1283–6.

Christensen JE, Dudley EG, Pederson JA, Steele JL. 1999. Peptidases and amino acid catabolism in lactic acid bacteria. *Ant Leeuwenhoek* 76:217–46.

Courtin P, Monnet V, Rul F. 2002. Cell-wall proteinases PrtS and PrtB have a different role in *Streptococcus thermophilus/Lactobacillus bulgaricus* mixed cultures in milk. *Microbiology* 148(Pt 11):3413–21.

Deguchi Y, Morishita T 1992. Nutritional requirements in multiple auxotrophic lactic acid bacteria; genetic lesions affecting amino acid biosynthetic pathways in *Lactococcus lactis, Enterococcus faecalis* and *Pediococcus acidilactici*. *Biosc Biotechn Biochem* 56:913–918.

Dellaglio F Felis GE. 2005. Taxonomy of Lactobacilli and Bifidobacteria. In: Tannock, GW (Ed). *Probiotics and prebiotics. Scientific aspects*. Norfolk: Caister Academic Press, pp. 25–49.

Detmers FJ, Kunji ER, Lanfermeijer FC, Poolman B, Konings WN. 1998. Kinetics and specificity of peptide uptake by the oligopeptide transport system of *Lactococcus lactis*. *Biochemistry* 37(47):16671–79.

Detmers FJM, Lanfermeijer FC, Poolman B. 2001. Peptides and ATP-binding cassette peptide transporters. *Res Microbiol* 152(3-4):245–58.

Deutsch SM, Molle D, Gagnaire V, Piot M, Atlan D, Lortal S. 2000. Hydrolysis of sequenced beta-casein peptides provides new insight into peptidase activity from thermophilic lactic acid bacteria and highlights intrinsic resistance of phosphopeptides. *Appl Environ Microbiol* 66(12):5360–67.

Dherbécourt J, Falentin H, Jardin J, Maillard MB, Baglinière F, Barloy-Hubler F, Thierry A. 2010. Identification of a secreted lipolytic esterase in *Propionibacterium freudenreichii*, a ripening process bacterium involved in Emmental cheese lipolysis. *Appl Environ Microbiol* 76(4):1181–8.

Díaz-Muñiz I, Banavara DS, Budinich MF, Rankin SA, Dudley EG, Steele JL. 2006. *Lactobacillus casei* metabolic potential to utilize citrate as an energy source in ripening cheese: a bioinformatics approach. *J Appl Microbiol* 101:872–82.

Di Cagno R, De Pasquale I, De Angelis M, Calasso MP, Buchin S, Gobbetti M. 2010. Contribution of selected adjunct or attenuated non-starter lactobacilli to ripening of Italian caciotta cheese. *Austr J Dairy Technol* 65:188–91.

Do Carmo AP, Borges AC, de Moraes CA, de Carvalho AF. 2012. Unraveling genomics of lactic acid bacteria and flavour formation in dairy products. In: Perez Campos AI, Leon Mena A (Eds). Lactobacillus. Classification. Uses and health implications. Hauppauge, NY: Nova Biomedical, pp 203–22.

Do Carmo AP, da Silva DF, de Oliveira MNV, Borges AC, de Carvalho AF, de Moraes CA. 2011. Genes involved in protein metabolism of the probiotic lactic acid bacterium *Lactobacillus delbrueckii* UVFH2b20. *Beneficial Microbes* 2:27–38.

Doeven MK, Kok J, Poolman B. 2005. Specificity and selectivity determinants of peptide transport in *Lactococcus lactis* and other microorganisms. *Mol Microbiol* 57(3):640–49.

Dudley E, Steele J. 2001. *Lactococcus lactis* LM0230 contains a single aminotransferase involved in aspartate biosynthesis, which is essential for growth in milk. *Microbiol* 147(Pt 1):215–24.

Dunn HC, Lindsay RC. 1985. Evaluation of the role of microbial Strecker-derived aroma compounds in unclean-type flavors of Cheddar cheese. *J Dairy Sci* 68:2859–74.

Engels WJM, Alting AC, Arntz MMTG, Gruppen H, Voragen AGJ, Smit G, Visser S. 2000. Partial purification and characterization of two aminotransferases from *Lactococcus lactis* subsp. *cremoris* B78 involved in the catabolism of methionine and branched-chain amino acids. *Int Dairy J* 10:443–452.

Fernandez M, Kleerebezem M, Kuipers OP, Siezen RJ, Van Kranenburg R. 2002. Regulation of the metC-cysK operon, involved in sulfur metabolism in *Lactococcus lactis*. *J Bacteriol* 184:82–90.

Ferrario C, Duranti S, Milani C, Mancabelli L, Lugli GA, Turroni F, Mangifesta M, Viappiani A, Ossiprandi MC, van Sinderen D, Ventura M. 2015. Exploring amino acid auxotrophy in *Bifidobacterium bifidum* PRL2010. *Front Microbiol* 6:1331.

Fox PF. 1989. Proteolysis during cheese manufacture and ripening. *J Dairy Sci* 72(6):1379–14.

Fox PF (Ed). 1993. *Cheese: chemistry, physics and microbiology*. London: Chapman & Hall.

Fox PF, Wallace JM, Morgan S, Lynch CM, Niland EJ, Tobin J. 1996. Acceleration of cheese ripening. *Ant Leeuwenhoek* 70:271–97.

Fukuda S, Toh H, Taylor TD, Ohno H, Hattori M. 2012. Acetate-producing bifidobacteria protect the host from enteropathogenic infection via carbohydrate transporters. *Gut Microbes* 3(5):449–54.

Fulyani F, Schuurman-Wolters GK, Slotboom DJ, Poolman B. 2015. Relative rates of amino acid import via the ABC Transporter GlnPQ determine the growth performance of *Lactococcus lactis*. *J Bacteriol* 198(3):477–85.

Gagnaire V, Piot M, Camier B, Vissers JP, Jan G, Léonil J. 2004. Survey of bacterial proteins released in cheese: a proteomic approach. *Int J Food Microbiol* 94(2):185–201.

Ganesan B, Weimer BC. 2004. Role of aminotransferase IlvE in production of branched-chain fatty acids by *Lactococcus lactis* subsp. *lactis*. *Appl Environ Microbiol* 70:638–41.

Gatti M, Bottari B, Lazzi C, Neviani E, Mucchetti G. 2014. Microbial evolution in raw-milk, long-ripened cheeses produced using undefined natural whey starters. *J Dairy Sci* 97:573–91.

Grossiord BP, Luesink EJ, Vaughan EE, Arnaud A, de Vos WM. 2003. Characterization, expression, and mutation of the *Lactococcus lactis* galPMKTE genes, involved in galactose utilization via the Leloir pathway. *J Bacteriol* 185(3):870–78.

Guédon E, Serror P, Ehrlich SD, Renault P, Delorme C. 2001. Pleiotropic transcriptional repressor CodY senses the intracellular pool of branched-chain amino acids in *Lactococcus lactis*. *Mol Microbiol* 40(5):1227–39.

Hammes WP, Hertel C. 2003. The genera Lactobacillus and Carnobacterium. In: Dworkin M. *The prokaryotes: an evolving electronic resource for the microbiological community*. New York: Springer-Verlag, pp 320–403.

Hayaloglu AA McSweeney PLH. 2015. Primary biochemical events during cheese ripening, Chapter 7. In: Ozer BH Akdemir-Evrendilek G (Eds). *Dairy microbiology and biochemistry: recent developments*. Boca Raton: CRC Press, pp 134–39.

Herve-Jimenez L, Guillouard I, Guedon E, Boudebbouze S, Hols P, Monnet V, Maguin E, Rul F. 2009. Postgenomic analysis of *Streptococcus thermophilus* cocultivated in milk with *Lactobacillus delbrueckii* subsp. *bulgaricus*: involvement of nitrogen, purine, and iron metabolism. *Appl Environ Microbiol* 75:2062–2073.

Kang TS, Korber DR, Tanaka T. 2013a. Regulation of dual glycolytic pathways for fructose metabolism in heterofermentative *Lactobacillus panis* PM1. *Appl Environ Microbiol* 79(24):7818–26.

Kang TS, Korber DR, Tanaka T. 2013b. Contributions of citrate in redox potential maintenance and ATP production: metabolic pathways and their regulation in *Lactobacillus panis* PM1. *Appl Microbiol Biotechnol* 97(19):8693–703.

Kieronczyk A, Skeie S, Langsrud T, Yvon M. 2003. Cooperation between *Lactococcus lactis* and nonstarter lactobacilli in the formation of cheese aroma from amino acids. *Appl Environ Microbiol* 69:734–39.

Kimoto H, Nomura M, Suzuki I. 1999. Growth energetics of *Lactococcus lactis* subsp. *lactis* biovar *diacetylactis* in cometabolism of citrate and glucose. *Int Dairy J* 9(12):857–63.

Klaenhammer TR, Azcarate-Peril MA, Altermann E, Barrangou R. 2007. Influence of the dairy environment on gene expression and substrate utilization in lactic acid bacteria. *J Nutr* 137(3 Suppl 2):748S–50S.

Kunji ERS, Mierau I, Hagting A, Poolman B, Konings WN. 1996. The proteolytic systems of lactic acid bacteria. *Ant Leeuwenhoek* 70:187–221.

Lamarque M, Charbonnel P, Aubel D, Piard J-C, le Atlan D, Juillard V. 2004. A multifunction ABC transporter (Opt) contributes to diversity of peptide uptake specificity within the genus Lactococcus. *J Bacteriology* 186(19):6492–6500.

Lane CN, Fox PF. 1996. Contribution of starter and adjunct lactobacilli to proteolysis in Cheddar cheese during ripening. *Int Dairy J* 6:728–51.

Larrouture C, Ardaillon V, Pepin M, Montel MC. 2000. Ability of meat starter cultures to catabolize leucine and evaluation of the degradation products by using an HPLC method. *Food Microbiology* 17:563–70.

Law BA, Wigmore A. 1983. Accelerated ripening of Cheddar cheese with a commercial proteinase and intracellular enzymes from starter streptococci. *J Dairy Science* 50(5):519–25.

Lemieux L,Simard RE. 1992. Bitter flavour in dairy products. II. A review of bitter peptides from caseins: their formation, isolation and identification, structure masking and inhibition. *Lait* 72:335–82.

Licitra G, Carpino S. 2014. The microfloras and sensory profiles of selected protected designation of origin Italian cheeses. *Microbiol Spectr* 2(1):CM-0007-2012.

Liu M, Nauta A, Francke C, Siezen RJ. 2008. Comparative genomics of enzymes in flavor-forming pathways from amino acids in lactic acid bacteria. *Appl Environ Microbiol* 74:4590–600.

Lorca G, Reddy L, Nguyen A, Sun EI, Tseng J, Yen M-R, Saier MHJr. 2010. Lactic acid bacteria: comparative genomic analyses of transport systems. In: Mozzi F, Raya RR, Vignolo GM (Eds). *Biotechnology of lactic acid bacteria. Novel applications*. Ames, IA: Wiley Blackwell, pp 73–87.

Lu WW, Wang Y, Wang T, Kong J. 2015. The global regulator CodY in *Streptococcus thermophilus* controls the metabolic network for escalating growth in the milk environment. *Appl Environ Microbiol* 81(7):2349–58.

Lynch CM, McSweeney PLH, Fox PF, Cogan TM, Drinan FD. 1996. Manufacture of Cheddar cheese with and without adjunct lactobacilli under controlled microbiological conditions. *Int Dairy J* 6:851–67.

Madrau MA, Mangia NP, Murgia MA, Sanna MG, Garau G, Leccis L, Caredda M, Deiana P. 2006. Employment of autochthonous microflora in Pecorino Sardo cheese manufacturing and evolution of physicochemical parameters during ripening. *Int Dairy J* 16:876–85.

Mangia NP, Fancello F, Deiana P. 2016. Microbiological characterization using combined culture dependent and independent approaches of Casizolu pasta filata cheese. *J Appl Microbiol* 120:329–45.

Mangia NP, Murgia MA, Garau G, Deiana P. 2011. Microbiological and physicochemical properties of Pecorino Romano cheese produced with a selected starter culture. *J Agric Sci Technol* 13:585–600.

Mangia NP, Murgia MA, Garau G, Sanna MG, Deiana P. 2008. Influence of selected LAB cultures on the evolution of free amino acids, free fatty acids and Fiore sardo cheese microflora during the ripening. *Food Microbiol* 25:366–377.

Marilley L, Casey MG. 2004. Flavours of cheese products: metabolic pathways, analytical tools and identification of producing strains. *Int J Food Microbiol* 90(2):139–59.

McSweeney PLH. 2004. Biochemistry and cheese ripening. *Int J Dairy Technol* 57:27–144.

McSweeney PLH, Sousa MJ. 2000. Biochemical pathways for the production of flavour compounds in cheeses during ripening. *Lait* 80:293–324.

McSweeney PLH, Wash EM, Fox PF, Cogan TM, Drinan FD, Castelo-Gonzalez M. 1994. A procedure for the manufacture of Cheddar cheese under controlled bacteriological conditions and the effect of adjunct lactobacilli on cheese quality. *Irish J Agric Food Res* 33:183–92.

Moio L, Langlois D, Etievant PX, Addeo F. 1993. Powerful odorants in water buffalo and bovine Mozzarella cheese by use of extract dilution sniffing analysis. *It J Food Sci* 3:227–37.

Moio L, Piombino P, Addeo F. 2000. Odour-impact compounds of Gorgonzola cheese. *J Dairy Res* 67:273–85.

Monnet V, Nardi M, Chopin A, Chopin MC, Gripon J-C. 1994. Biochemical and genetic characterization of PepF, an oligopeptidase from *Lactococcus lactis*. *J Biol Chem* 269:32070–73.

Nardi M, Fiez Vandal C, Tailliez P, Monnet V. 2002. The EstA esterase is responsible for the main capacity of *Lactococcus lactis* to synthesize short chain fatty acid esters in vitro. *J Appl Microbiol* 93:994–1002.

Nierop Groot MN, De Bont JAM. 1998. Conversion of phenylalanine to benzaldehyde initiated by an aminotransferase in *Lactobacillus plantarum*. *Appl Environ Microbiol* 64:3009–3013.

Oommen BS, McMahon DJ, Oberg CJ, Broadbent JR, Strickland M. 2002. Proteolytic specificity of *Lactobacillus delbrueckii* subsp. *bulgaricus:* influences functional properties of Mozzarella cheese. *J Dairy Sci* 85:2750–58.

Ortigosa M, Arizcun C, Irigoyen A, Oneca M, Torre P. 2006. Effect of lactobacillus adjunct cultures on the microbiological and physicochemical characteristics of Roncal-type ewes'-milk cheese. *Food Microbiol* 23:591–98.

Ozer BH, Akdemir-Evrendilek G (Eds). 2015. *Dairy microbiology and biochemistry: recent developments*. Boca Raton: CRC Press.

Pedersen TB, Vogensen FK, Ardo Y. 2016 Effect of heterofermentative lactic acid bacteria of DL-starters in initial ripening of semi-hard cheese. *Int Dairy J* 57:72–79.

Peltoniemi K, Vesanto E, Palva A. 2002. Genetic characterization of an oligopeptide transport system from *Lactobacillus delbrueckii* subsp. *bulgaricus*. *Arch Microbiol* 177:457–67.

Pereira CI, Neto DM, Capucho JC, Gião MS, Gomes AMP, Malcata FX. 2010. How three adventitious lactic acid bacteria affect proteolysis and organic acid production in model Portuguese cheeses manufactured from several milk sources and two alternative coagulants. *J Dairy Sci* 93:1335–44.

Poveda JM, Chicon R, Cabezas L. 2015. Biogenic amine content and proteolysis in Manchego cheese manufactured with *Lactobacillus paracasei* subsp. *paracasei* as adjunct and other autochthonous strains as starters. *Int Dairy J* 47:94–101.

Pritchard G, Coolbear T. 1993. The physiology and biochemistry of the proteolytic system in lactic acid bacteria. *FEMS Microbiol Rev* 12:179–206.

Pudlik AM, Lolkema JS. 2012. Rerouting citrate metabolism in *Lactococcus lactis* to citrate-driven transamination. *Appl Environ Microbiol* 78(18):6665–73.

Pudlik AM, Lolkema JS. 2013. Uptake of α-ketoglutarate by citrate transporter CitP drives transamination in *Lactococcus lactis*. *Appl Environ Microbiol* 79(4):1095–101.

Qian M, Reineccius G. 2002. Identification of aroma compounds in Parmigiano-Reggiano cheese by gas chromatography/olfactometry. *J Dairy Sci* 85:1362–69.

Rajagopal SN, Sandine WE. 1990. Associative growth and proteolysis of *Streptococcus thermophilus* and *Lactobacillus bulgaricus* in skim milk. *J Dairy Sci* 73:894–99.

Randazzo CL, De Luca S, Todaro A, Restuccia C, Lanza CM, Spagna G, Caggia C. 2007. Preliminary characterization of wild lactic acid bacteria and their abilities to produce flavour compounds in ripened model cheese system. *J Appl Microbiol* 103(2):427–35.

Ruggirello M, Dolci P, Cocolin L. 2014. Detection and viability of *Lactococcus lactis* throughout cheese ripening. *PLoS One* 9(12):e114280.

Rychlik M, Bosset JO. 2001. Flavour and off-flavour compounds of Swiss Gruyere cheese. Identification of key odorants by quantitative instrumental and sensory studies. *Int Dairy J* 11:903–10.

Sadat-Mekmene L, Genay M, Atlan D, Lortal S, Gagnaire V. 2011. Original features of cell-envelope proteinases of *Lactobacillus helveticus*. *A review*. *Int J Food Microbiol* 146(1):1–13.

Saier MH Jr, Ye JJ, Klinke S, Nino E. 1996. Identification of an anaerobically induced phosphoenolpyruvate-dependent fructose-specific phosphotransferase system and evidence for the Embden-Meyerhof glycolytic pathway in the heterofermentative bacterium *Lactobacillus brevis*. *J Bacteriol* 178(1): 314–6.

Sanna MG, Mangia NP, Garau G, Murgia MA, Massa T, Franco A, Deiana P. 2005. Selection of folate-producing lactic acid bacteria for improving fermented goat milk. *Ital J Food Sci* 17:143–54.

Santiago B, MacGilvray M, Faustoferri RC, Quivey RG Jr. 2012. The branched-chain amino acid aminotransferase encoded by *ilvE* is involved in acid tolerance in *Streptococcus mutans*. *J Bacteriol* 194(8):2010–19.

Sasaki Y, Horiuchi H, Kawashima H, Mukai T, Yamamoto Y. 2014. NADH oxidase of *Streptococcus thermophilus* 1131 is required for the effective yogurt fermentation with *Lactobacillus delbrueckii* subsp. *bulgaricus* 2038. *Biosci Microbiota Food Health* 33(1):31–40.

Savijoki K, Ingmer H, Varmanen P. Proteolytic systems of lactic acid bacteria. 2006. *Appl Microbiol Biotechnol* 71:394–406.

Schmitt P, Vasseur C, Palip V, Huang DQ, Divies C, Prevost H. 1997. Diacetyl and acetoin production from the co-metabolism of citrate and xylose by *Leuconostoc mesenteroides* subsp. *mesenteroides*. *Appl Microbiol Biotechnol* 47:715–718.

Schweikhard ES, Ziegler CM. 2012. Amino acid secondary transporters: toward a common transport mechanism. *Curr Topics Membranes* 70:1–28.

Sieuwerts S, Molenaar D, van Hijum SA, Beerthuyzen M, Stevens MJ, Janssen PW, Ingham CJ, de Bok FA, de Vos WM, van Hylckama Vlieg JE. 2010. Mixed-culture transcriptome analysis reveals the molecular basis of mixed-culture growth in *Streptococcus thermophilus* and *Lactobacillus bulgaricus*. *Appl Environ Microbiol* 76:7775–84.

Simov ZI, Ivanov GY. 2005. Proteolytic activity of *Lactobacillus delbrueckii* ssp. *bulgaricus* and *Streptococcus thermophilus* in frozen-stored Kashkaval cheese. *J Ind Microbiol Biotechnol* 32(10):449–54.

Smit BA, Engels WJM, Alewijn M, Lommerse GTCA, Kippersluijs E, Wouters JTM, Smit G. 2004c. Chemical conversion of alpha-ketoacids in relation to flavor formation in fermented foods. *J Agric Food Chem* 52:1263–1268.

Smit BA, Engels WJM, Bruinsma J, Van Hylckama Vlieg JET, Wouters JTM, Smit G. 2004a. Development of a high throughput-screening method to test flavour-forming capability of anaerobic microorganisms. *J Appl Microbiol* 97:306–313.

Smit BA, Engels WJM, Wouters JTM, Smit G. 2004b. Diversity of L-leucine catabolism in various microorganisms involved in dairy fermentations, and identification of the rate-controlling step in the formation of the potent flavour component 3-methylbutanal. *Appl Microbiol Biotechnol* 64:396–402.

Smit BA, Vlieg J, Engels WJM, Meijer L, Wouters JTM, Smit G. 2005a. Identification, cloning, and characterization of a *Lactococcus lactis* branched-chain alpha-ketoacid decarboxylase involved in flavor formation. *Appl Environ Microbiol* 71:303–311.

Smit G, Smit BA, Engels WJM. 2005b. Flavour formation by lactic acid bacteria and biochemical flavour profiling of cheese products. *FEMS Microbiol Rev* 29:591–610.

Smid EJ, Erkus O, Spus M, Wolkers-Rooijackers JCM, Alexeeva S, Kleerebezem M. 2014. Functional implications of the microbial community structure of undefined mesophilic starter cultures. *Microb Cell Factories* 13:S2.

Strahinic I, Kojic M, Tolinacki M, Fira D, Topisirovic L. 2010. The presence of prtP proteinase gene in natural isolate *Lactobacillus plantarum* BGSJ3e18. *Lett Appl Microbiol* 50:43–49.

Stressler T, Ewert J, Merz M, Funk J, Claaßen W, Lutz-Wahl S, Schmidt H, Kuhn A, Fischer L. 2016. A novel glutamyl (aspartyl)-specific aminopeptidase A from *Lactobacillus delbrueckii* with promising properties for application. *PLoS One* 11(3):e0152139.

Sun Z, Harris HM, McCann A, Guo C, Argimón S, Zhang W, Yang X, Jeffery IB, Cooney JC, Kagawa TF, Liu W, Song Y, Salvetti E, Wrobel A, Rasinkangas P, Parkhill J, Rea MC, O'Sullivan O, Ritari J, Douillard FP, Paul Ross R, Yang R, Briner AE, Felis GE, de Vos WM, Barrangou R, Klaenhammer TR, Caufield PW, Cui Y, Zhang H, O'Toole PW. 2015. Expanding the biotechnology potential of lactobacilli through comparative genomics of 213 strains and associated genera. *Nat Commun* 6:8322.

Sutula J, Coulthwaite L, Verran J. 2012. Culture media for differential isolation of *Lactobacillus casei* Shirota from oral samples. *J Microbiol Methods* 90:65–71.

Tabet E, Mangia NP, Mouannes E, Hassoun G, Helal Z, Deiana P. 2016. Characterization of goat milk from Lebanese Baladi breed and his suitability for setting up a ripened cheese using a selected starter culture. *Small Ruminant Research* 40:13–17.

Tanous C, Kieronczyk A, Helinck S, Chambellon E, Yvon M. 2002. Glutamate dehydrogenase activity: A major criterion for the selection of flavour-producing lactic acid bacteria strains. *Ant Leeuwenhoek* 82:271–278.

Thevenard B, Rasoava N, Fourcassié P, Monnet V, Boyaval P, Rul F. 2011. Characterization of *Streptococcus thermophilus* two-component systems: in silico analysis, functional analysis and expression of response regulator genes in pure or mixed culture with its yogurt partner, *Lactobacillus delbrueckii* subsp. *bulgaricus*. *Int J Food Microbiol* 151:171–81.

Trip H, Mulder NL, Lolkema JS. 2013. Cloning, expression, and functional characterization of secondary amino acid transporters of *Lactococcus lactis*. *J Bacteriol* 195(2):340–50.

Turner MS, Woodberry T, Hafner LM, Giffard PM. 1999. The bspA locus of *Lactobacillus fermentum* BR11 encodes an L-cystine uptake system. *J Bacteriology* 181(7):2192–8.

Urbach G. 1993. Relations between cheese flavour and chemical composition. *Int Dairy J* 3:389–422.

Van Kranenburg R, Kleerebezem M, Van Hylckama Vlieg JET, Ursing BM, Boekhorst J, Smit BA, Ayad EHE, Smit G, Siezen RJ. 2002. Flavour formation from amino acids by lactic acid bacteria: Predictions from genome sequence analysis. *Int Dairy J* 12:111–21.

Wagner N, Tran QH, Richter H, Selzer PM, Unden G. 2005. Pyruvate fermentation by *Oenococcus oeni* and *Leuconostoc mesenteroides* and role of pyruvate dehydrogenase in anaerobic fermentation. *Appl Environ Microbiol* 71:4966–71.

Wakai T, Yamamoto N. 2013. A novel branched chain amino acids responsive transcriptional regulator, BCARR, negatively acts on the proteolytic system in *Lactobacillus helveticus*. *PLoS One* 8(10) :e75976.

Wallace JM, Fox PF. 1997. Effect of adding free amino acids to cheddar cheese curd on proteolysis, flavour and texture development. *Int Dairy J* 7:157–67.

Weimer B, Seefeldt K, Dias B. 1999. Sulfur metabolism in bacteria associated with cheese. *Ant van Leeuwenhoek* 76:247–61.

Wolf IV, Peralta GH, Candioti M, Perotti MC. 2016. The role of propionibacteria in the volatile profile of Pategrás cheeses. *Dairy Sci Technol* 96:551–67.

Yvon M, Berthelot S, Gripon JC. 1998. Adding alpha-ketoglutarate to semi-hard cheese curd highly enhances the conversion of amino acids to aroma compounds. *Int Dairy J* 8:889–98.

Yvon M, Chambellon E, Bolotin A, Roudot-Algaron F. 2000. Characterization and role of the branched-chain aminotransferase (BcaT) isolated from *Lactococcus lactis* subsp. *cremoris* NCDO 763. *Appl Environ Microbiol* 66(2):571–7.

Zaunmüller T, Eichert M, Richter H, Unden G. 2006. Variations in the energy metabolism of biotechnologically relevant heterofermentative lactic acid bacteria during growth on sugars and organic acids. *Appl Microbiol Biotechnol* 72:421–29.

Yvon M, Chambellon E, Bolotin A, Roudot-Algaron F. 2000. Characterization and role of the branched-chain aminotransferase (BcaT) isolated from *Lactococcus lactis* subsp. *cremoris* NCDO 763. *Appl Environ Microbiol* 66(2):571–7.

Zaunmüller T., Eichert M., Richter H., Unden G. 2006. Variations in the energy metabolism of biotechnologically relevant heterofermentative lactic acid bacteria during growth on sugars and organic acids. *Appl Microbiol Biotechnol* 72:421–29.

7 Growth needs and culture media for LAB and dairy-associated species

Giuseppe Blaiotta[1], Maria Aponte[2] and Palmiro Poltronieri[3]

[1]*Department of Agricultural Sciences, Division of Grape and Wine Sciences, University of Naples Federico II, Avellino, Italy*
[2]*Department of Agricultural Sciences, Division of Microbiology, University of Naples Federico II, Portici, Italy*
[3]*Institute of Sciences of food Productions (CNR-ISPA), CNR, National Research Council of Italy, Lecce, Italy*

7.1 INTRODUCTION

Lactobacillus (*Lb.*), *Leuconostoc* (*Ln.*), *Lactococcus* (*Lc.*), *Pediococcus* (*Ped.*), *Streptococcus* (*St.*) and *Enterococcus* (*Ent.*) spp. are lactic acid bacteria (LAB) genera commonly retrieved in dairy environments where they may find compounds to sustain their growth. As described by several research groups, based on genome analyses and phenotypic characterisation, each species may show specific needs for amino acids as nitrogen source, whey protein utilization, preference for *N*-acetyl neuraminic acid (NANA)-modified whey proteins, lipids and oligosaccharides, exploited to promote their growth. In addition, this chapter will discuss dairy-associated species such as Bifidobacteria.

7.2 ESTABLISHED CULTURE MEDIA FOR LACTOBACILLI

Lactobacilli have been isolated from environments and niches to which members of the genus are well adapted to grow. Apart from dairy environment treated in this book, lactobacilli may be isolated from the mammary gland, the gut of livestock from which bacteria are transferred to the environment, milk used to breastfeed newborns, and milk-derived products. Technologies in dairy manufacturing, such as thermal treatments of milk, and mild thermization during the production of curd allow a selective recovery of thermoresistant species, that is, lactobacilli, enterococci and *St. thermophilus* strains. However, the isolation of single species or strain by mixed populations in co-culture is made problematic by the presence of other species that may often outgrow and outcompete the candidate to be isolated (Lima et al., 2009; Erkus et al., 2013). Numerous selective/differential media have been proposed based on the use of selective agents (Tharmaraj and Shah, 2003). Often, selective compounds exert high inhibition

Microbiology in Dairy Processing: Challenges and Opportunities, First Edition.
Edited by Palmiro Poltronieri.

Published 2018 by John Wiley & Sons Ltd and the Institute of Food Technologists.

power on injured or stressed cells (Reid et al., 2006; Reale et al., 2015), resulting in an underestimation of the actual number of LAB within the total microflora, or in a failure in the isolation of specific species in the sample.

For instance, in the context of the cooperative research project "Fermented Beverages in East Europe (Ferbev)", of the European Commission Framework Programme 6, the working groups studied the bacterial community in Ayran productions using two types of cultures: M17 with 0.5% lactose and de Man Rogosa Sharpe (MRS), supplemented with 0.1 g/l of cycloheximide to inhibit yeasts, for enumeration of mesophilic and thermophilic LAB with incubation under anaerobiosis for 48 h at 30 °C and 42 °C (Baruzzi et al., 2011; Baruzzi et al., 2016). With similar purposes, bifidobacteria were identified in milk-fed infants and dairy products, such as fermented milks of different origin, fermented milk products, yoghurts and other dairy productions (non-LAB dairy-associated species). To this aim, there are several standard procedures, such as ISO 7889:2003 to enumerate LAB and dairy-associated species from yoghurt (Davis, 2014). Among the LAB and dairy-associated bacteria present or added to these products, the strains with probiotic characteristics have been particularly isolated and enumerated (Shah, 2000).

In the following paragraphs specific attention will be paid to lactobacilli species and their growth requirements.

7.2.1 Rogosa agar

On Rogosa agar, lactobacilli appear as medium to large white colonies. As usual, if a quantitative plating has been performed, colonies may be counted, multiplied by the dilution factor of the sample, thus obtaining the colony-forming units (CFU) per gram or millilitre. Colonies presumptively belonging to *Lactobacillus* genus must be further confirmed by microscopic and biochemical tests. Rogosa composition is reported in Table 7.1, according to marketed products and bibliography (de Man et al., 1960).

Lactococci/streptococci do not form colonies on acidified Rogosa agar (Ravula and Shah, 1998). Growth of *Streptococcus thermophilus* is prevented by adjusting pH to 5.1.

Due to the high salt content, Rogosa is not a suitable medium for the isolation of *Lactobacillus delbrueckii* subsp. *lactis* and subsp. *bulgaricus*, a species commonly associated with dairy products. Rogosa medium with lower salt concentration and pH 5.4 could be used for the determination of the total lactobacilli counts after incubation at 37 °C, which enables the discrimination of the *Lb. bulgaricus* strains by difference.

Table 7.1 Rogosa composition (per litre of purified water).

Tryptone	10.0 g/l
Yeast extract	5.0 g/l
Glucose	20.0 g/l
Sorbitan mono-oleate (Tween 80)	1.0 ml/l
Potassium Dihydrogen Phosphate	6.0 g/l
Ammonium citrate	2.0 g/l
Sodium acetate anhydrous	17.0 g/l
Magnesium sulfate	0.575 g/l
Manganous sulfate	0.12 g/l
Ferrous sulfate	0.034 g/l
Acetic acid	1.32 ml/l
pH	5.4 ± 0.2

Table 7.2 MRS composition (per litre of purified water).

Peptone	10.0g/l
Lab-Lemco	8.0g/l
Yeast extract	4.0g/l
Glucose	20.0g/l
Sorbitan mono-oleate (Tween 80)	1.0ml/l
Potassium dihydrogen phosphate	2.0g/l
Sodium acetate * 3 H_2O	5.0g/l
Tri-ammonium citrate	2.0g/l
Magnesium sulfate * 7 H_2O	0.2g/l
Manganese sulfate * 4 H_2O	0.05g/l
pH	6.2±0.2

7.2.2 MRS medium

MRS medium is usually well suited for the growth of a large number of lactobacilli, as well as other bacteria, able to tolerate acidic pH (Smid et al., 2014). Optimal time and temperature for incubation are 2–3 days at 30–37 °C for mesophilic LAB. A longer incubation should be avoided, since the viability of the colonies may thereafter decrease. MRS composition, reported in Table 7.2, is similar to that of Rogosa agar, but pH is higher.

In MRS, casein peptone, yeast extract and ammonium salts provide nitrogen. Polysorbate 80 supplies fatty acids required for the growth of lactobacilli (Corcoran et al., 2007). Manganese and magnesium are growth factors. Glucose is a universal energy and carbon source. Ammonium and sodium acetate, acetic acid and ferrous sulfate act as inhibitors of streptococci and other contaminating microorganisms and provide the low pH, well endured by lactobacilli, but not by several other organisms. Phosphate and acetate stabilize pH.

Leuconostoc mesenteroides can be enumerated in parallel on MRS supplemented with vancomycin (de Man et al., 1960; Schillinger and Holzapfel, 2011; Smid et al., 2014).

MRS agar may be acidified by adding 1.23 ml of acetic acid per litre (MRSa), as previously reported for Rogosa agar. Plates may be incubated under anaerobic conditions at two temperatures, for mesophilic and thermophilic lactobacilli counts (Galat et al., 2016).

Alternatively MRS may be supplemented with calcium carbonate (0.3–0.5%), which confers a glassy colour to the agar. Colonies of highly acidifying LAB species exhibit a clear halo. This observation can be an alternative to the use of pH indicators, which usually turn red/yellow by pH lowering.

7.2.3 Skim milk and whey agar

Sterile, autoclaved 10% reconstituted skim milk (RSM) supplemented with 2% glucose and 1% yeast extract is often used for continuous culture of LAB strains. Liquid skim milk provides an anaerobic environment, but the clotting of proteins makes handling difficult.

For whey agar preparation, milk whey needs to be defatted by refrigerated centrifugation. Skimmed milk whey is then filtered (0.22 μm) and diluted 1:2 in 1% yeast extract; finally, 5% agar stock is added to final concentration 1.5%. At this point, agar needs to be rapidly poured onto plates.

7.3 M17 MEDIUM FOR SELECTION AND ENUMERATION OF LACTOCOCCI AND STREPTOCOCCI

M17 is used for lactococci and streptococci counting and isolation. M17 composition is reported in Table 7.3. Lactose is added after autoclave treatment as a filtered-sterilized solution. Mesophilic and thermophilic cocci can be differentiated by incubation at 30 and 45 °C, respectively.

Lactococci can be discriminated by colonies' appearance on differential media, such as Reddy's agar (Reddy et al., 1972). This medium allows the differentiation of *Lactococcus lactis* subsp. *cremoris* (no arginine deiminase activity, no citrate utilization), *Lc. lactis* subsp. *lactis* (arginine deiminase activity, no citrate utilization) and *Lc. lactis* subsp. *lactis* biovar *diacetylactis* (arginine deiminase activity, citrate utilization). Moreover, supplementation with sodium chloride (6.5% vs. 4%) may allow exploitation of the higher salt tolerance of *Lc. lactis* subsp. *lactis* versus subsp. *cremoris* (Al-Zoreky and Sandine, 1991; Fernández et al., 2011).

Smid and colleagues (2014) used different media in parallel, to recover all bacterial cells from a starter culture; such an approach allowed a collection of hundreds of cultures, that collectively represented the entire diversity in the sample (Erkus et al., 2013).

Table 7.3 M17 composition (per litre of purified water).

Tryptone	5.0 g/l
Soya peptone	5.0 g/l
Meat digest	5.0 g/l
Yeast extract	2.5 g/l
Ascorbic acid	0.5 g/l
Magnesium sulfate anhydrous	0.25 g/l
Disodium glycerophosphate	19.0 g/l
Lactose 10% (filtered solution)	10 g/l

7.3.1 St. thermophilus *agar*

St. thermophilus (ST) agar is a non-selective medium for the culture of thermophilic streptococci. In Table 7.4 the composition of ST is compared with that of Elliker broth (Elliker et al., 1956), and both have been used to isolate cocci (Gemelas et al., 2013).

Table 7.4 ST agar composition (per litre of purified water) and comparison with Elliker broth.

ST broth		Elliker	
Casein enzyme hydrolysate	20.0 g/l	Tryptone	20.0 g/l
Dextrose	5.0 g/l	Yeast extract	5.0 g/l
Lactose	5.0 g/l	Dextrose	5.0 g/l
Ascorbic acid	0.5 g/l	Gelatine	2.5 g/l
pH	6.8	Lactose	5.0 g/l
		Sucrose	5.0 g/l
		Ascorbic acid	0.5 g/l
		Sodium acetate	1.5 g/l
		Sodium chloride	4.0 g/l

ST agar at 37 °C for 24 h and in an aerobic condition is selective for *St. thermophilus* (Dave and Shah, 1996). *St. thermophilus* forms tiny colonies of around 0.1–0.5 mm diameter, on ST agar at 37 °C under aerobic incubation (Dave and Shah, 1996).

7.4 SELECTIVE MEDIA FOR LACTOBACILLI

Several authors described selective media supposedly able to quantitatively isolate strains in the *Lactobacillus casei* group reliably, regardless of their physiological conditions (Desai et al., 2006), from milk, milk curd and cheese (Ravula and Shah, 1998), such as Lc agar (Table 7.5) (Champagne et al., 1997).

Table 7.5 Lc agar composition (per litre of purified water).

Bacteriological peptone	10.0 g/l
Yeast extract	1.0 g/l
Lab-Lemco	4.0 g/l
KH_2PO_4	2.0 g/l
Sodium acetate	3.0 g/l
Tri-ammonium citrate	1.0 g/l
$MgSO_4$	0.2 g/l
$MnSO_4$	0.05 g/l
Acid casein hydrolysate	1.0 g/l
Tween 80	1.0 ml/l

LAB strains in a mixed population or outnumbered by other more abundant species may need some selective agents to support their growth. Among these agents, a unique carbohydrate source is often used to distinguish LAB at species or strain level.

7.4.1 MRS vancomycin

Many LAB are intrinsically resistant to vancomycin. Genera *Leuconostoc, Weissella* and *Pediococcus*, as well as most *Lactobacillus* species are resistant. Many thermophilic lactobacilli, including *Lactobacillus acidophilus* and *Lb. delbrueckii*, are sensitive (Vinderola and Reinheimer, 2000).

Gram negative bacteria are usually intrinsically resistant to vancomycin because of their outer membrane, which is impermeable to large glycopeptide molecules (Quintiliani et al., 1995). Additionally, a mechanism of resistance to vancomycin involves the alteration of the terminal amino acid residues of the peptidoglycan (under normal conditions the peptide is D-alanyl-D-alanine) to which vancomycin binds. The D-alanyl-D-lactate variation results in the loss of one hydrogen-bonding interaction (in respect to five bonds for D-alanyl-D-alanine) possibly between vancomycin and the peptide. The loss of one interaction results in a 1000-fold decrease in affinity. The D-alanyl-D-serine variation causes a six-fold loss of affinity between vancomycin and the peptide, likely due to steric hindrance.

In enterococci, such modification seems to be due to the expression of an enzyme that alters the terminal residue. To date, three main resistance variants have been characterised within resistant *Enterococcus faecium* and *faecalis* populations.

In 2011, a variant of vancomycin has been tested at the Scripps Research Institute that binds to the resistant D-lactic acid variation in vancomycin-resistant bacterial cell walls (Xie et al., 2011) and binds to the original target (vancomycin-susceptible bacteria).

7.4.2 Additional selective agents

Other selective agents that can be included in the MRS medium are cycloheximide, nalidixic acid, neomycin sulfate, paromomycin sulfate, bile salts, lithium chloride, metronidazole, sodium propionate, sodium lactate, 2,3,5-triphenyltetrazolium chloride (TTC) (Sakai et al., 2010; Sutula et al., 2012; Colombo et al., 2014). Moreover, selectivity may be increased by replacing glucose and Lab-Lemco in the MRS composition with a different carbon source, such as lactitole, rhamnose or maltose (Sakai et al., 2010). For instance, Sakai and co-workers (2010) selectively enumerated *Lb. casei* Shirota (LcS) on modified-rhamnose-2,3,5-triphenyltetrazolium chloride-LBS-vancomycin agar (M-RTLV agar; Table 7.6), by exploiting the selective use of L-rhamnose as a carbon source. In detail, on M-RTLV agar, *Lb. casei* and *Lactobacillus paracasei* formed red colonies, while *Lactobacillus rhamnosus* formed either pink-toned colonies or white colonies with a red spot.

MRS-S (pH 5.7) is MRS supplemented with 0.1% sorbic acid and is used as general selective medium for *Lactobacillus*, *Leuconostoc* and *Weissella* spp. (Reuter, 1985).

All-purpose Tween (APT) medium (pH 6.7) has been proposed for the enumeration and the cultivation of heterofermentative LAB including lactobacilli, leuconostocs and lactic streptococci, as well as other microorganisms with high requirements for thiamine in meat products, canned foods, fruit juices and other foodstuffs (Evans and Niven, 1951).

TPPY (tryptose proteose peptone yeast extract-eriochrome T agar) may be used for counting of *St. thermophilus*, *Lb. delbrueckii*. ssp. *bulgaricus* and *Lb. acidophilus*. *St. thermophilus* colonies appear circular or semi-circular, convex, opaque, white-violet, and often have a dark centre. *Lb. delbrueckii* ssp. *bulgaricus* colonies are flat, transparent, of undefined shape, with an irregular edge, while *Lb. acidophilus* colonies are small, white to violet, slightly elevated and somewhat fuzzy (Braquart, 1981).

Table 7.6 M-RTLV composition.

Trypticase peptone	10.0 g/l
Yeast extract	5.0 g/l
Potassium phosphate monobasic	6.0 g/l
Tri-ammonium citrate	2.0 g/l
Tween 80	1.0 g/l
Sodium acetate trihydrate	25.0 g/l
Vancomycin hydrochloride	1 mg/l
Metronidazole	10 mg/l
2,3,5 triphenyltetrazolium chloride (TTC)	30 mg/l
L-Rhamnose	20 g/l
Salt solution	5 ml
Distilled water for salts	100 ml
$MgSO_4$	11.5 g/l
$(Fe^{3+})_2(SO_4)_3$	0.68 g/l
$MnSO_4$	2.4 g/l
pH	6.0

L-S differential (LSD) agar is a selective medium that allows the differentiation of lactobacilli and streptococci species in yoghurt. The reduction of 2,3,5-TTC, together with the casein reaction, allows discrimination between lactobacilli and streptococci by means of colony morphology. LSD agar is composed of peptone from casein, peptone from soya, beef extract, yeast extract, dextrose, sodium chloride, 1-cysteine-hydrochloride, 2, 3, 5-TTC, milk powder, agar, and distilled or deionized water. Brilliant red colonies with clear halos are considered to be lactic streptococci, while pink colonies with a red centre are likely to be reported to *Lb. delbrueckii* ssp. *bulgaricus*. *Lb. acidophilus* and other lactobacilli from milk sources appear as red colonies with an opaque halo (Eloy and Lacrosse, 1976).

Briggs agar with addition of tomato juice is frequently used for cultivation of lactobacilli from dairy products (Mitsuoka et al., 1965).

M-RTLV composition, based on MRS, including TTC, vancomycin and L-rhamnose is reported in Table 7.6.

7.4.3 MRSV plus selective agents for *Lb. casei* group enumeration

MRS agar supplemented with nalidixic acid, bile, lithium chloride, metronidazole, sodium propionate and vancomycin may be used for the selective isolation of *Lactobacillus paracasei*. Based on the tolerance of members of the *Lb. casei* group species to TTC, Sutula and co-workers (2012) developed a differential medium, LcS Select, for the selective isolation and counting of the probiotic *Lb. casei* Shirota. In detail, MRS agar was supplemented with bromophenol blue as pH indicator, vancomycin and reducing agent L-cysteine hydrochloride for differential colony morphology development. *Lb. casei* Shirota cultured on the medium produced distinctive colony morphologies. The viable count of LcS colonies was correlated with those isolated on lactitol-LBS-vancomycin agar (LLV) (Sakai et al., 2010).

7.4.4 MRS-salicin, MRS-sorbitol, MRS-ribose, MRS gluconate agar

MRS, prepared without sugars according to Lankaputhra and Shah (1996), is supplemented with filter-sterilized solutions of salicin, sorbitol, ribose or gluconate (10%) after autoclaving (Dave and Shah, 1996). MRS-salicin or MRS-sorbitol agar can be used for the selective counting of *Lb. acidophilus,* provided that *Lb. casei* is not present in the product (in which case a total count for both species is obtained (Coeuret et al., 2003).

7.4.5 MRS-clindamycin-ciprofloxacin agar

MRS-clindamycin-ciprofloxacin (MRS-CC) agar proved to have a relatively good selectivity for *Lb. acidophilus*; however, it also promotes the growth of *Lb. casei* strains. MRS-CC agar can only be used as a selective medium for the enumeration of *Lb. acidophilus* if *Lb. casei* is not present in the sample at levels similar to or exceeding those of *Lb. acidophilus*.

According to Van de Casteele et al. (2006), MRS-CC agar has to be considered the elective medium for the selective enumeration of commercial probiotic *Lb. acidophilus* strains La-145 and Lafti L10. The International Organization for Standardization (ISO, 20128/IDF 192: 2006) (ISO, 2006) has recommended MRS-CC agar for the enumeration of presumptive *Lb. acidophilus* in milk products, including fermented and non-fermented milks, milk powders and infant formula, where presumptive *Lb. acidophilus* strains are present together with other LAB and bifidobacteria.

7.4.6 MMV medium for *Lb. casei* group enumeration

MMV, a basal medium with a mix containing all the components of MRS, except glucose and meat extract and supplemented with maltose as the only carbohydrate source, L-cysteine, vancomycin, and bromocresol purple is suggested for the isolation and enumeration of *Lb. casei* group members (Ludwig et al., 2009) (see Table 7.7).

The use of maltose as carbohydrate source is widely recognised for several lactobacilli, including the *Lb. casei* sharing the ability to metabolize maltose, thus acidifying the medium (Gänzle and Follador, 2012).

7.4.7 MRS containing fructose (MRSF)

Tabasco et al. (2007) discovered that a non-antibiotic medium based on the acidified MRS medium recommended for the enumeration of yoghurt characteristic microorganisms (ISO, 2002) was not selective for enumeration of *Lb. delbrueckii* subsp. *bulgaricus* since it also allowed growth of *Lb. acidophilus*, *Lb. paracasei* subsp. *paracasei*, and *Bifidobacterium (Bif.) lactis*. For this purpose, the authors proposed the use of MRSF: MRS without glucose or meat extract, supplemented with fructose, casein acid hydrolysate and cysteine. The medium allowed a morphological differentiation between lenticular colonies with 1–2 mm diameter of *Lb. delbrueckii* subsp. *bulgaricus* and fluffy colonies with 2–3 mm diameter of *Lb. acidophilus*. Moreover, the replacement of glucose by fructose coupled to the incubation at 45 °C allows exclusion of *Lb. paracasei* subsp. *paracasei* and *Bif. lactis* (Tabasco et al., 2007).

Table 7.7 MMV agar composition.

Peptone	10.0 g/l
Yeast extract	4.0 g/l
Dipotassium hydrogen phosphate	2.0 g/l
Sodium acetate $3H_2O$	5.0 g/l
Tri-ammonium citrate	2.0 g/l
Magnesium sulfate $7H_2O$	0.20 g/l
Manganese sulfate $4H_2O$	0.05 g/l
Tween 80	1.0 ml
Maltose	10.0 g/l
L-Cysteine HCl	0.5 g/l
pH	6.9–7.0
Bromocresol purple 1.6% (w/v, ethanol solution)	3.0 ml
Vancomycin 10 g/l (1 μg/ml final)	1 ml

7.4.8 mMRS-BPB

Lee and Lee (2008) developed a modified MRS medium (mMRS-BPB), namely MRS containing cysteine and bromophenol blue (pH 6.5), which, incubated in anaerobiosis, enables the differential enumeration of several species, including *Lb. plantarum, Lb. casei, Lb. paracasei* and *Lb. rhamnosus*, in fermented milk. According to Ricciardi et al. (2015), who tested the procedure on 461 LAB strains belonging to eight genera and 35 species, the medium correctly discriminated members of the *Lb. plantarum* and *Lb. casei* groups from other LAB species (Ricciardi et al., 2015).

7.4.9 MRS-NNLP agar and chromogenic agars for complex communities

MRS supplemented with nalidixic acid, neomycin sulfate, lithium chloride and paromomycin sulfate (MRS-NNLP) is useful to differentiate and enumerate LAB species in complex communities. Several bacteriological media have been evaluated for the selective enumeration of *Lb. delbrueckii* ssp. *bulgaricus, St. thermophilus, Lb. casei, Lb. rhamnosus, Lb. acidophilus*, bifidobacteria, and propionibacteria (Sakai et al., 2010; Sutula et al., 2012).

A culture-based method combining novel chromogenic agar media and appropriate incubation conditions was recently developed by Galat et al. (2016) to enumerate, according to our knowledge for the first time, LAB strains in fermented milk. M1 agar, containing chromogenic substrates for detection of β-galactosidase and β-glucosidase activities, allowed selective enumeration of *Lb. rhamnosus*, *Lb. paracasei* subsp. *paracasei* and *St. thermophilus*. Depending on the presence of some or all of the above species, M1 agar may be supplemented with L-rhamnose or vancomycin, and incubations can be carried out at 37 °C or 44 °C to increase selectivity. M2 agar, containing one chromogenic substrate, was used to selectively enumerate β-galactosidase producing *Lb. delbrueckii* subsp. *bulgaricus* strains after incubation at 47 °C. The chromogenic method allowed unambiguous enumeration of each species, with discrimination between two *Lb. paracasei* strains in fermented milk (Galat et al., 2016).

7.4.10 Homofermentative-heterofermentative differential medium

Heterofermentative LAB reduce a portion of fructose to mannitol producing CO_2, lactic acid, and acetic acid when fructose is the unique carbohydrate source. On the other hand, homofermentative LAB produce 3 moles of lactic acid independently by the fermentable hexose, including fructose. Since homofermentative LAB produce more acid from a fixed amount of fructose, if compared to heterofermentative LAB, a pH difference can be established and exploited to discriminate between homofermentative and heterofermentative LAB in a medium with a limited amount of fructose as the sole carbohydrate source.

Firstly, Sims (1968) designed a selective medium containing fusidic acid to grow homofermentative lactobacilli of dairy, oral, and vaginal origin, but both homofermentative and heterofermentative lactobacilli proved to be able to grow in fusidic

Table 7.8 Homofermentative-heterofermentative differential (HHD) medium.

Fructose	0.25% (14 mM)
KH_2PO_4	0.25% (18 mM)
Free amino acids	1.0 g/l
Trypticase peptone	10.0 g/l
Phytone peptone	1.5 g/l
Casamino acids	3.0 g/l
Yeast extract	1.0 g/l
Tween 80	1.0 g/l

acid-containing medium. Later, McDonald et al. (1987) designed homofermentative–heterofermentative (HHD) medium that allowed differential enumeration of the two physiological kinds of LAB. The medium, whose composition is reported in Table 7.8, is based on the use of bromocresol green as pH indicator. HHD broth inoculated with homofermentative LAB is green, while it remains blue in the case of heterofermentative LAB. In addition, the cells of the homofermentative species on the bottom of the tube are blue to green, whereas the heterofermentative cells remain white. So, the difference in pH lowering may be exploited to discriminate homofermentative and heterofermentative LAB in HHD broth. On HHD agar plates, after 3 days of incubation at 30 °C, homofermentative LAB produce blue to green colonies, while heterofermentative LAB show white colonies (McDonald et al., 1987).

Actually, HHD formula is based on the MD medium (Daeschel et al., 1984), containing fructose (14 mM, 0.25%), KH_2PO_4 (18 mM, 0.25%), and amino acid sources.

7.5 MEDIA FOR THE ISOLATION OF BIFIDOBACTERIA

Bifidobacteria can be enumerated using standard international microbiological reference methods (ISO 29981:2010). Several compounds can promote the growth of bifidobacteria (Ventura and Zink, 2002; Zitz et al., 2007), that is, *N*-acetyl-neuraminic acid (NANA)-containing whey proteins, fructooligosaccharides (FOS) (Gibson and Wang, 1994) and inulin (Roberfroid et al., 1998) have been used as bifidogenic compounds. Another bifidogenic compound, produced by *Propionibacterium freudenreichii* (PF) (Isawa et al., 2002; Furuichi et al., 2007) and strains of *Lb. casei* group (Kang et al., 2015), is 1,4-dihydroxy-2-naphthoic acid (DHNA). DHNA is a precursor of menaquinone (MK) and is transformed to MK by combination with an isoprenoid unit.

Bifidobacterium spp. can be propagated in modified GAM (Gifu Anaerobic Medium) broth (Table 7.9) supplemented with 1% (w/v) glucose. GAM broth is a liquid medium for anaerobic bacteria, containing hemin as essential growth factor, sodium thioglycollate and L-cysteine as reducing agents, and starch to absorb the toxic metabolites produced.

Table 7.9 GAM broth (g/l).

Peptone	5.0 g/l
Soya peptone	3.0
Proteose-peptone	5.0
Digested serum	10.0
Yeast extract	2.5
Meat extract	2.2
Liver extract	1.2
Dextrose	0.5
Soluble starch	5.0
L-tryptophan	0.2
L-cysteine	0.3
Na thioglycollate	0.3
L-arginine	1.0
Vitamin K	5 mg
Hemin	10 mg
KH_2PO_4	2.5
NaCl	3.0
pH	7.3

Table 7.10 NNLP medium composition.

Nalidixic acid	15.0 mg
Neomycin sulfate	100 mg
Lithium chloride	3.0 g
Paromomycin sulfate	200 mg
L-cysteine hydrochloride	0.5 g
Filter sterilized (0.22 µm) distilled water	100 ml
Added to MRS broth at 10%	

7.5.1 MRS-NNLP agar

MRS-NNLP (nalidixic acid, neomycin sulfate, lithium chloride, paromomycin sulfate) is used to enumerate bifidobacteria (Table 7.10) under anaerobic incubation at 37 °C for 72 h. Improvements to selective growth were obtained by adoption of new supplements.

For MRS-NNLP agar, filter-sterilized NNLP was added to autoclaved MRS and L-cysteine (0.05%) (Laroia and Martin, 1991).

7.5.2 BSM, WSP, TOS-MUP

Bifidobacteria-selective medium (BSM) containing arginine, sodium pyruvate, menadione and hemin and modified Wilkins-Chalgren anaerobe agar (Wilkins and Chalgren, 1976) with soya peptone (WSP) have both been applied for bifidobacteria enumeration (Bunesova et al., 2015).

MRS-CC agar, raffinose-propionate lithium mupirocin (RP-MUP), transgalactosylated oligosaccharide (TOS)-propionate agar base supplemented with MUP and lithium salts (ISO 29981:2010) are alternative bifidobacteria-selective growth media. TOS-MUP contains transgalactosylated oligosaccharides obtained by the transformation of lactose by the enzyme β-galactosidase. Specifically, TOS-MUP was suitable to selectively enumerate bifidobacteria and LAB in fermented milks (Miranda et al., 2014; Süle et al., 2014).

Raffinose-propionate (RP) base broth supplemented with MUP solution has been proposed as an alternative medium for enumerating bifidobacteria. For RP preparation, the TOS component is substituted by raffinose (5 g/l) in the TOS-propionate broth composition.

Comparison of MUP-based media for selective enumeration of bifidobacteria in probiotic supplements has been performed by Bunesova and co-workers (2015).

7.5.3 MRS-ABC

MRS-ABC (A: dicloxacillin solution, B: lithium chloride solution, C: cysteine solution) composition is reported in Table 7.11. MRS-NNLP and MRS-ABC were used in association with 3 M Petrifilm plating system (3 M Corporation, Food Safety Division, St. Paul, MN, USA) to enumerate bifidobacteria in fermented milk (Miranda et al., 2011). In other studies, Petrifilm systems were used for enumeration of *St. thermophilus* (Miranda et al., 2015) and with MRS vancomycin (MRS-V) for resolution of *Lb. casei* colonies incubated under anaerobic conditions (Di Lena et al., 2015).

RP-MUP reliability was assessed by comparison with reference protocols and culture media for bifidobacteria enumeration.

MRS-ABC is presented in Table 7.11. After incubation, the colonies formed in each culture medium composition were counted and the results expressed as colony-forming units per millilitre (CFU/ml).

Table 7.11 MRS-ABC medium. A, B and C solutions composition.

A	Dicloxacillin, 10 mg/100 ml distilled water	Filter sterilized (0.22 μm)	0.5% in MRS
B	LiCl, 2 g/18 ml distilled water	Filter sterilized (0.22 μm)	1.0% in MRS
C	L-cysteine, 10 g/100 ml distilled water	Autoclave (121 °C/15 min)	0.5% in MRS

7.6 PHENOTYPING

Phenotypical characterisation of LAB strains may be based on biochemical, serological and morphological features. Recently, Biolog's automated Phenotype Microarrays (Hayward, CA, USA) may provide up to 100 characters to be screened in one step. Based on the use of different carbon, nitrogen, phosphate, and sulfur sources, on the inclusion of nutritional additives or stress-inducing agents (salts, pH, antimicrobial agents), about 2000 phenotypes can be compared (Galat et al., 2016). These methods allow strain-specific phenotypical profiles for targeted strains. Differential carbohydrate fermentation patterns can be observed and selected for preliminary testing of the corresponding enzymatic activities.

Physiological and biochemical characteristics of LAB are based on general known properties and tentative assays. *Leuconostoc* spp. are mesophilic, while many others may grow at temperatures over 37 °C. Some species are grouped by homolactic fermentation, while others are heterofermentative and produce CO_2. This may be evidenced using an agar laid over the selective agar after inoculation of the bacteria. After few days, the gas will push the agar on the top of the vial.

Characterisation is performed by monitoring: acid, salt and temperature tolerance; the optical nature of the isomer of lactic acid produced, "D, L, or both (meso) lactic acid; enzymatic assays" (API20, Biolog, phenotyping assays); screening on media for nutrient requirement, and temperatures; growth on citrate, metabolism on different sugars; final pH of fermented broth; Gas chromatography or thin layer chromatography (TLC) or high-performance liquid chromatography (HPLC) for analysis of metabolites; identification of exopolysaccharides synthesised and viscous property of the fermented broth.

The exopolysaccharide (EPS) production in association with detection of the *eps* gene of *St. thermophilus* strains can be screened on ruthenium red milk plates (Stingele et al., 1996).

7.7 CONCLUSIONS

In dairy microbiology, the media used to selectively grow and enumerate lactococci, streptococci, lactobacilli and bifidobacteria from complex microbial communities have evolved significantly to support the studies on the biodiversity of strains occurring in natural dairy products and to characterise dairy food productions with natural microbiota or supplemented with functional strains, such as probiotic strains.

References

Al-Zoreky N, Sandine WE. 1991. Lactococcus genus: A selective and differential agar medium. *J Food Science* 56(6):1729–30.

Baruzzi F, Quero GM, Poltronieri P, Morea M, Morelli L. 2011. An in vitro protocol for direct isolation of potential probiotic lactobacilli from raw bovine milk and traditional fermented milks. *Appl Microbiol Biotechnol* 90:331–42.

Baruzzi F, Quintieri L, Caputo L, Cocconcelli PS, Borcakli M, Owczarek L, Jasinska UT, Skapska S, Morea M. 2016. Improvement of Ayran quality by the selection of autochthonous microbial cultures. *Food Microbiol* 60:92–103.

Braquart P. 1981. An agar medium for the differential enumeration of *Streptococcus thermophilus* and *Lactobacillus bulgaricus* in yoghurt. *J Appl Bacterio* 51:303–305.

Bunesova V, Musilova S, Geigerova M, Pechar R, Rada V. 2015. Comparison of mupirocin-based media for selective enumeration of bifidobacteria in probiotic supplements. *J Microbiol Meth* 109:106–109.

Corcoran BM, Stanton C, Fitzgerald GF, Ross RP. 2007. Growth of probiotic lactobacilli in the presence of oleic acid enhances subsequent survival in gastric juice. *Microbiol* 153:291–99.

Champagne CP, Roy D, Lafond A. 1997. Selective enumeration of *Lactobacillus casei* in yoghurt-type fermented milks based on a 15° C incubation temperature. *Biotechnol Technol* 11:567–69.

Coeuret V, Dubernet S, Bernardeau M, Gueguen M, Vernoux JP. 2003. Isolation, characterisation and identification of lactobacilli focusing mainly on cheeses and other dairy products. *Lait* 83:269–306.

Colombo M, Zimmermann de Oliveira AE, de Carvalho AF, Nero LA. 2014. Development of an alternative culture medium for the selective enumeration of *Lactobacillus casei* in fermented milk. *Food Microbiol* 39:89–95.

Daeschel MA, McFeeters RF, Fleming HP, Klaenhammer TR, Sanozky RB. 1984. Mutation and selection of *Lactobacillus plantarum* strains that do not produce carbon dioxide from malate. *Appl Environ Microbiol* 47, 419–420.

Dave RI, Shah NP. 1996. Evaluation of media for selective enumeration of Streptococcus thermophilus, *Lactobacillus delbrueckii* ssp. *bulgaricus*, *Lactobacillus acidophilus*, and bifidobacteria. *J Dairy Sci* 79:1529–36.

Davis C. 2014. Enumeration of probiotic strains: review of culture-dependent and alternative techniques to quantify viable bacteria. *J Microbiol Methods* 103:9–17.

de Man JC, Rogosa M, Sharpe ME. 1960. A medium for the cultivation of lactobacilli. *J Appl Bacteriol* 23:130–35.

Desai AR, Shah NP, Powell IB. 2006. Discrimination of dairy industry isolates of the *Lactobacillus casei* group. *J Dairy Sci* 89:3345–51.

Di Lena M, Quero GM, Santovito E, Verran J, De Angelis M, Fusco V. 2015. A selective medium for isolation and accurate enumeration of *Lactobacillus casei*-group lactobacilli in probiotic milks and dairy products. *Int Dairy J* 47:27–36.

Elliker PR, Anderson AW, Hannesson G. 1956. An agar culture medium for lactic acid Streptococci and Lactobacilli. *J Dairy Sci* 39:1611–12.

Eloy C, Lacrosse R. 1976. Lactobacillus Streptococcus differential medium. *Bullettin de Recherches Agronomiques Gembloux* 11:83.

Erkus O, de Jager VC, Spus M, van Alen-Boerrigter IJ, van Rijswijck IM, Hazelwood L, Janssen PW, van Hijum SA, Kleerebezem M, Smid EJ. 2013. Multifactorial diversity sustains microbial community stability. *ISME J* 7:2126–36.

Fernández E, Alegría A, Delgado S, Martín MC, Mayo B. 2011. Comparative phenotypic and molecular genetic profiling of wild *Lactococcus lactis* subsp. *lactis* strains of the *L. lactis* subsp. *lactis* and *L. lactis* subsp. *cremoris* genotypes, isolated from starter-free cheeses made of raw milk. *Appl Environ Microbiol* 77(15):5324–35.

Galat A, Dufresne J, Combrisson J, Thépaut J, Boumghar-Bourtchai L, Boyer M, Fourmestraux C. 2016. Novel method based on chromogenic media for discrimination and selective enumeration of lactic acid bacteria in fermented milk products. *Food Microbiol* 55:86–94.

Gänzle MG, Follador R. 2012. Metabolism of oligosaccharides and starch in lactobacilli: a review. *Frontiers Microbiol* 26:340.

Gemelas L, Rigobello V, Ly-Chatain MH, Demarigny Y. 2013. Selective Lactococcus enumeration in raw milk. *Food Nutr Sci* 4:49–58.

Gibson GR, Wang X. 1994. Bifidogenic properties of different types of fructooligosaccharides. *Food Microbiol* 11:491–8.

ISO 7889:2003. Yogurt – Enumeration of characteristic microorganisms – Colony-count technique at 37 °C.

ISO 29981:2010. Milk products. Enumeration of presumptive bifidobacteria. Colony count technique at 37 °C.

Laroia S, Martin JH. 1991. Methods for enumerating and propagating bifidobacteria. *Cult Dairy Products J* 26:32–33.

Lee HM, Lee Y. 2008. A differential medium for lactic acid-producing bacteria in a mixed culture. *Lett Appl Microbiol* 46:676–81. doi: 10.1111/j.1472-765X.2008.02371.x.

Lima KGC, Kruger MF, Behrens J, Destro MT, Landgraf MB, Franco DGM. 2009. Evaluation of culture media for enumeration of *Lactobacillus acidophilus*, *Lactobacillus casei* and *Bifidobacterium animalis* in the presence of *Lactobacillus delbrueckii* subsp. *bulgaricus* and *Streptococcus thermophilus*. *LWT-Food Sci Technol* 42:491–95.

Ludwig W, Schleifer K-H, Whitman WB. 2009. Lactobacillaceae. In: Vos P, Garrity G, Jones D, Krieg NR, Ludwig W, Rainey FA (Eds). *Bergey's manual of systematic bacteriology, the firmicutes*. New York: Springer.

McDonald LC, McFeeters RF, Daeschel MA, Fleming HP. 1987. A differential medium for the enumeration of homofermentative and heterofermentative lactic acid bacteria. *Appl Environ Microbiol* 33(6): 1382–84.

Miranda RO, de Carvalho AF, Nero LA. 2014. Development of a selective culture medium for bifidobacteria, raffinose-propionate lithium mupirocin (RP-MUP) and assessment of its usage with Petrifilm™ Aerobic Count plates. *Food Microbiol* 39:96–102.

Miranda RO, Gama Neto G, de Freitas R, de Carvalho AF, Nero LA. 2011. Enumeration of bifidobacteria using Petrifilm™ AC in pure cultures and in a fermented milk manufactured with a commercial culture of *Streptococcus thermophilus*. *Food Microbiol* 28:1509–13.

Mitsuoka T. 1969. Comparative studies on bifidobacteria isolated from the alimentary tract of man and animals, including descriptions of *Bifidobacterium thermophilum* nov spec, and *Bifidobacterium pseudolongum* nov. spec. *Infektionskr. Hyg Abt 1 Orig* 210:52–64.

Quintiliani R Jr, Courvalin P. Mechanisms of resistance to antimicrobial agents. 1995. In: Murray PR, Baron EJ, Pfaller MA, Tenover FC, Yolken RH. Manual of clinical microbiology, 6th ed.). Washington, DC: ASM Press, pp 1319–25.

Ravula RR, Shah NP. 1998. Selective enumeration of *Lactobacillus casei* from yogurts and fermented milk drinks. *Biotechnol Technol* 12:819–822.

Reale A, Di Renzo T, Rossi F, Zotta T, Iacumin L, Preziuso M, Parente E, Sorrentino E, Coppola R. 2015. Tolerance of *Lactobacillus casei, Lactobacillus paracasei* and *Lactobacillus rhamnosus* strains to stress factors encountered in food processing and in the gastro-intestinal tract. *LWT - Food Sci Technol* 60:721–28.

Reddy MS, Vedamuthu ER, Washam CJ, Reinbold GW. 1972. Agar medium for differential enumeration of lactic streptococci. *Appl Microbiol* 24:947–952.

Reid G, Kim SO, Köhler GA. 2006. Selecting, testing and understanding probiotic microorganisms. *FEMS Immunol Med Microbiol* 46:149–157.

Ricciardi A, Ianniello RG, Parente E, Zotta T. 2015. Modified chemically defined medium for enhanced respiratory growth of *Lactobacillus casei* and *Lactobacillus plantarum* groups. *J Appl Microbiol* 119:776–85.

Roberfroid MB, Van Loo JAE, Gibson GR. 1998. The bifidogenic nature of chicory inulin and its hydrolysis products. *J Nutr* 128:11–19.

Sakai T, Oishi K, Asahara T, Takada T, Yuki N, Matsumoto K, Nomoto K, Kushiro A. 2010. M-RTLV agar, a novel selective medium to distinguish *Lactobacillus casei* and *Lactobacillus paracasei* from *Lactobacillus rhamnosus*. *Int J Food Microbiol* 139(3), 154–160.

Schillinger U, Holzapflel WH. 2011. Culture media for lactic acid bacteria. In: Corry JEL, Curtis GDW, Baird RM (Eds). *Handbook of culture media for food and water microbiology*. Cambridge, UK: Leatherhead Publishing and RSC publishing, pp 174–92.

Shah NP. 2000. Probiotic bacteria: selective enumeration and survival in dairy products. *J Dairy Sci* 83:894–907.

Sims W. 1968. Use of fusidic acid for differentiating homofermentative from heterofermentative lactobacilli. *J Dent Res* 47:105–107.

Smid EJ, Erkus O, Spus M, Wolkers-Rooijackers JCM, Alexeeva S, Kleerebezem M. 2014. Functional implications of the microbial community structure of undefined mesophilic starter cultures. *Microbial Cell Factories* 13(Suppl 1):S2.

Stingele F, Neeser JR, Mollet B. 1996. Identification and characterization of the eps (exopolysaccharide) gene cluster from *Streptococcus thermophilus* Sfi6. *J Bacteriol* 178:1680–90.

Süle J, Kõrösi T, Hucker A, Varga L. 2014. Evaluation of culture media for selective enumeration of bifidobacteria and lactic acid bacteria. *Braz J Microbiol* 45(3):1023–1030.

Sutula J, Coulthwaite L, Verran J. 2012. Culture media for differential isolation of *Lactobacillus casei* Shirota from oral samples. *J Microbiol Meth* 90:65–71.

Tabasco R, Paarup T, Janer C, Peláez C, Requena T. 2007. Selective enumeration and identification of mixed cultures of *Streptococcus thermophilus, Lactobacillus delbrueckii* subsp. *bulgaricus, L. acidophilus, L. paracasei* subsp. *paracasei* and *Bifidobacterium lactis* in fermented milk. *Int Dairy J* 17:1107–14.

Tharmaraj N, Shah NP. 2003. Selective enumeration of *Lactobacillus delbrueckii* ssp. *bulgaricus, Streptococcus thermophilus, Lactobacillus acidophilus*, bifidobacteria, *Lactobacillus casei, Lactobacillus rhamnosus*, and propionibacteria. *J Dairy Sci* 86(7):2288–96.

Van de Casteele S, Vanheuverzwijn T, Ruyssen T, Van Assche P, Swings J, Huys G. 2006. Evaluation of culture media for selective enumeration of probiotic strains of lactobacilli and bifidobacteria in combination with yoghurt or cheese starters. *Int Dairy J* 16:1470–76.

Ventura M, Zink R. 2002. Rapid identification, differentiation, and proposed new taxonomic classification of *Bifidobacterium lactis*. *Appl Environ Microbiol* 68: 6429–34.

Vinderola C, Reinheimer J. 2000. Enumeration of *Lactobacillus casei* in the presence of *L. acidophilus*, bifidobacteria and lactic starter bacteria in fermented dairy products. *Int Dairy J* 10:271–75.

Wilkins TD, Chalgren S. 1976. Medium for use in antibiotic susceptibility testing of anaerobic bacteria. *Antimicrob Agents Chemother* 10:926–928.

Xie J, Pierce JG, James RC, Okano A, Boger DL. 2011. A redesigned vancomycin engineered for dual d-Ala-d-Ala and d-Ala-d-Lac binding exhibits potent antimicrobial activity against vancomycin-resistant bacteria. *J Am Chem Soc* 133 (35):13946–9.

Zitz U, Kneifel W, Weiss H, Wilrich PT. 2007. Selective enumeration of bifidobacteria in dairy products: development of a standard method. *Bull Int Dairy Fed* 411:3–20.

8 LAB species and strain identification

Cinzia Randazzo[1], Alessandra Pino[1], Koenraad Van Hoorde[2] and Cinzia Caggia[1]

[1]*Department of Agriculture, Food and Environment (Di3A), University of Catania, Catania, Italy*
[2]*Laboratory of Brewing and Biochemistry, Faculty of Bioscience Engineering, Ghent University, Gent, Belgium*

8.1 INTRODUCTION

Milk is converted into a large amount of dairy products, and its high water content and near neutral pH value support the growth of a variety of microorganisms. More than 100 genera and 400 microbial species have already been detected in raw milk (see review Montel et al., 2014). These are both yeasts and bacteria, mainly halophilic and/or alcalophilic Gram positive or negative bacteria, both of which can have considerable negative effects on the quality of milk and dairy products (Cousin, 1982). They can also be of technological relevance, such as lactic acid bacteria (LAB). LAB are naturally present in milk as contaminants from various sources such as: the udder surface, milking equipment, dairy factory environment, transport and filling operations, storage surfaces (Eneroth et al., 1998; McPhee and Griffiths, 2002). From a technological point of view, most of the studies have been focused on the detection and the identification of LAB species, which are mainly involved in the texture and flavour formation of dairy products. The rapid conversion of the lactose into lactic acid, coupled with the activity of their proteolytic enzymes, responsible for small peptides and amino acids formation, especially in cheeses, are considered the main contributions of LAB. For centuries, research has been focused on the exploration of the autochthonous LAB biodiversity from various ecological dairy niches in order to address its role in the texture and flavour formation and to select strains to be used as starter cultures. In this context, it is very useful to investigate the dairy microbial composition, addressing the questions "which microorganisms are present" and "what can the community do"?

Microbiology in Dairy Processing: Challenges and Opportunities, First Edition.
Edited by Palmiro Poltronieri.

Published 2018 by John Wiley & Sons Ltd and the Institute of Food Technologists.

8.2 GENOTYPIC FINGERPRINTING METHODS

Currently, a wide range of molecular genetic techniques is available for identifying members of dairy microbiota. When they are used in combination, they are referred to as a polyphasic identification approach (Vandamme, 1996). The choice of the best method depends on several factors such as:

1) the purpose or context of the research: building a taxonomic inventory of microbial species present in a dairy matrix or monitoring dynamics during production and/or storage, searching for microorganisms with interesting functional and/or technological properties, source tracking and microbial forensics, quality control;
2) the required taxonomical information level, which is strongly associated with the purpose and context: genus, species or strain level;
3) available time: long-term versus real-time analysis;
4) financial resources availability: small-scale or large-scale studies, cost-effective or expensive applications;
5) what methods are current state-of-the art.

As a result of the use of molecular genotypic approaches, our knowledge regarding the diversity of microbial communities of different complexity has improved tremendously. Many of these techniques rely on sequence information inferred from the 16S rRNA gene, from housekeeping genes and from partial or complete genomes. Since Woese (1987) proposed the concept of reconstructing bacterial phylogeny based on ribosomal RNA sequences, 16S rRNA gene sequencing, and to a much lesser extent the 23S rRNA gene, has been routinely used for bacteria identification, allowing the construction of a phylogenetic framework for bacterial taxonomy. Hence, identification through 16S rRNA gene sequencing of the various members thriving in milk and dairy products has been widely used (Delbes et al., 2007; Randazzo et al., 2010; Van Hoorde, 2010; Caro et al., 2013; Bautista-Gallego et al., 2014; Franciosi et al., 2015; Carafa et al., 2015, 2016; Tsafrakidou et al., 2016). Now, thanks to the continuous advancements and improvements in sequencing technologies, more and more high-throughput sequencing approaches using nucleic acids directly extracted from the dairy matrix are used. These so-called second- and third-generation sequencing techniques allowed a more in-depth microbial characterization and increased the numbers of samples analysed (Leite et al., 2012; O'Sullivan et al., 2013; De Pasquale et al., 2014a; Riquelme et al., 2015).

This chapter will give an overview on the culture-dependent and -independent approaches used to assess the microbial composition of dairy products. Moreover, recent advances in omics technologies, used to understand metabolic functionality and diversity of dairy microbiota, will be discussed. Figure 8.1 shows a flowchart of the current molecular techniques used singularly or in combination to study dairy microbiota.

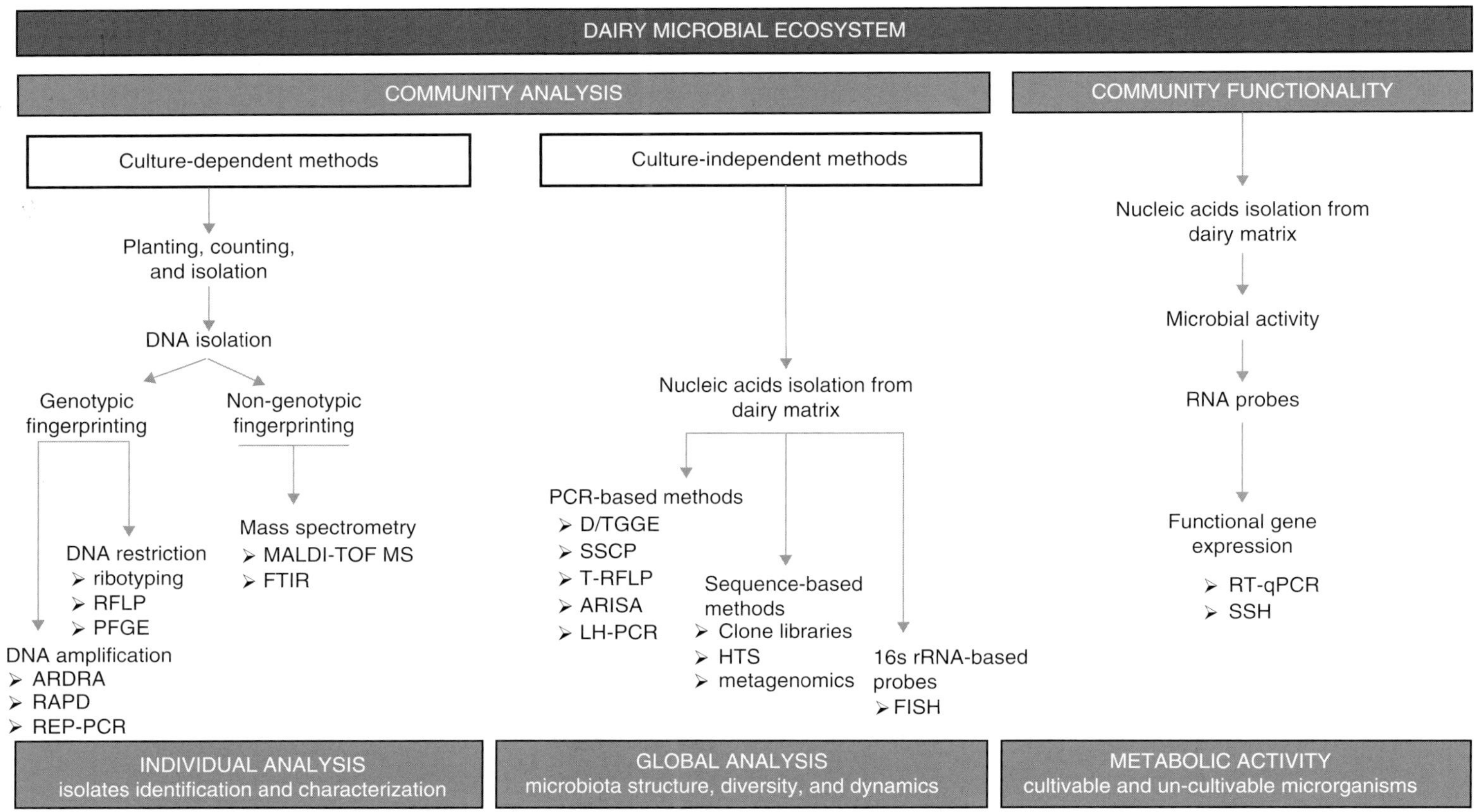

Figure 8.1 Flowchart of the current molecular techniques used singularly or in combination to study dairy microbiota.

8.3 CULTURE-DEPENDENT APPROACHES

Dairy products are generally populated by a great variety of cell types, including intact, viable, non-viable and partially or fully disintegrated (i.e. autolyzed) cells (Carraro et al., 2011), which contribute to the physico-chemical and microbiological characteristics of the final product. For centuries, culture-dependent techniques have been used to allow bacterial identification. The methods are based on the cultivation of the microorganisms, on specific media, and on the phenotypic and/or genotypic characterization of a fraction of the community (Ward and Roy, 2005; Jany and Barbier, 2008).

8.3.1 Random amplification of polymorphic DNA

Random amplified polymorphic DNA (RAPD) is a PCR-based technique able to discriminate microorganisms based on genetic variation. RAPD involves the use of short synthetic primers (generally 10 bp) of random sequence in a PCR reaction, resulting in the amplification of many discrete DNA products. These oligonucleotides serve as both forward and reverse primers and are usually able to amplify fragments from 1 to 10 genomic sites simultaneously. Amplified fragments, usually within the 0.5–5 kb size range, are separated by agarose gel electrophoresis, and polymorphisms are detected as the presence or absence of bands of particular sizes (Williams et al., 1990; Hadrys et al., 1992). RAPD-PCR is one of the most popular typing techniques applied to dairy ecosystems. Its use has been documented in various reports describing and typing the diversity of cheese and dairy ecosystems (Martín-Platero et al., 2009; Carafa et al., 2015; Mangia et al., 2016; Tsafrakidou et al., 2016), to monitor the dynamics of specific strains throughout cheese ripening (Bove et al., 2011; Feligini et al., 2012; Pogačić et al., 2013; Silva et al., 2015) and/or as a tool for characterization of dairy isolates with strain-specific features such as technological properties, proteolytic properties, γ-aminobutyric acid (GABA) production, production of antimicrobials or antibiotic susceptibility, in view of a possible future application in the development of new starter or adjunct cultures (Morandi and Brasca, 2012; Baruzzi et al., 2012; Franciosi et al., 2015; Carafa et al., 2016). In the Pecorino di Filiano PDO (Protected Denomination of Origin) cheese, a traditional southern Italian sheep's milk cheese, plate counting in combination with RAPD-PCR, and amplified ribosomal DNA restriction analysis (ARDRA) allowed identification of *Lactobacillus paracasei* subsp. *paracasei* as the dominant species during the ripening process (Bonomo and Salzano, 2012). Similar results were shown by Tsafrakidou and co-workers (2016), demonstrating the dominance of *L. paracasei*, followed by *Lactobacillus brevis*, in Graviera Kritis PDO cheese, a hard Greek cheese made from sheep's milk (Tsafrakidou et al., 2016). Moreover, applying the RAPD-PCR approach, it was possible to detect *Lactococcus lactis* subsp. *lactis* dominance both in Casizolu pasta filata cheese, made with raw cow's milk (Mangia et al., 2016), and in fresh and mature Feta PDO cheeses, made from ovine milk and from a mixture of ovine and caprine milks (Bozoudi et al., 2016). In addition, some reports included the use of RAPD-PCR analysis, combined with other molecular techniques, to determine the occurrence and genotypic biodiversity of *Geotricum candidum* in Armada cheese (Sacristán et al., 2013) and to provide a complete picture of the high biodiversity, at genetic and phenotypic levels, among *Kluyveromyces marxianus* population, isolated from Pecorino di Farindola (Tofalo et al., 2014).

8.3.2 ARDRA and RFLP

Next to their widespread applications in sequence-based approaches, ribosomal genes can also be used as genomic targets for DNA fingerprinting methods, such as ribotyping and amplified rDNA restriction analysis (ARDRA). Both techniques can be considered as a variation of conventional restriction fragment length polymorphism (RFLP) analysis. RFLP analysis is a profiling tool based on the banding patterns obtained from genomic or rRNA restriction digests. The crucial element of this technique is the correct selection of the restriction enzymes since the use of rare cutting enzymes reduces the number of DNA fragments obtained (McCartney, 2002). In fact, the generated DNA fingerprint depends both on the specificity of the restriction enzymes used and on the sequence of the bacterial genome that is characteristic for each bacterial species or strain. When the targeted (read as amplified and digested) region corresponds to the ribosomal DNA, this RFPL variant is referred to as ARDRA. However, ARDRA has only limited discriminative power close to the genus-species level (Marilley and Casey, 2004). The resolution of this technique, however, depends on and can be influence by the choice of restriction endonuclease used and can be increased using multiple restriction enzymes separately or in combination (Justé et al., 2008). Moreover, the choice of endonucleases can be adapted in view of the taxonomic group under study.

Recently, ARDRA, using *Taq*I and *Hae*III restriction enzymes, was applied in order to identify the *Lactobacillus* species of Dalameh cheeses and yoghurt produced in Iran. Results showed the dominance of strains belonging to *Lactobacillus delbrueckii*, *Lactobacillus acidophilus*, *L. brevis*, and *Lactobacillus delbrueckii* subsp. *lactis* (Ghafari et al., 2014). The same approach, coupled to 16S rDNA sequence analysis, was applied in order to characterize the LAB population isolated from mature Konya Kuflu cheese, a traditional Turkey blue cheese produced from raw milk without starter culture addition (Guley et al., 2014). Moreover, Khemariya and co-workers (2013) demonstrated that PCR-RFLP assay could be routinely used for the identification of subspecies of *L. lactis* from dairy products based on amplicons targeting the *yueF* and *htrA* genes subjected to restriction by *Alu*I and *Taq*I enzymes, respectively. Similarly, Jans and co-workers (2012b) developed a specific ARDRA for the *Streptococcus bovis*/*Streptococcus equinus* complex, targeting the 16S rRNA gene based on multiplex PCR followed by RFLP using *Mse*I and *Xba*I restriction enzymes.

An ARDRA approach, using *Hha*I and *Mse*I digestion, was part of a polyphasic study to inventory the non-starter lactic acid bacteria (NSLAB) community present in ripened Parmigiano Reggiano cheese (Solieri, 2012), whereas Blasco and co-workers successfully used a *Hae*III- and *Alu*I-based ARDRA for species designation of 107 strains belonging to the dairy genus *Propionibacterium* (Blasco, 2015).

8.3.3 Ribotyping

Ribotyping is a variation of the conventional RFLP and allows the generation of a less complex pattern since the genomic DNA is enzymatically digested, and the resulting fragments are later transferred to a membrane and hybridised with an rDNA probe. The probes used in ribotyping vary from partial sequences of the DNA genes or from their spacer regions to the whole rDNA operon, which could be used as a probe in a classic multi-step Southern blotting or in automated ribotyping. If the probe contains conserved regions of rDNA, it can be used for ribotyping of a wide range of bacteria, even phylogenetically distant. Evidently, more fragments will hybridise with probes that encompass a large region of the rDNA operon than with a shorter probe. Thus, the discriminatory power of the technique is not only dependent on the size of the probe, but also on the used restriction enzyme(s) (Randazzo et al., 2009). Caro and co-workers characterized LAB isolates from a raw milk cheese using only a single endonuclease *Eco*RI and single probe. Recently Mormile and co-workers (2016) used ribotyping to identify 169 LAB isolated from artisanal Pecorino di Tremonti cheese. The tecnique allowed to cluster the isolates into following species: *Enterococcus faecium, Lc. lactis* subsp. *lactis, Lc. lactis* subsp. *cremoris, Enterococcus faecalis* and *Enterococcus durans*. A combination of RFLP and ribotyping allowed discrimination among *Lactobacillus helveticus* strains, according to the cheese they were isolated from (Giraffa et al., 2000). It is, however, important to remember that whenever one wants to use the information within the ribosomal DNA, such approaches can fail in discriminating phylogenetically closely related species or subspecies (Felis and Dellaglio, 2007) and, as mentioned before, certainly do not allow focus at the strain level. Alternatives exist that provide a better discrimination of closely related species sharing highly similar, if not identical, rRNA gene sequences; some of these high-resolution fingerprint methods allow the delivery of a closer look at the strain level whenever necessary.

8.3.4 Repetitive element sequence-based PCR

Repetitive element sequence-based PCR (rep-PCR) is a typing method that takes advantage of the presence of interspersed short extragenic sequence elements (e.g. REP, ERIC, $(GTG)_5$ and BOX sequences) repetitively found throughout the bacterial chromosome. PCR with primers targeting these repetitive elements generates specific profiles that can be used to discriminate up to the strain level (Versalovic et al., 1994). The rep-PCR method utilizes primers that hybridise to non-coding sequences scattered across the genome. DNA between adjacent repetitive elements is amplified using PCR, and multiple amplicons can be produced. Amplicons are then characterized by electrophoresis, and the banding patterns are compared to determine the genetic relatedness between the analysed bacterial isolates.

Rep-PCR proved to be useful for rapid and reliable differentiation at species, subspecies and, potentially, strain level of lactobacilli and other LAB associated with the production of different kinds of cheeses (Van Hoorde et al., 2008b, 2010; Bove et al., 2011; Baruzzi et al., 2012; Solieri et al., 2012; Bautista-Gallego et al., 2014; Perin et al., 2015; Mangia et al., 2016) and other dairy products (Kesmen and Kacmaz, 2011; Jans et al., 2012a; Akabanda et al., 2013; Tormo et al., 2015). In particular, Terzić-Vidojević et al. (2015) used the REP-PCR combined with 16S rDNA sequencing, underlining the dominance of the *Leuconostoc pseudomesenteroides* and *Lc. lactis* species in 5-day-old Vlasina raw goat's milk. *Lactobacillus plantarum, E. durans* and *Pediococcus pentosaceus* species were dominant in the same cheese after 15 days of ripening (Terzic-Vidojevic et al., 2015). Recently, REP-PCR was applied in order to characterize the LAB population in Egyptian Karish cheese and Laban Zeer, a traditional fermented milk, showing the dominance of *Lc. lactis* subsp. *lactis, L. delbrueckii* subsp. *lactis* and *E. feacium* species (Awad, 2016).

8.3.5 Amplified fragment length polymorphism

Amplified fragment length polymorphism (AFLP) combines PCR amplification with a preceding DNA restriction digest whereby the restriction fragments are ligated to double-stranded adaptors providing the annealing sites for the primers. Following PCR, resulting fragments are analysed via electrophoresis, generating fingerprints with a highly discriminatory power. AFLP has proven to be a high-resolution fingerprinting technique for intraspecies identification and for genotyping of LAB and bifidobacteria from various fermented products as well as from the human gastrointestinal biota (Ben Amor et al., 2007). Although not frequently employed in combination with dairy products, AFLP was used in combination with $(GTG)_5$-PCR to monitor the persistence of LAB during the ripening of artisan Gouda-type cheeses (Van Hoorde et al., 2008a). Interestingly, AFLP proved to perform better compared to RAPD for assessing the genotypic diversity among different strains of the same species as was shown for *Streptococcus thermophilus* (Lazzi et al., 2009) and for *L. plantarum* (Di Cagno et al., 2010).

8.3.6 Pulsed field gel electrophoresis

Like rep-PCR and RAPD, pulsed field gel electrophoresis (PFGE) analyses the complete genome and thus has a high discriminatory power. However, PFGE is a very laborious and time-consuming technique as it involves the digestion of genomic DNA with a statistically rare cutting endonuclease generating large DNA fragments. These DNA fragments are then separated in an electrical field with direction-changing pulses. As a result, limited numbers of samples can be studied, making PFGE a less suitable identification tool, for example, for large-scale isolation campaigns with a descriptive taxonomic objective. On the other hand, as this method is highly discriminatory, PFGE is a powerful method for strain typing of dairy isolates (Blasco et al., 2015; Tsafrakidou et al., 2016), and it is often the method of choice for source-tracking applications (Freitas et al., 2015; Terzić-Vidojević et al., 2015; Acciari et al., 2016). Using a combination of RAPD and PFGE, Coppola and co-workers (2006) were able to trace back most isolates from ripened pasta filata cheese to the raw milk.

Every method has its strengths and its limitations. Hence, to trying to partly circumvent inherent drawbacks, often two or more techniques are combined to maximally benefit from each of the methods used. Whenever the information of multiple approaches, both phenotypic and/or genotypic, is integrated for a more robust differentiation, identification and classification, this is referred to as a 'polyphasic taxonomy strategy' (Vandamme et al., 1996). In most of the aforementioned studies, such a polyphasic strategy was implemented.

8.4 NON-GENOTYPIC FINGERPRINTING METHODS

Possibly new advances in microbial ecology can also give an additional answer to the aforementioned problems. A technique that recently emerged in the field of microbiology is matrix-assisted laser desorption ionization-time-of-flight mass spectrometry (MALDI-TOF MS). MALDI-TOF MS is a proteomics approach that generates fingerprints representing the molecular masses of peptides and small proteins. As these proteins confer genetic information, the profiles are considered suitable for identification purposes. It is a soft ionization method, and the generated ions are separated and detected according to their molecular mass and charge, generally a positive single charge. Its high sensitivity and high throughput have made it a very interesting method for the characterization and identification of (large numbers of) microbial isolates. To obtain the proteinaceous cell extracts for analysis, in general three approaches can be used: the shortest method relies on the direct application of cell material as a thin layer on the target plate (i.e. direct smear method); subsequently, the sample is dried and covered with the matrix before being analysed. Somewhat more elaborate is the on-target extraction in which formic acid is added to the cell smear in order to enhance protein extraction. The third option consists of performing an ethanol-formic acid-acetonitrile extraction, and, next, after centrifugation, 1 µl of the supernatants is applied on the target plate, allowed to dry, and covered with matrix prior to analysis. The nature of the analyte determines the type of matrix used: for microbial identification purposes that rely on proteins, generally α-cyano-4-hydroxycinnamic acid (CHCA) is used and, less frequently sinapinic acid (SA). Whereas the use of MALDI-TOF MS was initially limited to a clinical microbiology context (Carbonnelle et al., 2011), more and more frequently MALDI-TOF MS is being exploited for the characterization of microorganisms from food origin in general and dairy in particular, although its use is still limited (Pavlovic et al., 2013). Using MALDI-TOF MS fingerprinting, Dušková et al. (2012) successfully identified 93% of the 148 LAB isolated from various food (meat and dairy) sources. This success rate was remarkably higher compared to ARDRA, for which only 77% of the isolates were identified unambiguously. Several studies used MALDI-TOF MS to characterize beer spoilage microorganisms (Wieme et al., 2014a; Turvey et al., 2016), and MALDI-TOF MS was used for assessing the biodiversity of psychrotrophic microbial consortia in raw milk (Vithanage et al., 2014; 2016) and to identify *Enterococcus* species and *Escherichia coli* retrieved from Bryndza cheese (Vrabec et al., 2015). As with most fingerprinting strategies, a library-based approach containing spectra of known type and reference bacteria to identify unknowns is most common. However, additionally, it is also possible to identify a given species through unique biomarker(s) (i.e. mass peaks with a certain m/z ratio in the MS spectrum), and with this, important advantages are the comparative simplicity of the mass spectra and the high reproducibility of the technique (Sauer et al., 2008). In this way, Fernández-No and co-workers (2010) were able to unambiguously differentiate between 16 biogenic amine-producing enteric and marine bacteria via the assignment of specific protein marker peaks. Similarly, isolates of two subspecies of the probiotic microorganism *Bifidobacterium animalis*, subsp. *lactis* and subsp. *animalis*, could be discriminated on the basis of specific peaks (Ruiz-Moyano et al., 2012). This open perspectives towards a more accurate description and identification of the different members of a community up to strain level through the generation of strain-specific MALDI-TOF MS spectra whether or not in combination with strain-specific biomarker peaks (Sandrin et al., 2013). However, the potential of MALDI-TOF MS to classify beyond the subspecies level is still a matter of debate, with some studies being in favor (Kern et al., 2014; Wieme et al., 2014b) and others having encountered the often taxon-dependent (Ghyselinck et al., 2011) taxonomic limitations of the technique (Zeller-Péronnet et al., 2013).

Another not new (with the earliest projects dating back 20 to 30 years) (Naumann et al., 1991) but nowadays emerging technique in the field of microbiology, is Fourier transform infrared (FTIR) spectroscopy. Whereas MALDI-TOF MS records merely a protein-content mass spectrum of whole bacterial cells, FTIR spectroscopy generates overall biochemical fingerprints: absorbance of the infrared (IR) light by the cellular compounds of the cells results in a fingerprint-like spectrum containing signals of many different functional groups and biomolecules, including lipids, proteins, carbohydrates, nucleic acids and even water. Like MALDI-TOF MS, it is a rapid and straightforward technique that delivers reliable identification at genus, species, even up to strain, level (Wenning et al., 2014). In a 2015 study, for example, von Neubeck and his colleagues were able to assign 2906 raw milk isolates to 169 species of 61 genera using FTIR spectroscopy (von Neubeck et al., 2015).

Although these methods have been optimized to detect and quantify relevant microorganisms, they present a lower diversity coverage than culture-independent methods. Moreover, the conventional techniques are known to be biased because bacteria can only be cultivated if their metabolic and physiological requirements can be in vitro reproduced (Ndoye et al., 2011). Culture-dependent techniques currently used for dairy product microbiota characterization are reported in Figure 8.1.

8.5 CULTURE-INDEPENDENT APPROACHES

Developments of molecular methods allowing the identification of microbes that are difficult, or impossible, to grow on media have revolutionised our knowledge on the microbial ecology of dairy foods. This shift towards the use of culture-independent methods that no longer require a culturing step has facilitated the simultaneous study of viable, non-cultivable and stressed/injured microorganisms. In fact, their application to the dairy matrix has helped to give significant insight into specific isolates, microbial populations and into the evolution and the nature of the microbial groups into these ecosystems (Bonaïti et al., 2006). In addition, culture-independent methods tend to be more rapid, sensitive and less susceptible to bias than culture-dependent ones.

8.5.1 Culture-independent methods for qualitative analysis of dairy foods microbiota

The qualitative analysis of dairy foods microbiota to study microbial diversity and dynamics can be performed applying PCR-based tools such as denaturing gradient gel electrophoresis (PCR-DGGE), temperature gradient gel electrophoresis (PCR-TTGE), single-strand conformation polymorphism PCR (SSCP-PCR), terminal restriction fragment length polymorphism (T-RFLP), length heterogeneity PCR (LH-PCR), and automated ribosomal intergenic spacer analysis (ARISA) techniques (Ndoye et al., 2011). Other community fingerprinting techniques used in dairy food microbiology include denaturing high-performance liquid chromatography (DHPLC) (Ercolini et al., 2008; Mounier et al., 2010; Delavenne et al., 2011) and amplified ribosomal DNA restriction analysis (ARDRA) (Singh et al., 2009). Moreover, when particular interest lies in profiling the metabolically active community members, analysis can be performed on the reverse-transcribed (RT) RNA, e.g. RT-PCR-DGGE, RT-PCR-SSCP, and LH-RT-PCR.

DGGE and the related method TGGE are the most widely used community fingerprinting approaches for studying microbial population in dairy products. Since their first use in microbial ecology research in the early 1990s (Muyzer et al., 1993), DGGE and TGGE analysis of rRNA-encoding genes amplified by PCR have been widely applied to identify the dominant microbial populations and to show the diversity and the structure of the microbial community present without previous knowledge of its composition. DGGE and TGGE consist of an electrophoretic separation of PCR products in a polyacrylamide gel containing a gradient of chemical (urea and formamide in DGGE) or physical (temperature in TGGE) denaturants. As the DNA molecule migrates through the denaturant gradient, a sequence-dependent partial denaturation of the double strand occurs (Muyzer et al., 1993). Although DGGE and TGGE are not quantitative techniques, their application allows showing and monitoring of the spatial/temporal changes that occur during dairy food production and storage (Cocolin et al., 2013). In addition, reverse transcription (RT)-PCR-DGGE has also been performed to obtain a picture of the species metabolically active at a particular sampling time (Dolci et al., 2013; Alessandria et al., 2016). DGGE and TGGE methods, coupled with culturing and/or sequencing, have been successfully applied to several cheese types and dairy products (Leite et al., 2012; Delgado et al., 2013; Arcuri et al., 2013; Quero et al., 2014; Terzić-Vidojević et al., 2015; Chombo-Morales et al., 2016; Carpino et al., 2017). Recently, a new culture-independent method based on PCR-DGGE, was developed in order to detect and identify, at species level, the histaminogenic bacteria present in cheese. Specific primers were designed based on the *hdcA* gene sequences available for Gram positive bacteria, and PCR and DGGE were optimized in order to differentiate amplicons that correspond to different histamine-producing species (Diaz et al., 2016).

SSCP is a culture-independent method similar to PCR-D/TGGE as it allows the separation of different DNA fragments of similar length, also relying on electrophoretic separation due to sequence-dependent variation in migration speed of the PCR products. SSCP has been applied to analyse mutations or dynamics among microbial populations at the genus and species levels in cheese (Callon et al., 2006; Delbes et al., 2007; Saubusse et al., 2007; Mounier et al., 2009). The principle of SSCP is that the mobility of a single-stranded DNA fragment, in a polyacrylamide gel, is dependent on the DNA conformation (Orita et al., 1989; Hayashi, 1991). A typical SSCP profile consists of two single-stranded DNA fragments and one double-stranded DNA fragment, although different conformations from one strand are also possible. With respect to cheese, SSCP has not been extensively employed. Sofu and Ekinc (2016) used PCR-SSCP combined with PCR-DGGE to profile the community diversity and dynamics of milk and Ezine cheese, revealing the dominance of *L. lactis* and *S. thermophilus* spp.

Recently, Hermet and co-workers (2014) have developed a method based on capillary electrophoresis CE-SSCP analysis of nuclear ribosomal DNA ITS amplicons

(ITS1 and ITS2 conformers) in order to determine the fungal community composition of 36 cheeses, including blue-veined, pressed-cooked, pressed-uncooked, red-smear and surface-mould ripened cheeses. The method has proved to be reproducible and sensitive. It can be considered an effective tool to identify the fungi present in various cheese types and may be of interest for the cheese industry to rapidly describe the composition of cheese fungal communities (Hermet et al., 2014). As for DGGE, TGGE and TTGE, SSCP also provides community fingerprints that cannot be phylogenetically assigned directly. A database containing the migration profile of reference strains has to be created. One disadvantage of the SSCP technique, in contrast to DGGE, is that the labelled single-strand DNA fragments cannot be sequenced to confirm the database-derived species designations.

Terminal restriction fragment length polymorphism (T-RFLP) is a rapid and sensitive molecular approach with a high resolution and ability to assess subtle genetic differences among strains, providing a deeper insight into the structure of microbial communities. The technique combines selective PCR amplification of target genes with restriction enzyme digestion, high-resolution electrophoresis and fluorescent detection. The use of one fluorescently labelled primer restricts the analysis to the terminal fragment only. T-RFLP is increasingly used to analyse microbial communities due to its simplicity and reliability for bacterial 16S rRNA genes (Schütte et al., 2008; Arteau et al., 2010; Rasolofo et al., 2011). Moreover, it has become a valuable method for rapidly comparing the temporal changes and relationships of bacterial communities as well as for revealing the most dominant microbial sequences in dairy samples (Ndoye et al., 2011). Fuka and co-workers (2013) applied T-RFLP and 454 pyrosequencing of tagged 16S rRNA gene amplicons to obtain a complete overview of the bacterial community structure of Croatian raw ewe's milk cheeses.

Length heterogeneity PCR (LH-PCR) is a fingerprinting method able to study the microbial diversity, associated with a particular ecosystem, based on the variations in the length of sequences of the 16S rRNA gene or other genes (Giraffa and Neviani, 2001). The LH-PCR is performed using a fluorescent dye–labelled oligonucleotide as forward primer coupled with an unlabelled reverse primer to amplify hypervariable regions. Labelled fragment are subsequently separated, and the fluorescence is detected with an automated sequencer. LH-PCR is capable of discriminating amplicons originating from different organisms based on natural variation in the lengths of their DNA target regions. This method has been demonstrated to be easy, fast, reliable, and highly reproducible (Quigley et al., 2011). It was used in order to explore the structure and the dynamics of the Grana Padano PDO cheese microbiota as well as to evaluate the trend of microbial dynamics of LABs during the entire cheese-making process (Pogačić et al., 2013; Santarelli et al., 2013). Recently, Lazzi and co-workers (2016) applied the same approach in order to study the relationship between the dynamics of growth and lysis of LAB in Grana Padano cheese and their link with the volatile flavour compound formation during the ripening time.

Automated ribosomal intergenic spacer analysis (ARISA), introduced by Fisher and Triplett (1999), is a commonly used method for microbial community analysis. Based on the length heterogeneity of the bacterial rRNA operon 16S–23S intergenic spacer, ARISA provides estimates of microbial richness and diversity. This method has been frequently used to study a variety of habitats, including dairy products, becoming a very useful tool for comparing community structure across multiple samples based on pattern profiles. Moreover, the ARISA approach presents important perspectives in terms of applicability along the entire dairy production chain for diagnostic purposes as well as for the microbiological quality of milk evaluation (Feligini et al., 2014). A profiling approach based on ARISA, associated to T-RFLP, was used in combination with principal component analysis to assess fungal changes in Camembert cheese, according to process parameters (Arteau et al., 2010). This approach was sufficiently sensitive to detect differences in fungal populations between batches and types of cheese (Camembert and Brie). Additionally, Porcellato and co-workers (2014) evaluated the application of ARISA as a screening method to study cheese microbiota and to assess the microbial dynamics in cheddar cheeses differing in type and salt cation level. The results indicated that ARISA is a useful technique for the analysis of the microbial ecology of cheese and, due to its high resolution and similarity to other molecular methods used for community studies, may be used as a screening method for studying a large number of samples. The same authors also determined the dynamics of the microbiota during the ripening time of the cheddar cheeses applying the LAB-ARISA. The ARISA results indicated that the primary lactobacilli present in the cheese matrix during ripening were *L. helveticus* and *Lactobacillus curvatus*. An original application of the ARISA was developed by Panelli and co-workers (2013). In fact, relying on a multi-species–specific primer set, it was possible to simultaneously display distinctive fingerprints diagnostic for *Clostridium tyrobutyricum* and for the other species so far associated with the late blowing in Grana Padano PDO cheese. In addition, a species-specific primer set was designed and used for the ARISA in order to diagnose the presence of *C. tyrobutyricum* in raw milk before cheese making (Panelli et al., 2013).

Bacterial diversity may also be assessed by sequencing clone libraries generated from the 16S rRNA gene amplification of DNA extracted from dairy products (Rasolofo et al., 2010; Schornsteiner et al., 2014; Carafa et al., 2015; Avnİ Kırmacı et al., 2015). Despite the progress made in the field of high-throughput sequencing, this approach is expensive and time-consuming. Moreover, clone libraries are often constructed in a parallel approach to complement fingerprinting techniques such as DGGE, T-RFLP or SSCP.

8.5.2 Culture-independent methods for quantitative analysis of dairy foods microbiota

To obtain quantitative information of specific community members, PCR-based tools such as quantitative PCR (qPCR) and reverse transcription quantitative PCR (RT-qPCR) can be applied. The qPCR technique is used especially in studies of DNA, while RT-qPCR is devoted to the study of specific expressed genes (mRNA).

The qPCR technology is based on the detection and quantification of a fluorescent signal either produced by non-specific intercalating dyes, such as SYBR green, or sequence-specific probes such as molecular beacons and TaqMan probes. The intensity of the fluorescent signal is linked to the initial amount of DNA. qPCR is a useful tool for the effective detection, identification, and quantification of bacteria in different types of samples or products. Moreover, this technique does not require large quantities of sample (Postollec et al., 2011). The qPCR approach has been used to quantify different microbial species in dairy foods such as *S. thermophilus,* in commercial yoghurt samples (Ongol et al., 2009), *Propionibacterium freudenreichii*, *L. paracasei*, *S. thermophilus*, *L. helveticus* (Falentin et al., 2010; 2012) in experimental Emmental cheese; *Enterococcus gilvus* in artisanal raw milk cheeses (Zago et al., 2009); *E. faecium,* in Lebanese raw goat's milk cheeses (Serhan et al., 2009); *S. aureus* genotype B in milk (Boss et al., 2011); *Brucella* spp. in buffalo milk (Amoroso et al., 2011); *Penicillium roqueforti* and *Penicillium camemberti* in model and commercial Camembert-type cheeses (Le Dréan et al., 2010); and fungal microbiota in bloomy-rind cheeses, such as Camembert (Lessard et al., 2012).

The lack of differentiation between the DNA of viable and non-viable bacterial cells is a significant barrier to the use of qPCR (Nocker et al., 2006) that can be resolved by eliminating the amplification of DNA from dead cells, for example by complexing with agents such as propidium monoazide. In this context, RT-qPCR is able to exceed the above limitations since it targets the RNA molecules characterizing microorganisms and metabolic activities. RT-qPCR experiments involve the following steps: RNA extraction, evaluation of RNA integrity, DNase treatment, reverse transcription and qPCR (Nolan et al., 2006; Bustin et al., 2009). Up to now, RT-qPCR has been applied to dairy products mainly with the aim to detect, quantify, and study growth and metabolic activity of specific microbial species (Rantsiou et al., 2008; Falentin et al., 2010; Duquenne et al., 2010; Ruggirello et al., 2014).

8.6 NOVEL HIGH-THROUGHPUT TECHNIQUES: SEQUENCING AND METAGENOMICS

High-throughput sequencing (HTS) and metagenomics represent recent advances in the study of food microbial ecology and are widely applied to study food microbiota composition and to monitor microbial dynamics during food production and storage. Since their advent in the mid-2000s, HTS has revolutionised the field of microbial ecology, allowing for a more accurate identification of microbial taxa, including those that are difficult to culture and/or are present in low abundance (Sogin et al., 2006). HTS boasts higher sensitivity compared with traditional culture-independent methods, and, although PCR-dependent, the amplicon-based HTS is considered quantitative. Depending on the desired level of sample coverage, many food samples can be sequenced at the same time, and a reliable identification of the majority of microorganisms, in raw materials and during food production/storage, can be achieved. In addition, when microbiomes are studied by shotgun library sequencing, insights into microbial activities can be obtained from the sequences of microbial genes present in the food sample, which offers important advances in studying microbial ecology of foods (Ercolini, 2013; De Filippis et al., 2016). Dairy food and fermented milk were extensively studied through HTS in order to monitor the fermentative process and the microbiota dynamics during production and aging and to explore the spatial distribution of microorganisms in different parts of the same sample (Nalbantoglu et al., 2014; De Filippis et al., 2014; De Pasquale et al., 2014a, 2014b; Dolci et al., 2014; Wolfe et al., 2014; Riquelme et al., 2014; Sun et al., 2014; Delcenserie et al., 2014, 2016; Liu et al., 2015a, 2015b; Garofalo et al., 2015; Stellato et al., 2015; Guidone et al., 2015; Calasso et al., 2016).

In metagenomics or metatranscriptomics studies, total DNA or cDNA is sequenced, without the need for PCR, avoiding the possibility of amplification biases. By this way, a picture of the entire microbial community can be obtained, tracking and comparing the abundance of bacteria and other organisms at the same time. Therefore, besides the taxonomical composition of the community, this approach allows to obtain the abundance of all microbial genes. The output of metagenomics analysis, based on DNA sequencing, is, however, the potential activities of the microbial communities, because a specific gene may not be expressed in that condition or because DNA may arise from dead or metabolically inactive cells. In order to identify the genes actually expressed in a food sample, RNA sequencing (RNAseq) is the most appropriate path to take, which is what happens in a metatranscriptomic approach. Up to now, the application of metagenomics and metatranscriptomics is limited to only few studies focused mainly on dairy products. Lessard and co-workers (2014) conducted, to our knowledge, for the first time, a comprehensive metatranscriptome analysis of the Camembert cheese complex fungal ripening ecosystem, increasing the knowledge about the biological activities of the dominant ripening microbial population of Camembert cheese. Dugat-Bony and co-workers (2015), using an integrated metagenomic, metatranscriptomic and biochemical approach studied the cheese maturation process and the metabolic activities of the different community members as well as their possible interactions. Moreover, Monnet et al. (2016) reported differences in amino acid catabolism between the two yeasts inoculated on a Reblochon-type cheese: *Geotrichum candidum* was found to reach the maximum expression of amino acid-related genes in the first phase of ripening, while *Debaryomyces hansenii* took over after 30 days, highlighting a different role of these yeasts in flavour production.

8.7 CONCLUSIONS

Molecular fingerprinting methods are highly discriminatory at species and strain level, allowing a comprehensive diversity assessment of isolates recovered from cultivation plates. However, the choice of a typing method depends on its application, for example, compositional and/or dynamic studies, source tracking, monitoring and controlling food quality, food preservation and so on. Currently, several culture-independent approaches are used to increase knowledge on microbial composition and its functionality of dairy products and to monitor the fate of certain bacteria through the dairy process.

Further improvements can be expected by the application of new high-throughput techniques for dairy ecology studies.

References

Acciari VA, Iannetti L, Gattuso A, Sonnessa M, Scavia G, Montagna C, Addante N, Torresi M, Zocchi L, Scattolini S, Centorame P, Marfoglia C, Prencipe VA, Gianfranceschi MV. 2016. Tracing sources of Listeria contamination in traditional Italian cheese associated with a US outbreak: investigations in Italy. *Epidemiol Infect* 144:2719–27.

Akabanda F, Owusu-Kwarteng J, Tano-Debrah K, Glover RLK, Nielsen DS Jespersen L. 2013. Taxonomic and molecular characterization of lactic acid bacteria and yeasts in nunu, a Ghanaian fermented milk product. *Food Microbiol* 34:277–83.

Alessandria V, Ferrocino I, De Filippis F, Fontana M, Rantsiou K, Ercolini D, Cocolin L. 2016. Microbiota of an Italian Grana-like cheese during manufacture and ripening, unraveled by 16S rRNA-based approaches. *Appl Environ Microbiol* 82:3988–95.

Amoroso MG, Salzano C, Cioffi B, Napoletano M, Garofalo F, Guarino A, Fusco G. 2011. Validation of a real-time PCR assay for fast and sensitive quantification of Brucella spp.in water buffalo milk. *Food Control* 22:1466–70.

Arcuri EF, Sheikha AF, Rychlik T, Isabelle PM, Montet D. 2013. Determination of cheese origin by using 16S rDNA fingerprinting of bacteria communities by PCR-DGGE: preliminary application to traditional Minas cheese. *Food Control* 30:1–6.

Arteau M, Labrie S, Roy D. 2010. Terminal-restriction fragment length polymorphism and automated ribosomal intergenic spacer analysis profiling of fungal communities in Camembert cheese. *Int Dairy J* 20:545–54.

Avnİ Kırmacı H, Hamdİ Özer B, Akçelik M, Akçelİk N. 2015. Identification and characterisation of lactic acid bacteria isolated from traditional Urfa cheese. *Int J Dairy Technol* 69:301–307.

Awad S. 2016. Microbial safety criteria and quality of traditional Egyptian Karish cheese. *African J Microbiol Res* 10:804–12.

Baruzzi F, Lagonigro R, Quintieri L, Morea M, Caputo L. 2012. Occurrence of non-lactic acid bacteria populations involved in protein hydrolysis of cold-stored high moisture Mozzarella cheese. *Food Microbiol* 30:37–44.

Bautista-Gallego J, Alessandria V, Fontana M, Bisotti S, Taricco S, Dolci P, Cocolin L, Rantsiou K. 2014. Diversity and functional characterization of Lactobacillus spp. isolated throughout the ripening of a hard cheese. *Int J Food Microbiol* 181:60–66.

Ben Amor K, Vaughan EE, de Vos WM. 2007. Advanced molecular tools for the identification of lactic acid bacteria. J Nutrition 137:741S–47S.

Blasco L, Kahala M, Jatila H, Joutsjoki V. 2015. Application of 16S-ARDRA and RFLP-PFGE for improved genotypic characterisation of dairy propionibacteria and combination with characteristic phenotypes. *Int Dairy J* 50:66–71.

Bonaïti C, Parayre S, Irlinger F. 2006. Novel extraction strategy of ribosomal RNA and genomic DNA from cheese for PCR-based investigations. *Int J Food Microbiol* 107:171–79.

Bonomo MG, Salzano G. 2012. Microbial diversity and dynamics of Pecorino di Filiano PDO, a traditional cheese of Basilicata region (Southern Italy). *Int J Dairy Technol* 65:531–41.

Boss R, Naskova J, Steiner A, Graber HU. 2011. Mastitis diagnostics: quantitative PCR for *Staphylococcus aureus* genotype B in bulk tank milk. *J Dairy Sci* 94:128–37.

Bove CG, De Dea Lindner J, Lazzi C, Gatti M, Neviani E. 2011. Evaluation of genetic polymorphism among Lactobacillus rhamnosus non-starter Parmigiano Reggiano cheese strains. Int J Food Microbiol 144:569–72.

Bozoudi D, Kotzamanidis C, Hatzikamari M, Tzanetakis N, Menexes G, Litopoulou-Tzanetaki E. 2015. A comparison for acid production, proteolysis, autolysis and inhibitory properties of lactic acid bacteria from fresh and mature Feta PDO Greek cheese, made at three different mountainous areas. *Int J Food Microbiol* 200:87–96.

Bustin SA, Benes V, Garson JA, Hellemans J, Huggett J, Kubista M, Mueller R, Nolan T, Pfaffl MW, Shipley GL, Vandesompele J, Wittwer CT. 2009. The MIQE guidelines: minimum information for publication of quantitative real-time PCR experiments. *Clin Chem* 55(4):611–22.

Calasso M, Ercolini D, Mancini L, Stellato G, Minervini F, Di Cagno R, De Angelis M, Gobbetti M. 2016. Relationships among house, rind and core microbiotas during manufacture of traditional Italian cheeses at the same dairy plant. *Food Microbiol* 54:115–26.

Callon C, Delbès C, Duthoit F, Montel MC. 2006. Application of SSCP-PCR fingerprinting to profile the yeast community in raw milk Salers cheeses. *Syst Appl Microbiol* 29:172–80.

Carafa I, Clementi F, Tuohy K, Franciosi E. 2016. Microbial evolution of traditional mountain cheese and characterization of early fermentation cocci for selection of autochtonous dairy starter strains. *Food Microbiol* 53 (Part B):94–103.

Carafa I, Nardin T, Larcher R, Viola R, Tuohy K, Franciosi E. 2015. Identification and characterization of wild lactobacilli and pediococci from spontaneously fermented Mountain Cheese. *Food Microbiol* 48:123–32.

Carbonnelle E, Mesquita C, Bille E, Day N, Dauphin B, Beretti J-L, Ferroni A, Gutmann L, Nassif X. 2011. MALDI-TOF mass spectrometry tools for bacterial identification in clinical microbiology laboratory. *Clin Biochem* 44:104–109.

Caro I, Mateo J, Sandoval MH, Soto S, García-Armesto MR Castro JM. 2013. Characterization of Oaxaca raw milk cheese microbiota with particular interest in Lactobacillus strains. *J Dairy Sci* 96:3461–70.

Carpino S, Randazzo CL, Pino A, Russo N, Rapisarda T, Belvedere G, Caggia C. 2017. Influence of PDO Ragusano cheese biofilm microbiota on flavour compounds formation. Food Microbiol. 6: 126–35.

Carraro L, Maifreni M, Bartolomeoli I, Martino ME, Novelli E, Frigo F., et al. 2011. Comparison of culture-dependent and culture-independent methods for bacterial community monitoring during Montasio cheese manufacturing. *Res Microbiol* 162:231–39.

Chombo-Morales P, Kirchmayr M, Gschaedler A, Lugo-Cervantes E, Villanueva-Rodríguez S. 2016. Effects of controlling ripening conditions on the dynamics of the native microbial population of Mexican artisanal Cotija cheese assessed by PCR-DGGE. *LWT - Food Sci Technol* 65:1153–61.

Cocolin L, Alessandria V, Dolci P, Gorra R, Rantsiou K. 2013. Culture independent methods to assess the diversity and dynamics of microbiota during food fermentation. *Int J Food Microbiol* 167:29–43.

Coppola S, Fusco V, Andolfi R, Aponte M, Blaiotta G, Ercolini D, Moschetti G. 2006. Evaluation of microbial diversity during the manufacture of Fior di Latte di Agerola, a traditional raw milk pasta-filata cheese of the Naples area. *J Dairy Res* 73:264–72.

Cousin M A. 1982. Presence and activity of psychrotrophic microorganisms in milk and dairy products: A review. *J Food Protect* 36:108–207.

De Filippis F, Genovese A, Ferranti P, Gilbert JA, Ercolini D. 2016. Metatranscriptomics revels temperature-driven functional changes in microbiome impacting cheese maturation rate. *Sci Rep* 6:21871.

De Filippis F, La Storia A, Stellato G, Gatti M, Ercolini D. 2014. A selected core microbiome drives the early stages of three popular Italian cheese manufactures. *PLoS One* 9: e89680.

De Pasquale I, Calasso M, Mancini L, Ercolini D, La Storia A, De Angelis M, Di Cagno R, Gobbetti M. 2014b. Causal relationship between microbial ecology dynamics and proteolysis during manufacture and ripening of protected designation of origin (PDO) cheese Canestrato Pugliese. *Appl Environ Microbiol* 80:4085–94.

De Pasquale I, Di Cagno R, Buchin S, De Angelis M, Gobbetti M. 2014a. Microbial ecology dynamics reveal a succession in the core microbiota involved in the ripening of pasta filata caciocavallo pugliese cheese. *Appl Environ Microbiol* 80:6243–55.

De Pasquale I, Di Cagno R, Buchin S, De Angelis M, Gobbetti M. 2016. Spatial distribution of the metabolically active microbiota within Italian PDO ewes' milk cheeses. *PLoS One* 11:e0153213.

Delavenne E, Mounier J, Asmani K, Jany JL, Barbier G, Le Blay G. 2011. Fungal diversity in cow, goat and ewe milk. *Int J Food Microbiol* 151:247–51.

Delbès C, Ali-Mandjee L, Montel MC. 2007. Monitoring bacterial communities in raw milk and cheese by culture-dependent and -independent 16S rRNA gene-based analyses. *Appl Environ Microbiol* 73, 1882–1891.

Delcenserie V, Taminiau B, Delhalle L, Nezer C, Doyen P, Crevecoeur S, Rossey D, Korsak N, Daube G. 2014. Microbiota characterization of a Belgian protected designation of origin cheese, Herve cheese, using metagenomic analysis. *J Dairy Sci* 97:1–11.

Delgado S, Rachid CT, Fernández E, Rychlik T, Alegría A, Peixoto RS, Mayo B. 2013. Diversity of thermophilic bacteria in raw, pasteurized and selectively-cultured milk, as assessed by culturing, PCR-DGGE and pyrosequencing. *Food Microbiol* 36:103–11.

Di Cagno R, Minervini G, Sgarbi E, Lazzi C, Bernini V, Neviani E, Gobbetti M. 2010. Comparison of phenotypic (Biolog System) and genotypic (random amplified polymorphic DNA-polymerase chain reaction, RAPD-PCR, and amplified fragment length polymorphism, AFLP) methods for typing *Lactobacillus plantarum* isolates from raw vegetables and fruits. *Int J Food Microbiol* 143:246–53.

Diaz M, Ladero V, Del Rio B, Redruello B, Fernández M, Martin MC, Alvarez MA. 2016. Biofilm-forming capacity in biogenic amine-producing bacteria isolated from dairy products. *Front Microbiol* 7:591.

Dolci P, De Filippis F, La Storia A, Ercolini D, Cocolin L. 2014. rRNA-based monitoring of the microbiota involved in Fontina PDO cheese production in relation to different stages of cow lactation. *Int J Food Microbiol* 185:127–35.

Dolci P, Zenato S, Pramotton R, Barmaz A, Alessandria V, Rantsiou K, Cocolin L. 2013. Cheese surface microbiota complexity: RT-PCR-DGGE, a tool for a detailed picture? *Int J Food Microbiol* 162:8–12.

Dugat-Bony E, Straub C, Teissandier A, Onésime D, Loux V, Monnet C, Irlinger F, Landaud S, Leclercq-Perlat MN, Bento P, Fraud S, Gibrat JF, Aubert J, Fer F, Guédon E, Pons N, Kennedy S, Beckerich JM, Swennen D, Bonnarme P. 2015. Overview of a surface-ripened cheese community functioning by meta-omics analyses. *PLoS One* 10(4):e0124360.

Duquenne M, Fleurot I, Aigle M, Darrigo C, Borezée-Durant E, Derzelle S, Bouix M, Deperrois-Lafarge V, Delacroix-Buchet A. 2010. Tool for quantification of staphylococcal enterotoxin gene expression in cheese. *Appl Environ Microbiol* 76:1367–74.

Dušková M, Šedo O, Kšicová K, Zdráhal Z, Karpíšková R. 2012. Identification of lactobacilli isolated from food by genotypic methods and MALDI-TOF MS. *Int J Food Microbiol* 159:107–114.

Eneroth A, Christiansson A, Brendehaug J, Molin G. 1998. Critical contamination sites in the production line of pasteurised milk, with reference to the psychrotrophic spoilage flora. *Int Dairy J* 8:829–834.

Ercolini D, Frisso G, Mauriello G, Salvatore F, Coppola S. 2008. Microbial diversity in natural whey cultures used for the production of Caciocavallo Silano PDO cheese. *Int J Food Microbiol* 124:164–70.

Ercolini D. 2013. High-throughput sequencing and metagenomics: moving forward in the culture-independent analysis of food microbial ecology. *Appl Environ Microbiol* 79(10):3148–55.

Falentin H, Postollec F, Parayre S, Henaff N, Le Bivic P, Richoux R, Thierry A, Sohier D. 2010. Specific metabolic activity of ripening bacteria quantified by real-time reverse transcription PCR throughout Emmental cheese manufacture. *Int J Food Microbiol* 144:10–19.

Falentin H, Henaff N, Le Bivic P, Deutsch SM, Parayre S, Richoux R, Sohier D, Thierry A, Lortal S, Postollec F. 2012. Reverse transcription quantitative PCR revealed persistency of thermophilic lactic acid bacteria metabolic activity until the end of the ripening of Emmental cheese. *Food Microbiol* 29:132–40.

Feligini M, Brambati E, Panelli S, Ghitti M, Sacchi R, Capelli E, Bonacina C., 2014. One-year investigation of *Clostridium* spp. occurrence in raw milk and curd of Grana Padano cheese by the automated ribosomal intergenic spacer analysis. *Food Control* 42:71–77.

Feligini M, Panelli S, Buffoni JN, Bonacina C, Andrighetto C, Lombardi A. 2012. Identification of microbiota present on the surface of Taleggio cheese using PCR-DGGE and RAPD-PCR. *J Food Sci* 77:M609–15.

Felis GE, Dellaglio F. 2007. Taxonomy of lactobacilli and bifidobacteria. *Curr Issues Intestinal Microbiol* 8:44–61.

Fernández-No IC, Böhme K, Gallardo JM, Barros-Velázquez J, Cañas B, Calo-Mata P. 2010. Differential characterization of biogenic amine-producing bacteria involved in food poisoning using MALDI-TOF mass fingerprinting. *Electrophoresis* 31:1116–27.

Fisher MM, Triplett EW. 1999. Automated approach for ribosomal intergenic spacer analysis of microbial diversity and its application to freshwater bacterial communities. *Appl Environ Microbiol* 65:4630–36.

Franciosi E, Carafa I, Nardin T, Schiavon S, Poznanski E, Cavazza A, Larcher R, Tuohy, KM. 2015. Biodiversity and γ-aminobutyric acid production by lactic acid bacteria isolated from traditional alpine raw cow's milk cheeses. *BioMed Res Int* 1–11.

Freitas Rd, Chuat V, Madec M-N, Nero LA, Thierry A, Valence F, Carvalho AFd. 2015. Biodiversity of dairy Propionibacterium isolated from dairy farms in Minas Gerais, Brazil. *Int J Food Microbiol* 203:70–77.

Fuka MM, Wallisch S, Engel M, Welzl G, Havranek J, Schloter M. 2013. Dynamics of bacterial communities during the ripening process of different Croatian cheese types derived from raw ewe's milk cheeses. *PLoS ONE* 8:e80734.

Garofalo C, Osimani A, Milanović V, Aquilanti L, De Filippis F, Stellato G, Di Mauro S, Turchetti B, Buzzini P, Ercolini D, Clementi F. 2015. Bacteria and yeast microbiota in milk kefir grains from different Italian regions. *Food Microbiol* 49:123–33.

Ghafari E, Doosti A, Mokhtari-Farsani A, Ghasemi-Dehkordi P, Mahmoudi E, Kabiri H. 2014. Genotypic and phenotypic study of Lactobacillus species isolated from traditional yogurt and cheese in southwest Iran. *Int J Biosci* 5:155–62.

Ghyselinck J, Van Hoorde K, Hoste B, Heylen K, De Vos P. 2011. Evaluation of MALDI-TOF MS as a tool for high-throughput dereplication. *J Microbiol Methods* 86:327–36.

Giraffa G, Gatti M, Rossetti L, Senini L, Neviani E. 2000. Molecular diversity within *Lactobacillus helveticus* as revealed by genotypic characterization. *Applied Environ Microbiol* 66:1259–65.

Giraffa G, Neviani E. 2001. DNA-based, culture-independent strategies for evaluating microbial communities in food-associated ecosystems. *Int J Food Microbiol* 67:19–34.

Guidone A, Braghieri A, Cioffi S, Claps S, Genovese F, Morone G, Napolitano F, Parente E. 2015. Effect of adjuncts on microbiological and chemical properties of Scamorza cheese. *J Dairy Sci* 98:1467–78.

Guley Z, Uysal HR, Kilic S. 2014. Lactic acid bacteria flora of Konya Kuflu cheese: a traditional cheese from Konya province in Turkey. *J Microbiol Biotechnol Food Sci* 4:238–42.

Hadrys H, Balick M, Schierwater B. 1992. Applications of random amplified polymorphic DNA (RAPD) in molecular ecology. *Mol Ecol* 1:55–63.

Hayashi K. 1991. PCR-SSCP: a simple and sensitive method for detection of mutations in the genomic DNA. *PCR Methods Appl* 1:34–38.

Hermet A, Mounier J, Keravec M, Vasseur V, Barbier G, Jany JL. 2014. Application of capillary electrophoresis single-stranded conformation polymorphism (CE-SSCP) analysis for identification of fungal communities in cheese. *Food Microbiol* 41:82–90.

Jans C, Bugnard J, Njage PMK, Lacroix C, Meile L. 2012a. Lactic acid bacteria diversity of African raw and fermented camel milk products reveals a highly competitive, potentially health-threatening predominant microflora. *LWT - Food Sc Technol* 47:371–79.

Jans C, Lacroix C, Meile L. 2012b. A novel multiplex PCR/RFLP assay for the identification of *Streptococcus bovis/Streptococcus equinus* complex members from dairy microbial communities based on the 16S rRNA gene. *FEMS Microbiol Lett* 326(2):144–50.

Jany JL, Barbier G. 2008. Culture-independent methods for identifying microbial communities in cheese. *Food Microbiol* 25:839–48.

Justé A, Thomma BPHJ, Lievens B. 2008. Recent advances in molecular techniques to study microbial communities in food-associated matrices and processes. *Food Microbiol* 25:745–61.

Kern CC, Vogel RF, Behr J. 2014. Differentiation of *Lactobacillus brevis* strains using matrix-assisted-laser-desorption-ionization-time-of-flight mass spectrometry with respect to their beer spoilage potential. *Food Microbiol* 40:18–24.

Kesmen Z, Kacmaz N. 2011. Determination of lactic microflora of kefir grains and kefir beverage by using culture-dependent and culture-independent methods. *J Food Sc* 76:M276–M283.

Khemariya P, Singh S, Nath G, Gulati AK. 2013. Isolation, identification, and antibiotic susceptibility of nis+ *Lactococcus lactis* from dairy and non-dairy sources. *Czech J Food Sci*, 31:323–31.

Lazzi C, Bove CG, Sgarbi E, Monica G, La Gioia F, Sandra T, Neviani E. 2009. Application of AFLP fingerprint analysis for studying the biodiversity of *Streptococcus thermophilus*. *J Microbiol Methods* 79:48–54.

Lazzi C, Povolo M, Locci F, Bernini V, Neviani E, Gatti M. 2016. Can the development and autolysis of lactic acid bacteria influence the cheese volatile fraction? The case of Grana Padano. *Int J Food Microbiol* 233:20–28.

Le Dréan G, Mounier J, Vasseur V, Arzur D, Habrylo O, Barbier G. 2010. Quantification of *Penicillium camemberti* and *P. roqueforti* mycelium by real-time PCR to assess their growth dynamics during ripening cheese. *Int J Food Microbiol* 138:100–107.

Leite AMO, Mayo B, Rachid CTCC, Peixoto RS, Silva JT, Paschoalin VMF, Delgado S. 2012. Assessment of the microbial diversity of Brazilian kefir grains by PCR-DGGE and pyrosequencing analysis. *Food Microbiol* 31:215–221.

Lessard MH, Bélanger G, St-Gelais D, Labrie S. 2012. The composition of Camembert cheese-ripening cultures modulates both mycelial growth and appearance. *Appl Environ Microbiol* 78:1813–19.

Lessard MH, Viel C, Boyle B, St-Gelais D, Labrie S. 2014. Metatranscriptome analysis of fungal strains *Penicillium camemberti* and *Geotrichum candidum* reveal cheese matrix breakdown and potential development of sensory properties of ripened Camembert-type cheese. *BMC Genomics* 5:235.

Liu W, Zheng Y, Kwok LY, Sun Z, Zhang J, Guo Z, Hou Q, Menhe B, Zhang H. 2015b. High-throughput sequencing for the detection of the bacterial and fungal diversity in Mongolian naturally fermented cow's milk in Russia. *BMC Microbiol* 15:45.

Liu X-F, Liu C-J, Zhang H-Y, Gong F-M, Luo Y-Y, Li X-R. (2015a) The bacterial community structure of yond bap, a traditional fermented goat milk product, from distinct Chinese regions. *Dairy Sci Technol* 95:369–80.

Mangia NP, Fancello F, Deiana P 2016. Microbiological characterization using combined culture dependent and independent approaches of Casizolu pasta filata cheese. *J Applied Microbiol* 120, 329–345.

Marilley L, Casey MG. 2004. Flavours of cheese products: metabolic pathways, analytical tools and identification of producing strains. *Int J Food Microbiol* 90:139–159.

Martín-Platero AM, Maqueda M, Valdivia E, Purswani J, Martínez-Bueno M. 2009. Polyphasic study of microbial communities of two Spanish farmhouse goats' milk cheeses from Sierra de Aracena. *Food Microbiol* 26:294–304.

McCartney AL. 2002. Application of molecular biological methods for studying probiotics and the gut flora. *Br J Nutr* 88:S29–37.

McPhee JD, Griffiths MW. 2002. *Pseudomonas spp*. In: Roginski H, Fuquay WJ, Fox FP (Eds). *Encyclopedia of dairy sciences*, Vol. 4. Academic Press, pp 2340–50.

Monnet V, Juillard V, Gardan R. 2016. Peptide conversations in Gram-positive bacteria. *Crit Rev Microbiol* 42:339–51.

Montel MC, Buchin S, Mallet A, Delbes-Paus C, Vuitton DA, Desmasures N, Berthier F. 2014. Traditional cheeses: rich and diverse microbiota with associated benefits. *Int J Food Microbiol* 177:136–54.

Morandi S, Brasca M. 2012. Safety aspects, genetic diversity and technological characterisation of wild-type *Streptococcus thermophilus* strains isolated from north Italian traditional cheeses. *Food Control* 23:203–209.

Mormile A, Barile M, Mercogliano R, Johansson P, Björkroth KJ, Aponte M, Murru N. 2016. Dynamics of lactic acid bacteria in "Pecorino di Tramonti"—a ewe's milk cheese—with particular emphasis on enterococci: a preliminary study. *Ann Microbiol* 66:179–85.

Mounier J, Le Blay G, Vasseur V, Le Floch G, Jany JL, Barbier G. 2010. Application of denaturing high-performance liquid chromatography (DHPLC) for yeasts identification in red smear cheese surfaces. *Lett Appl Microbiol* 51:18–23.

Mounier J, Monnet C, Jacques N, Antoinette A, Irlinger F. 2009. Assessment of the microbial diversity at the surface of Livarot cheese using culture-dependent and independent approaches. *Int J Food Microbiol* 133:31–37.

Muyzer G, de Waal EC, Uitterlinden AG. 1993. Profiling of complex microbial populations by denaturing gradient gel electrophoresis analysis of polymerase chain reaction-amplified genes coding for 16S rRNA. *Appl Environ Microbiol* 59:695–700.

Nalbantoglu U, Cakar A, Dogan H, Abaci N, Ustek D, Sayood K, Can H. 2014. Metagenomic analysis of the microbial community in kefir grains. *Food Microbiol* 41:42–51.

Naumann D, Helm D, Labischinski H. 1991. Microbiological characterizations by FT-IR spectroscopy. *Nature* 351:81–82.

Ndoye B, Lessard MH, LaPointe G, Roy D. 2011. Exploring suppression subtractive hybridization (SSH) for discriminating *Lactococcus lactis* ssp. *cremoris* SK11 and ATCC 19257 in mixed culture based on the expression of strain-specific genes. *J Appl Microbiol* 110:499–512.

Nocker A, Cheung CY, Camper AK. 2006. Comparison of propidium monoazide with ethidium monoazide for differentiation of live vs. dead bacteria by selective removal of DNA from dead cells. *J Microbiol Methods* 67:310–20.

Nolan T, Hands RE, Bustin SA. 2006. Quantification of mRNA using real-time RT-PCR. *Nat Protoc* 1:1559–82.

Ongol MP, Tanaka M, Sone T, Asano K. 2009. A real-time PCR method targeting a gene sequence encoding 16S rRNA processing protein, rimM, for detection and enumeration of *Streptococcus thermophilus* in dairy products. *Food Res Int* 42:893–98.

Orita M, Iwahana H, Kanazawa H, Hayashi K, Sekiya T. 1989. Detection of polymorphisms of human DNA by gel electrophoresis as single-strand conformation polymorphisms. *Proc Natl Acad Sci USA* 86:2766–70.

O'Sullivan DJ, Giblin L, McSweeney PL, Sheehan JJ, Cotter PD. 2013. Nucleic acid-based approaches to investigate microbial-related cheese quality defects. *Frontiers in Microbiol* 4.

Panelli S, Brambati E, Bonacina C, Feligini M. 2013. Detection of *Clostridium tyrobutyricum* in milk to prevent late blowing in cheese by automated ribosomal intergenic spacer analysis. *J Food Sci* 78:M1569–74.

Pavlovic M, Huber I, Konrad R, Busch U. 2013. Application of MALDI-TOF MS for the identification of food borne bacteria. *Open Microbiol J* 7:135–41.

Perin LM, Dal Bello B, Belviso S, Zeppa G, Carvalho AFd, Cocolin L, Nero LA. 2015. Microbiota of Minas cheese as influenced by the nisin producer *Lactococcus lactis* subsp. *lactis* GLc05. *Int J Food Microbiol* 214:159–67.

Pogačić T, Mancini A, Santarelli M, Bottari B, Lazzi C, Neviani E, Gatti M. 2013. Diversity and dynamic of lactic acid bacteria strains during aging of a long ripened hard cheese produced from raw milk and undefined natural starter. *Food Microbiol* 6:207–215.

Porcellato D, Brighton C, McMahon DJ, Oberg CJ, Lefevre M, Broadbent JR, Steele JL. 2014. Application of ARISA to assess the influence of salt content and cation type on microbiological diversity of Cheddar cheese. *Lett Appl Microbiol* 59(2):207–16.

Postollec F, Falentin H, Pavan S, Combrisson J, Sohier D. 2011. Recent advances in quantitative PCR (qPCR) applications in food microbiology. *Food Microbiol* 28:848–61.

Quero GM, Fusco V, Cocconcelli PS, Owczarek L, Borcakli M, Fontana C, Skapska S, Jasinska UT, Ozturk T, Morea M. 2014. Microbiological, physico-chemical, nutritional and sensory characterization of traditional Matsoni: selection and use of autochthonous multiple strain cultures to extend its shelf-life. *Food Microbiol* 38:179–91.

Quigley L, O'Sullivan O, Beresford TP, Ross RP, Fitzgerald FG, Cotter PD. 2011. Molecular approaches to analyzing the microbial composition of raw milk and raw milk cheese. *Int J Food Microbiol* 150:81–94.

Randazzo CL, Caggia C, Neviani E. 2009. Application of molecular approaches to study lactic acid bacteria in artisanal cheeses. *J Microbiol Methods* 78:1–9.

Randazzo CL, Pitino I, Ribbera A, Caggia C. 2010. Pecorino Crotonese cheese: study of bacterial population and flavour compounds. *Food Microbiol* 27:363–74.

Rantsiou K, Urso R, Dolci P, Comi G, Cocolin L. 2008. Microflora of Feta cheese from four Greek manufacturers. *Int J Food Microbiol* 26(1-2):36–42.

Rasolofo EA, LaPointe G, Roy D. 2011. Assessment of the bacterial diversity of treated and untreated milk during cold storage by T-RFLP and PCR-DGGE methods. *Dairy Sci Technol* 91(5):573–97.

Rasolofo EA, St-Gelais D, LaPointe G, Roy D, 2010. Molecular analysis of bacterial population structure and dynamics during cold storage of untreated and treated milk. *Int J Food Microbiol* 138(1-2):108–118.

Riquelme C, Câmara S, Enes Dapkevicius MdLN, Vinuesa P, da Silva CCG, Malcata FX, Rego OA. 2015. Characterization of the bacterial biodiversity in Pico cheese (an artisanal Azorean food). *Int J Food Microbiol* 192:86–94.

Ruggirello M, Dolci P, Cocolin L. 2014. Detection and viability of Lactococcus lactis throughout cheese ripening. *PLoS One* 9(12):e114280. doi: 10.1371/journal.pone.0114280.

Ruiz-Moyano S, Tao N, Underwood MA, Mills DA. 2012. Rapid discrimination of *Bifidobacterium animalis* subspecies by matrix-assisted laser desorption ionization-time of flight mass spectrometry. *Food Microbiol* 30:432–37.

Sacristán N, Mayo B, Fernández E, Fresno JM, Tornadijo ME, Castro JM. 2013. Molecular study of Geotrichum strains isolated from Armada cheese. *Food Microbiol* 36(2):481–87.

Sandrin TR, Goldstein JE, Schumaker S. 2013. MALDI TOF MS profiling of bacteria at the strain level: a review. *Mass Spectrom Rev* 32:188–217.

Santarelli M, Bottari B, Lazzi C, Neviani E, Gatti M. 2013. Survey on the community and dynamics of lactic acid bacteria in Grana Padano cheese. *Syst Appl Microbiol* 36:593–600.

Saubusse M, Millet L, Delbès C, Callon C, Montel MC. 2007. Application of single strand conformation polymorphism—PCR method for distinguishing cheese bacterial communities that inhibit *Listeria monocytogenes*. *Int J Food Microbiol* 116:126–35.

Sauer S, Freiwald A, Maier T, Kube M, Reinhardt R, Kostrzewa M, Geider K. 2008. Classification and identification of bacteria by mass spectrometry and computational analysis. *PLoS ONE* 3:e2843.

Schornsteiner E, Mann E, Bereuter O, Wagner M, Schmitz-Esser S. 2014. Cultivation-independent analysis of microbial communities on Austrian raw milk hard cheese rinds. *Int J Food Microbiol* 180:88–97.

Schütte U, Abdo Z, Bent S, Shyu C, Williams C, Pierson J, Forney L. 2008. Advances in the use of terminal restriction fragment length polymorphism (T-RFLP) analysis of 16S rRNA genes to characterize microbial communities. *Appl Microbiol Biotechnol* 80:365–80.

Serhan M, Cailliez-Grimal C, Borges F, Revol-Junelles AM, Hosri C, Fanni J. 2009. Bacterial diversity of Darfiyeh, a Lebanese artisanal raw goat's milk cheese. *Food Microbiol* 26:645–52.

Silva LF, Casella T, Gomes ES, Nogueira MCL, De Dea LJ, Penna ALB. 2015. Diversity of lactic acid bacteria isolated from Brazilian water buffalo mozzarella cheese. *J Food Sci* 80:M411–M417.

Singh S Goswami P, Singh R Hellerc KJ. 2009. Application of molecular identification tools for Lactobacillus, with a focus on discrimination between closely related species: a review. *LWT - Food Sci Technol* 42:448–45.

Sofu A, Ekinc FY. 2016. Bacterial diversity dynamics of traditional Turkish Ezine cheese as evaluated by PCR-DGGE and SSCP analysis. *Int J Dairy Technol* 69:592–600.

Sogin ML, Morrison HG, Huber JA, Welch DM, Huse SM, Neal PR, et al. 2006. Microbial diversity in the deep sea and the underexplored 'rare biosphere'. *Proc Nat Acad Sci USA* 103:12115–20.

Solieri L, Bianchi A, Giudici P. 2012. Inventory of non starter lactic acid bacteria from ripened Parmigiano Reggiano cheese as assessed by a culture dependent multiphasic approach. *System Applied Microbiol* 35:270–77.

Stellato G, De Filippis F, La Storia A, Ercolini D. 2015. Coexistence of Lactic Acid Bacteria and potential spoilage microbiota in a dairy processing environment. *Appl Environ Microbiol* 81(22):7893–904. doi: 10.1128/AEM.02294-155.

Sun Z, Liu W, Bao Q, Zhang J, Hou Q, Kwok L, Sun T, Zhang H. 2014. Investigation of bacterial and fungal diversity in tarag using high-throughput sequencing. *J Dairy Sci* 97(10):6085–96. doi: 10.3168/jds.2014-8360.

Terzić-Vidojević A, Tonković K, Leboš Pavunc A, Beganović J, Strahinić I, Kojić M, Veljović K, Golić N, Kos B, Čadež N, Gregurek L, Šušković J, Raspor P, Topisirović L. 2015. Evaluation of autochthonous lactic acid bacteria as starter cultures for production of white pickled and fresh soft cheeses. *LWT - Food Sci Technol* 63:298–306.

Tofalo R, Fasoli G, Schirone M, Perpetuini G, Pepe A, Corsetti A, Suzzi G. 2014. The predominance, biodiversity and biotechnological properties of *Kluyveromyces marxianus* in the production of Pecorino di Farindola cheese. *Int J Food Microbiol* 187:41–49.

Tormo H, Ali Haimoud Lekhal D, Roques C. 2015. Phenotypic and genotypic characterization of lactic acid bacteria isolated from raw goat milk and effect of farming practices on the dominant species of lactic acid bacteria. *Int J Food Microbiol* 210:9–15.

Tsafrakidou P, Bozoudi D, Pavlidou S, Kotzamanidis C, Hatzikamari M, Zdragas A, Litopoulou-Tzanetaki E. 2016. Technological, phenotypic and genotypic characterization of lactobacilli from Graviera Kritis PDO Greek cheese, manufactured at two traditional dairies. *LWT - Food Sci Technol* 68:681–89.

Turvey ME, Weiland F, Meneses J, Sterenberg N, Hoffmann P. 2016. Identification of beer spoilage microorganisms using the MALDI Biotyper platform. *Applied Microbiol Biotechnol* 100:2761–73.

Van Hoorde K, Van Leuven I, Dirinck P, Heyndrickx M, Coudijzer K, Vandamme P, Huys G. 2010. Selection, application and monitoring of *Lactobacillus paracasei* strains as adjunct cultures in the production of Gouda-type cheeses. *Int J Food Microbiol* 144:226–35.

Van Hoorde K, Vandamme P, Huys G. 2008a. Molecular identification and typing of lactic acid bacteria associated with the production of two artisanal raw milk cheeses. *Dairy Sci Technol* 88:445–55.

Van Hoorde K, Verstraete T, Vandamme P, Huys G. 2008b. Diversity of lactic acid bacteria in two Flemish artisan raw milk Gouda-type cheeses. *Food Microbiol* 25:929–35.

Vandamme P, Pot B, Gillis M, de Vos P, Kersters K, Swings J. 1996. Polyphasic taxonomy, a consensus approach to bacterial systematics. *Microbiol Rev* 60:407–38.

Versalovic J, Schneider M, De Bruijn FJ, Lupski JR. 1994. Genomic fingerprinting of bacteria using repetitive sequence-based polymerase chain reaction. *Methods Molec Cell Biol* 5:25–40.

Vithanage NR, Dissanayake M, Bolge G, Palombo EA, Yeager TR, Datta N. 2016. Biodiversity of culturable psychrotrophic microbiota in raw milk attributable to refrigeration conditions, seasonality and their spoilage potential. *Int Dairy J* 57:80–90.

Vithanage NR, Yeager TR, Jadhav SR, Palombo EA, Datta N. 2014. Comparison of identification systems for psychrotrophic bacteria isolated from raw bovine milk. *Int J Food Microbiol* 189:26–38.

von Neubeck M, Baur C, Krewinkel M, Stoeckel M, Kranz B, Stressler T, Fischer L, Hinrichs J, Scherer S, Wenning M. 2015. Biodiversity of refrigerated raw milk microbiota and their enzymatic spoilage potential. *Int J Food Microbiol* 211:57–65.

Vrabec M, Lovayová V, Dudriková K, Gallo J, Dudriková E. 2015. Antibiotic resistance and prevalence of *Enterococcus* spp. and *Escherichia coli* isolated from Bryndza cheese. *Italian J Animal Sci* 14:3968.

Ward P, Roy D. 2005. Review of molecular methods for identification, characterization and detection of bifidobacteria. *Lait* 85:23–32.

Wenning M, Breitenwieser F, Konrad R, Huber I, Busch U, Scherer S. 2014. Identification and differentiation of food-related bacteria: a comparison of FTIR spectroscopy and MALDI-TOF mass spectrometry. *J Microbiol Methods* 103:44–52.

Wieme AD, Spitaels F, Aerts M, De Bruyne K, Van Landschoot A, Vandamme P. 2014a. Identification of beer-spoilage bacteria using matrix-assisted laser desorption/ionization time-of-flight mass spectrometry. *Int J Food Microbiol* 185:41–50.

Wieme AD, Spitaels F, Vandamme P, Van Landschoot A. 2014b. Application of matrix-assisted laser desorption/ionization time-of-flight mass spectrometry as a monitoring tool for in-house brewer's yeast contamination: a proof of concept. *J Inst Brewing* 120:438–43.

Williams DL, Gillis TP, Booth RJ, Looker D, Watson JD. 1990. The use of a specific DNA probe and polymerase chain reaction for the detection of *Mycobacterium leprae*. *J Infect Dis* 162:193–200.

Woese CR. 1987. Bacterial evolution. *Microbiol Rev* 51:221–71.

Wolfe BE, Button JE, Santarelli M, Dutton RJ. 2014. Cheese rind communities provide tractable systems for in situ and in vitro studies of microbial diversity. *Cell* 158:422–33.

Zago M, Bonvini B, Carminati D, Giraffa G. 2009. Detection and quantification of *Enterococcus gilvus* in cheese by real-time PCR. *Syst Appl Microbiol* 32(7):514–21.

Zeller-Péronnet V, Brockmann E, Pavlovic M, Timke M, Busch U, Huber I. 2013. Potential and limitations of MALDI-TOF MS for discrimination within the species *Leuconostoc mesenteroides* and *Leuconostoc pseudomesenteroides*. Journal für Verbraucherschutz und Lebensmittelsicherheit 8:205–214.

9 LAB strains with bacteriocin synthesis genes and their applications

Lorena Sacchini, Giacomo Migliorati, Elisabetta Di Giannatale, Francesco Pomilio and Franca Rossi

Istituto Zooprofilattico Sperimentale, "G. Caporale", Teramo, Italy

9.1 INTRODUCTION

Lactic acid bacteria (LAB) constitute a heterogeneous group of microorganisms that can produce bacteriocins, which are proteinaceous antimicrobial molecules with a diverse genetic origin, posttranslationally modified (PTM) or not, and exported to the extracellular medium, which can help the producer organism to outcompete other bacterial species. Bacteriocins produced by LAB are particularly interesting due to the long history of safe use of some of them and the generally regarded as safe (GRAS) and qualified presumption of safety (QPS) status that most LAB possess. LAB bacteriocins are diverse in terms of chemical structure, molecular weight, posttranslational modification, mechanism of action, and percentage of modified amino acids. Among bacteriocins, nisin is approved in the food industry as a food additive, but other bacteriocins have a great potential to be applied as biopreservatives for food safety interventions.

9.2 BACTERIOCINS FROM LAB

The first classification of LAB bacteriocins was proposed by Klaenhammer (1993). Cotter et al. (2005) suggested a more radical modification of the previous classification scheme, and in recent years a new scheme has been proposed based on previous ones (Heng et al., 2007; Nishant et al., 2011; Nishie et al., 2012). According to this scheme, bacteriocins are grouped into two categories (Table 9.1): lantibiotics (class I) and non-lanthionine-containing bacteriocins (class II), as opposed to the four classes of the Klaenhammer (1993) classification scheme. The most notable change in this scheme is that it reclassified the class III bacteriocins as bacteriolysins, since they are lytic enzymes rather than peptides.

Class I bacteriocins or lantibiotics (lanthionine-containing antibiotics) exhibit a cationic and amphiphilic structure with a low molecular weight, from approximately 3 to 10 kDa, and are subdivided in two classes, I and II. Class I lantibiotics are formed

Microbiology in Dairy Processing: Challenges and Opportunities, First Edition.
Edited by Palmiro Poltronieri.

Published 2018 by John Wiley & Sons Ltd and the Institute of Food Technologists.

Table 9.1 Classes of bacteriocins produced by Lactic Acid Bacteria (LAB).

Class	Subclass	Subgroup	Features	Examples
I Small peptides (<5 kDa) with posttranslational modification yielding unusual amino acid residues such as lanthionine and ß-methyllanthionine	Lantibiotics I		Unusual amino acids introduced by LanB and LanC	Nisin A/Z
	Lantibiotics II		Unusual amino acids introduced by LanM	Lacticin 3147
II Small (<10 kDa), no posttranslational modification, heat-stable, small cationic and amphiphilic	IIa	Subgroup I	Difference in *C*-terminal, antilisterial activity	Pediocin PA1/AcH, sakacin P, enterocin A
		Subgroup II		Leucocin A, mesentericin Y105
		Subgroup III		Curvacin A, carnobacteriocin B2
	IIb		Two-peptide system for activation	Lactococcin G, Q and ABP-118, lactacin F, plantaricin EF/JK, plantaricin S, lactocin 705, enterocin X, enterocin NKR-5-3A/Z, leucocin A, munditicin
	IIc		Circular structure with *N*- and *C*- covalently linked	Enterocin AS-48, gassericin A, uberolysin, carnocyclin A, lactocyclin Q, leucocyclicin Q, garvicin ML, acidocin B
	IId		Other class II bacteriocins	Lactococcin A and 972, enterocin EJ97 and L50, lacticin Q and Z, weissellicin Y and M, leucocin Q and N, garvieacin Q, bacteriocin 7293

by small peptides containing 19–38 amino acid moieties and contain unusual modified amino acids, such as lanthionine and 3-methyllanthionine and some dehydrated amino acids. These unusual residues form covalent bonds between amino acids, which result in internal 'rings' conferring to lantibiotics stability against heat, to a wide range of pH, against proteolysis and resistance to oxidation (Cotter et al., 2005; Nishie et al., 2012).

Based on posttranslationally modifying (PTM) enzymes involved in the maturation process, lanthipeptides can be divided into type I, which can be modified by enzymes LanB and LanC, and type II, modified by enzyme LanM (Knerr and van der Donk 2012; Nishie et al., 2012). Nisin, a type I lantibiotic produced by *Lactococcus lactis* is the best studied.

In Figure 9.1, the clusters of genes encoding the production of some of wide spectrum lantibiotics produced by LAB, i.e. nisin, lacticin 481 and lacticin 3147 and the cellular localization and function of proteins encoded by the nisin operon are shown.

Class II bacteriocins, or the non-lantibiotics, are the bacteriocins most frequently occurring in nature. They are small (<10 kDa), heat-stable, non-lanthionine-containing peptides, which, unlike lantibiotics, do not undergo extensive posttranslational modification. This group can be further subdivided into four subclasses: pediocin-like bacteriocins (class IIa), two-component bacteriocins (class IIb), circular

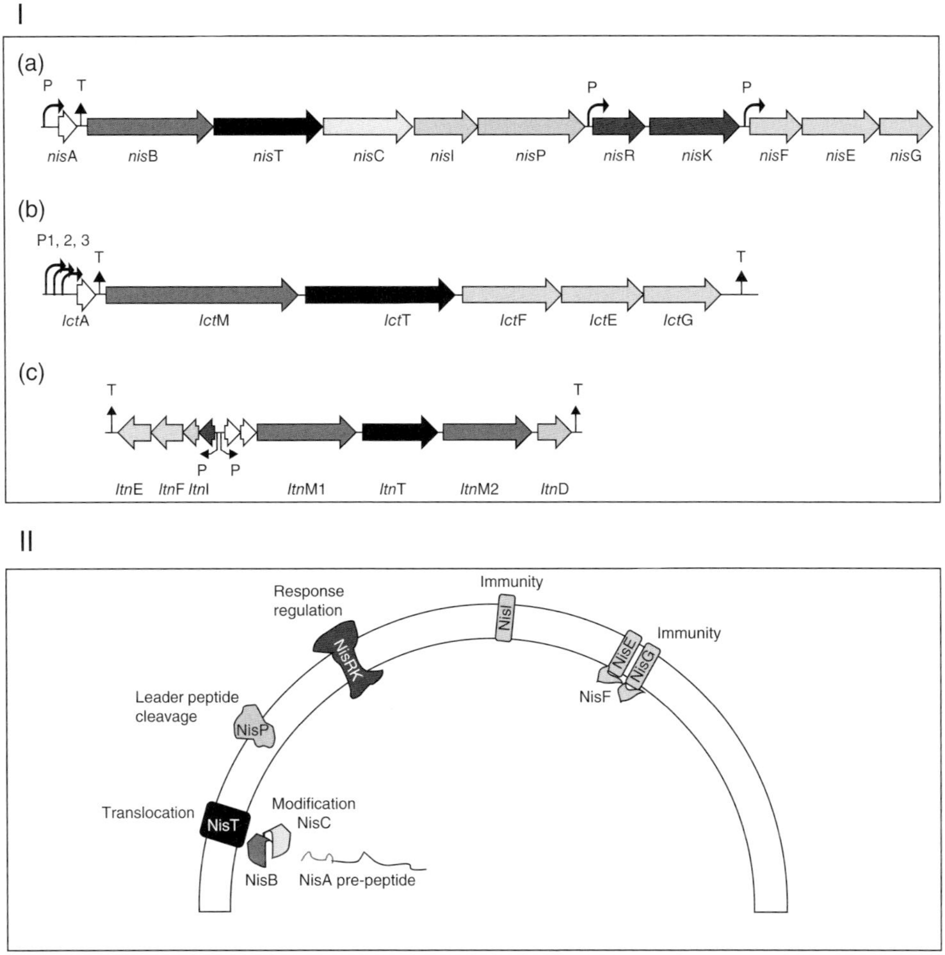

Figure 9.1 Structure of some lantibiotic encoding operons and function of the respective gene products. Panel I: gene arrangement in the nisin (a) (Lubelski et al., 2008), lacticin 481 (b) (Rincé et al., 1997), and lacticin 3147 (c) gene clusters (McAuliffe et al., 2001). T: terminator is often represented with a circle on the stem. Panel II: function and cellular localization of the nisin modification, export and immunity proteins.

bacteriocins (class IIc), and unmodified, linear, non-pediocin-like bacteriocins (class IId) (Cotter et al., 2005).

Class IIa bacteriocins have a distinct conserved sequence (YGNGVXC) in the cationic *N*-terminal region and contain two cysteine residues joined by a disulfide bridge that are responsible for their high potency against the food pathogen *Listeria monocytogenes* (Cui et al., 2102).

The class IIb bacteriocins are two-peptide bacteriocins that require both peptides to work synergistically to be fully active. Class IIc bacteriocins are grouped on the basis of their circular structural configuration. The *N*- and *C*-termini of class IIc bacteriocins are covalently linked, giving to the peptide an extremely stable structure. On the other hand, class IId bacteriocins comprise the remaining bacteriocins combined as a miscellaneous, one-peptide non-pediocin linear group (Cotter et al., 2005).

9.3 POTENTIAL FOR USE OF LAB BACTERIOCINS AS FOOD PRESERVATIVES

Given their capacity to kill pathogenic microorganisms, bacteriocins are considered a safe alternative to other antimicrobials as food preservatives and medical applications, and bacteriocins produced by food-grade organisms like LAB are of particular interest in this respect.

In food products, bacteriocins can replace chemical preservatives when used in purified or concentrated preparations or as *in situ* produced molecules by an inoculated bacteriocinogenic culture. An optimized use of these compounds in food could respond to consumers' demand for safe foods with a longer shelf-life that do not contain chemical preservatives.

Evidence for the lack of toxicity was achieved for nisin, which was found to have a very high not observed adverse effects level (NOAEL) in rats that was on average 3 g/kg per day (Hagiwara et al., 2010).

At present, only the wide spectrum bacteriocins nisin, produced by *Lactococcus lactis*, and pediocin PA1/AcH, produced by *Pediococcus acidilactici*, have found application in food products. The nisin preparation most widely used is Nisaplin (DuPont Danisco), while pediocin PA1 has been commercially exploited in the form of ALTA 2431 or 2351 that was produced by Quest International (Rodriguez et al., 2002). These bacteriocins are used mainly in canned food, processed cheeses and minimally processed foods (De Arauz et al., 2009). The reasons for the under-exploitation of bacteriocins in fermented food products is the inhibition exerted towards technologically relevant microbial components. Therefore, strategies to optimize use of bacteriocins in these products without affecting the microbiological processes for the transformation of fermented food products, for example by combining bacteriocins with narrow spectrum and active towards different target bacteria, should be pursued. The combinations of different bacteriocins were not extensively studied, and it was only shown that Pediocin AcH and nisin were more effective in combination than when they were used alone (Hanlin et al., 1993).

Dairy products represent a rich source of bacteriocin-producing bacteria, mainly LAB, so that it can be stated without any doubt that these antimicrobial compounds have been ingested for millennia, without known adverse effects, in this food category of wide consumption. The natural occurrence of bacteriocin-producing LAB in dairy products strongly suggests that an optimized addition of bacteriocinogenic cultures adapted to these food matrices would effectively increase safety. The possibility of adding cultures in these products, instead of the purified or semi-purified bacteriocin, facilitates use of these natural antimicrobials since, if produced *in situ* by food-grade bacterial strains, they are not food additives.

For some applications bacteriocins with a wide spectrum of activity are more appropriate, and factors including pH optima, solubility and stability have to be taken into account in choosing a particular inhibitor for a given food or target bacterium. Furthermore, the antimicrobial spectra of some LAB bacteriocins can be extended to encompass Gram-negative bacteria through their use in combination with compounds, such as chelating agents, or physical treatments that increase their sensitivity by destabilizing the outer membrane (Stevens et al., 1991). On the other hand, narrow-spectrum bacteriocins, e.g. the class IIa bacteriocins, which are specifically active against *Listeria* spp., can find application in ready to eat (RTE) foods where *L. monocytogenes* must be absent or not exceed 10^2 CFU/g during the shelf-life (EC, 2075). Narrow-spectrum bacteriocin-producing cultures could be used also in fermented foods, where these peptides do not impact the success of the fermentation (Eijsink et al., 1998). Bacteriocins formed *in situ* in dairy products have been shown to increase healthiness of these foods, for example by reducing biogenic amine formation by other bacteria (Tabanelli et al., 2014; Perin et al., 2015).

Since bacteriocins are increasingly gaining attention in the medical field as possible substitutes of antibiotics, given their capacity to specifically kill pathogens and their antibiotic resistant variants, use of bacteriocinogenic cultures in food products could present one advantage for consumer's health, which is the supply of beneficial bacteriocin-producing bacteria with the diet. The possibility of inhibiting pathogenic bacteria *in vivo* by ingested bacteriocin-producing bacteria has been demonstrated in animal models.

The advantages of administering narrow-spectrum bacteriocin-producing bacteria could be the selective killing of pathogens such as *L. monocytogenes* without affecting the beneficial components of the gut microbiome and the bacteriocin production *in situ* to counteract its possible degradation by proteases. On the other hand, it has been demonstrated that bacteriocin production by probiotic bacteria favours their gut-colonizing capacity (Cotter et al., 2013).

In the following section a brief description of the production, regulation and inhibitory activities of bacteriocins produced by LAB species most commonly found in dairy products is presented.

9.4 BACTERIOCINS PRODUCED BY DAIRY LAB

Lactococci are among the most important bacteriocin producers among LAB, being able to synthesize a range of antimicrobial peptides including broad- and narrow-spectrum ones. An extensive survey carried out by Geis et al. (1983), observed that about 5% of 280 lactococcal strains tested were bacteriocin producers. Moreover, conjugal transfer of bacteriocin plasmids among *L. lactis* strains occurred (Neve et al., 1984). A higher percentage of naturally occurring bacteriocin-producing lactococci was reported by Alegría et al. (2010), who identified 16 *L. lactis* subsp. *lactis* and 1 *L. lactis* subsp. *cremoris* able to produce different bacteriocins among 60 strains tested. The best characterized bacteriocin produced by *L. lactis* subsp. *lactis* is the lantibiotic nisin A (subclass I), which has been used in the food industry as a preservative since 1953 (Cotter et al., 2005). This is the sole bacteriocin in its semi-purified form to be currently permitted as a food additive (E234 in the European list of food preservatives) (European Parliament and Council, 1995). It is a 3353-Da, cationic, linear peptide of 34 amino acids that contains five intramolecular ring structures (Figure 9.2).

Nisin has a dual mode of action, i.e. inhibition of cell wall biosynthesis and pore formation in the cell membranes of susceptible cells (Wiedemann et al., 2001). Like other

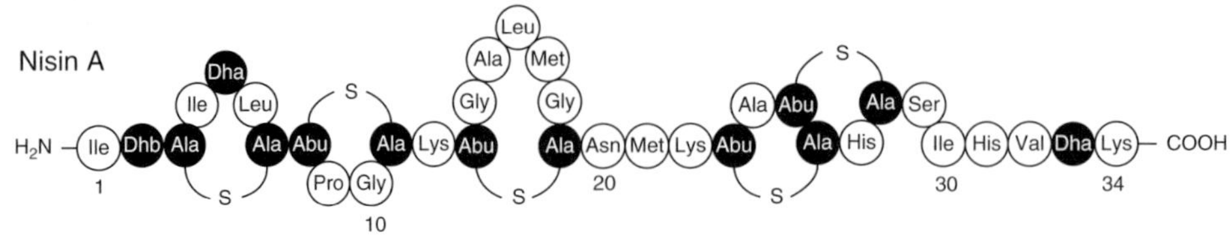

Figure 9.2 Structure of nisin A, a 3353-Da cationic, linear peptide of 34 amino acids, containing five intramolecular ring structures.

LAB lantibiotics, among which is lacticin 3147, it binds to a docking molecule, lipid II, at the membrane level, forming a complex (Brotz et al., 1998, Wiedemann et al., 2006). The formation of the lantibiotic-lipid II complex inhibits peptidoglycan synthesis. The phases of the nisin mechanism of action are as follows: (i) adsorption of nisin to the target cell surface allowed by electrostatic interactions of the positively charged bacteriocin and the negatively charged membrane phospholipids (Martin et al., 1996); (ii) binding to lipid II via two of its amino-terminal rings; (iii) pore formation, which involves a stable transmembrane orientation of nisin and the assembly of four nisins: lipid II complexes; (iv) four additional nisin molecules bound to form the pore complex.

The inhibitory spectrum of nisin is one of the most completely known. It includes the pathogens *Clostridium botulinum* and *L. monocytogenes* and the Gram-positive bacterial genera *Staphylococcus*, *Lactococcus*, *Lactobacillus*, *Leuconostoc*, *Listeria*, *Pediococcus* and *Micrococcus*, while it does not inhibit yeasts, filamentous fungi, viruses or Gram-negative bacteria (Gharsallaoui et al., 2016). It has been suggested that nisin is able to inhibit the germination of *Bacillus* and *Clostridium* spores beyond the vegetative forms of these spore-forming bacteria (Venema et al., 1995).

The nisin-producing operon is linked to genes responsible for sucrose catabolism on the nisin-sucrose transposon, Tn5276 (Kuiper et al., 1993).

Six natural nisin variants with a similar activity spectrum have been described so far: nisin A, nisin Z, nisin Q, a nisin variant described in *L. lactis* isolates from goat milk, and two variants of nisin U. Nisin U and U2 are produced by *Streptococcus uberis* (Lubelski et al., 2008; Perin et al., 2014; Yoneyama et al., 2008).

Lacticin 3147 is a wide spectrum lantibiotic, produced by *L. lactis* and encoded by a 63-kb plasmid, pMRC01, that can be easily conjugally transferred to commercial starter lactococci. It is more inhibitory than nisin against *L. monocytogenes*, *Bacillus cereus*, *Enterococcus faecium* and *Enterococcus faecalis*. It is mainly active against *Clostridium sporogenes* and *Clostridium tyrobutyricum* among spoilage microorganisms in dairy products and is moderately inhibitory for *Staphylococcus aureus* strains, while it is highly active against cheese-associated LAB, i.e. *Lactobacillus helveticus*, *Lactobacillus casei* and other *L. lactis* strains (Iancu et al., 2012). For this reason its suggested use was the inhibition of adventitious nonstarter LAB (NSLAB) in cheddar cheese to standardize quality (Ryan et al., 1996). Also, another bacteriocin produced by *L. lactis*, lacticin 481, is active against other LAB, and therefore it was used to speed up cheese ripening by lysis of starter cultures, increasing the amount of free intracellular enzymes (Garde et al., 2006). Lacticin 481 includes in its spectrum of inhibition *C. tyrobutyricum*, responsible for late blowing of hard cheeses (Thuault et al., 1991). Its bactericidal action consists in the formation of voltage-dependent pores through which an efflux of ions and small molecules occurs (Demele et al., 1996). Its production is stimulated by acid pH at the transcriptional level, as the two promoters P1 and P3, lying upstream of *lct*A, are acid-induced (Hindre et al., 2004). In addition, lacticin 481 production is stimulated by hyperosmotic stress (Uguen et al., 1999).

Several 'lactococcins' were described: among these, Lactococcin 972 (Lcn972) inhibits cell wall biosynthesis in *L. lactis* and is the first non-lantibiotic non-posttranslationally modified bacteriocin that was found to bind to lipid II, possibly for the presence of a novel lipid II-binding motif. It is encoded by a plasmid and is highly bactericidal to lactococci (Martínez et al., 2008).

Also, other starter LAB (SLAB) are capable of producing bacteriocins, and among these, *Streptococcus thermophilus* strains can synthesize a broad inhibitory spectrum lantibiotic, thermophilin 1277, active against *Clostridium butyricum*, *C. sporogenes* and *B. cereus* (Kabuki et al., 2009).

Thermophilin 13 is a nontypical anti-listerial poration complex bacteriocin with a broad spectrum that functions without a receptor and is composed of the antibacterial peptide ThmA and the enhancing factor ThmB, which, when present in equimolar ratio, give maximum activity (Marciset et al., 1997).

Moreover, *S. thermophilus* produces a multi-peptide bacteriocin, thermophilin 9, that is encoded by the *blp* gene cluster, composed of 23 genes, that inhibits related Gram-positive bacteria and *L. monocytogenes*. The antimicrobial peptide *blp*D is responsible for most of the antimicrobial activity of thermophilin 9 (Fontaine and Hols, 2008). The gene order, content and identity in the thermophilin 9 gene cluster varies greatly among *S. thermophilus* strains, and this determines differences in the inhibitory spectrum (Rossi et al., 2013). Rossi et al. (2013) observed that some strains possessing the structural genes of this bacteriocin were not all able to produce functional antimicrobial peptides. The ability to inhibit other *S. thermophilus* strains with a thermophilin 9 producer was proven advantageous to control the growth of a tyraminogenic and tetracycline resistant *S. thermophilus* strain when co-cultivated in skim milk.

Streptococcus macedonicus, another streptococcal species specifically adapted to dairy products, produces a lantibiotic of the lacticin 481 group, macedocin, slightly active against *S. aureus* and active against LAB and *C. tyrobutyricum* and induced by a casein fragment. Moreover, *S. macedonicus* produces a broad spectrum bacteriocin, macedovicin, identical to thermophilin 1277 (Georgalaki et al., 2013).

Starter lactobacilli have been found to produce narrow-spectrum bacteriocins active against strains of the same or closely related species. An example is helveticin J, a large (molecular weight 37 kDa), heat-labile protein that was the first bacteriocin from *Lactobacillus* spp. to be sequenced and has a bactericidal mode of action (Joerger and Klaenhammer, 1990).

L. helveticus PJ4, isolated from rat gut, produces bacteriocin PJ4, effective against both Gram-positive and Gram-negative bacteria involved in various diseases, including *Escherichia coli*, *Bacillus subtilis*, *Pseudomonas aeruginosa*, *E. faecalis* and *S. aureus*. The antimicrobial peptide is relatively heat resistant and is also active over a wide pH range (2–10); therefore, it represents a potential alternative agent to control important pathogens (Jena et al., 2013).

Among NSLAB mesophilic lactobacilli, *Lactobacillus plantarum* produces a high number of different bacteriocins, and more than 20 plantaricins have been described so far, some with great potential for use in food preservation. Those characterized most recently are plantaricin ZJ5 (Song et al., 2014) and plantaricin JLA-9, which has potential applications in the control of *Bacillus* spp. in the food industry (Zhao et al., 2016).

L. plantarum C11 produces six bacteriocin-like peptides, PlnA, PlnE, PlnF, PlnJ, PlnK, and PlnN, and two two-peptide bacteriocins (PlnEF and PlnJK.). A strain-specific antagonistic activity was detected at nanomolar concentrations when PlnE and PlnF were combined and when PlnJ and Plnk were combined in approximately equal amounts.

PlnA also acts as an extracellular signal (pheromone) that triggers bacteriocin production (Anderssen et al., 1998).

Plantaricin 423, which is bactericidal for many Gram-positive foodborne pathogens and spoilage bacteria, including *Listeria* spp. and *Staphylococcus* spp., has an operon structure similar to pediocin PA-1/AcH from *P. acidilactici* (Van Reenen et al., 2003).

L. plantarum NCIM 2084 produced an antibacterial substance with bactericidal and lytic effect against many Gram-positive and Gram-negative foodborne pathogenic and spoilage bacteria, while related mesophilic lactobacilli were not inhibited; thus, it has a good potential for application as a biopreservative in food (Suma et al., 1998).

Though some strains of enterococci possess virulence traits and can disseminate antibiotic resistance genes, these bacteria represent a component of the microbiota in many traditional cheese varieties where they exert positive roles in ripening. Enterocin A is a potent class IIa bacteriocin co-expressed with enterocin B (Nes et al., 2014).

A large-spectrum enterococcal bacteriocin is enterocin AS-48, which was originally isolated from *E. faecalis*; it is a circular bacteriocin and has been studied in great detail with respect to its genetics, biochemical features, and structure, as well as its mode of action (Maqueda et al., 2004).

Pediococci typically are unable to ferment lactose so that their applications in milk fermentations are restricted. Nevertheless, adjunct *Pediococcus* spp. impart desirable attributes to cheese.

Pediocins are among the most promising bacteriocins for use as food preservatives, and pediocin-like bacteriocins (36–48 residues) are produced by many LAB and share a 40–60% amino acid sequence similarity. The *N*-terminal region of all pediocins currently identified contains two cysteines, joined by a disulfide bond, in a motif known as the 'pediocin box': -Y-G-N-G-V-X1-C-X2-K/N-X3-X4-C-, with X1–4 representing polar uncharged or charged residues. The peptides of this group are known as antilisterial or *Listeria*-active peptides, and they are characterized by a -Y-G-N-G-V-N- terminus. Pediocin AcH production genes are encoded by four gene operons present on plasmids in *Pediococcus parvulus*, *P. pentosaceus* and *L. plantarum*, which appear to have spread in different bacterial genera (Papagianni and Anstasiadou, 2009).

9.5 IDENTIFICATION OF LAB-PRODUCING BACTERIOCINS

Identification and characterization of novel bacteriocins in novel strains of bacteria can lead to the discovery of new antimicrobial peptides with potential of application in food products.

On the other hand, Woraprayote et al. (2016) reported that among 185 LAB bacteriocins isolated, only 53% were well-characterized and sequenced at the protein or DNA levels.

The classical method to identify LAB-producing bacteriocins is represented by the spot-on-the-lawn technique (Harris et al., 1989; Lewus et al., 1991) in which fresh broth cultures of LAB isolates are spotted on the surface of MRS agar and incubated for 24h; an appropriate culture medium agarized with 0.5% (w/v) agar is inoculated with 0.1 v/v suspension of the bacteria to be used as indicators of the antimicrobial activity, overlaid on the plates where LAB have grown, and incubated for 24h. The presence of inhibition halos indicates that the LAB isolates exert antimicrobial activity. In an alternative to the LAB broth cultures, cell-free supernatant (CFS) can be used, thus reducing the time required for incubation.

A rapid method to search for the presence of different bacteriocins encoding genes is represented by PCR. Over the years, numerous primer pairs were designed to detect structural, transporter, modification or maturation genes belonging to bacteriocin gene clusters. Table 9.2 shows a list of references in which methods of bacteriocin detection by PCR are described.

Table 9.2 PCR references for the detection of bacteriocin gene clusters.

Producing bacteria	Bacteriocin gene	PCR references
L. lactis	Nisin	Ward et al., 1994; Rodriguez et al., 1995; Choi et al., 2000; Li et al., 2002; Wirawan et al., 2006
	Lacticin 481	Piard et al., 1993
	Lactococcin A	Martinez et al., 1998
	Lactococcin B	van Belkum et al., 1992
	Lactococcin 513	Villani et al., 2001
	Lacticin RM	
	Lactococcin 972	Martinez et al., 1999
L. sakei, L. curvatus	CurvacinA	Remiger et al., 1996
L. sakei, L. curvatus	Sakacin P	Remiger et al., 1996
L. plantarum	Plantaricin A	Remiger et al., 1996
	Plantaricin EF	
P. acidilactici	Pediocin PA-1	Rodriguez et al., 1997
E. faecium, P. pentosaceus	Pediocin PA-1-like	Devi and Halami, 2011
L. sakei, L. curvatus	Sakacin Q	Cocolin and Rantsiou, 2007
E. faecalis	Enterocin 4	Du Toit et al., 2000
	Bacteriocin 31	Du Toit et al., 2000; De Vuyst et al., 2003
	Enterocin AS48	
E. faecium	Enterocin A	Du Toit et al., 2000; De Vuyst et al., 2003
	Enterocin B	
	Enterocin L50A and L50B	
	Enterocin P	Toit et al., 2000; De Vuyst et al., 2003; Wieckowicz et al., 2011

9.6 A NOVEL APPROACH FOR SCREENING LAB BACTERIOCINS

Generally, identifying novel bacteriocins is necessary to obtain their purified form to perform studies and characterization. However, establishing a purification system is expensive, time-consuming, and tedious. Moreover, these systems often identify previously reported bacteriocins. Zendo (2013) developed a rapid system that can identify novel bacteriocins from various sources during the early stage of the screening and isolation of bacteriocin-producing strains. This system employs molecular mass analyses using electrospray ionization liquid chromatography/mass spectrometry (ESI-LC/MS) in combination with principal component analyses (PCA) of the antimicrobial activity spectrum of each bacteriocin-producing LAB strain (Figure 9.3). To determine molecular mass of bacteriocins, supernatants, with or without pretreatment of overnight culture with LAB strains, were injected to the ESI-LC/MS system for analysis. The molecular mass of the active compound in the supernatant was compared to the molecular mass of reported bacteriocins. When molecular weights of the bacteriocins did not resemble any previously reported strains, he considered them novel. For statistical analysis, antimicrobial spectra of the LAB strains are determined using the spot-on-the-lawn method against a set of indicator strains (usually about 10 strains) and a PCA graph. The data on the strains producing the reported bacteriocins were analyzed for comparison with those of the new isolates. In most cases, bacteriocins were clustered into three groups: nisin variants, class IIa (anti-*Listeria*) bacteriocins, and narrow-spectrum bacteriocins. The remaining bacteriocins, not classified into any of these three groups, have been proposed as potentially novel ones. Thus, data obtained from the PCA should be combined with data from the molecular mass analysis.

The continuous reporting of novel bacteriocins necessitates their reclassification. Several approaches have been proposed to classify them. Nowadays, a freely accessible database is available, known as Bactibase, which contains information on more than

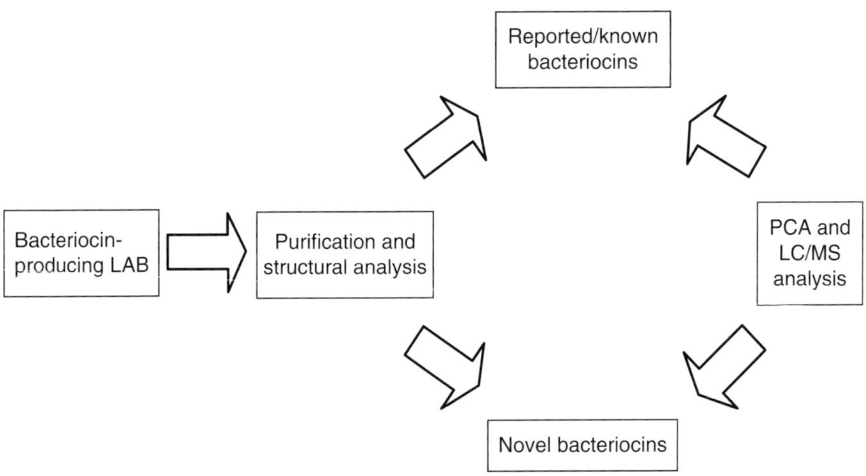

Figure 9.3 Rapid screening system for novel LAB bacteriocins.

Table 9.3 Revised classification for bacteriocins of LAB.

Heat-stable (<10 kDa)	Class I (modified)	Lanthipeptides
		Cyclizedpeptides
		Linear azol(in)e-containing peptides (LAPs)
		Sactibiotics (sulfur-to-α-carbon-containing peptides)
		Glycocins (glycosylated residues containing peptides)
		Lasso peptides (amide bond between the first amino acid in the core peptide and a negatively charged residue)
	Class II (Unmodified)	Pediocin-like
		Two-peptides
		Leaderless
		Non-pediocin-like, single-peptide
Thermo-labile (>10 kDa)	Class III	Bacteriolysins
		Non-lytic

Classes identified in silico are depicted in gray.

200 bacteriocins produced by many bacteria (Hammami et al., 2010). In addition to published bacteriocins, the repertoire of molecules hidden in the genomes that have not yet been isolated represents a valuable source of novel compounds with great potential. Diverse tools have been created that can be used for the automated screening of bacteriocin gene clusters (Blin et al. 2013; van Heel et al. 2013). The most recent classification scheme, extended to bacteriocins, even detected *in silico* in publicly available LAB genomes, has been reported (Alvarez-Sieiro et al., 2016). Alvarez-Sieiro et al. (2016) analyzed using Bagel3 (van Heel et al. 2013) a total of 238 complete LAB genomes deposited in public databases. This search resulted in a list of 785 putative bacteriocin gene clusters, including ribosomally produced and posttranslationally modified peptides (RiPPs) that were not previously identified in LAB. In this list, they observed previously characterized bacteriocins or natural variants, some of them spread among different species, and new putative bacteriocins with no significant homology to known peptides based on the BLAST results provided. Due to these results, they proposed a revised classification for LAB bacteriocins (Table 9.3) that can accommodate the novel subclasses that are appearing, based on the biosynthesis mechanism and biological activity, which is in agreement with previous proposals (Cotter et al. 2013; Arnison 2013). Although this new classification is innovative, it will take time to be introduced and accepted worldwide.

9.7 BIOTECHNOLOGICAL INTERVENTIONS FOR BACTERIOCIN ENGINEERING

Due to their long history of safe use and natural benefits to human health, LAB serve as promising candidates for therapeutic purposes. To exploit the full potential of LAB, one fundamental need is a powerful capacity for the engineering of complex gene networks, such as biosynthetic pathways and multicomponent artificial gene circuits. A couple of new methodologies have been established, including single-stranded DNA- (ssDNA-) based recombineering (van Pijkeren and Britton, 2014) and CRISPR–Cas9 assisted recombineering (Oh and van Pijkeren, 2014), which offer new strategies for manipulating

LAB chromosomes. These approaches are based on ssDNA-based recombination and on the assumption that chromosomes remain mostly in single copy. Recently, Kong et al. (2016) presented a synthetic biology platform for rapid engineering of complex gene networks in LAB. The platform involves a shuttle system and two associated engineering strategies. A complex pathway responsible for nisin biosynthesis was adopted for the demonstration and characterization of the strategies. This work offers a new solution to the engineering of complex gene networks in LAB, enabling a wide spectrum of LAB-based applications.

9.8 CONCLUSIONS

LAB strains producing bacteriocins are a valid and useful approach for pathogen control and food safety interventions and as an instrument to control dairy productions by means of starter lysis and ripening promoting factors. New methods for the detection of bacteriocin-producing strains are available that will expedite the screening of LAB strains to be applied to increase the safety of food products and to reduce use of chemical preservatives.

References

Alegría A, Delgado S, Roces C, López B, Mayo B. 2010. Bacteriocins produced by wild *Lactococcus lactis* strains isolated from traditional, starter-free cheeses made of raw milk. *Int. J Food Microbiol* 143:61–66.

Alvarez-Sieiro P, Montalbán-López M, Mu D, Kuipers OP. 2016. Bacteriocins of lactic acid bacteria: extending the family. *Appl Microbiol Biotechnol* 100:2939–51.

Anderssen EL, Diep BD, Nes IF, Eijsink VGH, Nissen-Meyer J. 1998. Antagonistic activity of *Lactobacillus plantarum* C11: Two new two-peptide bacteriocins, plantaricin EF and JK, and the induction factor plantaricin. *Appl Environ Microbiol* 6:2269–72.

Arnison PG, Bibb MJ, Bierbaum G, Bowers AA, Bugni TS, Bulaj G, Camarero JA, Campopiano DJ, Challis GL, Clardy J, Cotter PD, Craik DJ, Dawson M, Dittmann E, Donadio S, Dorrestein PC, Entian KD, Fischbach MA, Garavelli JS, Göransson U, Gruber CW, Haft DH, Hemscheidt TK, Hertweck C, Hill C, Horswill AR, Jaspars M, Kelly WL, Klinman JP, Kuipers OP, Link AJ, Liu W, Marahiel MA, Mitchell DA, Moll GN, Moore BS, Müller R, Nair SK, Nes IF, Norris GE, Olivera BM, Onaka H, Patchett ML, Piel J, Reaney MJ, Rebuffat S, Ross RP, Sahl HG, Schmidt EW, Selsted ME, Severinov K, Shen B, Sivonen K, Smith L, Stein T, Süssmuth RD, Tagg JR, Tang GL, Truman AW, Vederas JC, Walsh CT, Walton JD, Wenzel SC, Willey JM, van der Donk WA. 2013. Ribosomally synthesized and post-translationally modified peptide natural products: overview and recommendations for a universal nomenclature. *Nat Prod Rep* 30:108–160.

Blin K, Medema MH, Kazempour D, Fischbach MA, Breitling R, Takano E, Weber T. 2013. AntiSMASH 2.0—a versatile platform for genome mining of secondary metabolite producers. *Nucleic Acids Res* 41:W204–W212.

Brotz H, Josten M, Wiedemann I, Schneider U, Gotz F, Bierbaum G, Sahl HG. 1998. Role of lipid-bound peptidoglycan precursors in the formation of pores by nisin, epidermin and other lantibiotics. *Mol Microbiol* 30:317–27.

Choi HJ, Cheigh CI, Kim SB, Pyun YR. 2000. Production of a nisin-like bacteriocin by *Lactococcus lactis* subsp. *lactis* A164 isolated from kimchi. *J ApplMicrobiol* 88:563–71.

Cocolin L, Rantsiou K. 2007. Sequencing and expression analysis of sakacin genes in *Lactobacillus curvatus* strains. *Appl Microbiol Biotechnol* 76:1403e–1411e.

Cotter PD, Hill C, Ross RP. 2005. Bacteriocins: developing innate immunity for food. *Nat Rev Microbiol* 3:777–88.

Cotter PD, Ross RP, Hill C. 2013. Bacteriocins – a viable alternative to antibiotics? *Nat Rev Microbiol* 11:95–105.

Cui Y, Zhang C, Wang Y, Shi J, Zhang L, Ding Z, Qu X, Cui H. 2012. Class IIa bacteriocins: diversity and new developments. *Int J. Mol Sci* 13:16668–707.

de Arauz LJ, Jozala AF, Gava Mazzola P, Vessoni Penna TC. 2009. Nisin biotechnological production and application: a review. *Trends Food Sci Technol* 20:146–54.

De Arauz LJ, Jozala AF, Mazzola PG, Vessoni Penna TC. 2009. Nisin biotechnological production and application: a review. *Trends Food Sci Technol* 20:146–54.

De Vuyst L, Foulquie Moreno MR, Revets H. 2003. Screening for enterocins and detection of hemolysin and vancomycin resistance in enterococci of different origins. *Int J Food Microbiol* 84:299–18

Demel, RA, Peelen T, Siezen RJ, De Kruijff B, and Kuipers OP. 1996. Nisin Z, mutant nisin Z and lacticin 481 interactions with anionic lipids correlate with antimicrobial activity: a monolayer study. *Eur J Biochem* 235:267–74.

Devi SM, Halami PM. 2011. Detection and characterization of pediocin PA-1/AcH like bacteriocin producing lactic acid bacteria. *Curr Microbiol* 63:181–85.

Du Toit MD, Franz CMAP, Dicks LMT, Holzapfel WH. 2000. Preliminary characterization of bacteriocins produced by *Enterococcus faecium* and *Enterococcus faecalis* isolated from pig faeces. *J Appl Microbiol* 88:482–94.

EC 2073. 2005. European Commission Regulation N 2073/2005 on microbiological criteria for foodstuffs. *Official Journal Eur Un* L 338/1.

Eijsink VG, Skeie M, Middelhoven PH, Brurberg MB, Nes IF. 1998. Comparative studies of class IIa bacteriocins of lactic acid bacteria. *Appl Environ Microbiol* 64:3275–81.

European Parliament and Council Directive No 95/2/EC of 20 February 1995 on food additives other than colours and sweeteners. *Official Journal* L 61, 18. 3.

Fontaine L, Hols P. 2008. The inhibitory spectrum of thermophilin 9 from *Streptococcus thermophilus* LMD-9 depends on the production of multiple peptides and the activity of BlpG, a thiol-disulfide oxidase. *Appl Environ Microbiol* 74:1102–10.

Garde S, Avila M, Gaya P, Medina M, Nuñez M. 2006. Proteolysis of Hispanico cheese manufactured using lacticin 481-producing *Lactococcus lactis* ssp. *lactis* INIA 639. *J Dairy Sci* 89:840–849.

Geis A, Jasjit J, Teuber M. 1983. Potential of lactic streptococci to produce bacteriocins. *Appl Environ Microbiol* 45:205–211.

Georgalaki M, Papadimitriou K, Anastasiou R, Pot B, Van Driessche G, Devreese B, Tsakalidou E. 2013. Macedovicin, the second food-grade lantibiotic produced by *Streptococcus macedonicus* ACA-DC 198. *Food Microbiol* 33:124–30.

Gharsallaoui A, Oulahal N, Joly C, Degraeve P. 2016. Nisin as a food preservative: Part 1: Physicochemical properties, antimicrobial activity, and main uses. *Crit Rev Food Sci Nutr* 56:1262–74.

Hagiwara A, Imai N, Nakashima H, Toda Y, Kawabe M, Furukawa F, Delves-Broughton J, Yasuhara K, Hayashi S. 2010. A 90-day oral toxicity study of nisin A, an anti-microbial peptide derived from *Lactococcus lactis* subsp. *lactis*, in F344 rats. *Food Chem Toxicol* 48:2421–28.

Hammami R, Zouhir A, Le Lay C, Ben Hamida J, Fliss I. 2010. BACTIBASE second release: a database and tool platform for bacteriocin characterization. *BMC Microbiol* 10:22.

Hanlin MB, Kalchayand N, Ray P, Ray B. 1993. Bacteriocins of lactic acid bacteria in combination have greater antibacterial activity. *J Food Protection* 56:252–55.

Harris LJ, Daeschel MA, Stiles ME, Klaenhammer TR. 1989. Antimicrobial activity of lactic acid bacteria against *Listeria monocytogenes*. *J Food Protection* 52:384–87.

Heng NCK, Wescombe PA, Burton JP, Jack RW, Tagg JR. 2007. The diversity of bacteriocins. In Gram-positive bacteria. In: Riley MA, Chavan MA (Eds). *Bacteriocins: ecology and evolution*. Berlin: Springer.

Hindré T, Le Pennec J-P, Haras D, Dufour A. 2004. Regulation of lantibiotic lacticin 481 production at the transcriptional level by acid pH. *FEMS Microbiol Lett* 231:291–98.

Iancu C, Grainger A, Field D, Cotter PD, Hill C, Ross RP. 2012. Comparison of the potency of the lipid II targeting antimicrobials nisin, lacticin 3147 and vancomycin against Gram-positive bacteria. *Probiotics Antimicrob Proteins* 4:108–115.

Jena PK, Trivedi D, Chaudhary H, Sahoo TK, Seshadri S. 2013. Bacteriocin PJ4 active against enteric pathogen produced by *Lactobacillus helveticus* PJ4 isolated from gut microflora of wistar rat *Rattus norvegicus*: partial purification and characterization of bacteriocin. *Appl Biochem Biotechnol* 169:2088–2100.

Joerger MC, Klaenhammer TR. 1990. Cloning, expression, and nucleotide sequence of the *Lactobacillus helveticus* 481 gene encoding the bacteriocin helveticin. *J Bacteriol* 172:6339–47.

Kabuki T, Uenishi H, Seto Y, Yoshioka T, Nakajima H. 2009. A unique lantibiotic, thermophilin 1277. *J Appl Microbiol* 106:853–62.

Klaenhammer TR. 1993. Genetics of bacteriocins produced by lactic acid bacteria. *FEMS Microbiol Rev* 12:39–85.

Knerr PJ, van der Donk WA. 2012. Discovery, biosynthesis, and engineering of lantipeptides. *Annu Rev Biochem* 81:479–505.

Kong W, Kapuganti VS, Lu T. 2016. A gene network engineering platform for lactic acid bacteria. *Nucleic Acids Res* 44(4):e37.

Kuiper OP, Beerthuyzen MM, Siezen RJ, De Vos WM. 1993. Characterization of the nisin gene-cluster NisABTCIPR of *Lactococcus lactis*- requirement of expression of the *nis*A and *nis*I genes for development of immunity. *Eur J Biochem* 216:281–91.

Lewus CB, Kaiser A, Montville TJ. 1991. Inhibition of food-borne bacterial pathogens by bacteriocins from lactic acid bacteria isolated from meat. *Appl Environ Microbiol* 57:1683–88.

Li H, O'Sullivan DJ. 2002. Heterologous expression of the *Lactococcus lactis* bacteriocin, nisin, in a dairy Enterococcus strain. *Appl Environ Microbiol* 68:3392–3400.

Lorca G, Reddy L, Nguyen A, Sun EI, Tseng J, Yen M-R, Saier MH Jr. 2010. Lactic acid bacteria: Comparative genomic analyses of transport systems. In: Mozzi F, Raya RR, Vignolo GM (Eds). *Biotechnology of lactic acid bacteria: novel applications*. Ames, IA: Wiley-Blackwell, pp 73–87.

Lubelski J, Rink R, Khusainov R, Moll GN, Kuipers OP. 2008. Biosynthesis, immunity, regulation, mode of action and engineering of the model lantibiotic nisin. *Cell Mol Life Sci* 65:455–76.

Maqueda M, Gálvez A, Bueno MM, Sanchez-Barrena MJ, González C, Albert A, Rico M, Valdivia E. 2004. Peptide AS-48: prototype of a new class of cyclic bacteriocins. *Curr Prot Pept Sci* 5:399–416.

Marciset O, Jeronimus-Stratingh MC, Mollet B, Poolman B. 1997. Thermophilin 13, a nontypical antilisterial poration complex bacteriocin, that functions without a receptor. *J Biol Chem* 272:14277–284.

Martin I, Ruysschaert J-M, Sanders D, Giffard CJ. 1996. Interaction of the lantibiotic nisin with membranes revealed by fluorescence quenching of an introduced tryptophan. *Eur J Biochem* 239:156–64.

Martínez B, Böttiger T, Schneider T, Rodríguez A, Sahl HG, Wiedemann I. 2008. Specific interaction of the unmodified bacteriocin lactococcin 972 with the cell wall precursor lipid II. *Appl Environ Microbiol* 74:4666–70.

Martinez B, Fernandez M, Suarez JE, Rodriguez A. 1999. Synthesis of lactococcin 972, a bacteriocin produced by *Lactococcus lactis* IPLA 972, depends on the expression of a plasmid-encoded bicistronic operon. *Microbiology* 145:3155–61.

Martinez B, Rodriguez A, Medina M, Hernandez PE, Rodriguez JM. 1998. Detection of specific bacteriocin-producing lactic acid bacteria by colony hybridization. *J Appl Microbiol* 84:1099–1103.

McAuliffe O, O'Keeffe T, Hill C, Ross RP. 2001. Regulation of immunity to the two-component lantibiotic, lacticin 3147, by the transcriptional repressor LtnR. *Mol Microbiol* 39:982–93.

Nes IF, Diep DB, Ike Y. 2014. Enterococcal bacteriocins and antimicrobial proteins that contribute to niche control. In: Gilmore MS, Clewell DB, Ike Y, Shankar N. editors. *Enterococci: from commensals to leading causes of drug resistant infection*. Boston: Massachusetts Eye and Ear Infirmary.

Neve H, Geis A, Teuber M. 1984. Conjugal transfer and characterization of bacteriocin plasmids in group N (lactic acid) streptococci. *J Bacteriol* 157:833–838.

Nishant T, Sathish KD, Arun KR, Hima BK, Raviteja Y. 2011. Bacteriocin producing probiotic lactic acid bacteria. *J Microbial Biochem Technol* 3:121–24.

Nishie M, Nagao J, Sonomoto K. 2012. Antibacterial peptides "bacteriocins": an overview of their diverse characteristics and applications. *Biocontrol Sci* 17:1–16.

Papagianni M, Anastasiadou S. 2009. Pediocins: the bacteriocins of pediococci. Sources, production, properties and applications. *Microbial Cell Factories* 8:3.

Perin LM, Dal Bello B, Belviso S, Zeppa G, de Carvalho AF, Cocolin L, Nero LA. 2015. Microbiota of Minas cheese as influenced by the nisin producer *Lactococcus lactis* subsp. *lactis* GLc05. *Int J Food Microbiol* 214:159–67.

Perin LM, Nero LA. 2014. Antagonistic lactic acid bacteria isolated from goat milk and identification of a novel nisin variant *Lactococcus lactis*. *BMC Microbiol* 14:36.

Piard JC, Kuipers OP, Rollema HS, Desmazeaud MJ, De Vos WM. 1993. Structure, organization and expression of the *lct* gene for lacticin 481, a novel lantibiotic produced by *Lactococcus lactis*. *J Biol Chem* 268:16361–68.

Remiger A, Ehrmann MA, Vogel RF. 1996. Identification of bacteriocins encoding genes in lactobacilli by polymerase chain reaction PCR. *Syst Appl Microbiol* 19:28–34.

Riley MA, Wertz JE. 2002. Bacteriocin diversity: ecological and evolutionary perspectives. *Biochimie* 84:357–64.

Rincé A, Dufour A, Uguen P, Le Pennec JP, Haras D. 1997. Characterization of the lacticin 481 operon: the *Lactococcus lactis* genes *lctF*, *lctE*, and *lctG* encode a putative ABC transporter involved in bacteriocin immunity. *Appl Environ Microbiol* 63:4252–60.

Rodriguez JM, Cintas LM, Casaus P, Dodd HM, Hernandez PE, Gasson MJ. 1995. Isolation of nisin-producing *Lactococcus lactis* strains from dry fermented sausages. *J Appl Bacteriol* 78:109–15.

Rodriguez JM, Cintas LM, Casaus P, Martinez MI, Suarez A, Hernandez PE. 1997. Detection of pediocin PA-1-producing pediococci by rapid molecular biology techniques. *Food Microbiol* 14:363–71.

Rodriguez JM, Martinez MI, Kok J. 2002. Pediocin PA-1, a wide-spectrum bacteriocin from lactic acid bacteria. *Crit Rev Food Sci Nutr* 42:91–121.

Rossi F, Marzotto M, Cremonese S, Rizzotti L, Torriani S. 2013. Diversity of *Streptococcus thermophilus* in bacteriocin production; inhibitory spectrum and occurrence of thermophilin genes. *Food Microbiol* 35:27–33.

Ryan MP, Rea MC, Hill C, Ross RP. 1996. An application in cheddar cheese manufacture for a strain of *Lactococcus lactis* producing a novel broad-spectrum bacteriocin, lacticin 3147. *Appl Environ Microbiol* 62:612–19.

Song DF, Zhu MY, Gu Q. 2014. Purification and characterization of Plantaricin ZJ5, a new bacteriocin produced by *Lactobacillus plantarum* ZJ5. *PLoS One* 9:e105549.

Stevens KA, Sheldon BW, Klapes NA, Klaenhammer TR. 1991. Nisin treatment for inactivation of *Salmonella* species and other gram- negative bacteria. *Appl Environ Microbiol* 57:3613–15.

Suma K, Misra MC, Varadaraj MC. 1998. Plantaricin LP84, a broad spectrum heat-stable bacteriocin of *Lactobacillus plantarum* NCIM 2084 produced in a simple glucose broth medium. *Int J Food Microbiol* 40:17–25.

Tabanelli G, Montanari C, Bargossi E, Lanciotti R, Gatto V, Felis G, Torriani S, Gardini F. 2014. Control of tyramine and histamine accumulation by lactic acid bacteria using bacteriocin forming lactococci. *Int J Food Microbiol* 190:14–23.

Thuault D, Beliard E, Leguern J, Bourgeois CM. 1991. Inhibition of *Clostridium tyrobutyricum* by bacteriocin-like substances produced by lactic acid bacteria. *J Dairy Sci* 74:1145–50.

Uguen P, Hamelin J, Le Pennec J-P, Blanco C. 1999. Influence of osmolarity and the presence of an osmoprotectant on *Lactococcus lactis* growth and bacteriocin production. *Appl Environ Microbiol* 65:291–93.

van Belkum MJ, Kok J, Venema G. 1992. Cloning, sequencing, and expression in *Escherichia coli* of *lcn*B, a third bacteriocin determinant from the lactococcal bacteriocin plasmid p9B4-6. *Appl Environ Microbiol* 58:572–77.

Van Heel AJ, de Jong A, Montalbán-López M, Kok J, Kuipers OP 2013. BAGEL3: automated identification of genes encoding bacteriocins and non-bactericidal posttranslationally modified peptides. *Nucleic Acids Res* 41:W448–53.

van Pijkeren JP, Britton RA. 2014. Precision genome engineering in lactic acid bacteria. *Microb Cell Factories* 13:S10. doi: 10.1186/1475-2859-13-S1-S10.

Van Reenen CA, Chikindas ML, Van Zy WH, Dicks LM. 2003. Characterization and heterologous expression of a class IIa bacteriocin, plantaricin 423 from *Lactobacillus plantarum* 423, in *Saccharomyces cerevisiae*. *Int J Food Microbiol* 81:29–40.

Venema K, Venema G, Kok J. 1995. Lactococcal bacteriocins: mode of action and immunity. *Trends Microbiol* 3:299–304.

Villani F, Aponte M, Blaiotta G, Mauriello G, Pepe O, Moschetti G. 2001. Detection and characterization of a bacteriocin, garviecin L1-5, produced by *Lactococcus garvieae* isolated from raw cow's milk. *J Appl Microbiol* 90:430–39.

Ward LJH, Brown JCS, Davey GP. 1994. Application of the ligase chain reaction to the detection of nisin A and nisin Z genes in *Lactococcus lactis* ssp. *lactis*. *FEMS Microbiol Lett* 117:29–34.

Wieckowicz M, Schmidt M, Sip A, Grajek W. 2011. Development of a PCR based assay for rapid detection of class IIa bacteriocin genes. *Letters Appl Microbiol* 52:281–89.

Wiedemann I, Bottiger T, Bonelli RR, Wiese A, Hagge SO, Gutsmann T, Seydel U, Deegan L, Hill C, Ross P, Sahl HG. 2006. The mode of action of the lantibiotic lacticin 3147 – a complex mechanism involving specific interaction of two peptides and the cell wall precursor lipid II. *Mol Microbiol* 61:285–96.

Wiedemann I, Breukink E, van Kraaij C, Kuipers OP, Bierbaum G, de Kruijff B, Sahl HG. 2001. Specific binding of nisin to the peptidoglycan precursor lipid II combines pore formation and inhibition of cell wall biosynthesis for potent antibiotic activity. *J Biol Chem* 276:1772–79.

Wirawan RE, Klesse NA, Jack RW, Tagg JR. 2006. Molecular and genetic characterization of a novel nisin variant produced by *Streptococcus uberis*. *Appl Environ Microbiol* 72:1148–56

Woraprayote W, Malila Y, Sorapukdee S, Swetwiwathana A, Benjakul S, Visessanguan W. 2016. Bacteriocins from lactic acid bacteria and their applications in meat and meat products. *Meat Science* 120:118–32.

Yoneyama F, Fukao M, Zendo T, Nakayama J, Sonomoto K. 2008. Biosynthetic characterization and biochemical features of the third natural nisin variant, nisin Q, produced by *L. lactis* 61-14. *J Appl Microbiol* 105:1982–90.

Zendo T. 2013. Screening and characterization of novel bacteriocins from lactic acid bacteria. *Biosci Biotechnol Biochem* 77:893–9.

Zhao S, Han J, Bie X, Lu Z, Zhang C, Lv F. 2016. Purification and characterization of plantaricin JLA-9: a novel bacteriocin against Bacillus Spp. produced by *Lactobacillus plantarum* JLA-9 from Suan-Tsai, a traditional Chinese fermented cabbage. *J Agric Food Chem* 64:2754–64.

10 Starter strains and adjunct non-starter lactic acid bacteria (NSLAB) in dairy products

Paola Dolci and Luca Simone Cocolin

Settore di Microbiologia agraria e Tecnologie alimentari, Dipartimento di Scienze Agrarie, Forestali e Alimentari, Grugliasco, Italy

10.1 INTRODUCTION

Lactic acid bacteria (LAB) are either added intentionally at the beginning of cheese making or belong to the natural microbial population of the milk, as is the case in many artisanal dairy products made from raw milk. The benefits and risk of traditional cheeses have been the subject of debate from scientific communities. Some authors reported that microbial diversity in traditional raw milk cheeses is a key factor for sensory benefits, leading to a more intense and rich flavor than that found in industrial cheese products (Montel et al., 2014). In terms of hygienic risk, the same authors highlighted that the battles against pathogens have to be focused on the process upstream from the vat milk (Brooks et al., 2012; Montel et al., 2014). On the contrary, defenders of pasteurization advocate managing the pathogen risk by applying heat to reduce the microbial load on equipment and in milk and by standardizing production by inoculating a few selected strains into milk.

Lately, many studies have been focused on parameters for the selection of adapt strains for both cheese making, known as starter lactic acid bacteria (SLAB), and ripening (no starter lactic acid bacteria, NSLAB), based on the understanding of their metabolic activities along cheese production and the pros and cons of their use.

10.2 CONTROLLED FERMENTATION

Over the centuries, cheese makers have domesticated fermentation processes empirically, without a deep understanding of the multiple biochemical reactions governing the sensory qualities of their cheeses. Currently, the role of microbiota in terms of metabolic activities has been mostly understood, and many dairy productions are piloted and optimized by using starter cultures to standardize the fermentation process and reach the targeted properties for the final products.

Microbiology in Dairy Processing: Challenges and Opportunities, First Edition.
Edited by Palmiro Poltronieri.

Published 2018 by John Wiley & Sons Ltd and the Institute of Food Technologists.

The concept of primary or starter culture is well established in the dairy sector, referring to homofermenting lactic acid strains which produce lactic acid giving rise, with added rennin, to curd formation. The mesophilic lactococci generally make up around 90% of a mixed dairy starter population while thermophilic bacteria, mainly *Streptococcus thermophilus*, *Lactobacillus delbrueckii* subsp. *bulgaricus* and *Lactobacillus helveticus*, are employed in cooked curds because of their heat tolerance. Starting from these basic concepts, literature is now rich in debate and investigation regarding new tools and parameters for strain selection.

10.2.1 Natural versus selected lactic acid bacteria starters

Natural starter cultures (NSCs) are produced daily at cheese plants by some form of back slopping and/or by application of environmental selective pressure. Because no actions are used to prevent contamination from the cheese-making environment and raw milk, NSCs are continuously changing, undefined mixtures of several species and strains of LAB (Parente and Cogan, 2004).

Biodiversity is considered a strong point in NSCs, and producers of traditional cheeses, such as Grana Padano and Parmigiano Reggiano, accept slight variability in their products as a compromise with the more peculiar and individual characteristics coming from their ability to manage wild populations in raw milk and whey starters (Gatti et al., 2014). Moreover, richness in microbial consortium enhances the robustness of the starter. In fact, the presence of different species and as well as diverse biotypes make NSCs more resistant to lytic bacteriophage attack than selected starter cultures (SSCs) (Giraffa et al., 1997). The tolerance has been explained not only in terms of biodiversity but also by the fact that NSCs are grown in the presence of phages, which lead to dominance of resistant or tolerant strains (Marcò et al., 2012). Finally, NSCs show a higher level of adaptation to environmental stresses and microbial dynamics due to genetic diversity present in wild strains. Natural selection leads to characteristics relevant to the improvement of starter cultures (Smid and Hugenholtz, 2010). Thus, since the beginning of the last century, NSCs (together with spontaneous fermentation) have been an important source for isolation of new strains studied under defined conditions by industrial companies and research institutes, for their technological characteristics, with the aim to formulate performing SSC to add to milk for a better standardization of the process and of the final product; currently, use of SSC is state-of-the-art in the dairy industry.

Recently, Bassi et al. (2015), to better support the general and recognized belief regarding the pros and cons of using NSC and SSC, examined a number of studies comparing, in the same experimental conditions, NSC and SSC and found significant advantages in the use of SSC in terms of acidification, acceptability of final ripened products, and control of undesired microorganisms (Afzal et al., 2013; Murtaza et al., 2013; Renes et al., 2014; Van Hoorde et al., 2010). Moreover, these studies confirmed the minor impact on development of peculiar aroma when using SSC, supporting the importance of microbial biodiversity in terms of variety in cheese flavor. However, the risk of off-flavors is practically reduced to zero by using SSC.

Although much is known about the practice and impact on the final product of NSC and SSC use, further studies will be necessary in the future, with the help of next-generation sequencing (NGS) technologies, to better understand the interactions, at ecological level, between both SSC and NSC and the autochthonous microbial communities.

10.2.2 Starter strains: selection parameter approaches and strain concept

Within a bacterial species, a strain is defined as "the descendant of a single isolation in pure culture and usually made up of a succession of cultures ultimately derived from an initial single colony" (Staley and Krieg, 1984).

In food fermentation and, more specifically, in dairy processing, it is clear that the characteristics of the final products are modulated by the metabolic activities of specific strains. In fact, technological and functional properties are strain-dependent and strictly involved in the formation and quality of cheese flavor and texture. Thus, the selection of a starter culture has to be strain-dependent. The diversity in phenotypic traits of strains adapted to dairy environment has been explained in terms of genomic plasticity provided by both vertical and horizontal gene transfer (Thierry et al., 2015).

Traditionally, in order to screen and select starter cultures, microbiological techniques are employed to emphasize technological phenotypes of strains including growth and fermentation performance, phage robustness, flavor production and impact on matrix texture. As an example, the "Pearce activity test" is commonly used in the dairy industry as an indicator for growth and temperature-induced autolysis of starter strains (Feirtag and McKay, 1987), while the EPS production is checked by a common assay which utilizes ruthenium red stain in milk agar plates for the differentiation of ropy and non-ropy colonies (Hassan et al., 1995). Other traditional microbiological techniques include detection of prophage presence by phage induction assays whereby bacterial cells are submitted to stressful conditions (chemical treatment or exposure to UV light) in order to stimulate excision of the integrated phage, which will subsequently turn on its lytic life cycle and lyse the host cell (Chopin et al., 2001); in addition, culture-based methods are used in screening for bacteriocin producers, most commonly based on the principles of diffusion in agar plates and cell-free supernatants (Barefoot and Klaenhammer, 1983; Kékessy and Piguet, 1970; Yang et al., 1992).

Culture-based techniques are excellent for the study of LAB strains but are both labour- and time-intensive; thus, they are impractical for screening large strain collections. Recent advances in NGS platforms, is a promising approach to facilitate rapid identification and selection of such strains based on gene-trait matching (Kelleher et al., 2015). Molecular tools possess several advantages over classical approaches including a shorter assay period, higher throughput and a greater ability to discriminate at both species and strains level.

Due to the relatively small size of the genome of LAB species as *Lactococcus lactis* (~2.5 Mb chromosome) and *S. thermophilus* (~1.8 Mb chromosome), molecular NGS techniques can be applied to detect metabolic gene traits interesting for dairy industry. LAB encode a number of genes associated with key metabolic pathways which could assume the role of genetic markers whose presence, detected by molecular tools, could guarantee certain strain characteristics expressed at phenotypic level.

Liu et al. (2010) carried out comparative genomic studies to investigate proteolytic systems in LAB. Despite wide knowledge about the contribution of microbial proteolysis to cheese flavor development, Liu et al. (2010) enhanced information relative to two sub-groups within protein superfamilies involved in proteolysis. According to these authors, the goal should be accessing enough information from gene bacterial apparatus to predict proteolytic ability of different LAB strains. In 2012, Solopova et al. found that a plasmid-cured laboratory strain, *L. lactis* MG1363, harboured a cellobiose-specific phosphotransferase system (PTS) that allowed its adaptation and growth on lactose-supplemented media. The strain was lacking the lac operon, but the cellobiose-specific PTS behaved as an alternative to lactose utilization.

A central topic in the dairy industry is the problem of prophages and the selection of starter culture resistant to phage infection. The advent of modern technologies and the sequencing of whole bacterial genomes can readily identify the presence of temperate phages within the host genome. Moreover, plasmid and/or chromosomally encoded phage resistance mechanisms, which are desirable in starter strains, can be detected (Ainsworth et al., 2014a, 2014b; Allison and Klaenhammer, 1998) by genome sequencing (Mills et al., 2010; Sistla and Rao, 2004).

The recent advances in NGS technologies have provided new possibilities for the knowledge regarding metabolic traits of LAB strains. According to Kelleher et al. (2015), it will be necessary in the future to enforce the link between sequencing-based approaches with other "omics"-based technologies, such as transcriptomics and proteomics, in cheese studies in order to better confirm the genotype-phenotype connection. At the same time, culture-dependent methods will be never be replaced completely, and pilot trial fermentations will continue to be the only reliable test of starter culture performance within an industrial setting (Kelleher et al., 2015).

10.2.3 Starter culture formulation

The use of defined starters is state-of-the-art in the dairy industry where specialized companies produce and distribute the cultures worldwide. The main goal in dairy starter production is to get concentrated forms for direct inoculation in the milk vat. Thus, companies are oriented to reach the following targets: the production technology used does not have to negatively affect the viability and vitality of bacterial cells. Moreover, storage conditions, preferably at ambient temperature, also have to ensure minimal loss in terms of number of viable cells and metabolic activity level, especially as acidification rate. Finally, depending on the preservation technology, the handling of the cell concentration must be easy.

Currently, freezing and freeze-drying processes represent the most reliable technologies present on the market. For frozen cultures, the viability and vitality is generally high (Santivarangkna et al., 2008), but the transport has to be carried out at subzero temperatures on dry ice, which has a high impact on costs and consumed energy. On the contrary, transport of freeze drying is easier and cheaper because during the transfer the cold chain can be shortly interrupted. Moreover, among drying processes, freeze-drying is quite gentle and allows the obtaining of high-quality final products characterized by high survival, high metabolic activity, and good rehydratability. A major hurdle of freeze-drying is its high cost. For this reason, drying processing with lower costs have been considered, at experimental level, with good potentiality in terms of attractive processes for industrial companies. These promising processes are, mainly, spray drying and vacuum drying. Spray drying is a continuous process with high throughput, common in the dairy industry (Fu and Chen, 2011; Peighambardoust et al., 2011; Perdana et al., 2011, 2012), and for that reason it attracts attention at industrial levels. For spray drying, the cell suspension is pumped as liquid concentrate to the dryer and separated in small droplets by means of a pressure nozzle or a rotary disc (Foerst and Santivarangkna, 2015). Vacuum-drying process is currently evaluated by different research groups as an alternative to freeze-drying: the equipment is similar, but the setup is simpler for vacuum drying. Moreover, vacuum-dried starter cultures, with appropriate protectants, are more stable than spray- and freeze-dried cultures, especially at high storage temperature and relative humidity (Foerst and Santivarangkna, 2015), giving this technique a great potential in future applications.

10.3 ADJUNCT NON-STARTER LACTIC ACID BACTERIA

The NSLAB role in cheese maturation and, in particular, in flavor development, has been widely studied. High peptidase activity and catabolism of amino acids are strongly related to their presence in ripened cheeses from which comes the production of aromatic volatile compounds. The presence of moderate lipase and esterase activities have been also demonstrated in NSLAB, and their presence at high levels throughout ripening may strongly affect lipolysis of long-ripened cheeses.

The possibility of controlling flavor development and reduce fluctuations in the characteristics of the final product, besides, in some cases, accelerating cheese ripening, has led to the idea of using adjunct cultures of selected NSLAB strains. Moreover, recent evidence about their potential as protective cultures and health promoters, have oriented studies toward the selection and use of mixed cultures providing different technological and beneficial effects.

10.3.1 Biodiversity and adaptation to cheese environment

The transformation of curd in cheese is due to synergic action by microbial enzymes released by SLAB, NSLAB, other microorganisms naturally present in milk (McSweeney, 2004; Upadhyay et al., 2004), as well as those released by adventitious microorganisms contaminating cheese during ripening through salt, equipment, and maturing cells. Among NSLAB, mesophilic lactobacilli are the most important components affecting cheese ripening. Raw milk, natural whey cultures, and dairy environment, including equipment used in cheese manufacture, are considered sources for contaminating NSLAB.

The group of NSLAB is particularly heterogeneous, with lactobacilli being mostly represented rather than non-*Lactobacillus* species such as *Pediococcus acidilactici*, *Pediococcus pentosaceus*, *Enterococcus durans*, *Enterococcus faecalis*, *Enterococcus faecium* and also leuconostocs (Chamba and Irlinger, 2004). Among lactobacilli, the facultatively heterofermentative species *Lactobacillus casei*, *Lactobacillus paracasei*, *Lactobacillus plantarum*, *Lactobacillus curvatus*, *Lactobacillus rhamnosus* and *Lactobacillus pentosus* constitute the core microbiota of NSLAB; *Lactobacillus fermentum*, *Lactobacillus buchneri*, *Lactobacillus parabuchneri* and *Lactobacillus brevis* among obligately heterofermentative species; and *Lactobacillus farciminis* among obligately homofermentative species (Coeuret et al., 2004; Gobbetti et al., 2002; Svec et al., 2005). Gobbetti et al. (2015) found statistical positive correlations between technological treatments and the presence of some species. As an example, the presence of *L. casei*, *L. rhamnosus* and *L. parabuchneri* has been mainly related to cooking and stretching, indicating a specific tolerance of these species to high temperatures.

NSLAB do not grow well in milk due to their scarce proteolytic activities, and the level of residual lactose in cheese is fully depleted within a few days of ripening. Lack of lactose together with low pH values (~4.8–5.3), NaCl (1–3%), low moisture (~45–25%) and temperature (~8–12 °C), besides microbial competition, make the cheese environment very hostile, and lactobacilli adaptation varies depending on different species and strains. Thus, environmental and technological factors strongly affect their presence (Gobbetti et al., 2015).

Because of lactose depletion, NSLAB had to adapt to diverse energy sources. Free fatty acids (FFA), peptides and free amino acids (FAA), and ribose released from lysed cells are the main nutriments for NSLAB (Beresford and Williams, 2004). They come from microbial catabolism and are accumulated in cheese during ripening progress, where they are present in significant amounts in most dairy product varieties. Their ability to use those energy sources for their growth make them highly competitive in the hostile environment of ripened cheese. FAA represent not only a way for lactobacilli sustenance, but they also are a precursor for other catabolic reactions, giving rise to volatile aroma compounds. Richness in enzymatic tools make lactobacilli high producers of ketoacids, ammonia, amines, aldehydes, acids, and alcohols, which are essential contributors to cheese taste and aroma (Hemme et al., 1981). Thus, NSLAB are considered essential for developing the diversity and typical features of many cheese varieties (Coolbear et al., 2008; Montel et al., 2014).

Other adaptation mechanisms of NSLAB to the cheese environment have been ascribed to over-synthesis of molecular chaperones (De Angelis and Gobbetti, 2011), basic compounds (Heunis et al, 2014) and cold-induced proteins (De Angelis and Gobbetti, 2004).

Finally, hydrolase activities able to attack fat globules have been detected in NSLAB. Despite the presence of lipase and esterase activities in mesophilic lactobacilli having been demonstrated since 1990 (Khalid and Marth) and, subsequently, intracellular esterases from *L. plantarum* (Gobbetti et al., 1996) and *L. casei* (Castillo et al.,1999) having been purified and characterized, NSLAB contribution to lipolysis has been considered to have some significance only in long-ripened cheeses (Gobbetti et al., 2015).

10.3.2 Prospective in industrial application

Adventitious NSLAB cause variability in the ripening process and the cheese maker may have difficulty controlling them (Settanni and Moschetti, 2010). Fluctuations in the final characteristics of the product have been detected at different levels (Franciosi et al., 2008), among diverse dairy farms (Antonsson et al., 2003; De Angelis et al., 2001), but also between cheeses produced at the same farm at different times or coming from different vats from the same day (Fitzsimons et al., 1999; Williams et al., 2002).

Rarely, a single NSLAB species or biotype may predominate throughout the entire ripening time (Coolbear et al., 2008). Also, biotypes may vary between dairy factories, depending on season, and even vary depending on day of manufacture (Gobbetti et al., 2015).

In order to minimize variability in ripened products and defects that NSLAB rarely cause, it would be fundamental to pilot the dominance of selected NSLAB strains (Gobbetti et al., 2015). In this way, they would outcompete adventitious NSLAB and limit their influence on the final product quality. Selected NSLAB would be introduced as adjunct cultures (Coolbear et al., 2008), defined as those added to cheese for purposes other than lactic acid production (El Soda et al., 2000).

The inoculation of raw or pasteurized milk with mixtures of lactobacilli would increase secondary proteolysis and catabolism of FAA, positively affecting sensory cheese attributes, and in some cases, it would accelerate the ripening of a variety of products. Despite the undoubted advantages, the addition of adjunct NSLAB to cheese milk may cause over-acidification of the curd due to lactose fermentation in addition to SLAB metabolic activities, with repercussions on increased whey drainage, which, in turn, affects cheese yield and rheology, and on sensory cheese acceptability (Gobbetti et al., 2015).

The solution proposed could be the attenuation of adjunct NSLAB where the aim of the treatment is getting microbial cells unable to grow and thus produce lactic acid, but at the same time, keeping most of cellular enzymes active on peptides and FAA. Over the years, many treatments have been applied to get attenuated cultures, such as heating (Asensio et al., 1995; Farkye et al., 1995), freezing-thawing (Aly, 1990; Madkor et al., 2000), sonication (Di Cagno et al., 2011, 2012) and, less frequently applied, high-pressure (Lanciotti et al., 2007) and spray-drying (Johnson and Etzel, 1995; Madkor et al., 2000). Whatever the method used, various attenuated lactobacilli species increased proteolysis and lipolysis, shortened ripening time, improved flavour and reduced bitterness in studies on cheddar (Madkor et al., 2000), Caciocavallo Pugliese cheese (Di Cagno et al., 2012) and Caciotta-type cheese (Di Cagno et al., 2011). Despite the promising results, the industrial application of attenuated adjunct cultures is not as widespread as would be expected, and the reasons may be related mainly to the empirical nature of the methods employed, not easily adaptable at industrial level, and the different sensitivity to treatments of different NSLAB species and biotypes (Gobbetti et al. 2015).

10.3.3 Biopreservation and health benefits

Interest in NSLAB as adjunct cultures also has been due to the fact they are considered "protective cultures" with a positive impact on food stability. The capability to produce antifungal compounds and bacteriocins, besides conferring a competitive advantage, is of great importance in planning strategies of "biopreservation" in the logic of extending shelf-life and improving safety of foods.

Some papers describe the results obtained by employing bacteriocin producing NSLAB in cheese during ripening, and many of them are focused on the inhibition of *Listeria monocytogenes* whose growth can be limited or, in some cases, stopped (Dal Bello et al., 2012; Rossi and Veneri, 2016). However, many aspects have yet to be understood and solved for an efficient application of NSLAB as bioprotective agents (Giannou et al., 2009; Izquierdo et al., 2009). The main problems rely on the production of efficient amounts of bacteriocins either *in vitro* or in cheese.

Inhibitory activity of NSLAB over moulds may result from the production of metabolites such as organic acids (lactic, propionic, acetic acids and phenyllactic acid), carbon dioxide, ethanol, hydrogen peroxide, diacetyl, and reuterin, and other small molecular weight metabolites included peptide (Foerst and Santivarangkna, 2015), other than growth competition. Mostly, the application of lactobacilli as antifungal agents is focused on sourdoughs where they may delay growth of moulds such as *Aspergillus niger* (Lavermicocca et al., 2000) and *Fusarium* spp. (Dal Bello et al., 2007). In the dairy sector, noteworthy results were obtained with the addition of *L. rhamnosus* in commercial cottage cheese (Stiles et al., 2002) that reduced the growth of both *A. niger* and *Penicillium commune*. More recently, *L. plantarum* isolates were inoculated into cottage cheese prior to the addition of *P. commune*, and they were found to prevent the visible mould growth by between 14 and >25 days longer than cottage cheese that contained either no added LAB or LAB that did not have antifungal activity, such as *L. lactis*, *Weissella soli*, *Leuconostoc inhae* and *Leuconostoc mesenteroides* isolates (Cheong et al., 2014).

Other studies described the addition of *L. amylovorus* cultures in cheddar cheese, which resulted in a six-day delay in the appearance of *Penicillium expansum* mycelia on the cheese surface (Lynch et al., 2014), while a *L. plantarum* strain was effective in delaying the mycelia growth of mycotoxigenic *Aspergillus flavus* and *Aspergillus parasiticus* on fresh cheese surface by 19 and 22 days, respectively, at 4 °C (Sedaghat et al., 2016). Thus, the use of NSLAB as natural biopreservatives for controlling the problems caused by mould growth in cheese could be considered promising.

Health benefits have also been associated with NSLAB, and the possibilities to apply adjunct selected cultures have also been evaluated in terms of consumer health improvement. Settanni and Moschetti (2010) reviewed, among many aspects, the probiotic characteristics of some strains, the production of bioactive peptides and γ-aminobutyric acid, and antigenotoxic abilities.

Cheese can be considered an interesting matrix as a vehicle for living microorganisms due to its intrinsic characteristics of solid matrix, higher pH (compared to fermented milk), buffering capacity and fat content, which may protect vitality of probiotic lactobacilli during the transit through the gastrointestinal tract (Ross et al., 2002). In different studies, the addition of adjunct probiotics has been evaluated during cheese ripening in terms of growth and impact on cheese organoleptic characteristics (Da Cruz et al., 2009; Grattepanche et al., 2008). Technological interventions have been evaluated to reduce negative sensorial effects (Drake et al., 2007). Vinderola et al. (2009) highlighted the strong impact of storage temperature on sensorial properties of Argentinean cheeses added with probiotic lactobacilli.

The production by NSLAB of bioactive peptides with a positive impact on body functions has been widely demonstrated during cheese aging (Pripp et al., 2006). They mainly show hypotensive and/or angiotensin-I-converting enzyme (ACE) inhibitory activity (FitzGerald and Murray, 2006; Gobbetti et al., 2004), but their performance is strictly related to their bioavailability and their resistance to gastrointestinal tract conditions has yet to be demonstrated.

Similarly, studies on γ-aminobutyric acid-producing (GABA-producing) NSLAB have been carried out in recent years, and they showed the capacity of some NSLAB strains to produce GABA from L-glutamate through glutamate decarboxylase activity, positively affecting blood pressure regulation (FitzGerald and Murray, 2006). Studies demonstrated the ability of some *L. paracasei*, *L. plantarum* and *L. casei* strains to survive and synthesize GABA in cheese ripening and under simulated gastrointestinal conditions (Siragusa et al., 2007; Wang et al., 2010).

Finally, antitoxigenic ability has been referred to NSLAB in recent studies. Nowak et al. (2015) reported that strains belonging to the species *L. casei* showed strong antitoxigenic activity *in vitro* even if the mechanisms involved were not fully understood. They may be linked to the binding or lowering of the concentration of genotoxins as well as to the induction of DNA repair in exposed cells. *L. rhamnosus* also seems to have great potential (Pithva et al., 2015).

10.4 CONCLUSIONS

Despite the use of defined starters being widespread in the dairy industry and the many SLAB mixtures available on the market, in order to develop new dairy products, the necessity of new culture blends is increasing. Recent molecular tools are becoming effective for the screening of huge numbers of potential SLAB strains, allowing shortening of laboratory time. Pilot fermentations will continue to be the only final reliable test of starter culture performance before industrial application. Moreover, in the prospective of producing safe and functional foods, it is becoming fundamental, in strain selection, to take into account, in addition to well-known technological parameters, new characteristics such as antifungal properties, bacteriocin and bioactive peptide production, and probiotic attitude. With this purpose, SLAB and, especially, NSLAB strains are, currently, widely studied, and so far, they have proven to have great potential for future applications in dairy industry.

References

Afzal MI, Ariceaga CC, Lhomme E, Ali NK, Payot S, Burgain J, Gaiani C, Borges F, Revol-Junelles AM, Delaunay S. 2013. Natural and starter cultures in food fermentations: characterization of *Carnobacterium maltaromaticum* LMA 28 for its positive technological role in soft cheese making. *Food Microbiol* 36:223–30.

Ainsworth S, Mahony J, van Sinderen D. 2014a. The plasmid complement of *Lactococcus lactis* UC5099 encodes multiple bacteriophage resistance systems. *Appl Environ Microb* 80:4341–49.

Ainsworth S, Sadovskaya I, Vinogradov E, Courtin P, Guerardel Y, Mahony J, Grard T, Cambillau C, Chapot-Chartier MP, van Sinderen D. 2014b. Differences in lactococcal cell wall polysaccharide structure are major determining factors in bacteriophage sensitivity. *mBio* 5(3):e00880–14.

Allison GE. Klaenhammer TR. 1998. Phage resistance mechanisms in lactic acid bacteria. *Int Dairy J* 8(3):207–226.

Aly ME. 1990 Utilization of freeze-shocked lactobacilli for enhancing flavour development of Ras cheese. *Egypt J Dairy Sci* 18:143–56.

Antonsson M, Molin G, Ardö Y. 2003 *Lactobacillus* strains isolated from Danbo cheese as adjunct cultures in a cheese model system. *Int J Food Microbiol* 85:159–69.

Asensio C, Gomez R, Pelaez C. 1995. Effect of heat treatment on the proteolytic activity of mesophilic bacteria isolated from goats' milk cheese. *Lett Appl Microb* 21:25–30.

Barefoot SF, Klaenhammer TR. 1983. Detection and activity of lactacin B, a bacteriocin produced by *Lactobacillus acidophilus*. *Appl Environ Microbiol* 45 (6):1808–15.

Bassi D, Puglisi E, Cocconcelli PS. 2015. Comparing natural and selected starter cultures in meat and cheese fermentations. *Curr Opin Food Sci* 2:118–122.

Beresford T, Williams A. 2004. The microbiology of cheese ripening. In: Fox PF, McSweeney PLH, Cogan TM, Guinee TP (Eds). *Cheese: chemistry, physics and microbiology. General aspects*. Amsterdam: Elsevier, pp 287–317.

Brooks JC, Martinez B, Stratton J, Bianchini A, Krokstrom R, Hutkins R. 2012. Survey of raw milk cheeses for microbiological quality and prevalence of foodborne pathogens. *Food Microbiol* 31:154–58.

Castillo I, Requena T, Fernandez de Palencia P, Fontecha J, Gobbetti M. 1999. Isolation and characterization of intracellular esterase from *Lactobacillus casei* subsp *casei* IFPL731. *J Appl Microbiol* 86:653–59.

Chamba J-F, Irlinger F. 2004. Secondary and adjunct cultures In: Fox PF, McSweeney PLH, Cogan TM, Guinee TP (Eds). *Cheese: chemistry, physics and microbiology*. London: Elsevier, pp 191–206.

Cheong EYL, Sandhu A, Jayabalan J, Le TTK, Nhiep NT, Ho HTM, Zwielehner J, Bansal N, Turner MS. 2014. Isolation of lactic acid bacteria with antifungal activity against the common cheese spoilage mould *Penicillium commune* and their potential as biopreservatives in cheese. *Food Control* 46:91–97.

Chopin A, Bolotin A, Sorokin A, Ehrlich SD, Chopin MC. 2001. Analysis of six prophages in *Lactococcus lactis* IL1403: different genetic structure of temperate and virulent phage populations. *Nucleic Acids Res* 29 (3):644–51.

Coeuret V, Gueguen M, Vernoux JP. 2004. In vitro screening of potential probiotic activities of selected lactobacilli isolated from unpasteurized milk products for incorporation into soft cheese. *J Dairy Res* 71:451–60.

Coolbear T, Crow V, Harnett J, Harvey S, Holland R, Martley F. 2008. Developments in cheese microbiology in New Zealand – use of starter and nonstarter lactic acid bacteria and their enzymes in determining flavour. *Int Dairy J* 18:705–13.

da Cruz AG, Buriti FCA, de Souza CHB, Fonseca JAF, Saad SMI. 2009. Probiotic cheese: health benefits, technological and stability aspects. *Trends Food Sci Technol* 20:344–54.

Dal Bello F, Clarke CI, Ryan LAM, Ulmer H, Shober TJ, Strom K. 2007. Improvement of the quality and shelf life of wheat bread by fermentation with the antifungal strain *Lactobacillus plantarum* FST 17. *J Cereal Sci* 45(3):309–18.

Dal Bello B, Cocolin L, Zeppa G, Field D, Cotter PD, Hill C. 2012. Technological characterization of bacteriocin producing *Lactococcus lactis* strains employed to control *Listeria monocytogenes* in cottage cheese. *Int J Food Microbiol* 153:58–65.

De Angelis M, Corsetti A, Tosti N, Rossi J, Corbo MR, Gobbetti M. 2001. Characterization of non-starter lactic acid bacteria from Italian ewe cheeses based on phenotypic, genotypic, and cell wall protein analyses. *Appl Environ Microbiol* 67:2011–20.

De Angelis M, Gobbetti M. 2004. Environmental stress responses in *Lactobacillus*: a review. *Proteomics* 4:106–122.

De Angelis M, Gobbetti M. 2011. Stress responses of Lactobacilli. In: Tsakalidou E, Papadimitriou K (Eds). *Stress responses of lactic acid bacteria*. New York: Springer, pp 219–49.

Di Cagno R, De Pasquale I, De Angelis M, Buchin S, Calasso M, Fox PF. 2011. Manufacture of Italian Caciotta-type cheeses with adjuncts and attenuated adjuncts of selected non-starter lactobacilli. *Int Dairy J* 21:254–60.

Di Cagno R, De Pasquale I, De Angelis M, Gobbetti M. 2012. Accelerated ripening of Caciocavallo Pugliese cheese with attenuated adjuncts of selected nonstarter lactobacilli. *J Dairy Sci* 95:4784–95.

Drake M. 2007. Sensorial analysis of dairy foods. *J Dairy Sci* 90:4925–37.

El Soda M, Madkor SA, Tong PS. 2000. Adjunct cultures: recent developments and potential significance to the cheese industry. *Int J Dairy Tech* 83:609–19.

Farkye NY, Madkor SA, Atkins HG. 1995. Proteolytic abilities of some lactic acid bacteria in a model cheese system. *Int Dairy J* 5:715–25.

Feirtag JM, McKay LL. 1987. Thermoinducible lysis of temperature-sensitive *Streptococcus cremoris* strains. *J Dairy Sci* 70 (9):1779–84.

FitzGerald RJ, Murray BA. 2006. Bioactive peptides and lactic fermentations. *Int J Dairy Technol* 59:118–25.

Fitzsimons NA, Cogan TM, Condon S, Beresford T. 1999. Phenotypic and genotypic characterization of nonstarter lactic acid bacteria in mature cheddar cheese. *Appl Environ Microbiol* 65:3418–26.

Foerst P, Santivarangkna C. 2015. Advances in starter culture technology: focus on drying processes In: Holzapfel W (Ed). *Advances in fermented foods and beverages*. Cambridge, Elsevier, pp 249–70.

Franciosi E, Settanni L, Carlin S, Cavazza A, Poznanski E. 2008. A factory-scale application of secondary adjunct cultures selected from lactic acid bacteria during "Puzzone di Moena" cheese ripening. *J Dairy Sci* 91:2981–91.

Fu N, Chen XD. 2011. Towards a maximal cell survival in convective thermal drying process. *Food Res Int* 44(5):1127–49.

Gatti M, Bottari B, Lazzi C, Neviani E, Mucchetti G. 2014. Microbial evolution in raw-milk, long-ripened cheeses produced using undefined natural whey starters. *J Dairy Sci* 97:573–91.

Giannou E, Kakur A, Matijasic BB, Rogelj I, Samelin J. 2009. Fate of *Listeria monocytogenes* on fully ripened Greek Graviera cheese stored at 4, 12, or 25 °C in air or vacuum packages: in situ PCR detection of a cocktail of bacteriocins potentially contributing to pathogen inhibition. *J Food Prot* 72:531–38.

Giraffa G, De Vecchi P, Reinheimer J. 1997. Population dynamics of thermophilic lactobacilli in mixed starter whey cultures. *Food Res Int* 37:137–40.

Gobbetti M, De Angelis M, Di Cagno R, Mancini L, Fox PF. 2015. Pros and cons for using non-starter lactic acid bacteria (NSLAB) as secondary/adjunct starters for cheese ripening. *Trends Food Sci Tech* 45:167–78.

Gobbetti M, Fox PF, Smacchi E, Stepaniak L, Damiani P. 1996. Purification and characterization of a lipase from *Lactobacillus plantarum* 2739. *Food Biochem* 20:227–46.

Gobbetti M, Minervini F, Rizzello CG. 2004. Angiotensin-I-converting-enzyme inhibitory and antimicrobial bioactive peptides. *Int J Dairy Technol* 57:173–88.

Gobbetti M, Morea M, Baruzzi F, Corbo MR, Matarante A, Considine T, Di Cagno R, Guinee T, Fox PF. 2002. Microbiological, compositional, biochemical and textural characterisation of Caciocavallo Pugliese cheese during ripening. *Int Dairy J* 12:511–23.

Grattepanche F, Miescher-Schwenninger S, Meile L, Lacroix C. 2008. Recent developments in cheese cultures with protective and probiotic functionalities. *Dairy Sci Technol* 88:421–44.

Hassan AN, Frank JF, Farmer MA, Schmidt KA, Shalabi S. 1995. Formation of yogurt microstructure and three-dimensional visualization as determined by confocal scanning laser microscopy. *J Dairy Sci* 78 (12):2629–36.

Hemme DH, Schmal V, Auclair JE. 1981. Effect of the addition of extracts of thermophilic lactobacilli on acid production by *Streptococcus thermophilus* in milk. *J Dairy Res* 51:345–51.

Heunis T, Deane S, Smit S, Dicks LM. 2014. Proteomic profiling of the acid stress response in *Lactobacillus plantarum*. *J Proteome Res* 13:4028–39.

Johnson JAC, Etzel MR. 1995. Properties of *Lactobacillus helveticus* CNRZ-32 attenuated by spray-drying, freeze-drying or freezing. *J Dairy Sci* 78:761–68.

Izquierdo E, Marchioni E, Aoude-Werner D, Hasselmann C, Ennahar S. 2009. Smearing of soft cheese with *Enterococcus faecium* WHE 81, a multi-bacteriocin producer, against *Listeria monocytogenes*. *Food Microbiol* 26:16–20.

Kékessy DA, Piguet JD. 1970. New method for detecting bacteriocin production *Appl Microbiol* 20 (2):282–83.

Kelleher P, Murphy J, Mahony J, van Sinderen D. 2015. Next-generation sequencing as an approach to dairy starter selection. *Dairy Sci Technol* 95:545–68.

Khalid NM, Marth EH. 1990. Lactobacilli and their enzymes and role in ripening and spoilage of cheese: a review. *J Dairy Sci* 73:2669–84.

Lanciotti R, Patrignani F, Iucci L, Saracino P, Guerzoni ME. 2007. Potential of high pressure homogenization in the control and enhancement of proteolytic and fermentative activities of some *Lactobacillus* species. *Food Chem* 102:542–50.

Lavermicocca P, Valerio F, Evidente A, Lazzaroni S, Corsetti A, Gobbetti M. 2000. Purification and characterization of novel antifungal compounds from the sourdough *Lactobacillus plantarum* strain 21 B. *Appl Environ Microbiol* 66:4084–90.

Liu M, Bayjanov J, Renckens B, Nauta A, Siezen R. 2010. The proteolytic system of lactic acid bacteria revisited: a genomic comparison. *BMC Genomics* 11(1):36.

Lynch KM, Pawlowska AM, Brosnan B, Coffey A, Zannini E, Furey A, McSweeney PLH, Waters DM, Arendt EK. 2014. Application of *Lactobacillus amylovorus* as an antifungal adjunct to extend the shelf-life of Cheddar cheese. *Int Dairy J* 34:167–73.

Madkor S, Tong PS, El Soda M. 2000. Ripening of cheddar cheese with added attenuated adjunct cultures of lactobacilli. *J Dairy Sci* 83:1684–91.

Marcó MB, Moineau S, Quiberoni A. 2012. Bacteriophages and dairy fermentations Bacteriophage 2:149–58.

McSweeney PLH. 2004. Biochemistry of cheese ripening. *Int J Dairy Technol* 57:127–44.

Mills S, Griffin C, Coffey A, Meijer W, Hafkamp B, Ross R. 2010. CRISPR analysis of bacteriophage insensitive mutants (BIMs) of industrial *Streptococcus thermophilus*–implications for starter design. *J Appl Microbiol* 108(3):945–55.

Montel MC, Buchinb S, Mallet A, Delbes-Paus C, Vuitton DA, Desmasures N, Berthier F. 2014. Traditional cheeses: rich and diverse microbiota with associated benefits. *Int J Food Microbiol* 177:136–54.

Murtaza M, Rehman S, Anjum F, Huma N. 2013. Descriptive sensory profile of cow and buffalo milk Cheddar cheese prepared using indigenous cultures. *J Dairy Sci* 96:1380–86.

Nowak A, Slizewska K, Otlewska A. 2015. Antigenotoxic activity of lactic acid bacteria, prebiotics, and products of their fermentation against selected mutagens. *Regul Toxicol Pharmacol* 73:938–46.

Parente E, Cogan TM. 2004. Starter cultures: general aspects. In: Fox PF (Ed). *Cheese: chemistry, physics and microbiology*, vol 1. London: Elsevier, pp 123–47.

Peighambardoust S, Goshan Tafti A, Hesari J. 2011. Application of spray drying for preservation of lactic acid starter cultures: a review. *Trends Food Sci Tech* 22:215–24.

Perdana JA, Fox MB, Schutyser MAI, Boom RM. 2011. Single-droplet experimentation on spray drying: evaporation of a sessile droplet. *Chem Eng Technol*, 34 (7):1151–58.

Perdana JA, Fox MB, Schutyser MAI, Boom RM. 2012. Enzyme inactivation kinetics: coupled effects of temperature and moisture content. *Food Chem* 133 (1):116–23.

Pithva SP, Ambalam PS, Ramoliya JM, Dave JM, Vyas BRM. 2015. Antigenotoxic and antimutagenic activities of probiotic *Lactobacillus rhamnosus* Vc against N-Methyl-N′-Nitro-N-Nitrosoguanidine. *Nutr Cancer* 67(7):1142–50.

Pripp AH, Sorensen R, Stepamak L, Sorhaug T. 2006. Relationship between proteolysis and angiotensin-I-converting enzyme inhibition in different cheeses. *LWT-Food Sci Technol* 39:677–83.

Renes E, Diezhandino I, Fernaandez D, Ferrazza RE, Tornandijo ME, Fresno JM. 2014. Effect of autochthonous starter cultures on the biogenic amine content of ewe's milk cheese throughout ripening. *Food Microbiol* 44:271—77.

Ross RP, Morgan S, Hill C. 2002. Preservation and fermentation: past, present and future. *Int J Food Microbiol* 79:3–16.

Rossi F, Veneri G. 2016. Use of bacteriocinogenic cultures without inhibiting cheese associated nonstarter lactic acid bacteria: a trial with *Lactobacillus plantarum*. *Challenges* 7:4.

Santivarangkna C, Kulozik U, Foerst P. 2008. Inactivation mechanisms of lactic acid starter cultures preserved by drying processes. *J Appl Microbiol* 105:1–13.

Sedaghat H, Eskandari MH, Moosavi-Nasab M, Shekarforoush SS. 2016. Application of non-starter lactic acid bacteria as biopreservative agents to control fungal spoilage of fresh cheese. *Int Dairy J* 56:87–91.

Settanni L,Moschetti G. 2010. Non-starter lactic acid bacteria used to improve cheese quality and provide health benefits. *Food Microbiol* 27:691–97.

Siragusa S, De Angelis M, Di Cagno R, Rizzello CG, Coda R, Gobbetti M. 2007. Synthesis of g-aminobutyric acid by lactic acid bacteria isolated from a variety of Italian cheeses. *Appl Environ Microbiol* 22:7283–90.

Sistla S, Rao DN. 2004. S-Adenosyl-L-methionine-dependent restriction enzymes. *Crit Rev Biochem Mol Biol* 39(1): 1–19.

Smid EJ, Hugenholtz J. 2010. Functional genomics for food fermentation processes. *Annu Rev Food Sci Technol* 1:497–19.

Solopova A, Bachmann H, Teusink B, Kok J, Neves AR, Kuipers OP. 2012. A specific mutation in the promoter region of the silent cel cluster accounts for the appearance of lactose-utilizing *Lactococcus lactis* MG1363. *I* 78 (16):5612–21.

Staley JT, Krieg NR. 1984. Classification of procaryotic organisms: an overview. In: Krieg NR, Holt JG (eds). *Bergey's manual of systematic bacteriology*, vol 1. Baltimore: The Williams & Wilkins Co, pp 1–4.

Stiles J, Penkar S, Plockova M, Chumchalova J, Bullerman LB. 2002. Research note: antifungal activity of sodium acetate and *Lactobacillus rhamnosus*. *J Food Protect* 65 (7):1188–91.

Svec P, Dráb V, Sedlácek I. 2005. Ribotyping of *Lactobacillus casei* group strains isolated from dairy products. *Folia Microbiol* 50:223–28.

Thierry A, Valence F, Deutsch SM, Even S, Falentin H, Le Loir Y, Jan G, Gagnaire V. 2015. Strain-to-strain differences within lactic and propionic acid bacteria species strongly impact the properties of cheese–a review. *Dairy Sci Technol* 95:895–18.

Upadhyay VK, McSweeney PLH, Magboul AAA, Fox PF. 2004. Proteolysis in cheese during ripening. In: Fox PF, McSweeney PLH, Cogan TM, Guinee TP (Eds). *Cheese: chemistry, physics and microbiology*, London: Elsevier, pp 391–33.

Van Hoorde K, Van Leuven I, Dirinck P, Heyndrickx M, Coudijzer K, Vandamme P, Huys G. 2010. Selection, application and monitoring of *Lactobacillus paracasei* strains as adjunct cultures in the production of Gouda-type cheeses. *Int J Food Microbiol* 144:226–35.

Vinderola G, Prosello W, Molinari F, Ghiberto D, Reinheimer J. 2009. Growth of *Lactobacillus paracasei* A13 in Argentinian probiotic cheese and its impact on the characteristics of the product. *Int J Food Microbiol* 135:171–74.

Wang HK, Dong C, Chen YF, Cui LM, Zhang HP. 2010. A new probiotic Cheddar cheese with high ACE-inhibitory activity and gamma-aminobutyric acid content produced with Koumiss-derived *Lactobacillus casei. Zhang Food Technol Biotechnol* 48:62–70.

Williams AG, Choi SC, Banks JM. 2002. Variability of the species and strain phenotype composition of the non-starter lactic acid bacterial population of Cheddar cheese manufactured in a commercial creamery. *Food Rev Int* 35:483–93.

Yang R, Johnson MC, Ray B. 1992. Novel method to extract large amounts of bacteriocins from lactic acid bacteria. *Appl Environ Microbiol* 58 (10):3355–59.

11 Milk fat: stability, separation and technological transformation

Gianluigi Scolari

Istituto di Microbiologia degli alimenti, Università Cattolica del Sacro Cuore, Piacenza, Italy

11.1 INTRODUCTION

Cream, butter and sour cream are the food products derived from milk fat. Today, milk is separated in bulk into cream and skim milk by huge machines. The cream is processed to produce various consumer products, depending on its thickness, its suitability for culinary uses and consumer demand, which differs from place to place and country to country.

Some cream is dried and powdered, some is condensed (by evaporation) mixed with varying amounts of sugar and canned. Most cream is made into butter. This is done by churning the cream until the fat globules coagulate and form a monolithic mass. This butter mass is washed and, sometimes, salted to improve keeping qualities. The butter is packaged (25- to 50-kg boxes) and chilled for storage and sale. At a later stage, these packages are broken down into home-consumption-sized packs. Originally the cream was fermented using *Lactobacillus delbrueckii* subsp. *lactis* and *Lactococcus lactis* subsp. *diacetylactis* strains, producing butter flavours such as diacetyl, acetylpropionyl and acetoin. Nowadays, the natural flavours can be added during industrial manufacture. The residual buttermilk goes on to further processing. Cream is fermented by lactic acid bacteria into sour cream, a product mostly used in cooking recipes around the world.

As knowledge of cooling increased, it became possible to skim the cream before it had gone sour and make butter from the sweet cream. Sweet cream could be soured by the addition of naturally soured milk or acid buttermilk. It then became possible to make ripened cream butter under more controlled conditions.

When the cream is produced at the creamery, the whole milk is preheated to 63 °C in the pasteurizer before being separated. The warm cream is routed into an intermediate storage tank before being pumped to the cream pasteurization plant. The skim milk from the separator is pasteurized and cooled before being pumped to storage. When cultured butter is to be produced, part of the skim milk should be utilized for starter preparation.

Microbiology in Dairy Processing: Challenges and Opportunities, First Edition.
Edited by Palmiro Poltronieri.

Published 2018 by John Wiley & Sons Ltd and the Institute of Food Technologists.

From the intermediate storage tank(s) the cream continues to pasteurization at a temperature of 95 °C or higher. The high temperature is needed to destroy enzymes and microorganisms that would impair the keeping quality of the butter.

The destruction of unwanted microorganisms is also beneficial in the case of sour cream butter, as this creates perfect growth conditions for the bacteria culture. The heat treatment releases strongly anti-oxygenic sulfidryl compounds, which further reduce the risk of oxidation.

Butter made from sour cream has certain advantages over the sweet cream variety. The aroma is richer, the butter yield higher, and there is less risk of re-infection after temperature treatment, as the bacteria culture suppresses undesirable microorganisms. Buttermilk from sour cream butter has a lower pH than buttermilk from sweet cream butter. However, sour cream butter has drawbacks. A disadvantage of cultured cream butter is that it is more sensitive to oxidation, with production of metallic taste. This is due to presence of copper or other heavy metals, reducing the chemical keeping properties of the butter.

11.1.1 Composition and physical state of milk fat

The milk fat is a mixture of triglycerides (TGs) the complexity of which is responsible for some unique traits of milk fat's physical behaviour.Such an high number of triglycerides depends firstly from the wide variety of fatty acids (FAs) in milk fat as well as their esterification at the stereospecific positions of glycerol (*sn*1, *sn*2, *sn*3) during TG synthesis.

The main saturated and unsaturated FAs are reported in Table 11.1 and 11.2, respectively. The saturation level (about 70% in bovine milk fat), the dominance of the even number of carbon chains and short-chain FAs are characteristics of ruminant milk fat which result from both desaturation activity of rumen and the synthetic pathway starting from acetate, β-hydroxybutirate and propionate.

As the melting point of a TG depends on structural characteristics of molecule, even considering the FA at concentration still less than 0.1% of total (including: oxy- and keto-acids, branched-chain and odd carbon number chains) the melting point interval becomes significantly broad. In fact, if n is the number of FA to be distributed within the three stereospecific positions of glycerol, n^3 will be the number of possible combinations, and then the number of TG. Thus, assuming the presence of approximately 400 different FAs the possible TG will be 64×10^6 (400^3). Such a complexity justifies the statement: "*Among dietary fat matrices that of milk is by far the most complex*".

Actually, the biosynthesis of TG follows some preferential ways: the *sn*3 position is almost occupied by the C_4 and other short-chain FA, while the *sn*2 position is preferential for palmitic and stearic acids as well as *sn*1 for oleic acid. As a consequence, the TG number lessens to about 10^4, which still maintains a considerable level of complexity (Alais, 1984).

Incidentally, such a complexity is the basis for the detection of extraneous fats in butter (lard, palm oil and other vegetable oil), thanks to a characteristic bimodal distribution of milk fat TG according to the total number of carbon number.

Table 11.1 Composition in saturated fatty acids of bovine milk fat (I = *iso*; AI = *anteiso*).

	Linear fatty acids		
A		**% total**	**Physical state (melting point °C)**
a) Soluble volatile			
	C_4	3–6	Liquid (−8)
	C_6	1.5–3	Liquid (−3)
b) Insoluble volatile (extractable by steam distillation)			
	C_8	1–1.5	Liquid/solid (15)
	C_{10}	1.3–1.8	Solid (30)
	C_{12}	2.3–3.9	Solid (42)
c) Fixed (nonextractable by steam distillation)			
	C_{14}	8–13	Solid (54)
	C_{15}	0.2–1.2	
	C_{16}	25–34	Solid (62)
	C_{17}	0.3–0.7	Solid (70)
	C_{18}	10–14	Solid (75)
	C_{20}	0.2	
B	**Branched chain fattyacids**		
	$C_{14}I$	0.2	
	$C_{15}I$	0.3	
	$C_{15}AI$	0.6	
	$C_{16}I$	0.3	
	$C_{17}I$	0.3	
	$C_{17}AI$	0.5	
	$C_{18}I$	0.1	
		A + B = 65–68%	

Table 11.2 Composition in unsaturated fatty acids of bovine milk fat.

	Unsatured fatty acids		
		% Total	**Physical state (melting point °C)**
a) Monounsatured			
	$C_{10:1}$	0.2–0.3	Liquid
	$C_{14:1}$	0.3–1.2	"
	$C_{16:1}$	2–3	Liquid (0.0)
	$C_{17:1\ (9cis)}$	0.1–0.2	
	$C_{18:1\ (9cis)}$	18–30	liquid/solid (16)
	$C_{18:1\ (11trans)}$	2–3	solid (41)
b) Polyunsatured (non-conjugated)			
	Linoleic	1–3.5	Liquid (−5)
	α-Linolenic	0.2–1.4	Liquid (−12)
	Arachidonic C_{20}	0.2	Liquid (−47)
c) Polyunsatured (conjugated)	CLA	0.4–1.8	Liquid

The above mentioned aspects determining the physical behaviour and stability of milk triglyceride mixture, are of paramount importance in dairy processes regarding the milk fat transformation as churning and whipping of cream.

11.1.2 Melting point of milk fat

Observing the solid liquid phase transition of separated milk fat (butter) it is possible to detect a temperature interval between 23 and 25 °C, while the reverse occurs in the 22–19 °C range. Thus two anomalies could be registered: the lack of a unique value of both melting temperature (mT) and solidifying (sT) temperature and two different temperature of transition according to the melting or solidification processes. However, following the process by accurate measurements (differential thermal analysis, conductimetry, etc.), it is possible to detect a transition temperature interval ranging between –40 and +40 °C, with a maximum quantity of TG fusion or solidification at the aforementioned temperature intervals. Considering that the mT (or sT) of a single FATG are triple, it is possible to infer that the number of mT (or sT) of the complex mixture of milk fat is characterized by a continuous spread of transition temperatures (–40, +40 °C) as observed. So, the phase transition observable by the naked eye suggests that in the TG mix in milk fat there is a prevalent quantity of TG with transition points corresponding to those of mP and sP.[1]

Concluding this short exposition on physical properties of milk fat, the mixed crystal formation that is responsible for a little contraction of the melting and solidification point interval should also be mentioned.

Moreover the mixed crystal formation gives the milk fat a thermal memory effect.

In detail, an increase of the refrigeration speed to a final fixed temperature produces a major fraction of mixed crystals: the consequence is an accumulation of a greater quantity of solid fat. Also, a different quantity of solid fat is produced by different thermal cycles to which the milk cream will be subjected. As the mixed crystals are not thermodynamically stable, such a phenomenon will be reversible, and the time taken for the equilibrium achievement depends on the temperature of solid fat storage. Thus, due to the dependence of churning and whipping effectiveness from the ratio of liquid fat/ solid fat, the aforementioned phenomenon makes necessary a standardization procedure for this parameter (Ranjith and Rajah, 2001).

Finally, the explanation difference between the mP and sP rely on the nucleation phenomenon, the realization of which requires temperatures lower than melting points. Deeper insight regarding these phenomena is outside the scope of this chapter.

11.2 PHYSICAL INSTABILITY OF MILK FAT

Globular milk fat should be considered as a true emulsion, and then as lyophobic substance it must undergo segregation, whatever the density difference from the dispersing phase (milk plasma).

[1] Among the milk fat TGs prevailing oleic-stearic-butyric acid, oleic-palmitic-butyric acid and oleic-oleic-caproic acid, which in any case do not overcome 1.5–3.0% of total; this justifies the extreme spreading of mPs.

The different kinds of milk fat globule instability are as follows:

1) **Flocculation**. This is the mildest instability, which causes a very faint link among globule surfaces. The repulsive interactions provided by the glycoproteins of the globule membrane are sufficient to avoid a tighter contact, and then when flocculation occurs, it is easily dispersed by simple agitation. The mechanism by which globules come in touch is termed 'natural creaming' and is mediated by the so-called agglutinins. Such an event is utilized in the production of hard cheeses like Grana cheese and Parmigiano Reggiano cheese: the natural creaming allows a reduction of microbial population in raw milk (cold pasteurization) before renneting due to the ability of agglutinins (mainly IgE) to link either fat globules or bacterial cells. Thus the microbially enriched cream is removed from cheese milk and destined for butter production.
2) **Homogenization cluster.** This destabilization occurs after the homogenization process. It is observed during the restoration of membranes in homogenized globules. Due to the very short time of transit in the homogenization valve, for several homogenized globules it could be difficult to get surfactants (mainly casein micelles) in appropriate quantity to cover the new generated interfacial surface and then share membrane material with other globules. Such an aggregation is more difficult to be redispersed than in the previous case. Usually the redispersion is obtained by an improperly termed "second homogenization stage" (at 10% of nominal homogenization pressure).
3) **Partial coalescence** (churning clamps). The aggregation of globules follows the rupture of natural membrane so that the mixture of triglycerides (in the liquid state) is exchanged between two or more globules. Such a destabilization requires shear fields of $10^4/10^5\ s^{-1}$, which are intermediate between those of flocculation ($10^2\ s^{-1}$) and homogenization (up to $10^7\ s^{-1}$). Partial coalescence is the starting step common to butter and whipping technologies.
4) **Destruction.** This is the premise for the homogenization technology. As stated previously, the required energy is supplied by the strong acceleration that is imposed on milk (or cream) in order to create shear fields enough to strain the globule up to the rupture. Also, due to the very short duration of such a treatment (0.1 ms), the high density of mechanical energy is transferred to the globule by an adiabatic process (Mulder and Walstra, 1974; Fox, 2002).

11.3 MILK FAT SEPARATION

11.3.1 Flocculation or natural creaming

In thermally treated milk, the milk fat segregation can be satisfactorily interpreted by the well-known Stokes model, although with some approximation. By contrast, following the fat globules rise in raw milk, it can be observed that fat separation is significantly faster than one could expect by the Stokes law; such a disagreement must be attributed to the globule aggregation by the agglutinins, the aggregates simulating 'globules' of higher dimensions and then rising faster. The biological nature of the aggregation should impose the term "agglutination".

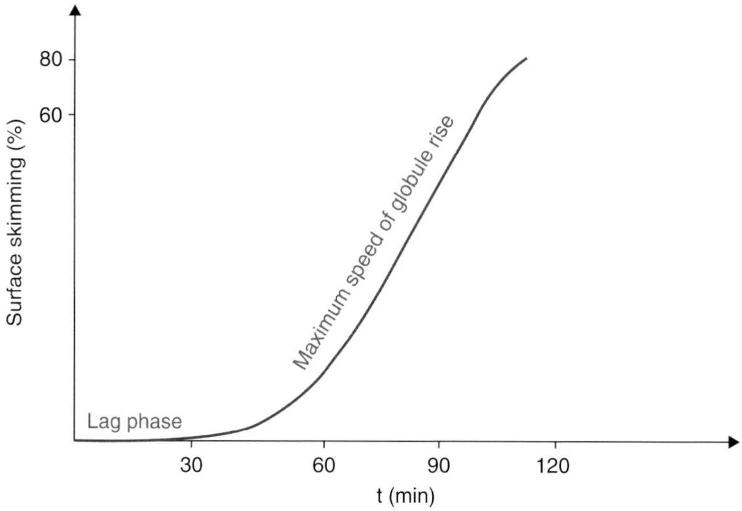

Figure 11.1 Percentage of surface skimming of milk fat over time. The geometry of surfacing basin is suited to lower the time needed in respect to the technological performance.

In Figure 11.1 is shown the percentage of surface skimming over time for milk fat (keeping in mind that the parameters are different from those of real technological plants).

Activation of agglutinins is obtained by lowering raw milk temperature to almost 10 °C. Then, after an induction time of approximately 30 min, IgM binds to the vessel and globule surfaces. In the next step, the large size of agglutinins allows the development of bridges between fat particles, overwhelming the stabilizing effect of the hydrophilic membrane glycoproteins as suggested by the Derjaguin, Landau, Verwey, Overbeek (DLVO) theory.

At this stage, as the repulsive interactions are impeded, the flocculation rate becomes linear with time, following the Smoluchowski kinetic model for association of particles without repulsive forces. The linkage of IgM to the fat globules is non-specific, while it is specific as established with some bacterial species, probably by polysaccharides on the bacterial surface. The final result is the segregation of a fraction of natural microbiota in raw milk, either by specific interactions or by its entrapment in fat globule aggregates.

The natural creaming can be compromised by:

- *unsuitable and prolonged agitation* of raw milk under refrigeration, probably due to the linkage of both sides of agglutinins on the same particle; in fact, the rewarming of milk to 37 °C (detachment of IgM) followed by appropriate refrigeration, completely restores the creaming ability;
- *homogenization*, as a consequence of an increased number of fat particles that make insufficient the agglutinin content;
- *raw milk at the extreme end of lactation*, during which IgG and IgM in whey protein fraction are almost substituted by BSA;
- *thermal treatments, like pasteurization,* which cause a significant precipitation of IgM.

The fat content of cream obtained by the natural creaming ranges between 21 and 24%, according to the time of creaming, such a cream require an adjustment of fat titre before its transformation to butter (Fox, 2002).

11.3.2 Milk fat separation by centrifugation

Both separation and titre adjustment of cream can be performed by centrifugal processes using cream separators. The parameters used in such a centrifugal process withstand some theoretical principles.

As previously mentioned, fat globule rising is satisfactorily described by the Stokes model, which is based on the equivalence of the friction and Archimedean forces to which particles are subjected:

- $\text{Farch} = 1/6\ \pi\ d^3\ (\rho_p - \rho_g)\ a$
- $\text{Ffric} = 3\ \pi\ \text{d}\ \eta_p \text{v}$

where: ρ_p and ρ_g are the plasma and fat density, respectively; η_p is plasma viscosity; d the globule diameter and v the globule speed of rise. Finally, $a = \omega^2 R$ in centrifugal gravitation field (ω = angular velocity of rotor; R = rotor radius).

After the rise starts, the two forces become equal, and the globule moves by straight uniform velocity. In such conditions:

$$\text{Farch} = \text{Fric and then } 1/6\,\pi\,\text{d}^3\left(\rho_p - \rho_g\right)a = 3\,\pi d \eta_p v$$

and then v can be obtained

$$v = ad^2\left(\rho_p - \rho_g\right)/18\eta_p$$

Therefore the separation speed depends directly on the gravitational intensity (a) and the square of d; also, the ratio enclosed in the ellipse plays an important role and is termed "efficiency factor".

Factors influencing the model effectiveness are:

- particles should be spherical (*true*)
- extraneous particles with diameter much smaller than the globule diameter (*roughly true*)
- mono dispersed system (*partially true after homogenization*)
- globule interactions absent (e.g. agglutination) (*true after pasteurization*)
- globules free from Brownian motion (*not true for globule* $d < 0.1\ \mu\text{m}$)
- volume fraction of fat < 0.01 (*scarcely true*)

Also, ρ_g is strongly affected by the crystallization rate as well as the quantity of casein micelles adsorbed on the globule surface, sometimes occurring after agitation in the cold or homogenization.

At the end, temperature can indirectly affect the separation process via its influence on ρ_p, η_p, ρ_g.

In spite of the above-cited limitation, in the first equation, established that at a fixed d, the cream separation can be accelerated, increasing either the speed of centrifuge

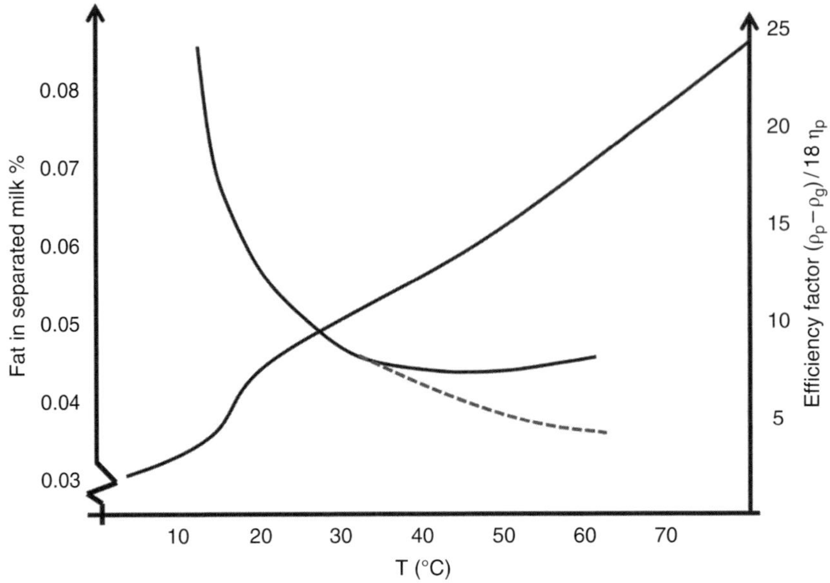

Figure 11.2 Efficiency factor and percentage of fat content in separated milk as a function of separation temperature. The dashed line corresponds to the theoretical fat percentage in separated milk, according to the formula of Stokes.

rotor ($a = \omega^2 R$) or the efficiency factor; the same holds for the reduction on milk fat in the separated milk.

From Figure 11.2 it is evident that the efficiency factor increases by increasing temperature until the temperature of approximately 35 °C, beyond which it decreases again, thus increasing fat percentage in separated milk. Such an unexpected trend is a function of rotation speed: by increasing ω values it becomes more and more evident.

Such behaviour can be explained considering the milk (or cream to be concentrated) path in the milk separator (Figure 11.3). Such equipment is constituted by a series of concentric conical disks (Figure 11.4) maintained in rotation by the axis of the motor. The disc alignment creates channels that are the way by which the milk enters in rotation; the number, the diameter and the distance between discs are characteristic of various commercial models. Usually, their rotation speed is maintained between 3000 and 4000 rpm as a function of disc diameter. In Figure 11.3 the point of milk inlet in the rotating section (black arrow) can be seen.

This is the most critical point: in fact, fat globules move suddenly from a non-rotating zone to the rotating section. This implies that fat globules are subjected to a strong acceleration that can produce a deformation of their shape according to both the intensity of the imposed shear rate and the physical state of fat. If the temperature is so high as to make prevalent the liquid fat, the rupture of the globule in smaller particles can occur, thus compromising fat separation.

Thus, the choice of the most proper centrifugation parameters is a compromise between temperature and rotor speed.

Currently, milk separators provide special geometries for the entry channel so that the risk of globule rupture is minimized, thus allowing either higher rotation velocity or separation temperatures as high as 50 °C (TetraPack, 2009).

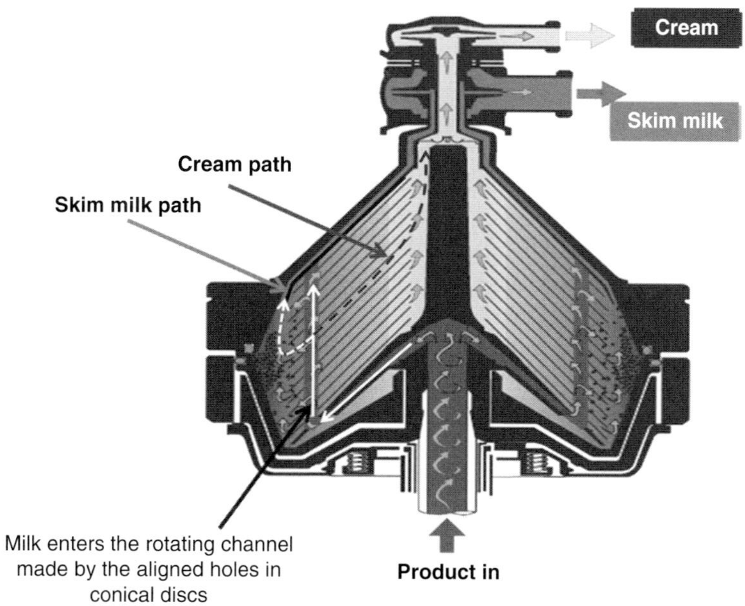

Figure 11.3 Section of a centrifugal separator.

Figure 11.4 Conical stainless steel discs of centrifugal separator.

11.4 PARTIAL COALESCENCE

11.4.1 General aspects

As previously described, partial coalescence occurs when two (or more) globules share the liquid fat after membrane rupture in a single point.

Factors influencing such a destabilization figure are:

- shear field intensity (s^{-1})
- globule concentration (%)
- diameter of particles

- solid TG fraction with respect to the total
- surfactant properties of membrane.

In detail, the coalescence per unit of time is linearly proportional to the shear rate (s^{-1}) at fixed values of other factors, while the same rate dependence from fat content (%) is more than cubic: in fact a shear rate of $1400\,s^{-1}$ (e.g. milk flowing in a pipeline) is enough to produce significant coalescence or none at all in 40 and 20% cream, respectively. As a consequence, in churning and whipping, the fat content is one of the key factors controlling the time to get the proper coalescence degree, this last being a prerequisite for these technologies.

As reported in Figure 11.5, the fraction of crystallized fat is of paramount importance in coalescence occurrence: only a given solid/liquid fat ratio (≈0.2–0.3) can maximize the coalescence and, consequently, produce visible butter granules, although there is little difference between the maxima of continuous and dashed curves. Additionally, it is noteworthy to emphasise how even a small deviation from the ideal value is more critical for butter granule appearance compared to coalescence; this demonstrates that the partial coalescence is a necessary, but not sufficient, condition to perform an appropriate churning.

In this respect, it assumes that it is of maximum importance to get a standard grade of fat crystallization by accurate thermal cycles before the churning or whipping process of cream.

Finally, Figure 11.5 shows the influence of globule dimension on the partial coalescence. It is interesting to observe that at a fixed shear rate the coalescence decreases until disappearance (e.g. globules with diameter <2 μm). This is the consequence of the internal Laplace overpressure that increases as the particle radius decreases, thus preventing the globule distortion up to membrane rupture. Therefore, churning or whipping homogenized cream requires the application of very high shear rates. A possible measure to overcome such a problem is to modify the membrane strength (e.g. adjunct of small-sized surfactant before homogenization). In this respect, some reasoning regarding the effects of shear stretching on globule membrane is possible after the examination of some aspects regarding the surfactants (present or added) at the TG–whey interface (Mulder and Walstra, 1974; Walstra, 2003).

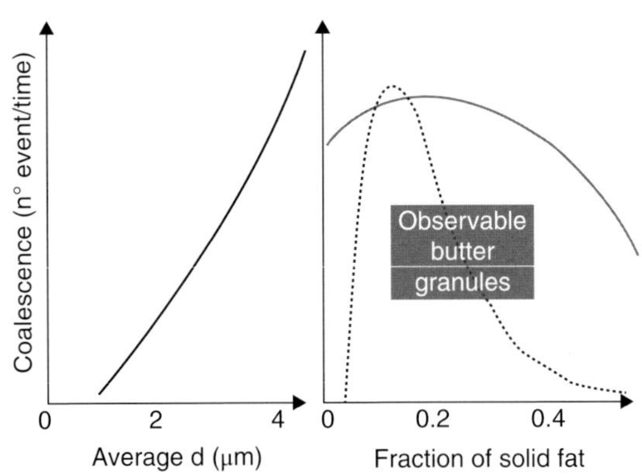

Figure 11.5 Grade of coalescence (number of events per time unit) as a function of mean globule diameter and of fraction of fat crystallized. (Approximate data from various sources.)

11.4.2 Barrier against coalescence

11.4.2.1 Low molecular mass surfactants

Mono- and di-glycerides of FA and phospholipids (P-lipids) belong to this class of surfactants and are contained in raw milk in small amounts. The mechanism or their possible milk fat stabilization is depicted in Figure 11.6, where two globules of different size are represented close to partial coalescence by collision. The bigger globule withstands an indentation by the pressure of the dispersing phase (milk plasma), while the smaller appears as non-deformable due to the high Laplace overpressure. The small-size surfactant (small bars) is evenly adsorbed onto the globule surface, so assuring the reduction of the interface tension (γ) over the whole surface. The stretching of the bigger globule immediately before the collision causes a transient surface increase in the incipient collision area and a lowering of surfactant concentration (molecules/μm^2). The resulting increase of surface tension γ causes a sudden migration of surfactant at the stretched site (B arrows) and consequently its new distribution on the whole globule surface; as the surfactant partially protrudes in the dispersing phase, its external portion drains milk plasma when moving towards the zone interposed between the two colliding particles (A arrows). Therefore, the density of milk plasma in this area increases, preventing the collision and partial coalescence (Marangoni effect).[2] Such a phenomenon is transient and completely restored when the globules come away.

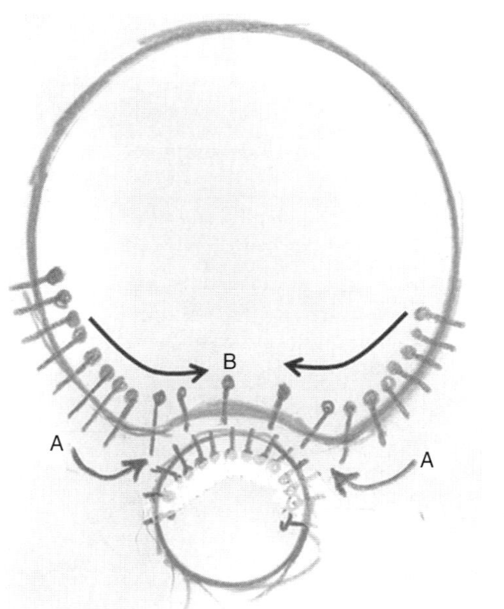

Figure 11.6 Representation of two fat globules with different sizes immediately before collision. The bars with circle distributed on the globule surfaces represent small-size surfactant (P-lipids, for instance).

[2] This suggests that the stabilizing effectiveness of surfactants depends on both the intrinsic ability to lower the interface energy and the stabilizing mechanism (affected by diffusivity and the orientation velocity once the interface is reached).

On the basis of this stabilization mechanism it can be deduced that the hydrophobic-hydrophilic balance (HHB) of the surfactant is of paramount importance: in fact, an excessive absorption into the hydrophobic phase (TGs) makes the milk plasma drainage impossible, while an excessive protrusion towards the dispersing phase exposes the stabilizing molecule to an easy desorption.

11.4.2.2 Large sized surfactants (casein micelle)

Due to the uneven distribution of hydrophobic and hydrophilic parts in caseins, the whole casein micelle behaves as a surfactant. Thus an interaction between the casein micelle and the milk fat globule can be expected in a way similar to that examined in the previous section. Gently shacking milk for a long time in the cold and then observing its particles by scanning electron microscopy reveals the presence of casein micelles adsorbed onto the fat globule surface; also, the contact angle results of 30° as represented in Figure 11.7. Here the angle α is determined by the resultant of the three interface tensions (force vectors) applied to the contact point T: $\boldsymbol{\gamma}_{SC}$ (serum/casein), $\boldsymbol{\gamma}_{SD}$ (serum/dispersed phase) and $\boldsymbol{\gamma}_{CD}$ (casein/dispersed phase) (dispersed phase = globular fat).

According to the intensity of the three interfacial forces, the resulting force vector determines the depth of casein immersion in the fat globule and then the contact angle, the value of which is determined by the formula

$$\alpha = (\gamma_{CD} - \gamma_{SC})/\gamma_{SD}$$

Taking into account the actual surface tension in the milk, the calculated value of α is 30°, coinciding with that observed. In the light of HHB significance, the maximum stabilizing effect is obtained by a contact angle between 30 and 90°; below the lower and above the upper limits, the micelle loses its stabilizing capacity.

It is noteworthy that in the true case, casein micelles are able to insert into the natural globule membrane, although this last is endowed with excellent stabilizing properties. One of the possible explanations is the concentration of caseins in milk: in fact, the

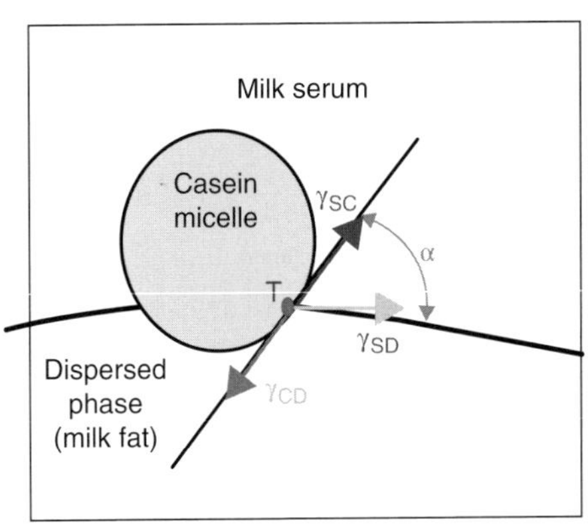

$$\gamma_{SC} + \gamma_{SD}\cos\alpha = \gamma_{CD}$$

$$\cos\alpha = (\gamma_{CD} - \gamma_{SC})/\gamma_{SD}$$

$$\alpha \approx 30°$$

Figure 11.7 Schematic representation of a casein micelle C adhering to a fat globule (D) immersed in the whey phase (S). (Arrows γ_{SC}, γ_{SD} and γ_{CD} are vectors corresponding to the three interface tensions).

absorption at an interface is also directed by the concentration (or rather activity) of the surfactant in one of the two phases, especially under agitation.

Therefore, under appropriate condition and concentration, surfactant as FAs and P-lipids can be partially adsorbed onto the globule surface, even in the presence of its natural membrane.

11.4.2.3 Polymeric surfactants (proteins and polysaccharides)

Polymeric molecules can behave as excellent surfactants provided they display an appropriate HHB. Among these, whey proteins are particularly suited. During the insertion into an interface, a deep unfolding occurs, so that the hydrophobic and hydrophilic groups dissolve in the hydrophobic and hydrophilic phases, respectively; in such a way, the energy for the unfolding of the protein (or polymer) is supplied by that stored at the interface, thus stabilizing it.

Moreover, for whey proteins, the possibility exists to realize intermolecular interactions or even chemical bonds, thus providing a film "adherent" (due to the wide surface of contact) and "coherent/elastic" (robustness due to the intermolecular interactions). Consequently, these interface "materials" are difficult to desorb (as could happen for the small surface active molecules), thus resembling the natural membrane of fat globule.

In some particular circumstances, caseins can also give rise to interfacial films, particularly after micelle spreading onto the contact surface. The same holds for peptides derived from casein proteolysis (better diffusivity and the orientation velocity once the interface is reached).

The stabilizing effect and properties of biopolymers is summarized in Table 11.3.

11.4.2.4 Mixed films

The addition of surfactants to the fat globule emulsion in raw milk (or cream) causes an impairment of the intrinsic elasticity-cohesiveness of natural membrane. The same is observed if artificial polymeric membranes are involved, as in the case of fat globules after homogenization with added small surfactants. This the alternation of polymeric units by small molecules results in an interruption of membrane structure. A not fully explained exception is the natural globule membrane that is composed of polymers, such as glycoproteins, and small molecules, such as P-lipids; the most probable reason is the presence of cholesterol, which is known to confer mechanical robustness

Table 11.3 Characteristics of polymeric films related to the polymer properties (data from various sources).

Film properties	Polymer properties
Membrane thickness	Solubility Diffusivity and orientation velocity
"Adherent" film (stability against desorption)	Partial "unfolding" and "spreading"
Film • "coherent" • viscoelastic (high Gibbs elasticity) • steric repulsion	Every intermolecular interactions

to eukaryotic membranes. Nevertheless, as stated before, appropriate concentration of small surfactants can reduce their robustness.

These aspects are fundamental in circumstance that imply milk fat stability:

- *Raw milk:* if the stability of natural globule is not well maintained, the surface activity of lipoprotein lipase could be responsible for an enzyme access to the TGs with an associated rapid rancidity development;
- *Partial coalescence (butter, whipped cream):* where allowed, the use of mono- and di-glycerides of FAs is beneficial for the partial coalescence. Such an advantage is particularly useful in ice cream, as the partial coalescence of the fat globules confer on the product the organoleptic trait of palatability; in this case, the destabilizing surfactants are provided by both the egg lecithins and mono- and di-glycerides (Walstra, 2003; Walstra et al., 1984; Fang and Dalgleish, 1993; Goff et al., 1999; Boode and Walstra, 1993).

11.5 FOAM IN MILK AND CREAM

11.5.1 General aspects

The substantial differences between the two kind of 'emulsion', foam and emulsion *sensu stricto*, can be deduced from the data from Table 11.4 on the basis of the energy stored at some interfaces. The water/air tension is as high as 74 mN/m. On the other hand, the interface tension between water and the hydrophobic *n*-octane is about 20 mN/m lower: such a reduction approximately corresponds to the tension at the air–*n*-octane interface (both mainly hydrophobic), thus demonstrating that gases are more hydrophobic than hydrophobic molecules.

As a consequence, it is not surprising that the same surfactant molecule could demonstrate different stabilizing properties of the emulsion with respect to the foam. For example, the better performance of casein (and its fragments) as emulsion stabilizer is recognized, while the whey proteins are excellent in foam stabilization.

In conclusion, it can be inferred that stability in foam is by far more critical than in emulsion.

11.5.1.1 Foam formation without surfactants

As previously seen, if air is beaten in water without surfactants, a foam forms, but the surface tension at the bubble–water interface is so high that it makes the foam very unstable: thus, it suddenly collapses, causing the two phases of separation. A different situation occurs when a hydrocarbon (or a pure TG) is vigorously stirred in water, producing an emulsion that is more stable. Either air cells or oil droplets are subjected to the internal overpressure of Laplace,[3] which is described by the following simple relation:

$$\Delta P \approx 2\gamma / r$$

[3] Laplace overpressure: pressure difference between the inner part of particles and the dispersing phase. It becomes negligible when the particle dimension is several micrometres.

where r is the particle radius and γ the interface tension. Therefore, air cell shows a major internal overpressure than oil droplet with the same radius, as its γ is higher.

From another point of view, the air cells in foam will maintain as larger than oil droplets in emulsion in order to get the same overpressure in both the cases.

The previous observations justify why the smaller air cells (high ΔP) will progressively coalesce with those of higher dimension (lower ΔP). This occurs in two ways:

1) due to a direct cell contact after removal of intercell dispersing medium (no added surfactants);
2) due to the high overpressure, the equilibrium between the gas inside the cell and that dissolved in dispersing phase (e.g. water) is strongly forced towards this last. The consequence is a progressive reduction of small cell radius and dramatic increase of internal ΔP; the end of this step is the total transfer of gas to the bigger cells via dispersing medium. The final result is the formation of a monodisperse foam before it collapses.

Compared to emulsion, instability of foam is accrued by the higher difference of density between dispersing and dispersed phases.

11.5.1.2 Foam formation with surfactants

From Table 11.4 it can be concluded that a surfactant dissolved in water before inflating air gives a wide range of γ reductions but always maintaining the aforementioned difference between foam and emulsion.

The surfactant nature determines the mechanism of foam stabilization. In this respect, the same considerations developed in Section 11.4.2 (Barrier Against Coalescence) are still true, although emphasised by the more critical conditions. In particular, adherent and coherent/elastic films are the best suited for the critical stabilization of foam.

Whey proteins, casein (isolate or micellar) and polysaccharides fulfil such requirements, the first being the best. In fact, the effect of the unavoidable reduction of bubble

Table 11.4 Interfacial tensions (γ) at pure substances contact and in some potential interface in milk.

Interfaces	Interfacial Tension γ (mN/m^{-1})
Water/air	74
n-octane/air	22
Water (with Na laurate)/air	40
Triglyceride mixture/air	34
Skimmed milk/air	48
Water/n-octane	51
Water/triglyceride mixture	19
Milk fat globule/skimmed milk	2
Liquid milk fat/skim milk	15

volume is the increase of film thickness, provided that surfactant molecules are adherent and coherent thus preventing their desorption. The γ reduction by surfactant is expressed by the Gibbs relation:

$$d\gamma = -\Gamma RT \, d \ln a$$

a = surfactant activity in the medium
Γ = surfactant film thickness[4]

Therefore, at a given instant, the γ increase deriving from cell volume reduction is counterbalanced by the film thickness increase: in such away, the bubbles become stable and, consequently, the foam, as well.

It now easy to understand the role of whey proteins as being the most appropriate foam stabilizers as a consequence of the possibility of intermolecular interactions, especially when a mild thermal treatment of foam is possible.

On the contrary, volatile surfactants (alcohol and esters, for example) are dramatically detrimental for the foam stability. Intuitively, once the interface is reached, during the bubble volume reduction they are desorbed into the bubble itself, producing a deleterious increase of the inner overpressure: the consequence is the rapid collapsing of the foam.

11.5.1.3 Drainage of dispersion liquid in foam

A particular kind of foam instability is the unavoidable dispersing air cells during the draining process of the aqueous medium. Due to its location between gas cells, this component can be considered to be a particular foam-stabilizing agent as well as a contributor to the foam characteristics. In fact, during the drainage, it creates a thin liquid barrier between bubbles that is called *lamella*: if the drainage is not completely prevented, the risk of *lamella* rupture exists that is the prerequisite for foam collapsing.

According to the *lamella* thickness, it is defined as the ratio l/d between the *lamella* thickness itself (l) and the cell diameter d, the level of which can be related to the physical characteristics of foam: for example, when $l \approx 1\%$ d, then the volume fraction of air in foam $\Phi = 0.95\%$, that is, the foam composition is 95% air and 5% dispersing liquid. Such a value is very close to that in foam from skim milk, which could be produced in the separated milk in a cream separator: such a foam is 'light', due to the high content of gas, and is efficiently stabilized by spreading of casein micelles on the air cell surface. The persistence of this foam can be a problem in cream separators, thus operating under pressure in order to prevent the foam formation.

The foam obtained from raw milk has a $\Phi = 0.85$, while in whipped cream $\Phi = 0.50$, thus indicating that it is composed of 50% milk plasma.

The volume fraction Φ is used to define the *overrun* of foam:

$$\text{Overrun } (\%) = \{\Phi / (1 - \Phi)\} \times 100$$

[4]Γ is the *specific stabilizing power* [dγ/RT d ln a] that is the ability to reduce γ at a given surfactant concentration *a*.

Therefore, skim milk produces a foam with an overrun = 1900 (%), while 33% cream gives a whipped cream with overrun = 100 (%).

In foam obtained from milk and, especially, cream, it is possible to observe the presence of small particles (micelles, fat globules) in the *lamella* between air cells; when present in appropriate quantities, they function as barrier to the liquid drainage, thus improving the foam stability in time.

In addition to the stability, the drainage delay is important for the organoleptic properties of foam; the dispersing liquid (milk plasma in whipped cream) strongly contributes to the product palatability due to the plasticizing effect of water. Therefore, when possible, technological measures should be taken to prevent drainage as much as possible.

In addition to fat globules and casein micelles, the addition of thickening polysaccharides is the method of choice, particularly those showing a thixotropic behaviour: in fact, they fulfil the fundamental requirement to keep the dispersing liquid viscosity low during the foam formation and increase such a rheological property at the end of the process, thus efficiently preventing deleterious drainage.

11.5.2 Foam from cream containing more than 30% milk fat

Before considering the cream whipping mechanism, it is noteworthy to observe the interaction that develops between the gas cell and fat globule.

Figure 11.8A shows three phases without any surfactant at their surfaces: a droplet of milk TGs dispersed in milk plasma and in incipient contact with an air cell. Once the contact has occurred, the droplet position relative to the air cell depends on the result of the three interfacial forces, each one attempting to close the corresponding surface; hence, the droplet is subjected to a spreading pressure determined by the following equation:

$$P_s = \gamma_{\text{milk plasma/air}} - \left(\gamma_{\text{milk plasma/oil}} + \gamma_{\text{air/butter oil}}\right)$$

If P_s is positive, then the oil droplet will spread into the air bubble and vice versa. In this case the $P_s > 0$, and the liquid portion of oil droplet will spread (Figure 11.8B).

In the real conditions (e.g. beating air in cream), bubbles come in touch with fat globules enveloped in their natural membranes (Figure 11.8C–D); also, as the surface tension air/water is the highest, air cells dispersed in milk plasma will suddenly be covered by surface active material present in the dispersing phase (mainly whey proteins and casein, to a lesser extent). As a consequence, even if once again $P_s > 0$, the spreading does not occur due to the mechanical resistance of the two films (Figure 11.8D). In other words, the energy stored at the interfaces is not enough to overcome the cohesion energy of membranes. Therefore, only when a fat globule collides with a still 'uncovered' air cell is the P_s positive enough that a membrane rupture occurs at the collision point followed by spreading of liquid TGs in the air cell (Figure 11.8E–F). The result of such an event is the fat globule membrane destabilization, which leads to the partial coalescence (the premise for whipped cream and butter production); the importance of a proper liquid/solid fat ratio is intuitive (see Section 11.1.2).

Continuous air beating in cream gives rise to air cells completely covered by milk fat globules, as shown by SEM microscopy. By the same technique, it has been possible to observe that the coalescence of such air cells produces the release of aggregates of

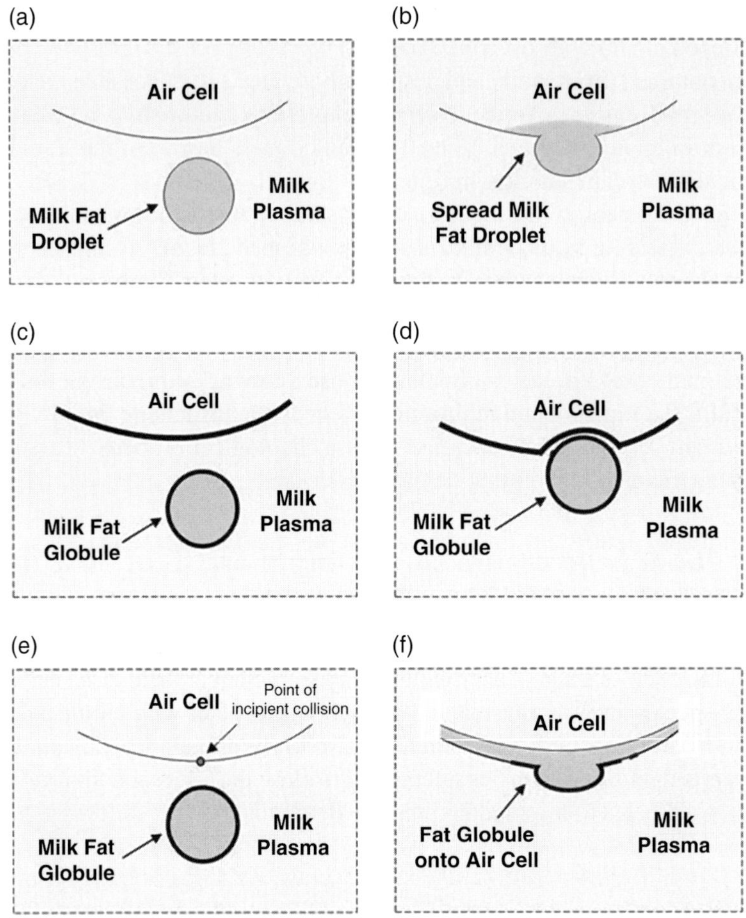

Figure 11.8 Oil droplet (A) and milk fat globule (C and E) interaction with air cell-milk plasma different conditions. Interfaces with adsorbed surface active substances are indicated by a thick line.

destabilized globules in the milk plasma due to the consequent reduction of total air cell surface: thus, in the milk plasma can be seen clumps of fat globules partially coalesced. Such a release can reach its maximum level after foam collapse.

In the light of the previous description, it can be evident that for whipped cream or butter technology the starting point is the same, as both require the partial coalescence step; then, some technological parameters (the beating speed, for example) will address the process towards the desired product.

An addition to the various parameters, the fat concentration in cream is of paramount importance. As a matter of fact, if an appropriate quantity of fat globules is available to give globule clumps of dimension (50–150 µm diameter) comparable to that of air cells, then a network is reliable, which can both link the air cells and efficiently delay the interstitial milk plasma drainage; such a result is attainable only if the fat content of cream is almost 33%. In Figure 11.9, a schematic picture of the network is depicted (Walstra et al., 1984; Goff et al., 1999; Boode and Walstra, 1993; Chen et al. 1999; McClements, 2005).

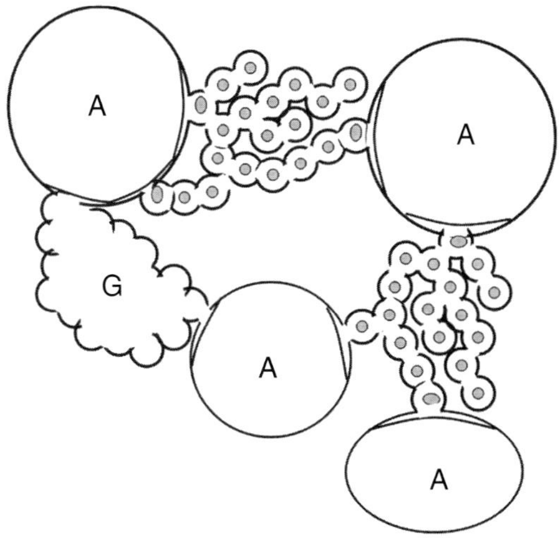

Figure 11.9 Foam-stabilizing network realized by clumps of partially coalescent fat globules linked to air cells. The represented structure is that of whipped cream. A: air cell; G: clump of globules.

11.6 WHIPPED CREAM AND BUTTER

11.6.1 Technological factors affecting whipped cream and butter production

The first steps in the technology of milk fat transformation in whipped cream or butter are common as described in the previous section; the surface and interfacial tensions in the presence of a gas (usually air) are the key to obtaining partial coalescence without employing high exogenous energies. However, in order to address the technology towards the foam or butter formation, it is necessary to examine the technological factors allowing the appropriate final result.

The following must be considered:

1) *Beating speed and effectiveness.* With other parameters fixed, the speed of cream beating is one of the most critical parameters for a successful process. At high speed the increasing dimension of air cells are constantly fractionated so avoiding the predominant tendency to the coalescence and the consequent release of globule clumps; also, air cells of minor dimensions require a lesser *clumping* grade to give satisfactory overrun to the whipped cream. Hence, slow beating forces globules to detach from the air cell, thus giving aggregates of visible butter grains (Figure 11.10).
2) *Globule stability.* Such a property substantially affects the formation of globule clumps that provide the network with air cells. When a fat globule is wrapped by the natural membrane, the only destabilization mechanism is air interaction (*flotation*). On the other hand, if such a membrane is impaired (adding small-sized surfactants), then globule aggregates occur also by the shear rate applied by beating, thus accelerating the whole process. The level of natural membrane destabilization can address the process towards whipping or churning, imposing the appropriate dosage of surfactants.

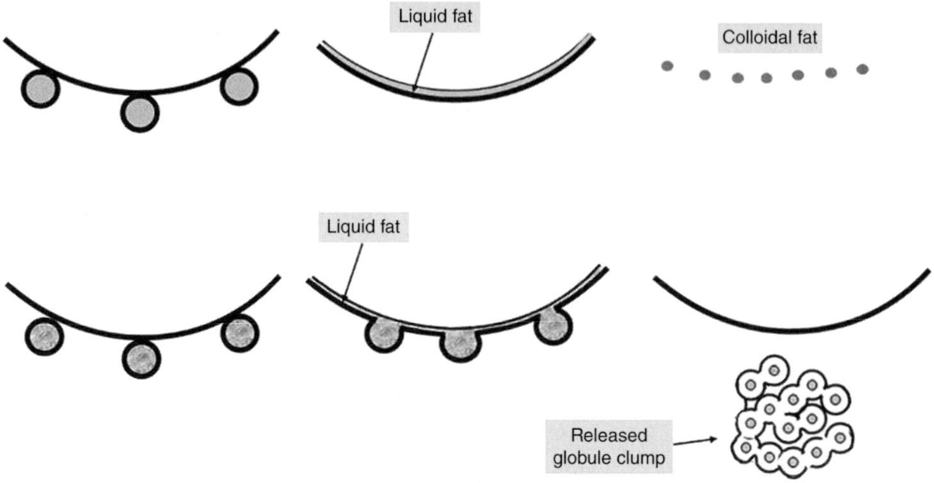

Figure 11.10 Schematic representation of fat globules and air cell during churning.

3) *Fat content.* As examined in Section 11.4.1, the fat content increases the partial coalescence. This could risk an excess of clumping, which reduces foam volume; therefore, the optimal cream concentration for whipping is about 33–35%, while churning can take an advantage by using more concentrated cream, although it is not so stringent.
4) *Liquid/solid fat ratio.* Such a parameter plays a role, the importance of which increases for cream less rich in fat, either a in whipping or churning process. An excessive presence of liquid fat is the cause of "colloidal" fat loosening after the air cell collapses, as depicted in the upper side of Figure 11.10, which is detrimental for successful results in both the technologies. Accordingly, attention must be paid to the differences of the melting point of winter or summer milk fat.
5) *Homogenization.* At the same fat content, homogenized fat globules delay the attainment of foam-stabilizing clumps. Also, due to the smaller dimension, homogenized globules are more stable. In this respect, it could be useful to leave the homogenization clusters.

11.7 CHURNING PROCESS

11.7.1 Type of cream

Cream used to produce butter belong to three typologies:

- Cream from centrifugal separator obtained during skim milk production or milk fat titration;
- Cream from natural creaming of milk for Grana production (only Italy);
- Whey cream obtained from centrifugal recovery of fat from cheese whey.

Table 11.5 Gross composition of different types of milk cream.

%	Natural cream	Cream from centrifuge	Cheese whey cream
Fat	20–24	32–48	44–48
Water	69–73	56–63	48–52
Protein	2.5–2.7	2.7–2.8	0.3–0.5
Lactose	3.5–3.8	3.5–3.8	2.7–3.0
pH/°SH[a]	5–6/6–10		

[a]Titratable acidity in Soxhlet-Henkel degree.

The gross composition of these products is reported in Table 11.5. It is evident the low-fat content of naturally separated cream which require a fat concentration before the churning process.

The lower protein and ash content in cheese whey cream is due to the lack of casein micelles and the associated CaP_i. Also, the higher titratable acidity is not surprising, as during the natural creaming, autochthonous microbiota of raw milk is captured by the rising globule aggregates. This suggest that such a cream is not a 'first quality' raw material for butter production, mainly due to the possible presence of psychrotrophic microbiota (mainly *Pseudomonas* genus) and its spoiling enzymes. Therefore, titratable acidity should not overcome 10 SH/50 ml.

In conclusion, the best quality cream for butter production is that obtained from raw milk by centrifuge (sweet cream).

Therefore, it can be concluded that cream from the milk separator will give the best quality butter, particularly when a biological maturation is provided by selected fermentation cultures.

The CEE n° 2991/94 and 257/97 Regulations (Spreadable fats, butter oil, concentrated butter) define as butter only that obtained from cow milk fat and can be labelled as 'traditional butter' if obtained from a cream content greater than 10%. For Italy only, the Ministry letter, Crc. MIPA n° 3, 1998, allows the use of natural cream for the production of 'traditional butter' when the requisites of CEE n° 2991/94 Regulations are satisfied.

The fat content of traditional butter must be between 82 and 90%, while water and dry matter must be below 16 and 2%, respectively.

Finally, a mention should be made of some kinds of butter that can be used in food products.

- Anhydrous milk fat or butter oil, which contains more than 99.8% fat matter (Reg CEE n° 2571/97) and is obtained from melting of butter, dilution by water, degassing, separating and finally solidified. It can be mainly employed in confectionery (Reg CE 1898/2005).
- Fractionated butter, which is obtained by the fractionated crystallization of milk fat TGs according to their melting point (m.p.). The main fractions are solid butter (8%, 52°Cm.p.), intermediate butter (15%, 38 °C m.p.) (substitute of cocoa butter) and butter oil (77%).

11.7.2 Physical (crystallization) and biological maturation of cream before churning

The physical maturation of cream fat is a particularly important step between cream pasteurization and churning; in fact, its function is to assure a reproducible solid/liquid milk fat ratio in order to standardize the churning process. Also, the appropriate value of such a parameter is necessary to get a coalescence that must be only partial, that is, two or more globules must maintain their shape after the membrane rupture thanks to the solid TGs fraction. On the other hand, an excessive portion of liquid fat causes 'colloidal' fat loss and then less yield in butter.

In Figure 11.11 is an example of thermal cycles of cream before the churning process. After cream pasteurization, the thermal 'memory' effect of fat matter is erased; then a cooling step at 7 °C is performed for 3 h. This allows the partial crystallization (tempering of fat) in controlled and reproducible conditions; such a step could be enough to get the appropriate solid/liquid fat ratio for the cream churning. However, in the profile of the figure, the temperature is raised in order to permit the growth of the starter bacteria for biological maturation: about 16 h of fermentation during which cream will be enriched with flavour compounds that will 'dissolve' in the fat phase.

A general scheme for the conditions useful for both physical and biological cream maturation are indicated in Figure 11.12. In the case of physical ripening alone (upper side of figure), the time and temperature differences between summer and winter cream are mainly due to the different saturation index of the TGs mixture; such a trait is sharply higher in the winter cream, with consequent presence of lower melting point TGs and vice versa for the summer fat.

However, the different conditions also apply to the same seasonal cream as a function of the acidity itself, with particular regard to the churning temperature (Figure 11.13A).

Concerning the fermented cream (lower side of figure), the profile of thermal cycles must be necessarily modified in order to create the proper conditions favorable to the starter cell growth, with the exception of resting time at the churning temperature.

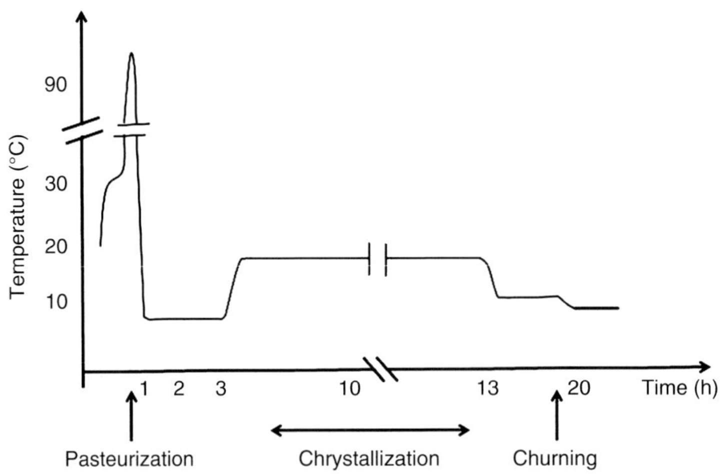

Figure 11.11 Example of thermal cycles to which a cream is subjected before churning.

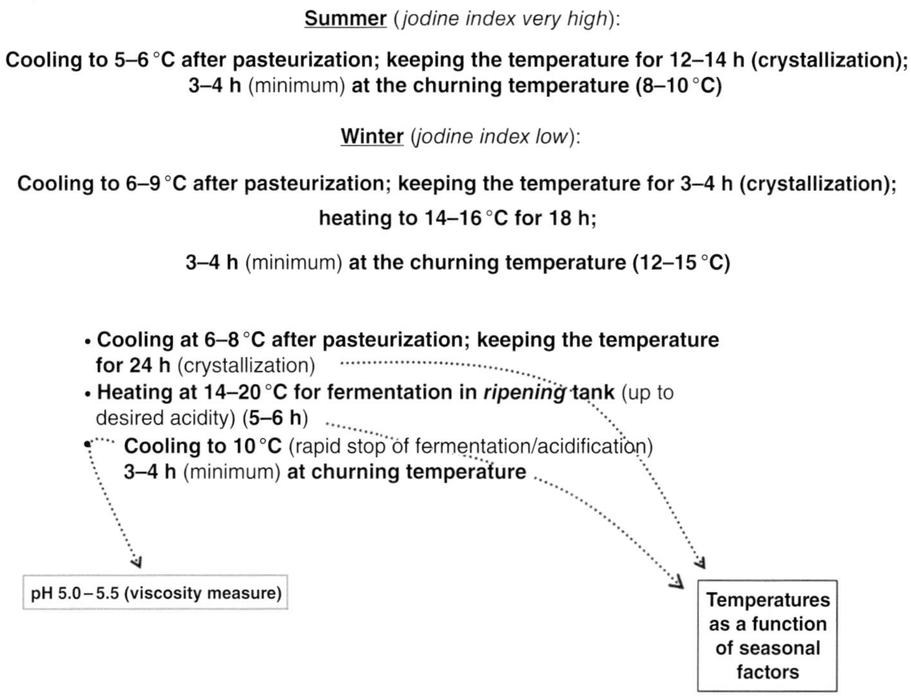

Figure 11.12 General scheme of physical and biological cream maturation before churning.

Therefore, in this case the choice of thermal cycles is a compromise between two necessities. In the biological maturation of cream of paramount importance are both the viscosity measure (indirect evaluation of pH) and the cooling velocity as critical to warrant either the optimal pH conditions for the best churning process or the "freezing" of the flavour profile; regarding this last aspect, it is well known that inhibitory function of residual citrate against the reductase activity converts the flavouring carbonyl compounds as diacetyl and acetoin in the flavourless 2-3, butylene glycol.

The cream fermentation is performed by starter of selected lactic acid bacteria mainly belonging to the following species: *Lactococcus lactis* subsp. *cremoris*, *Lactococcus lactis* subsp. *lactis*, *Lactococcus lactis* subsp. *lactis* biovar *diacetylactis* and *Leuconostoc mesenteroides* subsp. *cremoris*. These last two species constitute the association used worldwide (BD association) due to their excellent flavouring properties. Such an association permits a symbiotic effect for the two species: due to the scarce proteolytic and acidifying activity, *Leuconostoc* growth is supported by *Lactococcus,* whereas *Leuconostoc* is able to produce higher quantities of $CH_3COCOCH_3$, which is the di-ketone substantially contributing to the typical butter flavour. Moreover, *Leuconostoc mesenteroides* subsp. *cremoris*, plays the fundamental function of CH_3CHO oxidation to CH_3CH_2OH; the acetaldehyde is produced by *Lactococcus lactis* subsp. *lactis* biovar *diacetylactis* and stimulates *Leuconostoc* growth. Its accumulation is the main factor responsible for the 'green', 'unripe apple' flavour, whose pungency constitutes a deleterious off-flavour for fermented cream. Finally, both species are able to produce the organoleptic desirable flavour of $CH_3COCOCH_3$ via the co-metabolism of milk citrate

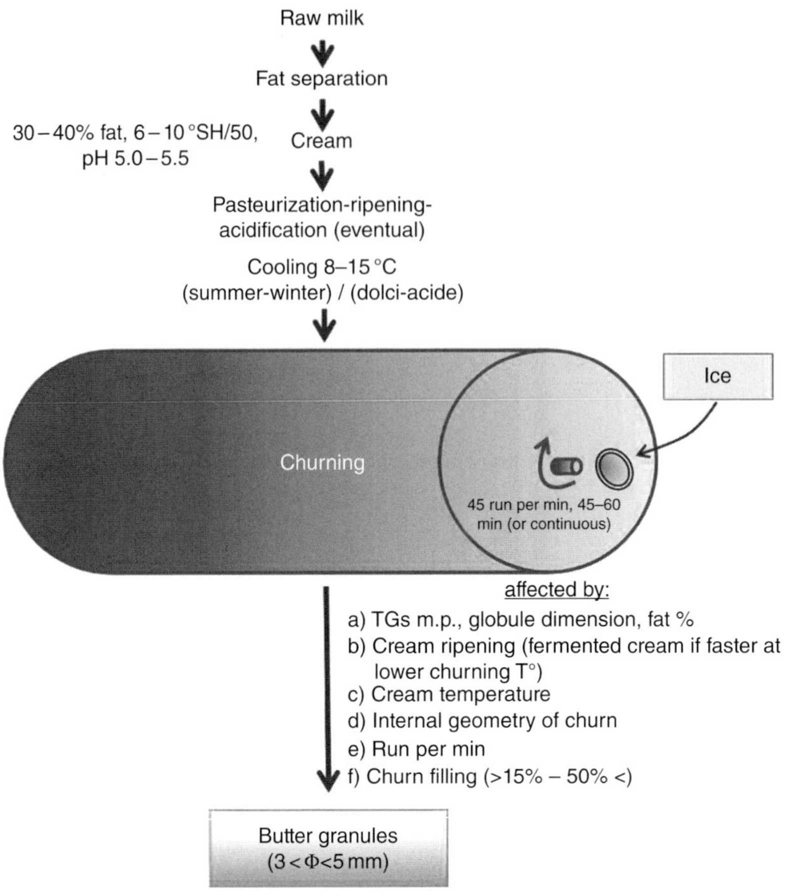

Figure 11.13A Initial steps of cream preparation and treatment for churning process up to the attainment of butter granules.

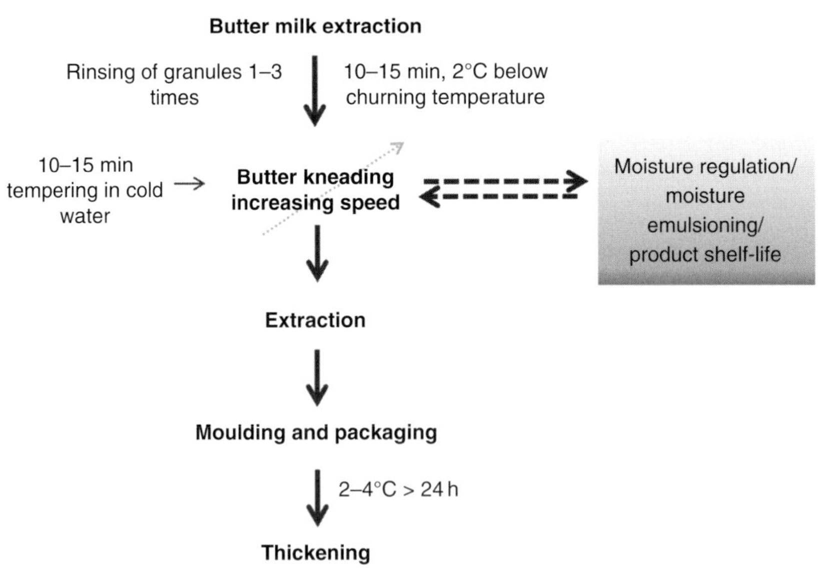

Figure 11.13B Steps following churning.

(at an *optimum* T of 20 °C). Such a production follows two different metabolic pathways: *Lactococcus* converts α-acetolactate produced by the action of α-acetolactate synthase on CH_3CH_2COOH; on the other hand, *Leuconostoc* follows the CH_3CHO-TPP pathway, obtained from $CH_3COCOCH_3$ reduction by CH_3CH_2COOH decarboxylation. Also, both species synthesize $CH_3COCOHCH_3$ via $CH_3COCOCH_3$ reduction conducted by the acetyl-methyl-carbinol dehydrogenase. It must be emphasised that $CH_3COCOHCH_3$ is the second determinant volatile molecule characterized by a typical butter flavour; particularly, the produced aroma is milder then that given by $CH_3COCOCH_3$ and, more importantly, gives the product a more persistent pleasant flavour note.

$CH_3COCOCH_3$ can be reduced to $CH_3COCOHCH_3$ by NAD(P)H, and in turn, $CH_3COCOHCH_3$ can be subjected to a further reduction to the flavourless $CH_3COHCOHCH_3$ by 2-3 butanediol dehydrogenase. The inhibiting activity of citrate explains the need to suddenly stop the fermentation as the lowering of pH is satisfactory for the churning process (5.0–5.5). In such a way, the milk citrate is not completely utilized, thus supplying the desired inhibitory activity; also, when the fermentation is stopped, an agitation of fermented cream can be beneficial in order to maintain an high red-ox potential. In order to obtain a pleasant butter flavour, the appropriate quantities of $CH_3COCOHCH_3$ and $CH_3COCOCH_3$ should be 500–530 and 5–12 ppm, respectively (Lopez et al., 2002; Keogh, 2006; Walstra et al., 1999; Starrenburg and Hughenoltz, 1991).

11.7.3 Churning technology

Any type of cream, before transformation, must be subjected to a pasteurization treatment, following some precautions:

- Reduced physical stress as centrifugal pumps, flow in pipeline, and so forth, in order to avoid as much as possible globule breaking from shear.
- Fat matter normalization to the desired fat percentage in order to make reproducible the partial coalescence during churning; this is particularly critical for naturally separated cream.
- Neutralization by two to three water rinses in order to eventually remove both lactic acid and short-chain FAs derived from fermentation and lipolysis by raw milk microbiota. Such a treatment is necessary when naturally separated cream is used. In fact, the cream acidified at pH <5 delays the churning phase and produces butter affected by off flavours.
- Pasteurization performed in 10–15 s at temperature from 90 to 96 °C up to 105 to 110 °C, according to the fat content.
- Deodorizing pasteurization in two steps (both in hot degasser): the first at 70–75 °C and the second at 90–95 °C. This rare intervention is necessary if cream has been subjected to an undesired spoiling microbial development.
- Decolourization, which is very rare and performed by vac-reactor procedure.

In Figures 11.13A and 11.13B, the various steps of churning are described; after preparation, the cream is introduced in the churn at a temperature of 8–15 °C, according to the season (different m.p. of TG). The rotation speed is usually maintained at 45 rpm for 1 hour.

The cream beating is warranted either by the inner helical rods in the cylinder or the filling factor, which is kept between 15 and 50% of churn volume, thus producing the intermediate foam. Also, during this step, the temperature of cream TG must be maintained constant by the addition of an appropriate quantity of ice in order to avoid the alteration of the initial fraction of solid/liquid fat.

Then, at the foam collapse, butter granules are released, and the process is stopped as their dimension reaches 3–5 mm; at this point, the buttermilk is removed together with the membrane material removed from fat globules. The composition of this by-product is similar to that of milk plasma, with the lipid fraction almost composed of P-lipids of globule membrane.

The mentioned end point is critical as it affects the water content of the final product via the dimension of butter granules.

The granule washing step must assure a decrease of temperature in order to:

- avoid an early granules fusion,
- keep low the level of microbiota (especially for butter from fermented cream) and
- efficient removal of membrane lipids, which can be source of oxidative rancidity.

However, it is advisable to avoid more than three washings in order to prevent the loss of flavour compounds of butter, particularly those obtained from fermented cream.

Finally, the kneading step is important in order to produce a well-distributed emulsion of residual water in fat; the dimension of water droplets must be appropriate to inhibit any bacterial growth.

11.7.4 Continuous churning

As the classic churning is a discontinuous process, continuous systems have been developed in order to increase production speed. Such a process can be used to produce anhydrous butter.

Besides the economic and the process advantages, an appreciable feature of the product is the finer dispersion of water-droplet emulsion ($d<0.3\,\mu m$), thus enhancing the microbiological durability of butter. Furthermore, organoleptic traits such as uniform texture of butter, more accurate control of composition, and water in product and an improved spread ability should be overlooked.

Briefly, the most employed systems belong to three types:

1) The Fritz system, also known as accelerated churning (Westfalia, Oelde, Germany, 1940), modifies the traditional churning by a higher churn rotation speed from 500 up to 3.200 rpm, with a consequent faster formation of butter granules; then the aggregates are worked and kneaded by an advancement device made up of cochlea (Ronholt et al., 2014).
2) The Alfa system, which is designed for continuous churning of heavy cream containing the same fat matter as the final product (82%). At such a concentration, fat globules touch each other, thus allowing a partial coalescence through shear rate

(gentle beating) and then rendering the phase inversion. The process is further ameliorated by a temperature lowering.

3) The Golden Flow system, in which the cream containing 80–90% fat is completely transformed in continuous molten phase.

In general, a continuous churning plant consists of six sections. The first is a churning cylinder slightly inclined towards the inlet of cream so that the product accumulates in such a side; then the cream advancement is obtained by the helicoidal beating devices that produce the foam and subsequent collapse, with butter granule separation. Granules and buttermilk are conveyed in the second section; at first, their separation occurs and buttermilk discharged, while butter grains are transported and washed and worked by a cochlea in the third conical section: the conical shape causes squeezing and drying of butter grains. A second cochlea consists of three sections: the fourth that performs the second step of working (kneading) before the product enters in the fifth section (injection, optional) in which salt (2% max) is added to partially kneaded butter; in the sixth section, butter is kneaded under vacuum by the same cochlea. At the end of this section, the automatic control of moisture in butter occurs.

11.7.5 Moulding and packaging

Whether butter is obtained by traditional or continuous churning, the final steps are moulding and packaging. The packaging material must be resistant to the fat and moisture, opaque to UV, and resistant to flavour substances such as ethers, esters, free FAs, etc. Also, it must be impermeable to water and humidity, otherwise the butter surface becomes dry and yellow.

Usually, butter to be sent to industry for further processing, is packed in 10–20 kg portions or in 200-kg barrels. Butter oil or concentrated butter is transported by cistern.

At the end of packaging, butter is still malleable, thus it must be completely solidified at 2–4 °C for almost 24–48 hours in order to obtain a satisfactory crystallization. Then, butter must be conserved in dry place (RH < 70%) and in the dark to prevent possible oxidative rancidity development. This is a particularly critical aspect when butter is obtained from fermented cream due to the rancidity-promoting pH (Keogh, 2006; Walstra et al., 1999).

11.8 CONCLUSIONS

Dairy products rely on whole milk, skim milk, or partially skimmed milk. Therefore, there are many opportunities for transformation of fat matter into butter, butter cream and fermented sour cream products. The technologies at the basis of the industrial transformation of these products have been summarised and will bring the attention of consumers and technologists for further advancements.

References

Alais C. 1984. *Sciences du lait. Principes des techniques laitieres*. 4th ed. Paris: Sepaic.

Boode K, Walstra P. 1993. Partial coalescence in oil-in-water emulsions. 1. Nature of the aggregation. *Colloids Surf A: Physicochem Eng Aspects* 3:81–121.

CEE 257/1997 Regulation. http://eur-lex.europa.eu/legal-content/EN/TXT/?uri=celex%3A31997R0577

CEE 2991/1994 Regulation. europa.eu/legal-content/IT/TXT/?uri=CELEX%3A31994R2991

Chen J, Dickinson E, Iveson G. 1999. Interfacial interactions, competitive adsorption and emulsion stability. *Food Struct* 3:12–135.

Commission Regulation (EC). 2005. Official Journal of the European Communities. No 1898/2005 of 9 November 2005 laying down detailed rules for implementing Council Regulation (EC) No 1255/1999 as regards measures for the disposal of cream, butter and concentrated butter on the Community market.

Fang Y, Dalgleish DG. 1993. Dimensions of the adsorbed layers in oil-in-water emulsions stabilized by caseins. *J Colloid Interface Sci* 3:329–334.

Fox PF. 2002. Fat globules in milk. In: Roginski H, Fuquay J, Fox PF. editors. *Encyclopaedia of dairy sciences*. London: Academic Press.

Goff HD, Vesperej E, Smith AK. 1999. A study of fat and air structures in ice cream. *Int Dairy J* 9:817–29.

Keogh MK. 2006. Chemistry and technology of butter and milk fat spreads. In: Fox PF, McSweeney PLH. editors. *Advanced dairy chemistry*, vol 2. *Lipids*. 3rd ed. New York: Springer.

Lopez C, Burgaux C, Bernadon S, Keller G, Ollivan M. 2002. Thermal and structural behavior of milk fat. 3. Influence of cooling rate and droplet size on cream crystallization. *J Colloid Interface Sci* 240: 150–61.

McClements DJ. 2005. *Food emulsions: principles. practices and techniques*. 2nd ed. London: CRC Press.

MIPA n°3, 1998. Ministry letter. http://www.gazzettaufficiale.it/atto/serie_generale/caricaDettaglioAtto/originario?atto.dataPubblicazioneGazzetta=1998-05-27&atto.codiceRedazionale=098A4403&elenco30giorni=false

Mulder H, Walstra P. 1974. *The milk fat globule*. Wagheninghen: PUDOC.

Ranjith HPM, Rajah KK. 2001. High fat content dairy products. In: *Mechanization and automation in dairy technology*. Sheffield: Academic Press.

Ronholt S, Madsen AS, Kirkensgaard JJK, Mortensen K, Knudsen JC. 2014. Effect of churning temperature on water content, rheology, microstructure and stability of butter during four weeks of storage. *Food Structure* 16:14–26.

Starrenburg MJC, Hughenoltz J. 1991. Citrate fermentation by *Lactococcus* and *Leuconostoc* spp. *Appl Environ Microbiol.* 57:3535–40.

TetraPack. 2009. Centrifugal separation and milk fat standardisation. In: *Dairy processing handbook*. Lund: TetraPack Processing Systems, AB.

Walstra P, Geurts TJ, Noomen A, Jellema A, van Boekel MAJS. 1999. Cream products. In: *Dairy technology*. New York: Marcel Dekker, Inc.

Walstra P, Jenness R, Badings HT. 1984. *Dairy chemistry and physics*, vol 5. *Lipids*. New York: John Wiley & Sons, pp 58–97.

Walstra P. 2003. *Physical chemistry of foods*. New York: Marcel Dekker, Inc.

12 Biological traits of lactic acid bacteria: industrial relevance and new perspectives in dairy applications

Diego Mora[1], Fabio Dal Bello[2] and Stefania Arioli[1]

[1]Department of Food Environmental and Nutritional Sciences (DeFENS), University of Milan, Milano, Italy
[2]Sacco System, Sacco srl, Cadonago (CO), Italy

12.1 INTRODUCTION

The market demand for dairy products with a high level of standardized quality in terms of nutritional values, sensory profiles, and safety, pushed the companies toward a continuous increase in applied research for the optimization of the technological processes and the selection of the microbial strains based on their taxonomy, safety aspects, genomics, and metabolic activities. The taxonomy issues related to the correct identification of microbial species to be introduced in the food chain have been rapidly solved owing to the development of molecular tools, mostly PCR-based and *16S rRNA* gene targeted, that were easily transferred from the academic research to the R&D laboratories of companies devoted to the production of starter cultures. The identification of strains at species level allowed their correct utilization based on the information available in the literature associated to their primary metabolism. All the other metabolic traits of dairy relevance can be classified as species-related or strain-dependent. The species-related characteristics are well known for most bacteria, yeast, and molds used so far in dairy, whereas the strain-dependent character can be investigated only by developing consistent screening protocols. As an example, lactose/galactose metabolism are species-related metabolic traits in the species *Streptococcus thermophilus* and *Lactobacillus delbrueckii* because all the strains that belong to these species use only the glucose moiety of lactose, whereas galactose is secreted in the medium in amounts equimolar to the lactose consumed. Another example of a species-specific metabolic trait is represented by urease activity, which is species-specific for *S. thermophilus* and *Lactobacillus fermentum* among dairy lactic acid bacteria. On the other hand, milk acidification rate, carbohydrate utilization pattern, types of endopeptidases, the presence of cell-envelope proteinases, the production of exopolysaccharides, the production of bacteriocins, the autolytic phenotype, the production of aroma compounds, the sensitivity to antibiotics, and the sensitivity to bacteriophages are strain-dependent metabolic traits

Microbiology in Dairy Processing: Challenges and Opportunities, First Edition.
Edited by Palmiro Poltronieri.

Published 2018 by John Wiley & Sons Ltd and the Institute of Food Technologists.

that very often represent the bottleneck for the selection of strain for dairy applications. In addition to the metabolic traits that are relevant for the obtainment of the desired dairy product, other characteristics are extremely important for the industrial production of microbial strains that are intended to be part of starter cultures. In this context, the biomass production yield and the retained metabolic activity after a freeze-drying process represents another set of characteristics that are used to select microbial strains for dairy applications. It follows that microbial strains, selected for one or some of the strain-dependent metabolic traits listed above, are then discarded because they do not reach sufficient production yield or they do not maintain enough metabolic activity after a freeze-drying process. Therefore, efforts should be undertaken both in the development of efficient screening protocols and to optimize in a strain-dependent way the biomass production processes. Some examples of technological traits to be considered for strain selection for dairy application will be discussed in the next paragraphs.

12.2 SELECTING FERMENTING BACTERIA FOR THEIR ABILITY TO HAVE A RESPIRATORY METABOLISM

Since the first observation of the respiratory metabolism induced by heme in several lactic acid bacteria species, several authors described this property (Bryan-Jones and Whittenbury, 1969; Sijpesteijn, 1970; Ritchey and Seeley, 1976). Historically known to have a strict fermentative metabolism, characterized by a high rate of sugar consumption but low yield of ATP production, some lactic acid bacteria species revealed the possibility to shift toward a respiratory metabolism characterized by a higher yield of ATP production associated to a high biomass production and robustness, a metabolic opportunity not dissimilar to that showed by the eukaryote *Saccharomyces cerevisiae* (Mora et al. 2013). Most of the studies related to the respiratory metabolism of lactic acid bacteria have been carried out on one of the most relevant dairy species, *Lactococcus lactis* (Duwat et al., 2001; Gaudu et al., 2002; Gaudu et al., 2003; Vido et al., 2004; Pedersen et al., 2008). In the presence of heme and oxygen, initial lactococcal growth occurs via fermentation, and when the external pH reduces to approximately 5.3, the transition to respiration occurs (Duwat et al., 2001). It was hypothesized that the physiological reprogramming that occurs during respiration might be a consequence of the shift in the NAD^+/NADH ratio that allosterically redirects metabolism in favor of enzymes that use either NAD or pyruvate as a substrate (Lechardeur et al., 2011). The industrial relevance of this metabolic shift is due to the increase in biomass production, growth efficiency, and an extraordinary increase in long-term survival when lactococcal cells were pushed toward a respiratory metabolism by the exogenous addition of a source of heme in the presence of oxygen (Gaudu et al., 2002). These remarkable phenotypes became of industrial relevance after the publication of a patent that specifically focused on improving the production of starter cultures by forcing the respiratory metabolism (Pedersen et al., 2005). In this context, it should be underlined that only *L. lactis* and *Lactobacillus plantarum* can improve, at a level relevant for industrial applications, their growth yield by shifting from the fermentative to the heme-dependent respiratory metabolism. In addition, the ability to shift toward a respiratory metabolism is a subspecies and or/strain dependent in both *L. lactis* and *L. plantarum*, and it is therefore a screening procedure is requested in order to select the best strain. After strain selection, the scale-up from laboratory to the production scale is necessary, in addition to the choice of a proper hemin food-grade source, considering that pure hemin cannot be economically sustainable at an industrial level. The respiratory metabolism of lactic acid bacteria is an interesting example of how the basic research, and not the applied research, has led to one of the most relevant revolutions in starter production at industrial level.

12.3 SELECTING GALACTOSE-POSITIVE YOGURT CULTURES: WORKING "AGAINST THE NATURAL EVOLUTION OF THE SPECIES"

Unfermented galactose in dairy products could represent a technological and health problem. Accumulation of galactose in dairy products may determine undesirable effects such as: i) browning of mozzarella cheese, ii) development of nonstarter lactic acid bacteria with undesirable production of CO_2, and iii) growth of spoilage and/or pathogenic microorganisms. In addition, excess galactose in dairy products may also adversely affect human health in individuals with galactosemia (Thomas et al. 1980; Tinson et al. 1982; Kindstedt and Fox 1993).

Looking at the scientific literature, the attempts to select or to genetically modify the lactic acid bacteria species involved in the yogurt consortium, are several, and all focused on *S. thermophilus* (Terence and Vaughan, 1984; Vaughan et al., 2001; Vaillancourt et al., 2004; Vaillancourt et al., 2008; Sorensen et al., 2016), but they all failed to reach the goal. The reason for the failed attempts to select or to construct a galactose *S. thermophilus* should be analyzed by taking into account the selective forces that have driven its speciation and adaptation to milk. The abundance of lactose in milk has probably driven the selection of highly adapted species able to ferment only the glucose moiety of the milk disaccharide using the remaining galactose, accumulating into the cell, as a molecule driving their highly specialized lactose/galactose antiporter transport system. In *S. thermophilus*, the LacS permease (the lactose/galactose antiporter system) was characterized in detail (Foucaud and Poolman, 1992; Poolman et al., 1992; Poolman et al., 1995; Gunnewijk and Poolman, 2000a, 2000b; Cochu et al., 2005), revealing a high affinity of this protein for the intracellular galactose coming from the activity of ß-galactosidase. Therefore, every attempt to force the galactose catabolism in *S. thermophilus* through the overexpression of enzymes of the Leloir pathway worked perfectly when *S. thermophilus* mutants were grown on galactose as a unique carbon source, but they dramatically failed when the same mutants grew in the presence of lactose, the favored carbon source that most probably has driven the evolution of this species. In absence of lactose, galactose can be catabolized by *S. thermophilus* through the Leloir pathway, whereas in presence of the milk disaccharide, galactose is probably disputed between the membrane permease LacS and the galactose-kinase (the first enzyme of the Leloir pathway), being favorite LacS. More recently, a new strategy for the selection of *S. thermophilus* and *L. delbrueckii* galactose positive have been developed by a company in the field (Sorensen et al, 2016). The authors developed a multiple serial selection in order to force cell metabolism to catabolize galactose by inhibiting the glucose catabolism. Using traditional bacterial genetic techniques, the authors redirected the metabolic pathways in *S. thermophilus* and *L. delbrueckii* subsp. *bulgaricus* so that the bacteria consume the nonsweet sugars lactose and galactose and secrete the much sweeter glucose. These goals were reached firstly by selecting mutants able to grow on galactose as a sole carbon source and secondly by using the galactose-positive mutants to select mutants resistant to glycolytic inhibitor 2-deoxyglucose. Final selected mutants were unable to grow on glucose as a carbon source, but they showed normal growth on lactose. Interestingly, whole genome sequencing revealed that *S. thermophilus* mutants were characterized by mutation in *gal* promoter region and missense mutation in the glucokinase gene (*glcK*) according to their ability to grow on galactose and the impaired growth on glucose, respectively. Some selected mutants showed an additional nonsense mutation in the *manM* gene, encoding the IIC^{man} protein of the glucose/mannose phosphotransferase system (PTS). All selected mutants showed a reduced acidification rate in milk compared to the wild type but a significant release of glucose in the medium between 0.8 and 22.9 mg/ml, depending on the mutant or the combination of mutants tested. The final result is an increased sweetness of yogurt, but the residual galactose content in yogurt was not significantly reduced, but rather quite often increased if compared to those measured in yogurt obtained by wild-type strains. In conclusion, an effective galactose consumption was not reached even using a multiple-selection strategy, thus confirming the difficulty in re-designing the evolution of energetic pathways. Interestingly, one combination of mutant strains left in the medium a lactose content below the detection limit, which was an extremely positive, unexpected side effect of the multiple serial selection strategy developed. All these metabolic effects have been observed prolonging the fermentation process for 30 h, which is a long incubation time compared to the 4–5 h requested for a standard yogurt fermentation. Therefore, the economic sustainability of the process that involve these kinds of mutants should be carefully evaluated.

12.4 ACCELERATING THE MILK ACIDIFICATION PROCESS BY SELECTING PROTEINASE-POSITIVE STRAINS

Besides lactose/galactose metabolism, other factors can affect the milk acidification rate, such as the efficacy of the proteolytic system and the urease activity. Most of the relevant lactic acid bacterial species used in dairy harbor a complex proteolytic system that is constituted of three main players: i) cell-envelope proteinases (CEP), located outside the cell and anchored to the cell wall, ii) oligopeptide transport systems across cell membrane, and iii) intracellular peptidases. The presence of the three players is a strain-dependent characteristic; therefore suitable screening protocols are necessary to select strains to be used in starter culture formulation. Several studies have been performed to characterize the proteolytic system in *Lactococcus lactis* (Tynkkynen et al. 1993; Hagting et al. 1994; Foucaud et al. 1995; Detmers et al. 1998; Guédon et al. 2001a, 2001b; den Hengst et al., 2005a, 2005b), *S. thermophilus* (Chapot-Chartier et al., 1994; Chavagnat et al., 1999; Chavagnat et al., 2000; Fernandez-Espla et al., 2000; Anastasiou et al., 2002; Garault et al., 2002) and *Lactobacillus delbrueckii* subsp. *bulgaricus* (Gilbert et al., 1994; Stefanitsi et al., 1995; Gilbert et al., 1996; Morel et al.,

1999; Morel et al., 2001; Germond et al., 2003). Other studies have focused attention between the presence of CEP and the rate of milk acidification. Moreover, other authors investigated the cooperative role of CEP in complex communities, such as the yogurt consortium (Zourari et al., 1992; Sasaki et al., 1995; Rul and Monnet, 1997; Courtin et al., 2002; Peltoniemi et al., 2002; Juilleet al., 2005). In all lactic acid bacteria species investigated, CEP plays a key role in milk acidification rate, ensuring the supply of amino acids during the exponential growth when nitrogen needs are more consistent due to the increased number of cells. In *S. thermophilus*, the induction of the CEP PrtS was related to the diauxic growth of this microorganism in milk (Letort et al., 2002). Using a recombinant *S. thermophilus* carrying a luminescent reporter system under the control of the *prtS* promoter, the authors elegantly showed that the second exponential growth phase of the diauxic growth was triggered by the induction of *prtS* as a consequence of the release of peptides and amino acids from caseins, which were able to sustain the growth of *S. thermophilus* until reaching the final stationary phase. Because the growth of *S. thermophilus* is associated with the production of lactic acid, the growth stimulation induced by the CEP PrtS resulted in a fast milk acidification. Considering that CEPs act by hydrolyzing the casein fractions in the medium, in a complex microbial community, not only the strains that harbor the CEP but also the strains that are free of such enzymes benefit from the availability of the generated casein-peptides. In this context, we should be advised that improvement of milk acidification by a starter culture could be achieved by introducing only some CEP-positive strains in the starter culture blend composition. In an interesting study, Delorme et al. (2010) focused on the evolution of *S. thermophilus* species; the authors hypothesized that the acquisition of *prtS* confers a fitness advantage for the milk adaptation of the species and realized that *prtS* gene is infrequent in historical microbiological collections, whereas it is quite common in recent industrial collections of dairy strains. These data clearly underlined that among the selection criteria used by companies, the ability to "grow well in milk" has favored the selection of *prtS*-positive strains. Nevertheless, an improvement could be requested for an already technological relevant strain that is *prtS*-negative and shows a low acidification in milk but displays industrially relevant properties such as the exopolysaccharide production or a high level of biomass yield and activity. In this case, the fast milk-acidifying phenotype of *S. thermophilus* can be elegantly acquired by natural transformation (*comS*-dependent) (Fontaine et al., 2010) of the genomic island encoding the cell-envelope proteinase PrtS, as described by Dandoy et al. (2011). For the *prtS* strainimprovement protocol developed, the insertion site ISS*th1* or IS*1167* element is mandatory in the genome of the receiving strains, but these IS elements are widespread among *S. thermophilus* genomes (Dandoy et al., 2011). Despite the industrial relevance of the methodological approach proposed, based on the *comS*-dependent induction of the natural transformation of *S. thermophilus,* it is still not clear if the derivative PrtS-positive mutants could be considered non-Genetically Modified Microorganisms according to the European Legislation (2001/18/EC; Fontaine et al., 2010). Despite all these considerations, it was recently observed that the acquisition of *prtS* is not the only requirement to be a fast milk-acidifying *S. thermophilus* strain (Galia et al, 2016). It was recently observed that *dtpT, amiF, ilvC, ilvB, bcaT, livJ, ackA, codY,* together with *prtS,* showed a high transcriptional level in fast-acidifying strains, thereby underlining the relevance of other metabolic pathways on the rate of milk acidification.

12.5 ACCELERATING THE MILK ACIDIFICATION PROCESS BY SELECTING UREASE-NEGATIVE *S. thermophilus* STRAINS

Among lactic acid bacteria, *S. thermophilus* shows some characteristics that are peculiar and rarely shared with other species used in dairy applications. *S. thermophilus*, is the only species of the genus *Streptococcus* to be generally recognized as safe (GRAS) and to receive the QPS status from the US FDA and from the European Food Safety Agency, respectively (Iyer et al., 2010). Genome analysis and comparison revealed the *S. thermophilus* emerged from closely related pathogens through a relevant genomic regression, together with the acquisition of relevant features that allowed the adaptation of the species to a defined and unique environmental niche, milk (Bolotin et al, 2004; Mora et al., 2005; Mora et al., 2013). Unlike *Lactococcus* and *Lactobacillus* species, *S. thermophilus* is auxotrophic toward few amino acids, essentially Glu, Cys, His, and Met, showing a relatively high conservation of functional amino acid biosynthetic genes (Bolotin et al, 2004; Hols et al., 2005; Arioli et al., 2007, 2008). More interestingly, *S. thermophilus* amino acid biosynthesis was shown to be modulated by the activity of urease, an enzyme peculiar to this species among all the other dairy species characterized by an homofermentative metabolism (Monnet et al., 2004; Mora et al., 2004; Mora et al., 2005; Pernoud et al., 2004; Monnet et al., 2005; Arioli et al., 2007). Urease is a urea amidohydrolase (EC 3.5.1.5) that catalyzes the hydrolysis of urea to yield ammonia and carbamate, which spontaneously decomposes to yield a second molecule of ammonia and carbonic acid. The released carbonic acid and the two molecules of ammonia are in equilibrium with their deprotonated and protonated forms, respectively, and the net effect of these reactions is an increase in the pH of the environment that surrounds the urease-positive microorganisms. Urea is present in milk in concentrations ranging from 0.1 to 0.4 g/l depending on the healthy status of animals, on diet, and on the season. Urease activity is considered a stress response that was developed by several bacteria to counteract a low environmental pH (Cotter and Hill, 2003). In *S. thermophilus*, urease is metabolically related to the biosynthetic pathways involved in aspartate, glutamine, arginine, and carbon dioxide metabolism (Monnet et al., 2005; Arioli et al., 2007; Arioli et al., 2009). Notably, urea hydrolysis increases the catabolic efficiency of *S. thermophilus* by modulating the intracellular pH and increasing the activity of ß-galactosidase, glycolytic enzymes, and lactate dehydrogenase (Arioli et al., 2010). Urea hydrolysis results in increases in both the pH_{in} and the pH_{out} due to the rapid diffusion of ammonia outside of the cell. Consequently, in the presence of urea and a urease-positive microorganism, urease-negative microorganisms share the environmental benefit derived from the transient local pH increase as recently observed for the yogurt consortium (Arioli et al., 2010). Despite the roles of urease in *S. thermophilus* physiology, roles that justify the presence of a huge urease operon (11 genes) in a regressed genomes, urease represents a negative technological trait in a dairy context. During the growth of *S. thermophilus*, urease activity and the consequent release of ammonia started after few hours when milk pH reaches values below 6. Ammonia generated by urea hydrolysis buffers lactic acid produced by the homolactic fermentation, thereby negatively affecting the rate of acidification (Monnet et al., 2004; Mora et al., 2004;

Monnet et al., 2005). Because *S. thermophilus* is irreplaceable by the other urease-negative lactic acid bacteria in several fermentation processes, a solution to the problem was found in the selection of natural urease-negative mutants or inducing mutants by chemical mutagenesis (Monnet et al., 2004). Unfortunately, urease-negative mutants are rare in nature, and mutants obtained by chemical mutagenesis could harbor other unpredictable mutations that can potentially affect their fitness in dairy fermentations. Moreover, due to the roles of urease activity in the physiology of the microorganism, quite often urease-negative mutants do not perform in milk as well as the wild type does. Therefore, other strategies are needed for the selection of strains that show a reduced, and not a complete absence of, urease activity.

12.6 PROTECTIVE CULTURES FOR DAIRY APPLICATIONS: "WORK BUT PLEASE DO NOT GROW AND DO NOT MODIFY THE SENSORY PROFILE OF THE PRODUCT"

Besides the criteria of strain selection focused on the increase of biomass yield and robustness and the improvement of the fermentation performance in milk, microbial cultures could be also involved as protective biological agents against spoilage and/or the development of pathogenic bacteria. According to the opinion of the Senate Commission on Food Safety (SKLM) of the German Research Foundation (DFG), "Protective cultures are preparations consisting of live microorganisms (pure cultures or culture concentrates) that are added to foods with the aim of reducing risks by pathogenic or toxigenic microorganisms" (Vogel et al., 2011). The protective cultures develop their biopreservation through their metabolic pathways, although they are not necessarily the same strains selected for the production of the dairy product. If protective cultures do not coincide with starter cultures, they should be metabolically active in the food matrix in order to promote their protection without altering the sensory profile of the product and, in some cases, without increasing their biomass. A protective culture may develop a protection through a competitive exclusion and/or by the production of antagonistically active substances. The competitive exclusion can be developed by competing for nutrients and/or through better adaptation to the micro-environmental condition that characterized the product. Antagonistic substances are typically organic acids (e.g. lactic, acetic, propionic, formic and benzoic acid), ethanol, H_2O_2, bacteriocins, as well as other antagonistic compounds active against bacteria or yeast and molds. While there is a lot of literature concerning the selection and characterization of lactic acid bacteria having activity against food pathogen bacteria, few data are available concerning the selection of protective strains showing antimycotic activity.

Despite fermented milks and yogurts not being subjected to microbiological contamination, acid-tolerant fungi can contaminate the product at all stages of food processing from raw materials to the finished products and can develop on the surface where oxygen is available even if at low concentration. *Candida parapsilosis*, *Candida diffluens*, *Debaryomyces hansenii*, *Kluyveromyces marxianus*, *Rhodotorula mucilaginosa*, *Yarrowia lipolytica*, *Zygosaccharomyces bailii* or *Penicillium brevicompactum* are among the most frequent fungal contaminants in yogurts and fermented milks, and their frequency increases when fruits, cereals or sugar are supplemented to the products (Mayoral et al., 2005). In addition, fungal development in yogurt or fermented milk can increase also when the cold chain is not correctly maintained, as could happen in country characterized by a warm climate. Fungal spoilage represents the main problem of conservation of yogurt and fermented milk, and it causes significant economic losses worldwide.

Due to their GRAS or QPS status, lactic acid bacteria are considered good candidates for the selection of protective cultures for dairy applications. For an exhaustive review on the literature on the antifungal lactic acid bacteria, the reader should refer to Crowley et al. (2013a). Lactic acid bacteria can produce several classes of molecules showing antifungal activity, among them the most characterized are several organic acids, phenyllactic acid, reuterin, some cyclic dipeptides and fatty acids. All these compounds have been isolated and described as the causative agent of fungal inhibition in a number of studies over the last decade and often play their role in a synergistic way. While some of the characterized antifungal compounds, such as phenyllactic acid and reuterin, do not interfere with the sensorial properties of the products, some organic acids or a mixture of them could affect considerably the quality of the product in a concentration-dependent manner. Acetic, formic, caproic, propionic, butyric, and n-valeric acids are active against several fungi and yeasts, but they also interfere with the sensorial characteristics of the product at low concentrations, whereas phenyllactic acid, reuterin and antimycotic peptides show absence of an apparent odour and are therefore considered the most promising candidates for the control of food spoilage (Crowley et al., 2013a).

The mechanism of action of the antimycotic molecules is in most cases not known or not completely understood. While it is known that organic acids act on bacteria by reducing the extracellular and intracellular pH to values that are incompatible with the growth and with the main metabolic functions as a consequence of their high level of dissociation inside the cell (Cotter and Hill, 2003), their mechanism of action on fungi remains incompletely characterized. Likewise, for fatty acids the mechanism of action is not known, but it is supposed that they can interfere with the composition and functionality of fungal cell membrane, increasing its fluidity and thereby determining an uncontrolled release of intracellular electrolytes (Avis and Belanger, 2001). Despite phenyllactic acid being one of the most characterized antimycotic molecules, its mechanism of action on fungal cells is far from being elucidated. Phenyllactic acid is an organic acid, and according to the weak acid theory, at low pH the acid is uncharged and can thus cross the phospholipid bilayer. Once inside the cell the organic acid dissociation leads to a decrease of intracellular pH, which causes metabolic and growth inhibition described previously. Moreover, D-phenyllactic acid is reported to display antibacterial activity (Dieuleveux 1998), and L-phenyllactic acid was demonstrated to be a competitive inhibitor of phenylalanine dehydrogenase in *Rhodococcus* sp. (Brunhuber et al., 2000). More recently Svanström et al. (2013) reported that the L isoform of phenyllactic acid is a more potent inhibitor of *Penicillium roqueforti* than the D form, even if the activity was as strongly modulated by the medium pH as for all other weak organic acids. Moreover, the addition of phenyllactic acid at sub-inhibitory concentrations determined a reduction of the fungal colonies' radial growth together with the inhibition of sporulation. The last effect of the phenyllactic acid should be considered beneficial as many foodborne fungi are spread by airborne conidia.

Concerning reuterin, in *Escherichia coli* this molecule is known to interact with its aldehyde group with thiol groups of small molecules and protein determining oxidative stress to the cell, a similar mechanism is hypothesized for its fungal growth inhibition activity. For antimycotic peptides no mechanism of action are described or hypothesized.

Most of the strains of lactic acid bacteria, characterized up to now for their ability to counteract fungal growth, belong to species isolated from vegetable materials rather than to those species which highly adapted to the milk environment. Antifungal compounds have been described and identified from *Lactobacillus sanfrancisciensis*, *Lactobacillus pentosus*, *Lactobacillus plantarum*, *Lactobacillus coryniformis*, *Lactobacillus rossiae*, *Lactobacillus amylovorus*, *Lactobacillus reuteri*, *Lactobacillus hammesii*, *Lactobacillus paracasei*, *Lactobacillus arizonensis*, *Propionibacterium jensenii*, *Weissella paramesenteroides,* and *Weissella cibaria* (Crowley et al., 2013a).

Among these lactic acid bacteria, only the species *Lactobacillus acidophilus*, *Lactobacillus harbinensis*, *L. paracasei* subsp. *paracasei*, *L. plantarum*, and *Lactobacillus rhamnosus* are currently selected for dairy applications both in yogurt and cheese and show activity against several yeasts including *Candida* species, *Debariomyces hansenii*, *Kluiveromyces marxianus,* and fungi such as *P. roqueforti* and *P. brevicompactum* (Schwenninger and Meile, 2004; Garcha and Natt, 2011; Muhialdin et al., 2011; Zhao, 2011; Delavenne et al., 2012; Crowley et al., 2013b).

In conclusion, despite several publications on antifungal activities of lactic acid bacteria being present in the literature, only few commercial protective cultures are available on the market, possibly due to the fact that the antimycotic activity of selected strains is dependent on the physico-chemical parameters of the dairy products and on the ability of such strains to produce the compounds *in situ* in the dairy matrix. In this context, screening tests in dairy matrix accompanied by challenge tests in a small-scale dairy production and followed with a sensory evaluation of the products is desirable.

12.7 SELECTION OF STARTER CULTURE FREE OF TRANSFERABLE ANTIBIOTIC-RESISTANCE MECHANISMS

Despite the absence of a national legislation, most of the companies involved in the production of starter culture have accepted the recommendation of transnational authorities. In Europe, the European Food Safety Authority has compiled a document in which the following recommendation is reported: "In principle, the selection of micro-organisms for use as feed additives should be oriented toward the least resistant organism whenever possible" (EFSA Journal, 2012). Despite this guidance document is intended to provide a method to identify resistance to antimicrobials of human and veterinary importance in bacterial strains intended for use as feed additives, it has been rapidly accepted and applied for bacterial strains intended for the preparation of starter cultures, and probiotics too. It follows that strains selected for dairy technological properties should undergo a further selection in order to limit the introduction to the food chain transmissible antibiotic-resistance mechanisms. The EFSA document proposes a detailed scheme for the antimicrobial resistance assessment of a bacterial strain used as a feed additive. The operative scheme started from the taxonomic identification of the strains, followed by the evaluation of the minimal inhibitory concentration (MIC) of a list of nine antibiotics. The MIC values obtained for the analyzed strain are then compared to the cutoff values identified by the FEEDAP Panel for one or more antimicrobials. When the MIC for an antimicrobial is above the cutoff values, further investigation is required to address the nature and the transferability of the resistance. In the absence of information on the genetic nature of a demonstrated resistance, the EFSA document recommends that the strain not be used as a feed additive. Moreover, any bacterial strain carrying an acquired resistance to antimicrobial(s) that is shown to be due to the acquisition of genetic determinant(s) presents the greatest potential for horizontal spread and should not be used as a feed additive. Considering that the development of resistance among bacteria to antimicrobials remains a serious concern in all countries, the aforementioned recommendation represents a good example for control of the spread of antibiotic-resistance genes in the food chain and should be implemented and applied by all starter producer companies worldwide.

12.8 CONCLUSIONS

Selection of starter cultures is still a difficult and complex task that needs to be considered using knowledge of microbial genetics, physiology, genomics, and microbial ecology, together with a deep knowledge of the physico-chemical parameter of the food product. While microbial genomic data are today available with a relatively low economic cost, microbial genetics and cell physiology still require substantial effort in terms of time and set up of analytical tools that need to be adapted from synthetic media to the more complex milk-based matrix.

Despite the recent advances in science technology, strain selection is often based on screening protocols that are still dependent on the measurement of simple physico-chemical parameters such as the pH, but a high throughput technology able to monitor the milk acidification process in 96- or 384-well plates is still not available. Besides the difficulties related to the screening tests, due to the lack of basic technology and automation, a further drawback is related to the complexity of several production processes, which are difficult to be carried out at small laboratory scale processes necessary to test new strains and/or new strain combination. Starter culture production still represents a uniqueness of the microbial-based market because microbes are selected and prepared in blend for the unique aim of activating their metabolism and the selected metabolic functions once inoculated in milk: in absence of microbiological activity, no product can be obtained. This statement underlies that the product "starter culture" is always under the control evaluation of the final user. This is a basic concept that can be understood if we compare the fate of starter cultures, commercialized for their activity, with that of probiotics, which are commercialized as "number of live cells" and not for the "probiotic activity" they exert once ingested and that nobody, as final user, can measure.

A further difficulty in the selection procedures is represented by the lack of biodiversity among strains of lactic acid bacteria intended for dairy applications. The worldwide use of starter cultures in dairy productions also determined the worldwide spreading of strains selected and characterized by the R&D section of the starter-producing company.

In this scenario, the higher the adaptation of lactic acid bacteria species to the milk environment, the lower the possibility to select new strains from milk environment.

This statement is particularly true for *L. delbrueckii* and *S. thermophilus,* both of which are described as highly adapted to the milk environment (Bolotin et al., 2004; Van de Guchte et al., 2006; Mora et al., 2013). Specifically, the selection of the best milk-performing strains to be a constituent of a starter culture has a risk: once inoculated in raw milk in high concentration, those selected strains are able to outcompete the autochthonous communities. The perpetuation of this practice inevitably could lead to the reduction of biodiversity in that environment. There is therefore a strong need for new strains isolated from natural environments in order to increase the genetic diversity and robustness of lactic acid bacteria collections intended for traditional and new dairy applications.

References

Anastasiou R, Papadelli M, Georgalaki MD, Kalantzopoulos G, Tsakalidou E. 2002. Cloning and sequencing of the gene encoding X-prolyl-dipeptidyl aminopeptidase (PepX) from *Streptococcus thermophilus* strain ACA-DC 4. *J Appl Microbiol* 93:52–59.

Arioli S, Ragg EM, Scaglioni L, Fessas D, Signorelli M, Karp M, Daffonchio D, De Noni I, Mulas L, Oggioni M, Guglielmetti S, Mora D. 2010. Alkalizing reactions streamline cellular metabolism in acidogenic microorganisms. *PLoS ONE* 5:e15520.

Arioli S, Roncada P, Salzano AM, Deriu F, Corona S, Guglielmetti S, Bonizzi L, Scaloni A, Mora D. 2009. The relevance of carbon dioxide metabolism in *Streptococcus thermophilus*. *Microbiol* 155:1953–65.

Avis TJ, Belanger RR. 2001. Specificity and mode of action of the antifungal fatty acid cis-9-heptadecenoic acid produced by *Pseudozyma flocculosa*. *Appl Environ Microbiol* 67:956–60.

Bolotin A, Quinquis B, Renault P, Sorokin A, Ehrlich SD, Kulakauskas S, Lapidus A, Goltsman E, Mazur M, Pusch GD, Fonstein M, Overbeek R, Kyprides N, Purnelle B, Prozzi D, Ngui K, Masuy D, Hancy F, Burteau S, Boutry M, Delcour J, Goffeau A, Hols P. 2004. Complete sequence and comparative genome analysis of the dairy bacterium *Streptococcus thermophilus*. *Nat Biotechnol* 22:1554–58.

Brunhuber N, Thoden JB, Blanchard J, Vanhooke JL. 2000. *Rhodococcus* L-Phenylalanine Dehydrogenase: kinetics, mechanism, and structural basis for catalytic specificity. *Biochemistry* 39:9174–87.

Bryan-Jones DG, Whittenbury R. 1969. Haematin-dependent oxidative phosphorylation in *Streptococcus faecalis*. *J Gen Microbiol* 58:247–60.

Chapot-Chartier MP, Rul F, Nardi M, Gripon JC. 1994. Gene cloning and characterization of PepC, a cysteine aminopeptidase from *Streptococcus thermophilus*, with sequence similarity to the eucaryotic bleomycin hydrolase. *Eur J Biochem* 224:497–506.

Chavagnat F, Casey MG, Meyer J. 1999. Purification, characterization, gene cloning, sequencing, and overexpression of aminopeptidase N from *Streptococcus thermophilus*. *Appl Environ Microbiol* 65:3001–3007.

Cochu A, Roy D, Vaillancourt K, Lemay JD, Casabon I, Frenette M, Moineau S, Vadeboncoeur C. 2005. The doubly phosphorylated form of HPr, HPr(Ser ~ P)(His-P), is abundant in exponentially growing cells of *Streptococcus thermophilus* and phosphorylates the lactose transporter LacSas efficiently as HPr(His ~ P). *Appl Environ Microbiol* 71(3):1364–72.

Courtin P, Monnet V, Rul F. 2002. Cell-wall proteinases PrtS and PrtB have a different role in *Streptococcus thermophilus/Lactobacillus bulgaricus* mixed cultures in milk. *Microbiology* 148:3413–21.

Cotter PD, Hill C. 2003 Surviving the acid test: responses of Gram-positive bacteria to low pH. *Microbiol MolBiol Rev* 67:429–53.

Crowley S, Mahony J, van Sinderen D. 2013a. Broad-spectrum antifungal-producing lactic acid bacteria and their application in fruit models. *Folia Microbiologica* (Praha) 58:291–99. doi:10.1007/s12223-012-0209-3.

Crowley S, Mahony J, van Sinderen D. 2013b. Current perspectives on antifungal lactic acid bacteria as natural bio-preservatives. *Trends in Food Sci Tecnol* 33:93–109.

Dandoy D, Fremaux C, de Frahan MH, Horvath P, Boyaval P, Hols P, Fontaine L. 2011. The fast milk acidifying phenotype of *Streptococcus thermophilus* can be acquired by natural transformation of the genomic island encoding the cell-envelope proteinase PrtS. *Microbial Cell Factories* 10(1):S21.

Delavenne E, Ismail R, Pawtowski A, Mounier J, Barbier B. 2012. Assessment of lactobacilli strains as yoghurt bioprotective cultures. *Food Control* 30:206e-13e.

Delorme C, Bartholini C, Bolotine A, Ehrlich SD, Renault P. 2010. Emergence of a cell wall protease in the *Streptococcus thermophilus* population. *Appl Environ Microbiol* 76(2):451–60.

den Hengst CD, Curley P, Larsen R, Buist G, Nauta A, van Sinderen D, Kuipers OP, Kok J. 2005b. Probing direct interactions between CodY and the *oppD* promoter of *Lactococcus lactis*. *J Bacteriol* 187:512–21.

den Hengst CD, van Hijum SA, Geurts JM, Nauta A, Kok J, Kuipers OP. 2005a. The *Lactococcus lactis* CodY regulon: identification of a conserved cis-regulatory element. *J Biol Chem* 280:34332–42.

Detmers FJ, Kunji ER, Lanfermeijer FC, Poolman B, Konings WN. 1998. Kinetics and specificity of peptide uptake by the oligopeptide transport system of *Lactococcus lactis*. *Biochemistry* 37:16671–179.

Dieuleveux V, Lemarinier S, Gueguen M. 1998.Antimicrobial spectrum and target site of D-3-phenyllactic acid. *Int J Food Microbiol* 40:177e–83e.

Duwat P, Sourice S, Cesselin B, Lamberet G, Vido K, Gaudu P, Le LoirY, Violet F, Loubière P, Gruss A. 2001. Respiration capacity of the fermenting bacterium *Lactococcus lactis* and its positive effect on growth and survival. *J Bacteriol* 183:4509–16.

EFSA. 2012. The European Union Summary Report on antimicrobial resistance in zoonotic and indicator bacteria from humans, animals and food in the European Union in 2010. *EFSA Journal* 10(3):2598.

Fernandez-Espla MD, Garault P, Monnet V, Rul F. 2000. *Streptococcus thermophilus* cell wall-anchored proteinase: release, purification, and biochemical and genetic characterization. *Appl Environ Microbiol* 66:4772–78.

Fontaine L, Dandoy D, Boutry C, Delplace B, de Frahan MH, Fremaux C, Horvath P, Boyaval P, Hols P. 2010. Development of a versatile procedure based on natural transformation for marker-free targeted genetic modification in *Streptococcus thermophilus*. *Appl Environ Microbiol*, 76:7870–77.

Garault P, Le Bars D, Besset C, Monnet V. 2002. Three oligopeptide-binding proteins are involved in the oligopeptide transport of *Streptococcus thermophilus*. *J Biol Chem* 277:32–39.

Gaudu P, Vido K, Cesselin B, Kulakauskas S, Tremblay J, Rezaiki L, Lamberret G, Sourice S, Duwat P, Gruss A. 2002. Respiration capacity and consequences in *Lactococcus lactis*. *Ant Van Leeuwenhoek* 82:263–69.

Fontaine L, Boutry C, de Frahan MH, Delplace B, Fremaux C, Horvath P, Boyaval P, Hols P. 2010. A novel pheromone quorum-sensing system controls the development of natural competence in *Streptococcus thermophilus* and *Streptococcus salivarius*. *J Bacteriol* 5:1444–54.

Foucaud C, Poolman B. 1992. Lactose transport system of *Streptococcus thermophilus*. Functional reconstitution of the protein and characterization of the kinetic mechanism of transport. *J Biol Chem* 267(31):22087–94.

Foucaud C, Kunji ER, Hagting A, Richard J, Konings WN, Desmazeaud M, Poolman B. 1995. Specificity of peptide transport systems in *Lactococcus lactis*: evidence for a third system which transports hydrophobic di- and tripeptides. *J Bacteriol* 177:4652–57.

Galia W, Jameh N, Perrin C, Genay M, Dary-Mourot A. 2016. Acquisition of PrtS in *Streptococcus thermophilus* is not enough in certain strains to achieve rapid milk acidification. *Dairy Sci Technol* 96:623–36.

Garcha S, Natt NK. 2011. In situ control of food spoilage fungus using *Lactobacillus acidophilus* NCDC 291. *J Food Sci Technol* 49:643–48.

Gaudu P, Lamberet G, Poncet S, Gruss A. 2003. CcpA regulation of aerobic and respiration growth in *Lactococcus lactis*. *Mol Microbiol* 50:183–92.

Germond JE, Delley M, Gilbert C, Atlan D. 2003. Determination of the domain of the *Lactobacillus delbrueckii* subsp. *bulgaricus* cell surface proteinase PrtB involved in attachment to the cell wall after heterologous expression of the prtB gene in *Lactococcus lactis*. *Appl Environ Microbiol* 69:3377–84.

Gilbert C, Atlan D, Blanc B, Portalier R. 1994. Proline iminopeptidase from *Lactobacillus delbrueckii* subsp. *bulgaricus* CNRZ 397: purification and characterization. *Microbiology* 140: 537–42.

Gilbert C, Atlan D, Blanc B, Portalier R, Germond GJ, Lapierre L, Mollet B. 1996. A new cell surface proteinase: sequencing and analysis of the *prtB* gene from *Lactobacillus debrueckii* subsp. *bulgaricus*. *J Bacteriol* 178:3059–65.

Guédon E, Renault P, Ehrlich D, Delorme C. 2001a. Transcriptional pattern of genes coding for the proteolytic system of *Lactococcus lactis* and evidence for coordinated regulation of key enzymes by peptide supply. *J Bacteriol* 183:3614–22.

Guédon E, Serror P, Ehrlich SD, Renault P, Delorme C. 2001b. Pleiotropic transcriptional repressor CodY senses the intracellular pool of branched-chain amino acids in *Lactococcus lactis*. *Mol Microbiol* 40:1227–39.

Gunnewijk MG, Poolman B. 2000a. Phosphorylation state of HPr determines the level of expression and the extent of phosphorylation of the lactose transport protein of *Streptococcus thermophilus*. *J Biol Chem* 275(44):34073–9.

Gunnewijk MG, Poolman B. 2000b. HPr(His approximately P)-mediated phosphorylation differently affects counterflow and proton motive force-driven uptake via the lactose transport protein of *Streptococcus thermophilus*. *J Biol Chem* 275(44):34080–5.

Hagting A, Kunji E, Leenhouts K, Poolman B, Konings W. 1994. The di- and tripeptide transport protein of *Lactococcus lactis*. A new type of bacterial peptide transporter. *J Biol Chem* 269:11391–99.

Hols P, Hancy F, Fontaine L, Grossiord B, Prozzi D, Leblond-Bourget N, Decaris B, Bolotin A, Delorme C, Ehrlich SD, Guedon E, Monnet V, Renault P, Kleerebezem M. 2005. New insights in the molecular biology and physiology of *Streptococcus thermophilus* revealed by comparative genomics. *FEMS Microbiol Rev* 29:435–63.

Iyer R, Tomar SK, Maheswari TU, Singh R. 2010. *Streptococcus thermophilus* strains: multifunctional lactic acid bacteria. *Int Dairy J* 20:133–41.

Kindstedt PS, Fox P. 1993. Effect of manufacturing factors, composition, and proteolysis on the functional characteristics of Mozzarella cheese. *Crit Rev Food Sci Nutr* 33:167–87.

Juille O, Le Bars O, Juillard V. 2005. The specificity of oligopeptide transport by *Streptococcus thermophilus* resembles that of *Lactococcus lactis* and not that of pathogenic streptococci. *Microbiology* 151:1987–94.

Lechardeur D, Cesselin B, Fernandez A, Lamberet G, Garrigues C, Pedersen M, Gaudu P, Gruss A. 2011. Using heme as an energy boost for lactic acid bacteria. *Curr Opin Biotechnol* 22:143–49.

Letort C, Nardi M, Garault P, Monnet V, Juillard V. 2002. Casein utilization by *Streptococcus thermophilus* results in a diauxic growth in milk. *Appl Environ Microbiol* 3162–65.

Mayoral MB, Martín R, Sanz A, Hernández PE, González I, García T. 2005. Detection of *Kluyveromyces marxianus* and other spoilage yeasts in yoghurt using a PCR-culture technique. *Int J Food Microbiol* 105:27e34.

Monnet C, Mora D, Corrieu G. 2005. Glutamine synthesis is essential for growth of *Streptococcus thermophilus* in milk and is linked to urea catabolism. *Appl Environ Microbiol* 71(6):3376–8.

Monnet C, Pernoud S, Sepulchre A, Fremaux C, Corrieu G. 2004. Selection and properties of *Streptococcus thermophilus* mutants deficient in urease. *J Dairy Sci* 87(6):1634–40.

Mora D, Arioli S, Compagno C. 2013. Food environments select microorganisms based on selfish energetic behavior. *Frontiers Microbiol* 4:348.

Mora D, Maguin E, Masiero M, Parini C, Ricci G, Manachini PL, Daffonchio D. 2004. Characterization of urease genes cluster of *Streptococcus thermophilus*. *J Appl Microbiol* 96:209–19.

Mora D, Monnet C, Daffonchio D. 2005. Balancing the loss and acquisition of pathogenic traits in food-associated bacteria. *Microbiology-SGM* 151:3814–16.

Mora D, Monnet C, Parini C, Guglielmetti S, Mariani A, Pintus P, Molinari F, Daffonchio D, Manachini PL. 2005. Urease biogenesis in *Streptococcus thermophilus*. *Res Microbiol* 156(9):897–903.

Morel F, Frot-Coutaz J, Aubel D, Portalier R, Atlan D. 1999. Characterization of a prolidase from *Lactobacillus delbrueckii* subsp. *bulgaricus* CNRZ 397 with an unusual regulation of biosynthesis. *Microbiology* 145:437–46.

Morel F, Lamarque M, Bissardon I, Atlan D, Galinier A. 2001. Autoregulation of the biosynthesis of the CcpA-like protein, PepR1, in *Lactobacillus delbrueckii* subsp. *bulgaricus*. *J Mol Microbiol Biotechnol* 3:63–66.

Muhialdini BJ, Hassan Z, Sadon SK, Zulkifli NA, Azfari AA. 2011. Effect of pH and heat treatment on antifungal activity of *Lactobacillus fermentum* Te007, *Lactobacillus pentosus* G004 and *Pediococcus pentosaceus* Te010. *Innovative Romanian Food Biotechnology* 8:41–53.

Official Journal of the European Communities. 2001. Directive 2001/18/EC of the European Parliament and of the Council of 12 March 2001 on the deliberate release into the environment of genetically modified organisms and repealing Council Directive 90/220/EEC. *Off J Eur Communities L* 106:1–38.

Pedersen MB, Garrigues C, Tuphile K, Brun C, Vido K, Bennedsen M, Mollgaard H, Gaudu P, Gruss A. 2008. Impact of aeration and heme-activated respiration on *Lactococcus lactis* gene expression: identification of a heme-responsive operon. *J Bacteriol* 190:4903–11.

Pedersen MB, Iversen SL, Sørensen KI, Johansen E. 2005. The long and winding road from the research laboratory to industrial applications of lactic acid bacteria. *FEMS Microbiol Rev* 29:611–24.

Peltoniemi K, Vesanto E, Palva A. 2002. Genetic characterization of an oligopeptide transport system from *Lactobacillus delbrueckii* subsp. *bulgaricus*. *Arch Microbiol* 177:457–67.

Pernoud S, Fremaux C, Sepulchre A, Corrieu G, Monnet C. 2004. Effect of the metabolism of urea on the acidifying activity of *Streptococcus thermophilus*. *J Dairy Sci* 87(3):550–5.

Poolman B, Knol J, Lolkema JS. 1995. Kinetic analysis of lactose and proton coupling in Glu379 mutants of the lactose transport protein of *Streptococcus thermophilus*. *J Biol Chem* 70(22):12995–33003.

Poolman B, Modderman R, Reizer J. 1992. Lactose transport system of *Streptococcus thermophilus*. The role of histidine residues. *J Biol Chem* 267(13):9150–7.

Rezaïki L, Cesselin B, Yamamoto Y, Vido K, van West E, Gaudu P, Gruss A. 2004. Respiration metabolism reduces oxidative and acid stress to improve long-term survival of *Lactococcus lactis*. *Mol Microbiol* 53:1331–42.

Ritchey TW, Seeley HW. 1976. Distribution of cytochrome-like respiration in streptococci. *J Gen Microbiol* 93:195–203.

Rul F, Monnet V. 1997. Presence of additional peptidases in *Streptococcus thermophilus* CNRZ 302 compared to *Lactococcus lactis*. *J Appl Microbiol* 82:695–704.

Sasaki M, Bosman BW, Tan PS. 1995. Comparison of proteolytic activities in various lactobacilli. *J Dairy Res* 62:601–610.

Schwenninger SM, Meile L. 2004. A mixed culture of *Propionibacterium jensenii* and *Lactobacillus paracasei* subsp. *paracasei* inhibits food spoilage yeasts. *Syst Appl Microbiol* 27:229–37.

Sijpesteijn AK. 1970. Induction of cytochrome formation and stimulation of oxidative dissimilation by hemin in *Streptococcus lactis* and *Leuconostoc mesenteroides*. *Ant Van Leeuwenhoek* 36:335–348.

Sørensen KM, Curic-Bawden M, Junge MP, Janzen T, Johansenc E. 2016. Enhancing the sweetness of yoghurt through metabolic remodeling of carbohydrate metabolism in *Streptococcus thermophilus* and *Lactobacillus delbrueckii subsp. bulgaricus*. *Appl Environ Microbiol* 82 (12):3683.

Stefanitsi D, Sakellaris G, Garel JR. 1995. The presence of two proteinases associated with the cell wall of *Lactobacillus bulgaricus*. *FEMS Microbiol Lett* 128:53–58.

Svanström A, Boveri S, Boström E, Melin P. 2013. The lactic acid bacteria metabolite phenyllactic acid inhibits both radial growth and sporulation of filamentous fungi. *BMC Research Notes* 6:464.

Terence DT, Vaughan LC. 1984. Selection of galactose-fermenting *Streptococus thermophilus* in lactose-limited chemostat cultures. *Appl Environ Microbiol* 186–91.

Thomas TD, Turner KW, Crow VL. 1980. Galactose fermentation by *Streptococcus lactis* and *Streptococcus cremoris*: pathways, products, and regulation. *J Bacteriol* 144:672–82.

Tinson W, Hillier A, Jago G. 1982. Metabolism of *Streptococcus thermophilus*. Utilization of lactose, glucose and galactose. *Aust J Dairy Technol* 37:8–13.

Tynkkynen S, Buist G, Kunji E, Kok J, Poolman B, Venema G, Haandrikman A. 1993. Genetic and biochemical characterization of the oligopeptide transport system of *Lactococcus lactis*. *J Bacteriol* 175:7523–32.

Vaillancourt K, Bédard N, Bart C, Tessier M, Robitaille G, Turgeon N, Frenette M, Moineau S, Vadeboncoeur C. 2008. Role of *galK* and *galM* in galactose metabolism by *Streptococcus thermophilus*. *Appl Environ Microbiol* 1264–67.

Vaillancourt K, LeMay JD, Lamoureux M, Frenette M, Moineau S, Vadeboncoeur C. 2004. Characterization of a galactokinase-positive recombinant strain of *Streptococcus thermophilus*. *Appl Environ Microbiol* 4596–603.

Van de Guchte M, Pernaud S, Grimaldi C, Barbe V, Bryson K, Nicolas P, et al. 2006. The complete genome sequence of *Lactobacillus bulgaricus* reveals extensive and ongoing reductive evolution. *Proc Natl Acad Sci USA* 24:9274–79.

Vaughan EE, van den Boogard P, Catzeddu P, Kuipers O, de Vos W. 2001. Activation of silent *gal* genes in *lac-gal* regulon of *Streptococcus thermophilus*. *J Bacteriol* 1184–94.

Vido K, le Bars D, Mistou M-Y, Anglade P, Gruss A, Gaudu P. 2004. Proteome analyses of heme-dependent respiration in *Lactococcus lactis*: involvement of the proteolytic system. *J Bacteriol* 186:1648–57.

Vogel R, Hammes WP, Habermeyer M, Engel KH, Knorr D, Eisenbrand G. 2011. Microbial food cultures – opinion of the Senate Commission on Food Safety (SKLM) of the German Research Foundation (DFG). *Mol Nutr Food Res* 55:654–62.

Zhao D. 2011. *Isolation of antifungal lactic acid bacteria from food sources and their use to inhibit mould growth in cheese*. PhD thesis. California Polytechnic State University.

Zourari A, Accolas JP, Desmazeaud MP. 1992. Metabolism and biochemical characteristics of yogurt bacteria—a review. *Lait* 72:1–34.

13 Lactic acid bacteria bacteriophages in dairy products: problems and solutions

Giorgio Giraffa, Miriam Zago and Domenico Carminati

CREA-ZA, Centro di ricerca Zootecnia e Acquacoltura, sede di Lodi, Italy

13.1 INTRODUCTION

Dairy starters are actively growing cultures of lactic acid bacteria (LAB), which are used in the production of a variety of dairy products, including cheese, fermented milks, and cream butter. These microorganisms produce lactic acid via lactose fermentation, which leads to a rapid decrease in pH. Cheese and fermented milk manufacture depend largely on this acidification, which is also crucial for efficient milk clotting, whey drainage and product structuring, to prevent the growth of pathogenic and spoilage microorganisms. Several factors can affect the activity of dairy starters, such as the composition of milk as growth medium, bacterial interactions, presence of antibiotics or sanitizers, chemical inhibitors and bacteriophage attack.

Problems due to the presence of phages were reported in the food, chemical, pharmaceutical, feed and pesticide industries. However, the dairy industry is probably the industry in which phage problems are the most well-documented (Garneau and Moineau, 2011). Phages of LAB may originate from both raw milk and lysogenic starter bacteria. Facilitated from the progress of knowledge on the biology, physiology and genetics of phages, a plethora of control measures have been introduced in the industrial practice to avoid or minimize phage-related problems. However, despite LAB bacteriophages now being among the most studied phages, they still remain a high-risk issue for the dairy industry because their complete removal is not realistically possible. LAB-infecting phages have been claimed as one of the principal sources of fermentation failure (spoilage or delay) in the manufacture of many dairy products (Brüssow et al., 1998; Josephsen and Neve, 1998; Garneau and Moineau, 2011). Some estimates assume that virulent phages are the primary reason for the largest economic loss of dairy factories, since they affect negatively up to the 10% of all milk fermentations, especially cheese (Moineau and Levesque, 2005). Phages can also cause problems in the fast-growing field of probiotic foods, where the consequences of phage infection may be even more detrimental because the cultures employed are generally composed of individual strains.

Microbiology in Dairy Processing: Challenges and Opportunities, First Edition.
Edited by Palmiro Poltronieri.

Published 2018 by John Wiley & Sons Ltd and the Institute of Food Technologists.

Emerging data suggest an increasing occurrence of phages for specific probiotic LAB species, especially *Lactobacillus plantarum*, *Lactobacillus acidophilus*, and *Lactobacillus casei/paracasei*, which are widely used in several fermented dairy products (Briggiler Marcò et al., 2012). Therefore, it is imperative not only to be able to easily recognize and quantify phages, but also to find even more effective ways to fight them, or at least maintain their numbers under a critical threshold, thus preventing huge economic losses for the industry of fermented dairy products.

Phages of dairy LAB have been reviewed many times in recent years in several reviews or book chapters, thanks to the explosion of knowledge on phage physiology, ecology, and genetics. For obvious reasons, being large volumes of raw milk daily fermented by LAB starters belonging to *Lactococcus lactis*, *Streptococcus thermophilus*, *Lactobacillus delbrueckii* and its two dairy subspecies *lactis* and *bulgaricus*, and *Lactobacillus helveticus*, cheese and yoghurt manufacture are the most affected processes. The aim of this chapter is to give an overview of most recent knowledge on phages infecting the main dairy LAB species used in the dairy industry, paying attention to the practical aspects of their detection and control.

13.2 PHAGE CLASSIFICATION

Phage classification schemes are often a matter of debate, but they do provide critical information to better control phage infections. Classification schemes based on traditional methods, including electron microscopy, serotyping and structural protein profiling, are presently available for *Lc. lactis*, *S. thermophilus*, *Leuconostoc* spp. and *Lactobacillus* phages (Mahony and Van Sinderen, 2014). The majority of these phages belong to the family of *Siphoviridae*, a vast group of phages with long, non-contractile tails and prolate or isometric capsides. Additionally, phages with short non-contractile tails (*Podoviridae*) and those displaying long contractile tails (*Myoviridae*) have also been described for some LAB genera. However, traditional classification approaches are used along with or being completely replaced by DNA sequencing. This has led to a significant increase in the amount of phage genomic data available, with which a rapid comparison and a genetic classification of phages is now possible. Phages infecting LAB possess double-stranded linear DNA, their genome size ranges from 14 to 141 kb, and their G+C content (33–48%) is similar to that of their hosts. Presently, full genome sequences of over 100 virulent LAB phages (with approximately 70% belonging to *Lactococcus* spp.), as well as several prophages, are available in public databases (Mahony and Van Sinderen, 2014). Comparative genome analysis has confirmed that phage diversification is most often due to the accumulation of point mutations, gene disruption and recombination (Rousseau and Moineau, 2009; Garneau and Moineau, 2011).

Undoubtedly, the most intensively researched and studied LAB-infecting phages are those of the dairy starter bacteria *Lc. lactis* and *S. thermophilus*. *Lactococcus* spp. phages are currently classified into 10 groups based on morphology and DNA homology (Villion and Moineau, 2009). At least one genome from each lactococcal phage group is available. Furthermore, two new species of lactococcal phages emerged, such as the Q54 and 1706, which were previously unknown or unclassified (Deveau et al., 2006).

However, most lactococcal phages isolated from dairy fermentations belong to 936 (45 sequenced genomes) and P335 (10 sequenced genomes) phage species, whereas the c2 phages, which dominated in early phage isolation studies, became less problematic recently (Briggiler Marcò et al., 2012; Mahony et al., 2014). However, while lactococcal strain diversity may be limited, their infecting phages have proven to co-evolve with their hosts and resident prophages by genomic rearrangements. This coevolution is the driving force supporting the ever-increasing genetic diversity of lactococcal phages within 936 and P335 groups (Mahony and van Sinderen, 2014).

All *S. thermophilus* phages reported to date are members of the *Siphoviridae* family and can be assembled into two distinct groups based on the number of major structural proteins (MSPs) and the mode of DNA packaging. The first group comprises phages with two MSP and cohesive genome extremities (*cos*-type). The second cluster includes phages with three MSP and a DNA packaging scheme that proceeds via a headful mechanism (*pac*-type) (Le Marrec et al., 1997). Interestingly, a third group has emerged according to phage 5093, isolated from Mozzarella starter. The genome of this phage was sequenced and revealed a novel genotype bearing little or no genetic similarity to previously identified *S. thermophilus* phages, yet retaining the conserved genomic modular arrangement found in the *pac* and *cos* phages (Mills et al., 2011). *S. thermophilus* phages have a linear double-stranded DNA genome, ranging in size from 30 to 45 kb (Brussow et al., 1998). DNA-DNA hybridization studies showed homology between all *S. thermophilus* phages, including virulent and temperate phages. Thirteen fully sequenced genomes, 12 of them sharing high homology, are available (Mahony and Van Sinderen, 2014). A great diversity of streptococcal phages is often observed in cheese making, in contrast to a more homogeneous phage population in yoghurt production facilities (Zago et al., 2003). This higher genetic diversity within cheese-associated *S. thermophilus* phages has been attributed to a continuous entry of new phages from environmental sources (i.e. raw milk), rather than genetic changes of a single phage invading the factory (Brüssow et al., 1998).

Lactobacillus phages belong to the families *Siphoviridae* and *Myoviridae* (Villion and Moineau, 2009). Classification of *Lactobacillus* phages is complex due to their wide genetic diversity. Before the availability of genomic sequences, *Lactobacillus* phages were typed by host range, morphology, and DNA homology. Later, several completely sequenced Lactobacillus genome phages were assigned, and to date, 24 Lactobacillus phages with genome sizes in the range ~31–42 kb are presently available (www.ebi.ac.uk/genomes/phage.html). Within the species of dairy interest, the phage genomes available include those of *L. plantarum*, *Lactobacillus gasseri*, *L. casei/paracasei*, *Lactobacillus rhamnosus*, *L. delbrueckii* and *L. helveticus*. Phages of *L. delbrueckii* are among the most widely studied. They belong to the *Siphoviridae,* and to date, six full genome sequences are available (Villon and Moineau, 2009). In *L. helveticus* the only full genomic sequence available is that of ϕAQ113, a *Myoviridae* bacteriophage with an isometric capsid, a contractile tail and a genome size of 36,566 bp. The sequence of ϕAQ113 is closely related to those of *L. gasseri* ϕKC5a and *Lactobacillus johnsonii* ϕLj771, two phages infecting gut species. The identification of genes involved in the establishment of lysogeny suggests that ϕAQ113 may have originated as a temperate phage (Zago et al., 2013).

Limited knowledge exists for dairy *Leuconostoc* phages. Since *Leuconostoc* strains do not contribute significantly to lactic acid production in milk, infections of the phages infecting *Leuconostoc* cultures usually go unnoticed during processing, but manifest their effects subsequently, altering the flavor profile of the final products. A recent and complete survey on 83 lytic phages infecting *Leuconostoc pseudomesenteroides* and *Leuconostoc mesenteroides* flavor-producing dairy starter strains showed that they belong to the *Siphoviridae* family with isometric heads and non-contractile tails. All phages are members of two (host species-specific) distinct genotypes but share a limited conserved DNA region according to restriction enzyme and Southern blot analysis (Ali et al., 2013).

13.3 PHAGE-HOST INTERACTIONS

Because of the economic impact of the phage problems on the food industry, phage resistance in LAB starter cultures has been intensively studied for decades. Bacteria have developed a plethora of strategies to defend themselves against the continuous attacks of phages. These systems have been generally classified according to the phage multiplication steps they target, including those that inhibit adsorption and block DNA injection (Ali et al., 2014; Mahony et al., 2008) as well as restriction and modification (R/M) (Suarez et al., 2009), abortive infection (Abi) (Samson et al., 2013), and clustered regularly interspaced short palindromic repeats (CRISPR) systems (Horvath et al., 2009). The vast majority of natural defense mechanisms have been found in *Lc. lactis* (Coffey and Ross, 2002). Adsorption represents the initial and obligate step of any phage infection process, and it is determined by a molecular interaction between a phage structural protein and a receptor present on the surface of the bacterial host cell. In *L. helveticus* the phage receptor has been identified as the S-layer (Ventura et al., 1999). Phage receptor masking, through the production of cell-wall-shielding components, has also been reported to confer phage resistance in lactococci (Trotter et al., 2002). As a matter of fact, phage receptor mutation is a frequently the outcome of the selection of spontaneous bacteriophage-insensitive mutants (BIMs) following phage challenge (Dupont et al., 2004).

Restriction/modification systems involve two distinct enzymatic activities: a site-specific restriction endonuclease that cleaves the incoming, unprotected phage DNA and a cognate methylase that, by methylation, protects the host genomic DNA from restriction (Blumenthal and Cheng, 2002). There are at least four types of R/M systems based on the co-factors needed, the subunit composition, the recognition sequence and the cutting site. R/M systems are the most studied intracellular phage defense mechanisms. Phage resistance based on R/M has been frequently identified in *Lactococcus* (Nyengaard et al., 1995) and also demonstrated in *L. helveticus* by de los Reyes-Gavilan et al. (1990) who observed a reduced efficiency of plating (EOP) for the resistant strain and identified the plasmid encoding the R/M system. By studying the lactic acid microflora of natural whey starters used in Argentina, Reinheimer et al. (1996) also demonstrated resistance mechanisms involving R/M systems for some *L. helveticus* strains exhibiting high phage adsorption rates. Also recent studies indicated the presence in a dairy *L. helveticus* strain of two phage resistance mechanisms, the Type I R/M system and the inhibition of phage adsorption (Zago et al., 2017). The vast majority of R/M systems that have been characterized so far in LAB species belong to Type II (which involves separate restriction and modification enzymes), whereas Type I and Type III R/M systems have occasionally been found (Sturino and Klaenhammer, 2004). In particular, Type I R-M systems, initially identified in and thought to be exclusive to members of the *Enterobacteriaceae,* such as *Salmonella* and *Escherichia coli*, have also been detected in *Lc. lactis*, *S. thermophilus*, and *L. delbrueckii* (Forde and Fitzgerald, 1999; Suarez et al., 2009). Despite their high efficacy (because phage DNA is degraded), R/M systems are also the most leaky because phages have several escape strategies that allow them to avoid this first line of intracellular defense (Labrie and Moineau, 2010). However, if combined with other defense mechanisms, they are even more powerful (Josephsen and Neve, 2004).

The Abi mechanisms refer to antiphage systems that inhibit phage multiplication between DNA entry and cell lysis. They are considered the most powerful among the natural mechanisms because they activate a premature cell death, thus preventing phage multiplication in the host population (Chopin et al., 2005). In LAB, most of the known Abi systems have been described in lactococci, with at least 23 distinct systems, and at least one in *S. thermophilus* (Emond and Moineau, 2007). The exact mode of action of Abi systems is not fully elucidated; they are conventionally categorized according to their ability to act early (before or upon DNA replication) or late (after replication). There is also little or no sequence similarity between Abi systems (Josephsen and Neve, 2004; Emond and Moineau, 2007).

Clustered regularly interspaced short palindromic repeats (CRISPR) are a novel adaptive bacterial immune system that has recently been described in *S. thermophilus*. The CRISPR-Cas (CRISPR associated proteins) defense includes a multistep process by which small and specific fragments of foreign nucleic acids are firstly recognized as being nonself and then incorporated into the host genome between short DNA repeats. This system is present in the genome of many bacteria (~40%) and in almost all archaebacteria (Horvath and Barrangou, 2010). CRISPR loci are usually composed of 23 to 50 bp direct DNA repeats interspaced by non-repetitive nucleotides of 26-72 bp called spacers. In a given locus, the size of the repeats and the spacers is conserved (Deveau et al., 2010). Although multiple CRISPR families have been identified within *Bifidobacterium* spp. and in several *Lactobacillus* genomes, such as *L. rhamnosus, L. casei, Lactobacillus acidophilus* and *L. helveticus* (Horvath et al., 2009), the functionality of these systems as phage defense mechanisms has not been yet demonstrated (Mills et al., 2010a).

As listed before, bacterial antiphage systems include the inhibition of phage attachment to cell surface receptors, the cleavage of the invading phage genome and even the induction of a cell suicide to abort phage infection. However, despite this arsenal, a large proportion of bacteria succumb to phage infection. Owing to their genomic plasticity and rapid multiplication rates, phages have evolved equally diversified strategies to thrive in apparently well-protected bacterial cells. These strategies include mechanisms such as point mutations in specific genes, genome rearrangements, and genomic exchange with other viral or microbial genomes to acquire new traits. Considering the abundance of phages on the planet (estimated at $>10^{31}$) and the fact that phages and hosts coexist in every possible ecosystem (Suttle, 2007), there are plenty of opportunities for phages to win or lose a battle against bacteria. Through a well-known biological process in which phages and bacterial hosts are locked in a continuous state of coevolution, phages have accumulated an impressive array of tactics to circumvent any given antiviral barrier that they face (Buckling and Brockhurst, 2012). This process influences the diversity of both bacteria and phage populations and is one of the driving forces generating genetic variability (Bono et al., 2013). Indeed, the pressure of phage predation selects for resistant bacteria, but the ability of some phages to circumvent resistant cells ensures persistence of host and phage populations. Understanding the complex dynamics of phage–host interactions is of interest for the dairy industry, which relies on phage-resistant bacteria to assure correct and reproducible fermentations. Coevolution between lytic phages and bacterial strains increases diversity within a bacterial population, thus increasing the probability to select strains with multiple phage resistances (Koskella and Brockhurst, 2014).

13.4 SOURCES OF CONTAMINATION

A phage attack does not always imply a total failure of fermentation batches, and thus a product loss; nevertheless, delays in production and variations in product quality are frequently encountered. As dairy phages are safe for human consumption, failed batches could be partially recovered and used to manufacture derivative products. In any case, elimination of phage problems are always costly and time-consuming for the industry because investigations have to be carried out to detect phage sources and introduce corrective actions, which can only control but not completely eradicate phages. It is therefore of paramount importance to know the potential sources of phages to at least limit their entry within the manufacturing facilities, which could be deleterious to the fermentation process.

13.4.1 Milk and cheese whey

Phages can come from various sources. Milk and derived products (fermented milk, cheese whey, cheese curd) are well-established and documented sources of phages. Milk is the first carrier and probably the most important phage contamination source in dairy factories, with concentrations ranging between 10^1 and 10^4 phages/ml (Briggiler Marcò et al., 2012). Given their increasing thermal resistance, both thermized or pasteurized milk can contain phages. Their numbers can increase in whey samples or final products, especially if no precaution is taken to avoid using starter cultures containing phage-sensitive strains. Therefore, phages may easily enter the manufacturing process and accumulate rapidly during fermentation, reaching values up to 10^9 phages/ml cheese whey (or per gram of product) (Briggiler Marcò et al., 2012). Lactococcal and streptococcal phages have been detected in 37% of the milk samples used for yogurt production in Spain (del Rio et al., 2007). An extensive and detailed study carried out on 900 samples of raw milk, each derived from a different farms located in northern Spain, showed that approximately 10% of samples contained lactococcal phages belonging to c2 (72% of all phages), 936 (24%), and P335 (4%) families (Madera et al., 2004). Taken together, these data underline the importance of analysing milk (or other ingredients) for the presence of phages, before beginning the dairy processing, to evaluate if the initial phage load might be risky for the regular fermentation process and, in that case, to reduce it to levels lower than 10 phages/ml.

However, phages can also reach milk through external routes, for example, brines, environmental air, plant facilities and starter cultures. Given the abundant volumes produced during dairy processing, residual dairy by-products are probably the most important indirect sources of milk contamination by phages. In poor hygiene situations, phages are dispersed into the plant environment through spreading and the aerosolising in the form of little droplets generated by the splashing of whey. The cheese whey is the largest source of phage in a dairy plant, since the host bacteria continue to grow in whey after it has been separated from the curd; consequently phages continue to increase (Carminati et al., 2012; Atamer et al., 2013). The recycling of whey proteins and their inclusion in the milk to be transformed is often applied in the modern dairy industry to improve the economic competitiveness of dairy production and the nutritional value of products. If not adequately controlled, the use of native or spray-dried whey preparations as an ingredient of fermented milk products, or as growth substrate for starter cultivations, may represent a serious and concrete source of phage contamination in dairy environments (Atamer et al., 2013).

13.4.2 Dairy cultures

As stated before, possible sources of milk contamination by phages are the starter cultures used in dairy fermentations. According to the way they are reproduced, starter cultures can be classified as defined (multi)strain cultures and undefined strain cultures. Defined strain cultures are used in fermented milks (including probiotics) and industrial cheese production. Although prepared under aseptic conditions from specialized industries, strains included in these cultures could be a source of phages if they hosted temperate phages. Undefined cultures are starters of undefined composition which contain strains of many LAB species, such as *S. thermophilus*, *L. helveticus*, *L. delbrueckii* (thermophilic cultures) and lactococci (mesophilic cultures). Whereas the latter have been left and progressively substituted by defined cultures, the former are still widely used for the manufacture of many Italian, Swiss and French hard cheeses. The production of such cultures is often derived from the traditional back-slopping practice (the use of a previous fermented product to inoculate a new one) and/or by application of selective pressures (heat treatment, incubation temperature, low pH). Although it is believed that the richness and heterogeneity of the microbial composition may represent a natural barrier against phage infection, phage contamination of undefined starter cultures may frequently occur because these cultures are propagated without special precautions to prevent contamination of the raw milk or the cheese-making environment. Consequently, phage contamination is considered a common trait of these cultures (Carminati et al., 2012).

13.4.2.1 The lysogenic state

When a temperate phage enters a strain, it can either start the lytic cycle or its genome can integrate into the bacterial chromosome and follow bacterial multiplication. When bacteria carry such a prophage, the cell is called a lysogen. The lysogen is resistant to its own prophage and similar phages, a phenomenon called "superinfection immunity" and observed in lactococci and *S. thermophilus* (Garneau and Moineau, 2011). Temperate phages can disturb the normal process of fermentations when, by mutation, they become virulent phages capable of overcoming superinfection immunity. Another problem derives from the induction of prophages that can be spontaneous or activated through external factors, such as heat, salt or antimicrobials, leading to the release of new virulent phages that can potentially infect phage sensitive strains present in the culture (Briggiler Marcò et al., 2012). The consequences of this fact are evident when commercial starters are used, because they are composed of a low number of strains. Prophages can also give rise to new virulent phages, with enhanced host range, through recombination with phages infecting the lysogenic strain (Labrie and Moineau, 2007).

Genome-sequencing studies have confirmed that many LAB species contain prophages (Canchaya et al., 2003). Lysogeny is widespread within most of the dairy-associated and probiotic LAB species, such as *Lactococcus* spp., *L. delbrueckii*, *L. helveticus*, *L. casei/paracasei* and *L. rhamnosus*, whereas it is rare in *S. thermophilus* (Carminati et al., 1997; Bruttin et al., 1997; Suarez et al., 2008; Garneau and Moineau, 2011). From a study on 74 strains of *L. helveticus* isolated from natural whey cultures, 47% of them hosted defective prophages, that is, incomplete phages able to kill the sensitive cells but unable to propagate within the host and liberate newly synthesized phage particles (Carminati et al., 1997). A recent study showed that 25 out of 30 probiotic strains of *Lactobacillus* contained inducible prophages (Mercanti et al., 2011). Starter culture suppliers must check their strains for the presence of prophage, excluding, in most cases, lysogenic strains from a commercial application.

13.5 PHAGE DETECTION AND QUANTIFICATION

Effective phage detection methods are necessary to identify phage contamination and monitor phage dynamics within the factory. These methods must be (i) sensitive enough to detect low levels of phage; (ii) fast, to allow rapid application of the appropriate control procedures; (iii) applicable to food matrices; (iv) easy to perform by the staff in place; and (v) affordable. Standard microbiological methods, that is, plaque assays, spot tests and activity tests, are usually applied to milk or dairy fermented products. One of the advantages of this type of technique is the discrimination between phage and non-phage inhibitors. Disadvantages include the requirement for a sensitive indicator strain and the relatively long time needed to obtain results. The incubation time is long (~18 h), and thus the presence of phages is confirmed after the manufacturing process is completed. Some phages are not able to form plaques or the plaques are difficult to observe (Anderson et al., 2011). Despite these negative aspects, the plaque assay is still a day-to-day tool in research and industrial laboratories, since it is relatively simple and does not require special equipment. Standard microbiological methods for phage detection and quantification are still used for routine controls in the dairy plant laboratory, as recommended by the International Dairy Federation (IDF, 1980, 1991). These methods can be used for single-strain cultures and include the turbidity test, which is based on the clearance of a broth culture caused by the presence of phages in a sterile filtrate, and the agar spot test or the double-layer plaque titration, which are based on the lysis of sensitive cells in solid media. These methods provide information on phage titers in the filtrate and represent the traditional choice to analyse industrial samples (IDF, 1991).

With the increasing number of phage DNA sequences available and the lower cost of molecular biology materials, many DNA-based methods for phage detection were developed. Several assays based on the polymerase chain reaction (PCR) have been designed and successfully applied to detect, or even classify, *Lactobacillus, Lactococcus* and *Streptococcus* phages in different dairy matrices (Binetti et al., 2008). The detection limit of a classical one-step PCR method is usually in the range 10^4–10^7 PFU/ml, depending on the phage type and sample, but an additional phage concentration step allows the detection up to 10^3 PFU/ml. A benefit of this technique is that neither phage particle concentration (e.g. purification steps) nor any other procedure is required to enrich the samples. PCR methods, however, cannot distinguish viable phage particles from DNA, but the presence of viral DNA could be an indication of the potential presence of infective viruses (Zago et al., 2008). Amplification of specific regions in phage genomes can be used to detect and identify different phages simultaneously. Multiplex PCR systems have been developed to discriminate specifically between the main lactococcal phage groups present in the dairy industry (Labrie and Moineau, 2000). A similar approach has been used to distinguish between phages of *S. thermophilus* (del Rio et al., 2008), and a multiplex PCR has also been designed to detect phages infecting *Lc. lactis, S. thermophilus,* and *L. delbrueckii* in a single reaction (del Rio et al., 2007).

Classical PCR amplification of phage DNA is not quantitative, thus methods based on quantitative real-time PCR (qPCR) are interesting because they provide highly sensitive, rapid and real-time monitoring of specific phages. Rapid detection of *L. delbrueckii* and *S. thermophilus* phages were recently reported, with 10^4 PFU/ml and 10^5 PFU/ml of milk as quantification limits, respectively (del Rio et al., 2008; Martin et al., 2008). In a recent study, the detection and enumeration (with detection limit as low as 10^2 PFU/ml) of three groups (c2, 936 and P335) of lactococcal phages in goat's raw milk and whey was possible by qPCR (Ly-Chatain et al., 2011). However, these molecular methods do not discriminate between active and non-active phage particles. Consequently, PCR-based methods and classical microbiological assays might be used together to obtain a more precise picture of the phages contained in the sample (titers, host range, phage type) (Garneau and Moineau, 2011).

Other methods based on microscopy and flow cytometry have been developed to study phages, but the need for specific and expensive equipment as well as specialized staff has restricted the practical application at the industrial level. *L. helveticus* phage φAQ113, isolated from cheese whey, was morphologically described (shape, dimensions) and quantified using transmission electron microscopy and epifluorescence microscopy, respectively (Zago et al., 2012). Detection of lactococcal phage infection with limits comparable to classical PCR methods (10^5 PFU/ml) was recently carried out by flow cytometry, a technique based on the detection of cells with low mass that are found in the later lytic cycle. During phage infection, the typical lactococcal chains are broken up while cells with low density appear, thus allowing

Table 13.1 Methods of bacteriophage detection.

Microbiological methods	• Turbidity test • Plaque assay
PCR-based methods	• Simplex PCR • Multiplex PCR • qPCR • MLST • RAPD
Microscopy	• Electron microscopy • Fluorescent microscopy • Atomic Force microscopy
Others	• Flow cytometry • Biosensors • Impedimetric • Immunodetection

MLST: multilocus sequence typing; RAPD: polymorphic DNA analysis.

detection of very early phage attacks (Michelsen et al., 2007). Table 13.1 reports a brief summary of the most widely used methods of LAB phage detection and quantification.

13.6 METHODS TO CONTROL PHAGE CONTAMINATION

The control of phage contamination in dairy factories is fundamental to maintaining phage levels as low as possible. Phages are difficult to eliminate because they rapidly disseminate in dairy environment, can persist on various surfaces, and can spread through the production facilities as airborne particles (Moineau and Levesque, 2005; Verreault, et al., 2011; Briggiler Marcó et al., 2012). The main control measures are physical and chemical treatments to improve the hygiene of dairy environment and the microbiological quality of raw milk. Moreover, biological approaches for starter cultures through the rotation of strains with different phage sensitivity and the use of phage-insensitive mutants contribute to preventing the negative consequences of phage infection (Coffey et al., 1998; Coffey and Ross, 2002; Sing and Klaenhammer, 1993) (Table 13.2).

13.6.1 Phage inactivation by physical treatments

The efficiency of physical and chemical treatments applied in the dairy industry for the inactivation of phages present in the dairy environment has been object of a number of reports (Atamer et al., 2012, 2013; Guglielmotti et al., 2012; Muller-Merbach et al., 2005a). Efficiency depends on phage susceptibility, phage initial load and suspension media. The susceptibility of LAB phages to thermal treatments usually applied to sanitize the raw milk and to assure the quality and the safety of dairy products varies between phages. Although the majority of dairy phages are known to be resistant to standard pasteurization procedures, the level of heat resistance is phage specific rather than species specific. Phages of *Lc. lactis* are known for their resistance to pasteurization (Deveau et al., 2006). Among the three *Lc. lactis* phage species commonly found

Table 13.2 Phage-control strategies in dairy plants.

Environment	• Adequate factory design (physical separation of processing areas, i.e. cheese manufacture, starter production, whey handling and storage systems) • Not using the same tools and equipment for milk and whey • Sanitation and disinfection of equipment, facilities, surfaces, etc. • Air control (filtration, physical or chemical disinfection)
Raw material, by-products and wastes	• Thermal treatments of raw materials • Thermal treatments of by-products before recycling • Not using by-products such as whey to rinse the vats
Starter cultures	• Starter culture rotations (with different phage sensitivity patterns) • Selection of phage-resistant starter cultures (phage-insensitive mutants, phage-resistant derivatives) • Aseptic propagation of starter cultures • Use of direct-to-vat concentrated starter cultures

in dairy plants, both the 936 and c2 species were more resistant than P335 phages to pasteurization conditions (i.e. 62 °C for 30 min or 72 °C for 15 s) (Madera et al., 2004). In a phage survey on virulent *Lc. lactis* phages collected in Germany, Atamer et al. (2009) found that phages of the 936 species are dominant, and 50% of them are thermo-resistant, surviving after a heat treatment at 80 °C for 5 min. The most resistant phage belonging to 936 phage species was still detectable in skim milk even at 90 °C for 20 min or 97 °C for 5 min. The D-value of this phage at pasteurization temperature of 72 °C was112 min. At 90–95 °C, 30–45 min were necessary to ensure 3-log inactivation, which is still insufficient considering that the concentration of lactococcal phages in raw milk can reach 10^4 PFU/ml and phages with an extremely high heat tolerance may survive at low numbers (Atamer et al., 2009). Also *Leuconostoc* phages show a phage-specific heat resistance. Heat inactivation experiments on dairy *Leuconostoc* phages showed that HTST pasteurization was not sufficient to ensure complete inactivation, and some phages also survived after treatment at 85 °C for 1 min (Atamer et al., 2011; Pujato et al., 2014). Some phages of *S. thermophilus* were rapidly inactivated at 63 °C, while others revealed high resistance to the same or even harsher treatments, and only heating at 90 °C for 5 min was effective for destruction of some autochthonous phages (Binetti and Reinheimer, 2000; Guglielmotti et al., 2012). In spite of the high heat sensitivity of some *Lactobacillus* phages, which showed 99% of inactivation after 2–3 min at 63 °C, more than 5 min at 90 °C were necessary for a complete inactivation of the most heat-resistant phages (Quiberoni et al., 2003; Capra et al., 2009; Guglielmotti et al., 2012). Studying thermo-resistant phages infecting LAB species, Capra et al. (2013) proposed a longer heating step at 90 °C (i.e. 45 min instead of 15 min) to ensure the absence of any phage particles in a sample.

The suspension media were also confirmed to affect the thermal susceptibility of phages. Milk seems to be protective to viral particles, as compared to media composed of either salt or carbohydrate solutions. The increased heat resistance in milk was mainly attributed to casein, while fat did not show a protective effect (Atamer et al., 2010; Guglielmotti et al., 2012). This is particularly important in recycling of milk by-products as cheese whey for production of whey protein concentrates. The concentration of whey by filtration techniques increases the risk of concentration of phages due to the possible retention of virions by the membranes. Moreover, whey proteins and other compounds may protect phages from thermal damage, thus increasing the risk of this recycling practice. Prior to recycling, an adequate heat treatment of by-products to reduce the risk of recontamination is recommended (Briggeler Marcò et al., 2012).

As stated previously, the spreading of by-products in the environment, as well as the movements of personnel or the transport of equipment or raw materials, increases the risk of aerosolization of phage particles. The generation of bioaerosols should be avoided, and the air must be filtered to remove dust particles carrying the phages. To inactivate phages suspended in the air, photocatalysis technology as alternative to the disinfection treatment by fumigation systems or UV light irradiation has been investigated. After photocatalysis exposure for 1 or 2 h, 6-log reductions of *Lactobacillus* and *Lc. lactis* phages were obtained (Briggeler Marcò et al., 2011; 2012).

Among non-thermal food preservation technologies, efficiency of high-pressure processing (i.e. HPH, high pressure homogenization, and HHP, high hydrostatic pressure) was demonstrated. Phages of different morphology, isometric vs. prolate, showed a different stability and resistance to high pressures (Moroni et al., 2002). Phage inactivation by HPH treatments were shown to be proportional to both applied pressure and number of passes. Phages of different LAB species suspended in skim milk were reduced from 2 to 6 logs after five successive passes at 100 MPa (Mercanti et al., 2012). By HHP treatments, isometric-headed *Lc. lactis* phages treated at up to 600 MPa showed a 5-log reduction after treatment for 2 h, whereas prolate-headed phages were less resistant (Muller-Merbach et al., 2005b). A non-thermal process based on membrane technology to eliminate bacteriophages from cheese whey was applied. Filtration studies were carried out with *Lc. lactis* phages to assess the efficiency on phage retention, avoiding interferences on whey protein permeation. Membranes with the cutoff of 300 kDa showed a significant phage retention (3.7 log units) with an acceptable whey protein permeation. To reduce or eliminate thermo-resistant phages in cheese whey, a combination of membrane filtration and thermal treatment was suggested (Atamer et al., 2013; Samtlebe et al., 2015).

13.6.2 Phage inactivation by chemical treatments

An essential requirement of biocides used in the food industry is the efficiency to inactivate the phages. Biocides are products and preparations containing disinfectants authorized for use and active on any harmful organism. The phage-killing efficacy of biocides is based on a reduction of at least 4 log units of a phage population after a given contact time in the presence of interfering material (Morin et al., 2015). Many studies for testing the effectiveness of sanitizers used by the dairy industries on phages infecting different LAB species were carried out. Peracetic acid-containing products are often the most effective for inactivation of phage particles. The strong oxidizing effects of peroxy acids, the low toxicity of their by-products and the minimal residues increased their use. In the conditions of European standard EN 13610:2003 (i.e. 15 min contact time at 20 °C with 1% of whey), 0.7 mmol/l (product concentration of 0.1%) of a commercial peracetic acid determined ≥4-log reduction of *Lc. lactis* P001 phage (Morin et al., 2015). With peracetic acid at concentrations applied in the dairy industry (about 0.15%), similar rates of inactivation were obtained for phages infecting *S. thermophilus*,

Lc. lactis and several dairy lactobacilli (Quiberoni et al., 1999, 2003; Suarez and Reinheimer 2002; Capra et al., 2004; Briggiler Marcó et al., 2009; Ebrecht et al., 2010). Murphy et al. (2014) showed the inactivation in less than 1 min of more than 99% of lactococcal 936-type phage particles by 0.015% peracetic acid at room temperature. Sodium hypochlorite is less efficient, and its effectiveness is phage dependent. Concentrations of 100–200 ppm of residual-free chlorine were enough to inactivate some phages of *L. delbrueckii*, *S. thermophilus*, *L. helveticus* and *Lc. lactis* (Guglielmotti et al., 2012). For other phages of *L. delbrueckii*, *Lc. lactis* and some phages of mesophilic lactobacilli, concentrations ≥400–800 ppm, not allowed in the food industry, were necessary for a total phage inactivation (Quiberoni et al., 2003; Capra et al., 2004). Also, the efficiency of quaternary ammonium-based biocide was phage dependent. The concentrations recommended by suppliers were effective against *L. paracasei* phages, while *L. delbrueckii* phages showed markedly different levels of resistance (Ebrecht et al., 2010; Mercanti et al., 2012). Ethanol and isopropanol, typically used for cleaning laboratory surfaces and utensils, showed a weak and industrially insufficient rate of phage inactivation (Guglielmotti et al., 2012). Among new disinfectants available for food factory sanitation, biocides at extreme pH, such as alkaline chloride foam or ethoxylated nonylphenol with phosphoric acid (pH values >12 and <2, respectively), were highly efficient (Briggiler Marcó et al., 2012).

13.6.3 Phage control by biological approaches

The first biological approach to counteract the proliferation of phages is the limitation of on-site preparation of bulk starter and the use of phage-inhibitory media designed to contain components (e.g. cation-scavenging compounds) that inhibit or delay phage adsorption and propagation. The replacement of bulk starters with concentrated starter cultures to be added directly to the cheese vat, associated with a rotation strategy, allow a decrease in the risk of phage propagation (Sturino and Klaenhammer, 2004). The rotation strategy, which is based on the alternated use of starter cultures composed of strains with different phage-resistance profiles but similar fermentative properties, is the basis for an efficient biological control of phage proliferation and a simple way to reduce fermentation failures and extend the longevity of strains in the dairy environment (Briggiler Marcó et al., 2012). Since this strategy is not suitable for all dairy manufacturing processes, the search for new pro-technological and phage-insensitive strains from raw materials (i.e. raw milk, natural starter cultures, or raw milk fermented dairy products) is a constant task for the dairy industry. Moreover, the ability of bacteria to face the rapid evolution of phage particles by developing various mechanisms to evade the infection and propagation of phages is exploited in the laboratory to select bacteriophage-insensitive mutants (BIMs). Several studies describe the isolation of spontaneous phage-resistant variants from sensitive strains of various LAB species (Neviani et al., 1992; Quiberoni et al., 1998; Guglielmotti et al., 2006; Binetti et al., 2007; Capra et al., 2011). The BIMs' selection strategy is exploited to increase the resistance in the phage sensitive starter strains, thus increasing the robustness of strain rotation (Mills et al., 2007; Chirico et al., 2014).

Phage resistance in dairy starter strains has attracted scientific interest worldwide since the early 1980s, when the first information on the genetic basis of this property in

LAB were acquired (Coffey and Ross, 2002). Among the defence mechanisms previously described (Section 13.3), the role of CRISPR/Cas systems in phage resistance is the most exciting breakthrough. The CRISPR/Cas system provides adaptive immunity against phages and invasive genetic elements. In *S. thermophilus* it has been demonstrated that the phage-resistance phenotype observed for the majority of the BIMs resulted from the integration of new spacers into the previously identified active CRISPR loci (Mills et al., 2010b; Barrangou and Horwath, 2012). A number of phage-resistant derivatives were also constructed by the use of recombinant DNA technology, but currently this is not an option for restrictive food legislation and consumer acceptance issues for genetically modified food ingredients. For improving the phage resistance of specific strains for industrial application, methods based on natural strategies (i.e. random mutagenesis, directed evolution and dominant selection) are applied (Derkx et al., 2014). Despite all these efforts to find and characterize new anti-phage mechanisms, the industrial use of phage-resistant bacteria will eventually lead to the emergence of phage mutants able to circumvent the resistance systems. Continuing research on anti-phage systems is still needed to stay one step ahead of phage evolution.

13.7 CONCLUSIONS

Despite significant progress made over the past decades to reduce the overall problem associated with phage contaminations, virulent LAB phages are still today a serious industrial concern, and dairy manufacturers are cyclically fighting against these viruses to keep them under control. The dairy industry has relied on an array of measures to control this natural phenomenon, including improved sanitation, process and starter production modifications, strain rotation, and the use of phage-resistant strains. In spite of these efforts, phages are rapidly co-evolving with the host according to a well-known biological mechanism, and new lytic variants are continuously emerging. Therefore, to avoid the battle against bacteriophages being lost, it is important that the search to find new control and antiviral strategies to keep pace with phage evolution has never stopped. In this regard, integrated phage biology approaches will help to further unravel phage diversity and phage-host interactions with the aim of improving the LAB strain selection process and developing bacterial strains for long-term industrial use in phage-contaminated environments, thus optimizing antiphage mechanisms in industrial practice.

References

Ali Y, Koberg S, Hessner S, Sun X, Rabe B, Back A, Neve H, Heller KJ. 2014. Temperate *Streptococcus thermophilus* phages expressing superinfection exclusion proteins of the Ltp type. *Front Microbiol* 5:98.

Ali Y, Kot W, Atamer Z, Hinrichs J, Vogensen FK, Heller KJ, et al. 2013. Classification of lytic bacteriophages attacking dairy *Leuconostoc* starter strains. *Appl Environ Microbiol* 79:3628–36.

Anderson B, Rashid MH, Carter C, Pasternack G, Rajanna C, Revazishvili T, Dean T, Senecal A, Sulakvelidze A. 2011. Enumeration of bacteriophage particles: comparative analysis of the traditional plaque assay and real-time QPCR- and nanosight-based assays. *Bacteriophage* 1:86–93.

Atamer Z, Ali Y, Neve H, Heller KJ, Hinrichs J. 2011. Thermal resistance of bacteriophages attacking flavour-producing dairy *Leuconostoc* starter cultures. *Int Dairy J* 21 327–34.

Atamer Z, Dietrich J, Müller-Merbach M, Neve H, Heller KJ, Hinrichs J. 2009. Screening for and characterization of *Lactococcus lactis* bacteriophages with high thermal resistance. *Int Dairy J* 19:228–35.

Atamer Z, Dietrich J, Neve H, Heller KJ, Hinrichs J. 2010. Influence of the suspension media on the thermal treatment of mesophilic lactococcal bacteriophages. *Int Dairy J* 20:408–14.

Atamer Z, Neve H, Heller K, Hinrichs J. 2012. Thermal resistance of bacteriophages in the dairy industry. In: Quiberoni AL, Reinheimer JA (Eds). *Bacteriophages in dairy processing*. New York: Nova Science Publishers, pp 195–214.

Atamer Z, Samtlebe M, Neve H, Heller KJ, Hinrichs J. 2013. Review: elimination of bacteriophages in whey and whey products. *Front Microbiol* 4:191.

Barrangou R, Horvath P. 2012. CRISPR: new horizons in phage resistance and strain identification. *Annu Rev Food Sci Technol* 3:19.1–19.20.

Binetti AG, Bailo NB, Reinheimer JA. 2007. Spontaneous phage-resistant mutants of *Streptococcus thermophilus*: isolation and technological characteristics. *Int Dairy J* 17:343–49.

Binetti AG, Reinheimer JA. 2000. Thermal and chemical inactivation of indigenous *Streptococcus thermophilus* bacteriophages isolated from Argentinian dairy plants. *J Food Prot* 63:509–15.

Binetti A, Reinheimer JA, Álvarez MA. 2008. Dairy bacteriophages and bacteriophage resistance. In: Mayo B, López P, Pérez-Martínez G (Eds). *Molecular aspects of lactic acid bacteria for traditional and new applications*: Kerala, India: Research Signpost, pp 209–34.

Blumenthal RM, Cheng X. 2002. Restriction-modification systems. In: Yasbin RE, Streips UN (Eds). *Modern microbial genetics*, 2nd ed. New York: Wiley, pp 177–225.

Bono LM, Gensel CL, Pfennig DW, Burch, CL. 2013. Competition and the origins of novelty: experimental evolution of niche-width expansion in a virus. *Biol Lett* 9:DOI: 10.1098/rsbl.2012.0616.

Briggiler Marcó M, De Antoni G, Reinheimer JA, Quiberoni A. 2009. Thermal, chemical and photocatalytic inactivation of *Lactobacillus plantarum* bacteriophages. *J Food Prot* 72:1012–19.

Briggiler Marcó M, Moineau S, Quiberoni A. 2012. Bacteriophages and dairy fermentations. *Bacteriophage* 2:149–58.

Briggiler Marcó M, Quiberoni AL, Negro AC, Reinheimer JA, Alfano OM. 2011. Evaluation of the photocatalytic inactivation efficiency of dairy bacteriophages. *Chem Eng J* 172:987–93.

Brussow H, Bruttin A, Desiére F, Lucchini S, Foley S. 1998. Molecular ecology and evolution of *Streptococcus thermophilus* bacteriophages. *Virus Genes* 16:95–109.

Bruttin A, DesièreF, Lucchini S, Foley S, Brussow H. 1997. Characterization of the lysogeny DNA module from the temperate *Streptococcus thermophilus* bateriophage Sfi21. *Virology* 233:136–48.

Buckling A, Brockhurst M. 2012. Bacteria–virus coevolution. *Adv Exp Med Biol* 751:347–70.

Canchaya C, Proux C, Fournous G, Bruttin A, Brüssow H. 2003. *Prophage genomics* Microbiol Mol Biol Rev 67:238–76.

Capra ML, Binetti AG, Mercanti DJ, Quiberoni A, Reinheimer, JA. 2009. Diversity among *Lactobacillus paracasei* phages isolated from a probiotic dairy product plant. *J Appl Microbiol* 107:1350–57.

Capra ML, Mercanti DJ, Rossetti LC, Reinheimer JA, Quiberoni A. 2011. Isolation and phenotypic characterization of *Lactobacillus casei* and *Lactobacillus paracasei* bacteriophage-resistant mutants. *J Appl Microbiol* 111:371–81.

Capra ML, Neve H, Sorati PC, Atamer Z, Hinrichs J, Heller KJ, Quiberoni A. 2013. Extreme thermal resistance of phages isolated from dairy samples: updating traditional phage detection methodologies. *Int Dairy J* 30:59–63.

Capra ML, Quiberoni A, Reinheimer JA. 2004. Thermal and chemical resistance of *Lactobacillus casei* and *Lactobacillus paracasei* bacteriophages. *Lett Appl Microbiol* 38:499–504.

Carminati D, Mazzucotelli L, Giraffa G, Neviani E. 1997. Incidence of inducible bacteriophage in *Lactobacillus helveticus* strains isolated from natural whey starter cultures. *J Dairy Sci* 80:1505–11.

Carminati D, Zago M, Giraffa G. 2012. Ecological aspects of phage contamination in natural whey and milk starters. In: Quiberoni A, Reinheimer NY (Eds). *Bacteriophages in Dairy Processing*, ch. 4, Hauppauge, NY: Nova Science Publishers, pp 79–98.

Chirico D, Gorla A, Verga V, Pedersen PD, Polgatti E, Cava A, Dal Bello F. 2014. Bacteriophage-insensitive mutants for high quality Crescenza manufacture. *Front Microbiol* 5:201.

Chopin M-C, Chopin A, Bidnenko E. 2005. Phage abortive infection in lactococci: variations on a theme. *Curr Opin Microbiol* 8:473–79.

Coffey A, Coakley M, McGarry A, Fitzgerald GF, Ross RP. 1998. Increasing phage resistance of cheese starters: a case study using *Lactococcus lactis* DPC4268. *Lett Appl Microbiol* 26:51–55.

Coffey A, Ross RP. 2002. Bacteriophage-resistance systems in dairy starter strains: molecular analysis to application. *Ant Van Leeuwenhoek* 82:303–321.

de los Reyes-Gavilan C, Linsowtin G, Sechaud L, Veaux M, Accolas JP. 1990. Evidence for a plasmid – linked to restriction – modification system in *Lactobacillus delbrueckii. Appl Environ Microbiol* 56:3412–19.

del Rio B, Binetti AG, Martin MC, Fernandez M, Magadan AH, Alvarez MA. 2007. Multiplex PCR for the detection and identification of dairy bacteriophages in milk. *Food Microbiol* 24:75–81.

del Rio B, Martin MC, Martinez N, Magadan AH, Alvarez MA. 2008. Multiplex fast real-time PCR for quantitative detection and identification of *cos-* and *pac*-type *Streptococcus thermophilus* bacteriophages. *Appl Environ Microbiol* 74:4779–81.

Derkx PMF, Janzen T, Sørensen KI, Christensen JE, Stuer-Lauridsen B, Johansen E. 2014. The art of strain improvement of industrial lactic acid bacteria without the use of recombinant DNA technology. *Microbial Cell Factories* 13 (Suppl 1):S5.

Deveau H, Garneau JE, Moineau S. 2010. CRISPR/Cas system and its role in phage-bacteria interactions. *Ann Rev Microbiol* 64:475–93.

Deveau H, Labrie SJ, Chopin M-C, Moineau S. 2006. Biodiversity and classification of lactococcal phages. *Appl Environ Microbiol* 72:4338–46.

Dupont K, Janzen T, Vogensen FK, Josephsen J, Stuer-Lauridsen B. 2004. Identification of *Lactococcus lactis* genes required for bacteriophage adsorption. *Appl Environ Microbiol* 70:5825–32.

Ebrecht AC, Guglielmotti DM, Tremmel G, Reinheimer JA, Suárez VB. 2010. Temperate and virulent *Lactobacillus delbrueckii* bacteriophages: comparison of their thermal and chemical resistance. *Food Microbiol* 27:515–20.

Emond E, Moineau S. 2007. Bacteriophages and food fermentations. In: McGrath S, van Sinderen (Eds.). *Bacteriophage genetics and molecular biology*, 1st ed. Norfolk: Horizon Scientific Press/Caister Academic Press, pp 93–124.

Forde A, Fitzgerald GF. 1999. Bacteriophage defense systems in lactic acid bacteria. *Ant Van Leeuwenhoek* 76:89–113.

Garneau JE, Moineau S. 2011. Bacteriophages of lactic acid bacteria and their impact on milk fermentations. *Microb Cell Fact* 10 (Suppl 1):S20.

Guglielmotti DM, Mercanti DJ, Reinheimer JA, Quiberoni AL. 2012. Review: efficiency of physical and chemical treatments on the inactivation of dairy bacteriophages. *Front Microbiol* 2:282.

Guglielmotti DM, Reinheimer JA, Binetti AG, Giraffa G, Carminati D, Quiberoni A. 2006. Characterization of spontaneous phage-resistant derivatives of *Lactobacillus delbrueckii* commercial strains. *Int J Food Microbiol* 111:126–33

Horvath P, Barrangou R. 2010. CRISPR/Cas, the immune system of bacteria and archaea. *Science* 327:167–70.

Horvath P, Coute-Monvoisin AC, Romero DA, Boyaval P, Fremaux C, Barrangou R. 2009. Comparative analysis of CRISPR loci in lactic acid bacteria genomes. *Int J Food Microbiol* 131:62–70.

IDF Bulletin N°129. 1980. Starters in the manufacture of cheese, 5–15.

IDF Bulletin N°263. 1991. Methods for phage monitoring, 29–39.

Josephsen J, Neve H. 1998. Bacteriophages and lactic acid bacteria. In: Salminen S, von Wright A (Eds.). *Lactic acid bacteria microbiology and functional aspects*. New York: Marcel Dekker Inc., pp 385–436.

Josephsen J, Neve H. 2004. Bacteriophage and antiphage mechanisms of lactic acid bacteria. In: In: Salminen S, von Wright A, Ouwehand A (Eds.). *Lactic acid bacteria: microbiological and functional aspect*, 3rd ed. New York: Marcel Dekker Inc., pp, 295–350.

Koskella B, Brockhurst M. 2014. Bacteria-phage coevolution as a driver of ecological and evolutionary processes in microbial communities. *FEMS Microbiol Rev* 385:916–31.

Labrie S, Moineau S. 2000. Multiplex PCR for detection and identification of lactococcal bacteriophages. *Appl Environ Microbiol* 66:987–94.

Labrie S, Moineau S. 2010. Bacteriophages in industrial food processing: incidence and control in industrial fermentation. In: Sabour PM, Griffiths MW (Ed.). *Bacteriophages, 1st*. Washington: ASM Press, pp, 199–216.

Labrie SJ, Moineau S. 2007. Abortive infection mechanisms and prophage sequences significantly influence the genetic makeup of emerging lytic lactococcal phages. *J Bacteriol* 189:1482–87.

Le Marrec C, van Sinderen D, Walsh L, Stanley E, Vlegels E, et al. 1997. Two groups of bacteriophages infecting *Streptococcus thermophilus* can be distinguished on the basis of mode of packaging and genetic determinants for major structural proteins. *Appl Environ Microbiol* 63:3246–53.

Ly-Chatain MH, Durand L, Rigobello V, Vera A, Demarigny Y. 2011. Direct quantitative detection and identification of lactococcal bacteriophages from milk and whey by Real-Time PCR: Application for the detection of Lactococcal bacteriophages in goat's raw milk whey in France. *Int J Microbiol* 59:43–51.

Madera C, Monjardín C, Suárez JE. 2004. Milk contamination and resistance to processing conditions determine the fate of *Lactococcus lactis* bacteriophages in dairies. *Appl Environ Microbiol* 70:7365–71.

Mahony J, Bottacini F, van Sinderen D, Fitzgerald GF. 2014. Progress in lactic acid bacterial phage research. *Microbial Cell Factories* 13 (Suppl 1):S1.

Mahony J, McGrath S, Fitzgerald G F, van Sinderen D. 2008. Identification and characterization of lactococcal-prophage-carried superinfection exclusion genes. *Appl Environ Microbiol* 74:6206–6215.

Mahony J, van Sinderen D. 2014. Current taxonomy of phages infecting lactic acid bacteria. *Front Microbiol* 5:1.7.

Martín MC, del Rio B, Martínez N, Magadán AH, Alvarez MA. 2008. Fast real-time polymerase chain reaction for quantitative detection of *Lactobacillus delbrueckii* bacteriophages in milk. *Food Microbiol* 25:978–82.

Mercanti DJ, Carminati D, Reinheimer JA, Quiberoni A. 2011. Widely distributed lysogeny in probiotic lactobacilli represents a potentially high risk for the fermentative dairy industry. *Int J Food Microbiol* 144:503–510.

Mercanti DJ, Guglielmotti DM, Patrignani F, Reinheimer JA, Quiberoni A. 2012. Resistance of two temperate *Lactobacillus paracasei* bacteriophages to high pressure homogenization, thermal treatments and chemical biocides of industrial application. *Food Microbiol* 29:99–104.

Michelsen O, Cuesta-Dominguez A, Albrechtsen B, Jensen PR. 2007. Detection of bacteriophage-infected cells of *Lactococcus lactis* by using flow cytometry. *Appl Environ Microbiol* 73:7575–81.

Mills S, Coffey A, McAuliffe OE, Meijer WC, Hafkamp B, Ross RP. 2007. Efficient method for generation of bacteriophage insensitive mutants of *Streptococcus thermophilus* yoghurt and mozzarella strains. *J Microbiol Meth* 70:159–64.

Mills S, Griffin C, Coffey A, Meijer WC, Hafkamp B, Ross RP. 2010a. CRISPR analysis of bacteriophage insensitive mutants BIMs of industrial *Streptococcus thermophilus* – implications for starter design. *J Appl Microbiol* 108:945–55.

Mills S, Griffin C, O'Sullivan O, Coffey A, McAuliffe O, et al. 2011. A new phage on the "Mozzarella" block: bacteriophage 5093 shares a low level of homology with other *Streptococcus thermophilus* phages. *Int Dairy J* 21:963–69.

Mills S, O'Sullivan O, Hill C, Fitzgerald G, Ross RP. 2010b. The changing face of dairy starter culture research: from genomics to economics. *Int J Dairy Technol* 63:149–70.

Moineau S, Lévesque C. 2005. Control of bacteriophages in industrial fermentations. In: Kutter R, Sulakvelidze (Eds.). *Bacteriophages: biology and applications*, ch. 10. Boca Raton: CRC Press, pp 285–96.

Morin T, Martin H, Soumet C, Fresnel R, Lamaudière S, Le Sauvage AL, Deleurme K, Maris P. 2015. Comparison of the virucidal efficacy of peracetic acid, potassium monopersulphate and sodium hypochlorite on bacteriophages P001 and MS2. *J Appl Microbiol* 119:655–65.

Moroni O, Jean J, Autret J, Fliss I. 2002. Inactivation of lactococcal bacteriophages in liquid media using dynamic high pressure. *Int Dairy J* 12:907–13.

Müller-Merbach M, Neve H, Hinrichs J. 2005. Kinetics of the thermal inactivation of the *Lactococcus lactis* bacteriophage P008. *J Dairy Res* 72:281–86.

Müller-Merbach M, Rauscher T, Hinrichs J. 2005. Inactivation of bacteriophages by thermal and high-pressure treatment. *Int Dairy J* 15:777–84.

Murphy J, Mahony J, Bonestroo M, Nauta A, van Sinderen D. 2014. Impact of thermal and biocidal treatments on lactococcal 936-type phages. *Int Dairy J* 34:56–61.

Neviani E, Carminati D, Giraffa G. 1992. Selection of some bacteriophages- and lysozyme-resistant variants of *Lactobacillus helveticus* CNRZ 892. *J Dairy Sci* 75:905–913.

Nyengaard N, Vogensen FK, Josephsen J. 1995. Restriction-modification systems in *Lactococcus lactis*. *Gene* 157:13–18.

Pujato SA, Guglielmotti DM, Ackermann H-W, Patrignani F, Lanciotti R, Reinheimer JA, Quiberoni A. 2014. *Leuconostoc* bacteriophages from blue cheese manufacture: long-term survival, resistance to thermal treatments, high pressure homogenization and chemical biocides of industrial application. *Int J Food Microbiol* 177:81–88.

Quiberoni A, Guglielmotti DM, Reinheimer JA. 2003. Inactivation of *Lactobacillus delbrueckii* bacteriophages by heat and biocides. *Int J Food Microbiol* 84:51–62.

Quiberoni A, Reinheimer JA, Tailliez P. 1998. Characterization of *Lactobacillus helveticus* phage-resistant mutants by RAPD fingerprints and phenotypic parameters. *Food Res Int* 31:537–42.

Quiberoni A, Suarez VB, Reinheimer JA. 1999. Inactivation of *Lactobacillus helveticus* bacteriophages by thermal and chemical treatments. *J Food Prot* 62:894–98.

Reinheimer JA, Quiberoni A, Tailliez P, Binetti AG, Suarez VB. 1996. The lactic acid microflora of natural whey starters used in Argentina for hard cheese production. *Int Dairy J* 6:869–79.

Rousseau GM, Moineau S. 2009. Evolution of *Lactococcus lactis* phages within a cheese factory. *Appl Environ Microbiol* 75:5336–44.

Samson JE, Belanger M, Moineau S. 2013. Effect of the abortive infection mechanism and type III toxin/antitoxin system *AbiQ* on the lytic cycle of *Lactococcus lactis* phages. *J Bacteriol* 195:3947–56.

Samtlebe M, Wagner N, Neve H, Heller KJ, Hinrichs J, Atamer Z. 2015. Application of a membrane technology to remove bacteriophages from whey. *In Dairy J* 48:38–45.

Sing WD, Klaenhammer TR. 1993. A strategy for rotation of different bacteriophage defences in a lactococcal single-strain starter culture system. *Appl Environ Microbiol* 59:365–72.

Sturino JM, Klaenhammer TR. 2004. Bacteriophage defense systems and strategies for lactic acid bacteria. *Adv Appl Microbiol* 56:331–78.

Suárez VB, Reinheimer JA. 2002. Effectiveness of thermal treatments and biocides in the inactivation of Argentinian *Lactococcus lactis* phages. *J Food Prot* 65:1756–59.

Suárez V, Zago M, Giraffa G, Reinheimer J, Quiberoni A. 2009. Evidence for the presence of restriction/modification systems in *Lactobacillus delbrueckii*. *J Dairy Res* 76:433–40.

Suárez V, Zago M, Quiberoni A, Carminati D, Giraffa G, Reinheimer J. 2008. Lysogeny in *Lactobacillus delbrueckii* strains and characterization of two new temperate prolate-headed bacteriophages. *J Appl Microbiol* 105:1402–11.

Suttle CA. 2007. Marine viruses — major players in the global ecosystem. *Nature Rev Microbiol* 5:801–812.

Trotter M, Ross RP, Fitzgerald GF, Coffey A. 2002. *Lactococcus lactis* DPC5598, a plasmid-free derivative of a commercial starter, provides a valuable alternative host for culture improvement studies. *J Appl Microbiol* 93:134–43.

Ventura M, Callegari ML, Morelli L. 1999. Surface layer variation affecting phage adsorption on seven *Lactobacillus helveticus* strains. *Annali Microbiol Enzimol* 49:45–53.

Verreault D, Gendron L, Rousseau GM, Veillette M, Massé D, Lindsley WG, Moineau S, Duchaine C. 2011. Detection of airborne lactococcal bacteriophages in cheese manufacturing plants. *Appl Environ Microbiol* 77:491–97.

Villion M, Chopin MC, Deveau H, Ehrlich SD, Moineau S, Chopin A. 2009. P087, a lactococcal phage with a morphogenesis module similar to an *Enterococcus faecalis* prophage. *Virology* 388:49–56.

Villion M, Moineau S. 2009. Bacteriophages of lactobacillus. *Front Biosci* 14:1661–83.

Zago M, Carminati D, Giraffa G. 2003. Characterization of *Streptococcus thermophilus* phages from cheese. *Ann Microbiol* 53:51–58.

Zago M, Rossetti L, Reinheimer J, Carminati D, Giraffa G. 2008. Detection and identification of *Lactobacillus helveticus* bacteriophages by PCR. *J Dairy Res* 75:196–201.

Zago M, Scaltriti E, Fornasari ME, Rivetti C, Grolli S, Giraffa G, Ramoni R, Carminati, D. 2012. Epifluorescence and atomic force microscopy: two innovative applications for studying phage-host interactions in *Lactobacillus helveticus*. *J Microbiol Methods* 88:41–46.

Zago M, Scaltriti E, Rossetti L, Guffanti A, Armiento A, Fornasari ME, Grolli S, Carminati D, Brini E, Pavan P, Felsani A, D'Urzo A, Moles A, Claude J, Grandori R, Ramoni R, Giraffa G. 2013. Characterization of the genome of the dairy *Lactobacillus helveticus* bacteriophage φAQ113. *Appl Environ Microbiol* 79:4712–18.

Zago M, Orrù L, Rossetti L, Lamontanara A, Fornasari ME, Bonvini B, Meucci A, Carminati D, Cattivelli L, Giraffa G. 2017. Survey on the phage resistance mechanisms displayed by a dairy Lactobacillus helveticus strain. *Food Micro* 66:110–16.

14 Lactic acid bacteria: a cell factory for delivering functional biomolecules in dairy products

Tiziana Silvetti, Stefano Morandi and Milena Brasca

ISPA-CNR, Institute of Sciences of Food Productions, Milano, Italy

14.1 INTRODUCTION

The functional food and nutraceuticals market is growing worldwide as a consequence of changing consumer behaviour and demands, coupled with scientific evidence for beneficial effects of particular ingredients and commercially driven interests to add value and competitiveness to existing products. Primarily, the ageing, health-conscious society desires food alternatives to increase health and well-being, preferably delivered in a conventional form. A food marketed as functional contains at least a component, whether it be a nutrient or not, that affects a target function of the organism in a specific, positive way, thereby generating a medical or physiological benefit beyond its basic nutritional functions. The term "nutraceutical", which was coined in 1989 as a contraction of the words "nutrition" and "pharmaceutical", refers to a food compound that not only supplements the diet but also aids in the prevention and/or treatment of disease and/or disorder (Burgess et al., 2006; Siró et al., 2008). The functional food and nutraceuticals sector will boom in coming years and become a major driver for the global food industry.

In this perspective, fermentation has been revalued as a pioneering approach for product innovation. Over the years, the practice of fermentation has expanded and improved to preserve and fortify the available food resources, providing nutritional and health-promoting attributes besides safety, texture and unique flavour in a natural and cost-effective way (Hugenholtz, 2013). Starter cultures with industrially important functionalities are continuously being developed. The diversity of bacterial species used in food fermentations has doubled in the past decade, increasing from 82 different species in 2002 to 195 in 2011 (Bourdichon et al., 2012). Ideally, a multifunctional strain is targeted or a combination of different strains endowed with multiple features is used (Ravyts et al., 2012). These cultures possess inherent, functional properties, aiming at improving the quality of the end product by offering sensory, organoleptic, technological, safety or nutritional and health advantages (Leroy and De Vuyst, 2004; Hati et al., 2013). Functional microorganisms and health benefits especially represent a binomial with great potential for functional foods through the indirect ingestion of

Microbiology in Dairy Processing: Challenges and Opportunities, First Edition.
Edited by Palmiro Poltronieri.

Published 2018 by John Wiley & Sons Ltd and the Institute of Food Technologists.

microbial metabolites synthesised during fermentation (biogenic effect) or the direct interactions of ingested live microorganisms with the host (probiotic effect) (Gobbetti et al., 2010). According to the Food and Agricultural Organization/World Health Organization definition, probiotics are live microorganisms that, when supplied in adequate amounts, may confer a health benefit on the host (FAO/WHO, 2006). The vitality and viability at high cell densities (at least 10^7 cfu g^{-1} of end product) is an essential requirement for probiotics (EFSA, 2010; Health Canada, 2009). Adhesion to the intestinal surface and the subsequent colonization of the human gastrointestinal tract are important prerequisites for probiotic action (Bermudez-Brito et al., 2013). Safety for human consumption and ability to reinforce the intestinal epithelial barrier and to modulate gut lymphoid tissue are other probiotic traits (Arena et al., 2014).

However, there is consistent evidence to support the immunoregulatory activity of dead probiotic cells (intact or broken). Biological responses, comprising immune modulation and carcinogen binding, have also been reported from crude cell extracts, fragments and cellular components, metabolites (cell homogenates, β-glucans, teichoic and lipoteichoic acids, peptidoglycans, lipopolysaccharides, extracellular phospho-polysaccharides and DNA/RNA) (Adams, 2010; Shigwedha et al., 2014).

Functional starter cultures are commonly represented by lactic acid bacteria (LAB), which have a long and safe history of application in a large variety of food fermentations, particularly in the dairy industry. In addition to *Bifidobacteria*, the most widely used probiotics include *Lactobacillus* and *Enterococcus* species. The use of probiotic LAB provides routes to deliver these microorganisms to the gut, where they can synthesise nutrients *in vivo*, besides secondary metabolites produced during fermentation and released directly in food. LAB represent ideal cell factories thanks to their generally limited biosynthetic capacity and metabolic versatility, their relative simple physiology, and their almost completely separated energy metabolism and biosynthesis processes (Burgess et al., 2006; Hati et al., 2013; Corbo et al., 2014; Tamang et al., 2016). In particular, LAB metabolism is chiefly geared towards production of a single metabolite, that is, lactic acid. Such a focused metabolism seems to be a perfect basis to make LAB efficient cell factories for the production of biomolecules and food ingredients, comprising macronutrients, micronutrients and non-nutritive compounds (Hugenholtz, 2008; Russo et al., 2012).

Within this scenario, the dairy industry is an excellent field to exploit the functional food market as confirmed by the recent, intense diversification of dairy products (Shortt and O'Brien, 2016). Milk and dairy products are a natural and highly nutritive part of a balanced daily diet, widely consumed from childhood to elder age, hence, their successful involvement in formulation of new lines with positive health impacts (Varzakas et al., 2010; González-Arrojo et al., 2016).

Nevertheless, the world of functional foods may still be considered an experimental research field showing some confirmed products, where microorganisms have a well-established capacity to markedly influence the quality and functionality, but with many other promising aspects to investigate. Dedicated research is essential in order to further clarify the relationship among starter cultures, food microbiota, metabolites, food matrix and processing conditions and to optimally implement functional starter in the existing production technology. Novel insights into the metabolism of LAB and quantitative data on their production offer perspectives for the application of a new generation of novel starter cultures and the exploitation of their full potential (Ravyts et al., 2012; Hati et al., 2013; Patel and Shah, 2016).

14.2 VITAMINS

Vitamins constitute a group of organic compounds required for supporting normal physiological functioning, but not endogenously synthesised by mammalian cells. Consequently, they have to be sequestered in small quantities from the diet. In total, 13 vitamins are essential for humans: four fat-soluble vitamins (A, D, E, K) and nine water-soluble vitamins, which comprise vitamin C and the B vitamins—thiamine (B1), riboflavin (B2), niacin (B3), pantothenic acid (B5), vitamin B6, folate (B9) and vitamin B12. Even though most vitamins occur in numerous foods and the required daily amounts are very small, vitamin deficiency in humans still persists because of insufficient food or unbalanced dietary intakes. The risk of vitamin deficiency exists in some population brackets, especially in the elderly due to their low food intake, in children for their limited variety of foods, and in pregnant women because of the critical role of folate in early embryonic development (LeBlanc et al., 2013; Rad et al., 2016).

The nutritional value of dairy products is indisputable and milk, cheese, and yoghurt have been recognized to contribute to the intake of many nutrients and vitamins. In the American diet, dairy consumption ensures more than 50% of total vitamin D and more than 25% of vitamin A and vitamin B12 (Rice et al., 2013). In a study aimed at determining if consumers have specific dietary and nutritional patterns, large consumers of fresh dairy products exhibited better adequacy in the dietary recommended allowances, including vitamins B1, B2, B5, B6, B9 (Lecerf et al., 2016). Dairy products may be a predominant source of dietary vitamin K2 in many regions of the world, too (Walther et al., 2013). Fermented milk products such as yoghurt naturally provide vitamins. Yet, vitamin content in yoghurt differs depending on the starter culture used, since fermentation microorganisms show a different ability to synthesise or destroy vitamins, even causing losses of these nutrients through their metabolism (Punnagaiarasi et al., 2016). In addition, dairy products emerged as appropriate vehicles for vitamin fortification using chemically synthesised pseudovitamins. Among the limited number of foods suitable for vitamin D fortification are dairy products; indeed, some yoghurt products, particularly those for children, currently contain added vitamin D (Bonjour et al., 2015; Leskauskaite et al., 2016). A consumer-friendly alternative to chemical fortification is represented by LAB bio-enrichment. The process of fermentation has been recognized as an easy, natural technology to raise vitamin levels in food and has prevailed over chemical synthesis due to economic and environmental issues. Numerous examples, especially involving dairy applications, have been proposed showing natural enrichment with riboflavin (del Valle et al., 2014), folate, vitamin B12, vitamin K2, and sometimes several of these vitamins, simultaneously (Hugenholtz, 2013; Thakur and Tomar, 2015). Secondary metabolism of LAB is targeted also to vitamin biosynthesis as these nutrients are essential cofactors in vital metabolic activities in LAB, such as the biosynthesis of amino acids and of nucleic acids (Hugenholtz, 2008).

There is copious scientific literature regarding the B-vitamin family produced by LAB in food. These micronutrients work closely in concert at a cellular level, acting as coenzymes in a vast array of catabolic and anabolic enzymatic reactions. Their collective effects are related to numerous aspects of brain function, including energy production, DNA/RNA synthesis/repair, genomic and non-genomic methylation, and the synthesis of numerous neurochemicals and signalling molecules. As water-soluble nutrients, they are generally safe at levels of consumption exceeding the recommended minimum consumption levels, except for folic acid (Kennedy, 2016). Dairy products have been recognized to contribute a small proportion to total vitamin K intake, hence vitamin K–producing bacteria in dairy fermentations represent effective tool for vitamin K supplementation in the food supply (Walther et al., 2013).

14.2.1 Vitamin B2 or Riboflavin

Riboflavin is the precursor of the coenzymes flavin mononucleotide (FMN) and flavin adenine dinucleotide (FAD), both of which act as hydrogen carriers in the transfer of electrons in oxidation-reduction reactions, functioning as cofactors for hundreds of FMN- or FAD-dependent enzymes called flavoproteins. Riboflavin deficiency, which persists in individuals whose diet is devoid of dairy and meat products, leads to impaired vision, reduced growth rate, increased levels of homocysteine with consequent cardiac risk, pre-eclampsia, oxidative stress and anaemia, liver and skin damage, and changes in cerebral glucose metabolism with symptoms such as hyperaemia, sore throat, oedema of oral and mucous membranes, cheilosis and glossitis. Milk and dairy products are well-known sources of riboflavin. However, its concentrations vary due to processing technologies and microbial activities during food processing. Bacterial requirement of this nutrient is uncommon, but riboflavin represents an essential growth factor for *Enterococcus faecalis*, *Streptococcus pyogenes*, *Listeria monocytogenes*, and some lactobacilli. Microbial biosynthesis from the precursors guanosine triphosphate (GTP) and D-ribulose 5-phosphate entails seven enzymatic steps. Most yoghurt starter cultures reduce riboflavin levels, while others can increase levels of this vitamin up to 160% of the concentration present in unfermented milk. The biosynthesis capability in LAB is species specific and/or strain specific. For instance, *Lactobacillus acidophilus* has been reported as a higher riboflavin producer compared with *Lactococcus lactis*. Beyond genetic engineering, riboflavin overproducing in riboflavin-consuming strains or in other strains that produce it at low level and are physiologically inactive can be achieved through exposure to chemical analogues: purine analogues and/or the toxic riboflavin analogue roseoflavin. Selection of natural riboflavin-overproducing strains of *Lc. lactis, Lactobacillus plantarum* and other food-grade microorganisms was pursued by putting the cells in contact with toxic concentrations of the flavin analogue roseoflavin. This method is unfailing and admissible from a consumer and regulatory perspective as it does not involve genetic engineering. Furthermore, possible elevated levels of the vitamin would not have any harmful effects as there is no consistent evidence of adverse health consequences in humans and no upper limit of intake has been established so far. Roseoflavin-resistant mutants of *Leuconostoc mesenteroides* overproduced up to 0.5 $mg\,l^{-1}$ of riboflavin, whereas riboflavin-overproducing *Lb. plantarum* was able to synthesise up to about 0.6 $mg\,l^{-1}$ (LeBlanc et al., 2013; Thakur and Tomar, 2015; Thakur et al., 2016).

14.2.2 Vitamin B9 or Folate

Folate is a general term for a large number of folic acid derivatives that differ by their state of oxidation, one-carbon substitution of the pteridine ring, and the number of glutamate residues. All these derivatives, called "vitamers", possess their own physico-chemical properties that may influence vitamin bioavailability.

The synthetic form of the vitamin is much more bioavailable than is natural food folate, and after ingestion, it is converted by dihydrofolate reductase to the partially reduced vitamer dihydrofolate and the fully reduced vitamer tetrahydrofolate form of folate, which corresponds to those that would arise from ingestion of natural folate (Sybesma et al., 2003; Bailey et al., 2015; Sabaude et al., 2016).

Folate is constituted by a pteridine moiety joined by a methylene bridge to para-aminobenzoic acid and linked to one or more molecules of L-glutamic acid. It serves as donors and acceptors of one-carbon units in a variety of reactions involved in the *de novo* biosynthesis of amino acids, purines and pyrimidines. Thus, it is involved in DNA synthesis, stability, and repair as well as alteration of DNA methylation, which is an important epigenetic determinant in gene expression, in the maintenance of DNA integrity, and in the development of mutations. Other folate functions include antioxidant properties which protect the genome against free radical damage. Folate deficiency implicates megaloblastic or macrocytic anaemia and augments the risk of congenital malformations such as neural tube defects, comprising spina bifida and anencephaly, in pregnancy. Low birth weight, preterm delivery, and fatal growth retardation are often related to folate deficiency during pregnancy, too (Bailey et al., 2015; Rad et al., 2016). This represents a health threat worldwide, both in developing and industrialized countries and concerns different age groups. Thus, daily folic acid supplementation in pregnant women is recommended. Yet, fortification programs have also been instituted, with obligatory flour fortification with folic acid in 63 countries and authorized in 4 countries in 2015.

Requirements for this vitamin are normally not satisfied through daily food intake, even by consuming folate-rich foods like liver and green leafy vegetables. Milk contains 20–50 μg/l of folate, while content in fermented milk reaches up to 150 μg/l (Divya and Nampoothiri, 2015). Being widely consumed (43% of total dairy products), fermented dairy products may be a useful means to increase folate levels. It has been demonstrated that the consumption of a portion (about 225 ml) of some fermented dairy products could cover 10–15% of daily recommended intake for adults. The possible side effects of chemically synthesised form of folate, the masking of B12 deficiency with consequent, permanent neurological damage, cause the need for choosing alternatives to fortification. To the contrary, natural folate produced by LAB has no adverse effects. The majority of LAB are unable to produce folate because they are missing the genes coding for enzymes necessary for folate biosynthesis. In this case, if precursors (para-aminobenzoic acid, guanosine triphosphate and glutamate) are available in culture medium or in food, many LAB can produce folate.

Folate biosynthesis is a highly variable and strain-dependent process. Growth kinetics and culture conditions play an important role in this metabolic activity of LAB. For instance, folate synthesis seems to occur during the exponential growth phase or at the beginning of the stationary phase; afterward, folate is consumed. To overcome this variability, the application of mixed LAB cultures seems to give best results compared to single cultures. Various combinations of strains have been tested. *Lb. acidophilus*, *Lb. plantarum*, *Lactobacillus delbrueckii* subsp. *bulgaricus*, and *Streptococcus thermophilus* are species active in folate biosynthesis. Among the *Lactobacillus* species, there are species which consume folate during growth. The mix of one *Lb. delbrueckii* subsp. *bulgaricus* strain with two *St. thermophilus* strains and the combination of strains of *Lc. lactis* subsp. *lactis* and *Lc. lactis* subsp. *cremoris* gave satisfactory results in milk and Domiati cheese, respectively. Through folate-producing starter cultures, Laiño et al. (2013) obtained in yoghurt an increase in folate content of about 250 and 150% in comparison to non-fermented milk and commercial yoghurt, respectively. Combinations of LAB with other folate-producing microorganisms such as yeasts or folate-producing probiotic bacteria may be an alternative strategy to enhance folate content in fermented milk. The main goal in this field is selecting strains, also in co-cultures, able to produce high amounts of folate, especially folate vitamers that are resistant to oxidation, acid pH, and heat treatments (Hati et al., 2013; Rad et al., 2016; Sabaude et al., 2016).

14.2.3 Vitamin B12 or cobalamin

Vitamin B12 is unique among vitamins. Firstly, it has the largest and most complicated structure. It consists of cobalt as the central atom and has a corrin ring that encloses the metal atom. The most frequent and active forms of B12 are adenosylcobalamin (also known as coenzyme B12) and methylcobalamin. Hydroxycobalamin and industrially produced cyanocobalamin are inactive forms of vitamin B12 that need to be metabolized in order to be used by humans. Moreover, no plant has the enzymes necessary for vitamin B12 synthesis, and it cannot be chemically synthesised. It can be industrially produced only through bacterial biosynthesis. Bacteria, mainly anaerobe, or archaea-bacteria synthesise vitamin B12 in the presence of cobalt. Microorganisms in the human gut are able to synthesise vitamin B12; nonetheless, humans are not able to absorb it since the location of synthesis (the colon) and absorption (the small intestine) are distant. Foods of ruminant origin represent excellent sources of vitamin B12 due to the production of cobalamin by their own gastrointestinal tract microbiota. Vitamin B12 combines important biological functions correlated to metabolism of fatty acids, amino acids, nucleic acids and carbohydrates. Methylcobalamin acts as an important carrier of methyl groups, and it is involved in the transformation of homocysteine to the metabolite methionine in the cytosol. Adenosylcobalamin acts as a cofactor of methylmalonyl-coenzyme A (CoA) mutase for the conversion of methylmalonyl-CoA to succinyl-CoA in the mitochondria, necessary for the catabolism of cholesterol, fatty acids and amino acids. Deficiency of this vitamin can potentially cause severe and irreversible damage, especially to the brain and nervous system, and haematopoietic (pernicious anaemia). A subclinical deficiency leads to symptoms such as fatigue, depression, and poor memory. Furthermore, this deficiency in male animal models influenced the number of offspring through growth retardation and decrease in blood parameters. Only food of animal origin contains adequate amount of vitamin B12, and milk and derivatives represent an excellent source.

Fermenting bacteria are often B12 consumers. The common starter cultures for yoghurt, *Lb. delbrueckii* subsp. *bulgaricus* and *St. thermophilus*, are very efficient vitamin B12 consumers for growth. A few bacteria have been found to produce vitamin

B12. Species of the food-grade propionibacteria such as *Propionibacterium shermani* showed vitamin B12 release in certain cheese varieties. Some members of the genus *Lactobacillus* have the ability to produce vitamin B12. In particular, a probiotic strain of *Lactobacillus reuteri* exhibiting hypocholesterolaemic activity in animals can produce B12, even if grown in B12-free cultivation medium. A corrinoid resembling B12, namely pseudovitamin B12, with biological activity analogue to vitamin B12, has been revealed, and a vitamin B12 biosynthesis gene cluster containing 25 genes has been identified and sequenced in this strain. *Lb. plantarum* and *Lactobacillus rossiae* strains isolated from diverse fermented foods have also been reported for B12 production (Hugenholtz, 2008; Patel et al., 2013; Gille and Schmid, 2015).

Among strategies screened for the selection of cobalamin-producing *Lactobacillus* strains is the B12-free growth condition. According to this hypothesis, only B12 producers could successfully thrive in cobalamin-deficient growth conditions. This theory stumbles, however, because of the presence of a number of B12-independent alternative metabolic pathways in bacteria as lactobacilli. In addition, high cobalt supplementation of growth medium was used for selection of potent B12-producing strains. Since cobalt has been supposed to generate oxidative stress in bacteria, it was assessed that only B12-producing strains have an edge to thrive owing to their ability to use more cobalt for vitamin bio-production.

Finally, the *cbiK* gene is one of the characteristic genes of anaerobic B12 biosynthesis as it codes for cobalt chelatase, which specifically incorporates cobalt ions as a metallic centre into tetrapyrrolic ring structure of cobalamin molecule (Bhushan et al., 2016). Enhanced levels of vitamin B12 in fermented milk have been reported as a consequence of the addition of prebiotic compounds lactulose and inulin at different concentrations during milk fermentation by *Bifidobacterium lactis* (Patel et al., 2013).

14.2.4 Vitamin K: menaquinone

Vitamin K is an essential fat-soluble vitamin existing in multiple dietary forms: the predominant dietary compound is phylloquinone, also known as vitamin K1, and is present in green leafy vegetables and their oils, whereas menaquinones, commonly called vitamin K2, are produced by intestinal bacteria, especially belonging to the genera *Lactococcus, Lactobacillus, Enterococcus, Leuconostoc* and *Streptococcus*, and found in much lower amounts in meat, dairy and fermented food products. There are, in turn, multiple forms of menaquinones, which, although clustered under the term vitamin K2, are different in origin and function. Vitamin K acts as an enzyme cofactor for γ-carboxylation of peptide-bound glutamate residues, thus proven to contribute to normal coagulation. Active roles in bone and kidney function, tissue calcification, cardiovascular and metabolic health was claimed, too. Its deficiency implicates clinical disorders like intracranial haemorrhage in newborn infants and possible bone fracture as a consequence of osteoporosis. Upper levels of menaquinone intake have not been defined currently since ingestion of large amount of menaquinone has not be shown to cause toxicity. In prokaryotes, menaquinones contribute to respiratory electron transport chains and serve as electron carriers in the cytoplasmic membrane. They play a role in the active transport of molecules across the cell membrane and in *Bacillus subtilis* sporulation. Reduced forms of menaquinones also exhibit antioxidant activities against lipid oxidation of cellular membranes. This is of utmost importance for industrial applications, considering that respiration is more efficient in energy production than fermentation. In fact, the addition of heme, oxygen and menaquinones to the culture media for LAB determines aerobic growth, and, consequently, higher biomass, lower acid production, greater amounts of minor end products, and the best response to the stressors encountered during processing. The capability of microbial biosynthesis of vitamin K is restricted to some naphthoquinones, which share the same 2-methyl-1,4-naphthoquinone ring (also known as menadione or vitamin K3), but diverge in the length of an isoprenoid side chain attached at the 3-position, generally ranging from 5 to 13 prenyl units (each having five carbons) in length. The metabolic pathway used by LAB consists of a series of enzymatic reactions by means of peptides encoded by *men* genes: the napthoquinone ring is first synthesised from chorismate derived from the shikimate pathway. The isoprenoid side chain is synthesised separately and joined to the napthoquinone ring to form dimethylmenaquinone, which is subsequently methylated to complete menaquinone biosynthesis. In general, Gram-positive bacteria like LAB primarily produce menaquinone; Gram-negative bacteria produce dimethylmenaquinone and ubiquinones. The concentrations and forms of menaquinone found in fermented foods depend on the specific bacterial strains used and production conditions during fermentation (pH, temperature, duration). Actually, some genera of obligate fermentative starters widely used in the dairy industry, such as *Lactobacillus* and *Streptococcus*, have lost this ability. Other genera, on the contrary, stand out as important sources of dietary vitamin K. High concentrations of menaquinone-8 and menaquinone-9 detected in Edam-type cheese derive from the addition as starter cultures of *Lc. lactis* subsp. *lactis* and *Lc. lactis* subsp. *cremoris* endowed with menaquinone-8 and menaquinone-9 synthesis capability. These really represent the most common menaquinones of microbial origin found in fermented dairy products. Other *Lactococcus* and *Leuconostoc* strains were revealed to produce long-chain menaquinones, more than 230 nmol/g of dried cells when grown in a synthetic medium, and up to 123 μg/l when cultured in reconstituted nonfat dry milk. In addition, a strain of faecal *Ent. faecium* was able to produce various forms of menaquinones. Unfortunately, most of bacterially synthesized menaquinones remain bound to the cell membrane and therefore are not accessible for absorption (Patel et al., 2013; Walther et al., 2013).

14.2.5 Other B-group vitamins

Higher contents of niacin, biotin, and pantothenic acid were observed in cultured dairy products compared to milk as a result of the biological activities of LAB (Hou et al., 2000; Le Blanc et al., 2013).

14.3 MINERALS

Like vitamins, minerals are micronutrients, too, that is, one of the five essential groups of nutrients necessary to sustain life. Regrettably, the mineral and trace element content in food is affected by the depletion in soil resulting from the increase in food production. Hence, prevalence of micronutrient malnutrition, also referred to as "hidden hunger", is a widespread public health concern, affecting more than a half of the global population (Mogna et al., 2012; Mrvčić at al., 2012). Since trace elements positively influence the physiological state of all organisms, the relevance of their adequate dietary intake or supplementation becomes obvious. On the other hand, LAB interaction with metal ions has also been proved successful in detoxification processes of food and water contaminated by heavy metals. Heavy metals are more and more present in food owing to the increasing environmental pollution, and their ingestion can lead to toxic effects, even at lower levels of exposure. In the body, they readily form complexes and accumulate in the tissues. After contact with the cells, they damage the cell membrane and alter its permeability. Within this context, LAB offer a strategy for absorption of essential or toxic metal ions with improvement of food safety and quality (Mrvčić et al., 2012). In fact, among the multidimensional functionalities owned by LAB, the ability to bind, uptake and biotransform metal ions from the substrate implies physiological, technological and nutritional added values for both LAB and humans. Several studies indicate that physiochemical properties of bacteria enriched with selected metals modify, improving their probiotic and health-promoting attributes (Pophaly et al., 2014). LAB are able to accumulate and biotransform inorganic metals into organic (metal-protein complexes) and elemental forms, which are much better absorbed by the body than if they are taken in the inorganic form. Se and Cu are the main essential metal ions tested in this research field. These selectively enriched biomasses have been developed as dietary supplements. In bacteria, getting trace elements in the organic form seems to follow two major routes: their binding to the microbial biomass or biotransformation during the food fermentation. The metal ion binding can happen through two different mechanisms: the metabolically passive binding of metal ions to the LAB cell wall and extracellular polysaccharides (biosorption) or the metabolically active process in which metal ions enter the cell membrane and accumulate into the cytosol (bioaccumulation). The high surface-area-to-volume ratio of bacteria enhances this process of accumulation because it makes available a large area for heavy metals to interact with microbial cell wall, making uptake of nutrients very efficient. Furthermore, the microbial cell envelope with a net negative charge makes microbes more prone to accumulating metal cations without differentiating between essential trace elements and toxic heavy metals (Gall et al., 2015).

Biosorption is a fast and reversible process, performed by both living and inactive cells, that is useful in exploiting microbial nutraceutical production and foodstuffs detoxification. Research about the binding of metal ions dependent on LAB metabolism mostly refers to the essential metal Se. A tiny body of research deals with other metals as Zn, Cr, Cu, Mn and Fe (Mrvčić et al., 2012). Se is a vital trace element for all higher eukaryotes as well as for some prokaryotes. In humans it plays an important antioxidant role, acts as chemopreventive agent for some cancers (prostate, lung and colon) and exhibits preventive effects against infections, pregnancy complications, male infertility and cardiovascular diseases, while in prokaryotes it serves as growth stimulator. Plants, in particular cereals and nuts, are the primary sources of Se, originally present in soil, while content in milk is generally too low to cover the recommended daily intake (RDA), that is, 20–100 μg/day (González-Arrojo et al., 2016). In southern and eastern Europe, Se intake does not fulfil this RDA. Se biological and toxicological effects are correlated to the type and concentration of Se species. Toxic reactions can rapidly progress, with chronic exposure leading to selenosis. However, the probability of Se deficiency exceeds that of toxicity. This is partially attributable to the Se bioavailability that is lower for tetravalent dietary Se. The microbial detoxification converts inorganic Se, potentially toxic and poorly bioavailable, to elemental selenium [Se(0)] and organic forms like selenocysteine (SeCys), selenomethionine (SeMet), and methylselenocysteine (MeSeCys). These amino acids could be randomly inserted during protein synthesis generating selenylated proteins. SeCys incorporated into proteins through the so-called "Selenocysteine insertion assembly" ultimately generates selenoproteins (glutathione peroxidase, thioredoxin reductase, iodothyronine deiodinase and selenoprotein P), endowed with antioxidant protection activity in all organisms (Mrvčić et al., 2012; Pophaly et al., 2014; González-Arrojo et al., 2016). Moreover, like peptides and proteins, these Se–protein complexes undergo absorption in the small intestine, where micronutrients can penetrate the intestinal wall. A variety of Se-rich products has been launched on the market, including fruits and vegetables cultured in Se-enriched soil (Kurek et al., 2016). Among dairy products, yoghurt and kefir were successfully enriched with Se thanks to the fermentation of Se-enriched milk by *Lactobacillus* strains and (only for kefir) also yeasts. Especially interesting are studies on application of Se-enriched lactobacilli in yoghurt production, since fermentation originated exclusively the most bioaccessible amino acids SeCys and methylSeCys (MeSeCys), which were also stable for at least 15 days (Alzate et al., 2008). A further work regarding Se-enriched yoghurt proved the relevant role of LAB in the incorporation of Se into proteins. Remarkably, the detection of chaperones in the controls, but not in the presence of Se, possibly indicates that Se might affect chaperone expression in *Lactobacillus* and consequently decrease the stress factor affecting this genus. In fact, chaperones are the key proteins in the early response to acid and heat shock in *Lactobacillus* (Palomo et al., 2014). This ability of Se-biotransformation has been detected in strains of *Lb. plantarum*, *Lb. delbrueckii* subsp. *bulgaricus*, *Lactobacillus casei*, *Lactobacillus paracasei*, *Lactobacillus rhamnosus* LB3, *Lactobacillus fermentum* LB7, *Lb. reuteri*, *Lactobacillus brevis*, *Lactobacillus sanfranciscensis* and *Lactobacillus buchneri* (Pophaly et al., 2014; Kurek et al., 2016). Some researchers studying behaviour of Se during yoghurt production discovered that LAB are able to reduce the selenite in toxic concentrations and excrete Se intracellularly into nano-sized, elemental Se spheres for its own detoxification purposes. These are relatively regular and uniform-sized, high-purity Se nanospheres that may have functional implications in the food industry when produced by food-grade microorganisms, such as *Lb. delbrueckii* subsp. *bulgaricus*, *Lb. acidophilus*, *St. thermophilus*, *Lb. casei*, *Lb. rhamnosus* and some Bifidobacteria. The size is genus-dependent, ranging as follows: 100–200 nm for *Lactobacillus* spp., 400–500 nm for *Bifidobacterium* spp., and 50–100 nm for *St. thermophilus*. A suspension or a powder containing Se nano-spheres possesses unique characteristics; a relevant one is that nano size helps the better absorption of nutritives into plants, animals, humans and microorganisms. A preliminary example of a product containing Se nano-particles is lactoselenium (LactoMicroSel), which is produced during the yoghurt-making process and might be used for safer Se supplementation in the future (Eszenyi et al., 2011). Evidence of accumulation and enhanced bioavailability of minerals such as Cu, Mn, Zn and Fe by LAB emerged, too (Mrvčić et al., 2012). Mrvčić and colleagues (2011) investigated the influence of Zn, Cu and Mn on the growth, acid production and binding capacity in strains of *Ln. mesenteroides*, *Lb. brevis* and *Lb. plantarum*. While *Ln. mesenteroides* was the most efficient in Zn-binding processes, *Lb. plantarum* accumulated high concentrations of Mn. *Lb. brevis* was the most Cu resistant and internalized high Cu levels. The influence of Cu ions on LAB is of great importance for cheeses produced in copper vats, where Cu ions can leach into milk and negatively influence LAB growth and acid production. In some LAB strains, Cu homeostasis, affected by the *cop* operon, enables resistance under copper-limiting conditions, as seen for *Enterococcus hirae*, which was able to grow in up to 8 mM Cu.

14.4 BIOACTIVE COMPOUNDS

Numerous definitions have been given for bioactive compounds, and one of the most appropriate is: components of consumption-ready foods which may exert a regulatory activity in the human organism, irrespective of their nutritive functions (Diplock et al., 2000). Milk of all different mammalian species contains a myriad of serum and glandular proteinaceous derivatives which are heterogeneous in molecular size, concentration and functionality as a result of the diverse milk composition (Park et al., 2015). Many of these milk-borne components lack activity when protein encrypted, but acquire biological effects when proteolytically released as a consequence of human digestive proteases and peptidases or hydrolysis by proteolytic microorganisms or activity of the pool of microbial enzymes. Actually, LAB fermentation contributes to enrich the diet with these bioactive components having several phenotypic effects after absorption in the gut and accumulation in the peripheral organs. In dairy products, LAB proteolytic system is very efficient in liberating peptides with a broad spectrum of biological activities from milk proteic matrix made up of alpha- and beta-caseins, albumin and globulin. In this modulation of body physiology, nitrogen-bearing molecules such as amino acids, amino acid derivatives and oligopeptides are the key players. They are able to control nutrition (mineral absorption and oxidative stress protection), metabolism (blood glucose and cholesterol lowering), cardiovascular function (antithrombotic and hypotensive action), infection (microbial inhibition and immunomodulation) and gut–brain axis (opioids and anti-opioids controlling mood and food intake) in humans (Pessione and Cirrincione, 2016). The biosynthetic abilities of LAB are very limited, especially in amino acid synthesis, thus leading LAB to develop a sophisticated proteolytic system that allows them to get amino acids from the proteins present in the environment. In general, lactobacilli possess a larger number of proteases, peptidases, and amino acid permeases than lactococci. Extracellular protein hydrolysis by a cell-envelope proteinase generates oligopeptides that are subsequently taken up by cells *via* specific transport systems and undergo further degradation into shorter peptides and amino acids in the cytoplasm. Only occasionally, the longer peptides obtained in the first hydrolysis step show biological activity. The most bioactive peptides derive from cell autolysis, which is a rare event because of short milk fermentation time. Actually, the cell-envelope proteinases from *Lb. delbrueckii* subsp. *lactis* are suitable to release both metal-chelating and anti-hypertensive peptides directly from caseins, independently from prior cell autolysis. Milk whey proteins (α-lactalbumin, β-lactoglobulin, lactoferrin, and immunoglobulins) are other interesting sources of bioactive peptides displaying a greater cholesterol-lowering activity than those from casein (Choi et al., 2016; Pessione and Cirrincione, 2016).

The study of bioactive compounds is difficult in milk because of the low content and the biochemical complexity of active peptides, even those endowed with a very strong action, the influence of components of the system and the lack of techniques for their quantification (Park et al., 2015). Differently, their presence has been detected and documented in different cheeses (Choi et al., 2012). The discovery of bioactive peptides has experienced recent, exponential progress. An emerging field called nutritional proteomics and a number of analytical techniques are now available to detect and identify various types of analytes over a wide range of concentrations, also in complex protein mixtures. The combination of shotgun proteomics and high-performance liquid chromatography plus high resolution mass spectrometry seems to be the most efficient approach applied so far (Capriotti et al., 2016).

14.4.1 Anti-hypertensive peptides

The renin–angiotensin system is the principal metabolic pathway in the regulation of arterial blood pressure *via* contraction of smooth muscles of the blood vessel. The angiotensin-converting enzyme (ACE) is a key factor in this process, as it converts angiotensin I into angiotensin II, a powerful vasoconstrictor, and inactivates the vasodilator peptides bradykinin and kallidin. The inhibition of ACE promotes a vasodilation response, which lowers blood pressure. Fermented milk and cheese enclose ACE inhibitory substances useful in the prevention and treatment of hypertension in a non-invasive manner as an alternative to pharmacological therapies. Anti-hypertensive peptides derived from milk proteins are the most investigated, although their actual physiological mechanism remains unclear and evidence of benefits is still too scarce (Beltrán-Barrientos et al., 2016). Nonetheless, the anti-hypertensive capacity of milk-borne peptides has been extensively demonstrated in the rat model and human studies. Particularly, *Lb. helveticus* strains are recognized for their production of the most hypotensive peptides, the best known of which are tripeptides Val-Pro-Pro (VPP) and Ile-Pro-Pro (IPP) (Bhat et al., 2016). *Lb. helveticus* strains are known for their fastidious nutritional requirements as they have multiple amino acid auxothrophies. When grown in milk, *Lb. helveticus* relies on its very potent proteolytic system that allows this species to break down more casein in culture media than other LAB species (Griffiths and Tellez, 2013). VPP and IPP peptides were found in fermented milk by *Lb. helveticus* CP790 in combination with *Saccharomyces cerevisiae*. These peptides decreased the systolic blood pressure of hypertensive rats after 6–8 h of administration. Moreover, they were resistant to digestive enzymes and underwent direct absorption in the gut. Using *Lb. helveticus* CHCC637 and *Lb. helveticus* CHCC641, fermented milk rich in ACE-inhibitors (ACEI) with a clinically relevant evidence was obtained (Choi et al., 2012). Fermented milk contained ACEI peptides other than VPP and IPP; for instance, Gly-Thr-Trp and Gly-Val-Trp revealed strong anti-hypertensive effects in an animal model study using spontaneously hypertensive rats. A tetrapeptide (Tyr-Gly-Leu-Phe) called α-lactorphin was proved to reduce blood pressure dose-dependently and possesses opioid-like activity (Beltrán-Barrientos et al., 2016). Novel ACE inhibitory peptides were identified in milk fermented with *Ent. faecalis* strains isolated from raw milk (Bhat et al., 2015), but also milk fermented by *Lb. delbrueckii* subsp. *bulgaricus* SS1 presented active fractions, which showed resistance to proteolysis either during dairy processing or by trypsin and chymotrypsin. Low-molecular-weight ACE inhibitory peptides were also reported in ripened cheeses, where their effects last until degradation to inactive fragments during ripening process. This negative impact of ripening on the ACE inhibitory activity was detected both in Gouda and Parmesan cheese, while other Italian cheeses characterised by short (Crescenza and Italico) and medium (Gorgonzola) ripening periods contained ACE inhibitory active compounds (Choi et al., 2012). Another study evaluated ACEI activity during refrigerated storage in yoghurt containing additional probiotic strains (*Lb. acidophilus* L10, *Lb. casei* L26 or *Bif. lactis* B94) besides the yoghurt culture, and evidenced that all probiotic yoghurts possessed greater ACEI activity during the initial stage of storage (first 3 weeks); however, activity decreased afterward. Thus, ACEI stability during shelf life is a critical factor in the development of functional fermented milk (Beltrán-Barrientos et al., 2016).

14.4.2 Antioxidative peptides

Among bioactive peptides decrypted by the proteolytic activity of LAB there are peptides exhibiting antioxidative activity, as those described in LAB-fermented milk whey and derived from bovine casein, mainly $\alpha_{s\text{-}}$casein. Hydrolysates attributed with antioxidant activity exert free radical scavengers, non-radical oxygen quenchers or metal chelators, but they also can eliminate radical precursors, hence controlling reactive oxygen species damage to cells and tissues. They act also as inhibitors of enzymatic and non-enzymatic peroxidation of fatty acids, thus preventing membrane lipid peroxidation (Beermann and Hartung, 2013; Pessione and Cirrincione, 2016). Oxidative processes are undesirable not only for their health consequences, but also for the deterioration of food quality. The formation of free radicals causes rancidity, unacceptable taste and reduction of shelf life, while the consumption of foods containing oxidized lipids has been correlated to cancers, diabetes and cardiovascular diseases (Bhat et al., 2015). The effectiveness in antioxidant activity relies on the amino acid sequence. It is generally accepted that the sulfur-containing and aromatic amino acids play important roles in the antioxidant activity. Antioxidant properties have been demonstrated for cysteine, methionine, tyrosine, tryptophan, histidine, and lysine. Tyrosine and tryptophan behave as antioxidant molecules since their indolic and phenolic groups can act as hydrogen donors, in particular when tyrosine and tryptophan are in the C-terminus. Basic amino acids can chelate metallic ions, and cysteine acts as a proton donor thanks to its thiol group. Leucine or proline residues to the N-terminus of a His-His dipeptide enhance antioxidant activity and facilitate synergy with synthetic antioxidants, like butylated hydroxyanisole, commonly used in the food industry (Beermann and Hartung, 2013; Pessione and Cirrincione, 2016). A κ-casein-derived peptide from milk fermented with *Lb. delbrueckii* subsp. *bulgaricus*, the peptide SKVLPVPQ present in Spanish milk beverages fermented using *Lb. helveticus* in combination with *S. cerevisiae* or the peptide PYVRYL from ovine α_{S2}-casein all exhibited oxygen radical absorbance capacity (Choi et al., 2012). As with many other LAB bioactivities, the radical scavenging activity is a highly species- and strain-specific feature. Beside *Lb. helveticus*, *Lactobacillus* strains, particularly *Lb. rhamnosus*, *Lb. casei* subsp. *casei* and *Lb. paracasei* subsp. *paracasei*, have been shown to be promising for developing functional foods to overcome oxidative stress (Ramesh et al., 2012; Namdari and Nejati, 2016). Milk-protein hydrolysates also act as potential anti-ageing agent. The dietary intake of *Lb. helveticus* fermented milk whey has been found to promote the epidermal barrier function and prevent dermal ageing processes and the onset of dermatitis in hairless mouse skin exposed to sodium dodecyl sulfate (Beermann and Hartung, 2013). Whey protein contribution in nullifying the negative effects of the oxidative stress through the supplying of bioactive molecules is substantial (Patel, 2015). Glutamylcysteine found in high levels in whey proteins subjected to a low temperature process promotes the synthesis of glutathione (GSH), a potent intracellular antioxidant with a key role in cellular detoxification, having the ability to bind heavy metals and other toxins and eliminate them from the body (Mrvčić et al., 2012; Park et al., 2015). Most of GSH biological functions are based on the activity of two enzymes: glutathione peroxidase, which converts reduced GSH to its oxidized form (GSSG), and glutathione reductase, which transforms GSSK back to GSH. The ratio GSH/GSSG permits the maintenance of the intracellular redox homeostasis on which the oxidative status of the cells depends. GSH is a thiol tripeptide made of the three amino acids, glutamate, cysteine and glycine. It is ubiquitous in eukaryotes, and in humans it is vital for immune system and health maintenance. As a result of low glutathione level in the body, a number of disease states such as cancer, AIDS, Alzheimer's disease and Parkinson's disease can arise. GSH is widely distributed in Gram-negative bacteria, while some members of Gram-positive bacteria are suggested either to synthesize or import it from the medium. In particular, some LAB species have evolved metabolic mechanisms to endure the stress conditions they commonly bump into in particular niches. GSH synthesis has been confirmed in strains belonging to the genus *Streptococcus,* where synthesis is carried out by a multidomain bifunctional fusion protein (GshF or GshAB) in which the N-terminal domain is responsible for γ-glutamylcysteine synthetase activity and the C-terminal domain carries glutathione synthetase activity. Similar putative fusion proteins have been detected in *St. thermophilus, Streptococcus suis, Streptococcus sanguinis, Streptococcus mutans, Ent. faecalis* and *Ent. faecium.* Among lactobacilli, GshF homologs have been described in *Lb. casei, Lb. rhamnosus, Lb. plantarum, Lactobacillus sakei* and *Lactobacillus ruminis*; however, GSH synthesis has not been established in these organisms. The probiotic *Lb. fermentum* ME-3 was recently reported to have a complete glutathione system able to perform GSH synthesis, uptake and redox turnover. Low GSH accumulation is observed in *Lb. helveticus* and *Lb. acidophilus.*

During industrial processing, unfavourable conditions can affect strain performance and viability of the starters. The stresses encountered by LAB because of industrial treatments and co-culturing in fermented foods or during transit through the gastrointestinal environment include oxidative, cold, acidic and osmotic stress. GSH supplementation guarantees multi-stress tolerance for LAB strains. Such a property is crucial for commercial starters prepared using treatments that load the cells with stress conditions or probiotic strains that need to withstand hostile conditions in the gut. GSH should also serve to deliver added functionalities like flavour generation and promote the growth of starter bacteria since it acts as sulfur and nutrient source. Supplemented GSH has also been reported to accelerate the ripening process even if the cellular reactions involved have not been ascertained. GSH seems to enhance the metabolic activities in cheese starters such as those of esterase and lipase that directly impact on the flavour development, hence accelerating ripening process. Addition of GSH increases the production of H_2S and methanethiol, both important contributors to the overall cheese flavour. Moreover, it helps maintain a low redox potential favourable for ripening process, too. In addition, protection offered by GSH against stress conditions, which would lead to higher viability of metabolically active starter bacteria in cheese, is a complementary factor in early ripening. In conclusion, dairy products supply all the necessary precursors for GSH synthesis, thus making LAB supported for glutathione biosynthesis and useful for delivering this molecule in the human system. Possessing unique properties and functions, glutathione can add a new dimension to the functional applications of LAB (Pophaly et al., 2012).

14.4.3 Bioactive amines

Bioactive amines—gamma-aminobutyric acid (GABA), β-phenylethylamine and tryptamine—are released by both eukaryotes and LAB *via* decarboxylation of their precursor amino acid glutamate, phenylalanine and tryptophan, respectively. LAB exploit amino acid decarboxylation as strategy for providing supplementary energy to cells as well as a defence system against environmental acidity, since the decarboxylated compounds are less acidic than the substrate (Mazzoli et al., 2014).

In recent years, various studies have been focused on bioactive substances derived from foods. Among the definitions given for bioactive compounds, one of the most appropriate could be components of consumption-ready foods which may exert a regulatory activity in the human organism, irrespective of their nutritive functions (Diplock et al., 2000).

GABA is a four-carbon, non-protein amino acid that is widely distributed in nature, from microorganisms to plants and animals (Ueno, 2000). GABA possesses well-known physiological functions such as neurotransmission, induction of hypotension and diuretic and tranquilizer effects (Jakobs et al., 1993). In addition, this compound promotes secretion of insulin by the pancreas and may prevent diabetic conditions (Adeghate and Ponery, 2002; Hiraga et al., 2008) and recent research has shown that it can enhance the level of plasma growth hormone and the rate of protein synthesis in the brain (Tujioka et al., 2009), and possibly contribute to the inhibition of certain lung adenocarcinomas (Schuller et al., 2008). GABA has also shown beneficial effects in alcoholism treatment, depression, and stimulation of immune cells (Okada et al., 2000; Oh and Oh, 2003; Oh and Oh, 2006).

In microbes, GABA has been reported to increase acid tolerance and ATP production (Schuller et al., 2008). During cell growth, the intracellular pH decreases as a result of the accumulation of organic acids. Glutamic acid decarboxylase (GAD) system converts by the irreversible α-decarboxylation of L-glutamic acid (monosodium glutamate), a molecule of glutamate to GABA consuming an intracellular proton, thus shifting the cytosolic environment pH towards neutrality. GABA is then exported in extracellular environment, thereby contributing to alkalinisation (Cotter et al., 2003).

GABA production has been found in different species of LAB such as *Lc. lactis, Lb. brevis*, *Lb. buchneri* (Cho et al., 2007), *Lb. helveticus*, *Lb. paracasei* (Komatsuzaki et al., 2005), *Lb. plantarum, Lb. sakei* and *St. thermophilus*. The ability to produce GABA depends on the degree of activation of GAD and availability of free glutamate (Oh and Yu, 2011). Moreover, the consumption of 10 μg/100 ml GABA containing fermented milk has been reported to provide a significant decrease in blood pressure in mildly hypertensive humans (Inoue et al., 2003). During milk fermentation and proteolysis, a high level of L-glutamate may theoretically be released, since native caseins contain a high proportion of this amino acid. Some studies have proved that fermented milk and cheeses are potential vehicles for GABA (Table 14.1) (Inoue et al., 2003; Komatsuzaki et al., 2005, Huang et al., 2007, Cho et al., 2007). Glutamate content, NaCl concentration, culture temperature (30–50 °C), low pH (4.0–5.0) and anaerobiosis are the most important factors affecting GABA production by LAB (Li and Cao, 2010). Cheese-ripening conditions (release of glutamate from casein proteolysis, relatively high NaCl content, low pH and anaerobiosis) are favourable to GABA production up to a level associated with health benefits (Hayakawa et al., 2004). Some studies evidenced that GABA-producing strains used in co-culture with highly proteolytic strains are an effective strategy to provide higher GABA-enriched dairy products (Inoue et al., 2003; Li and Cao, 2010).

Like an endogenous amphetamine, β-phenylethylamine acts as a mood elevator and regulates appetite–satiety balance. This molecule is applied in treatment of obesity and in weight-control diets, although some controversial effects appear (i.e. insomnia, anxiety). Even if the evidence of a specific phenylalanine decarboxylase enzyme has not yet been reported in LAB, β-phenylethylamine accumulation in *Lactobacilli* and *Enterococci* has been verified. The β-phenylethylamine found in fermented food seems to be associated with

Table 14.1 Comparison of lactic acid bacteria, GABA production conditions and final content in milk or dairy products.

LAB species	Dairy product	Production conditions	Final GABA content	References
Lactobacillus brevis	Skim milk	L-MSG 20 mM; 30 °C; 24 h	15 mg/kg	Siragusa et al., 2007
Lactobacillus delbrueckii subsp. *bulgaricus*	Skim milk	L-MSG 20 mM; 30 °C; 24 h	63.0 mg/kg	Siragusa et al., 2007
Lactobacillus helveticus	Skim milk	37 °C; 30 h	113 mg/l	Sun et al., 2008
Lactobacillus paracasei	Cheese	L-MSG 20 mM; 30 °C; 24 h	99.9 mg/kg	Siragusa et al., 2007
Lactobacillus plantarum	Skim milk Yogurt Skim milk	L-MSG 20 mM; 30 °C; 24 h L-MSG 80 mM; 36 °C	16.0 mg/kg 289 mg/100 g 314.56 mg/100 g	Siragusa et al. 2007 Park et al., 2013; Shan et al., 2015
Lactobacillus crispatus	12% skim milk	0.1% MSG; 40 °C; 18 h	74.56 mg/100 ml	Oh, 2006
Lactococcus lactis	Cheese Cheese Fermented milk Skimmed milk	pH 4.86; 16 °C; 2 wk pH 4.8; 3% S/M 2% NaCl; 367 mg/100 mL MSG L-MSG 20 mM; 30 °C; 24 h	197.9 mg/100 g 23 mg/30 g 500 mg/100 ml 36 mg/kg	Nomura et al., 1998 Gardner-Fortier et al., 2013 Lacroix et al., 2013 Siragusa et al., 2007
Streptococcus thermophilus	Fermented milk Milk	100 mM MSG; pH 4,87; 41 °C; 6 h 41 °C; 24 h	648 μM 100 mg/l	Somkuti et al., 2012 Brasca et al., 2016

MSG: monosodium glutamate; S/M: salt to moisture ratio.

the activity of tyrosine decarboxylating bacteria. In *Ent. faecalis* exists a bi-phasic kinetic of tyramine and β-phenylethylamine production based on the double catalytic activity of the same enzyme, known as tyrosine decarboxylase (EC 4.1.1.25), having higher specificity for tyrosine and only low affinity for phenylalanine. During logarithmic growth, tyrosine is consumed and completely converted into tyramine, then phenylalanine is decarboxylated to β-phenylethylamine, although with minor conversion rate and yield (Pessione and Cirrincione, 2016). Finally, *Lb. delbrueckii* subsp. *bulgaricus, Lb. acidophilus, Lb. casei* and *Lb. plantarum* strains can synthesise melatonin, another bioactive compound produced in the tryptophan/serotonin pathway, regulating sleep and the reproductive system but also controlling immunity, inflammation and carcinogenesis (Pessione and Cirrincione, 2016).

14.4.4 Immune system affecting peptides

The immune system is a complex defence system based on antigen-recognition and cell signalling by peptides. Among them, milk-borne peptides affect humoral and cell-mediated immune response, in particular lymphocyte proliferation, antibody synthesis and intercellular cytokine regulation (Beermann and Hartung, 2013). β-casein medium fermented with LAB generates bioactive peptides acting on monocytes, macrophages and T helper cells, particularly Th1-like cells. Different di- and tripeptides, like Tyr-Gly and Tyr-Gly-Gly, present in the primary structure of bovine κ-casein and α-lactalbumin significantly induce the proliferation of human peripheral blood lymphocytes *in vivo*. Casein hydrolysated by *Lb. rhamnosus* GG modulates allergic reactions through lymphocyte proliferation, enhancement of anti-inflammatory cytokines from Th1 lymphocytes and decrease of pro-inflammatory cytokines and immunoglobulins from Th2 lymphocytes, while hydrolysis of β-lactoglobulin by *Lb. paracasei* releases peptides stimulating interleukin 10 (IL-10) production and downregulation of IL-4 and gamma interferon secretion. Peptides from *Lb. helveticus* proteolysis downregulate cytokine production, stimulate macrophage phagocytosis and show protective effects against enteric infection by pathogens. To the contrary, all the non-proteolytic strains of *Lb. helveticus* lack immune system modulation capability. Immune-active peptides improve the effects of probiotic bacteria, too. In animal studies the peptidic fractions from milk fermented with *Lb. helveticus* inhibited tumour progression by stimulating cellular apoptosis, and yoghurt and peptides present in yoghurt have been demonstrated to inhibit tumour cell proliferation. Immune-modulatory peptides are often multifunctional, exhibiting ACE inhibitory or anti-microbial activities (Beermann and Hartung, 2013; Pessione and Cirrincione, 2016).

14.4.5 Opioid peptides

The hydrolysis of αs_1- and β-casein and α-lactalbumin in *Lactobacillus* GG fermented ultra-high-temperature milk by pepsin and trypsin has been shown to release several opioid peptides (Park and Nam, 2015). Opioids are substances that act on opioid receptors principally located in the central and peripheral nervous system and are involved in blood pressure regulation (Bhat et al., 2015). Other functions of these molecules comprise appetite-satiety balance regulation and control of the gut–brain axis at several levels, making them a promising way for treating stress-related behaviours such as anxiety, depression and mood disorder. Besides the best-studied opioid peptides β-casomorphins, αs_1 casein-derived peptide, casoxins and α-casozepine, exorphins and casoxins are peptides produced as means of LAB protease system on dairy proteins, the former having opioid-like and the latter opioid-antagonist activity (Pessione and Cirrincione, 2016).

14.4.6 Metal-binding peptides

Phosphopeptides phosphorylated on serine residues derived from casein digestion (CPP) possess a strong anionic character and hence a strong resistance to further proteolytic degradation. The cluster sequence Ser(P)-Ser(P)-Ser(P)-Glu-Glu is crucial for metal binding, but phosphorylated cysteines can be alternatively present, revealing significant differences in this capability. Since they can form soluble complexes with calcium, even at alkaline pH, calcium is absorbed in the intestine, and this helps in the treatment of osteoporosis, dental enamel recalcification and prevention of dental caries. However, calcium bioavailability varies as a function of meal composition and complex interactions between different foods ingested. Similarly, trace elements such as Fe, Zn, Cu, Se, Mg and Mn are more available and better absorbed thanks to the CPP high affinity for these molecules. They have been detected during cheese ripening due to microbial protease activity. *Lb. helveticus* LA can decrypt from α-casein, a peptide showing calcium-binding activity, and *Lb. helveticus* LBK16H can decrypt tripeptides (IPP and VPP) with bone anabolic action (Pessione and Cirrincione, 2016).

14.4.7 Conjugated linoleic acid and conjugated linolenic acid

Conjugated linoleic acid (CLA) and conjugated linolenic acid (CLNA) are conjugated fatty acids whose biological significance on human health has been well documented (Del Guire et al., 2012).

Many studies showed that dietary intervention with CLA could lead to anti-carcinogenic, anti-atherogenic, anti-inflammatory, and anti-diabetic activities and ability to reduce body fat deposition (Bassaganya-Riera at al., 2012; Chen et al., 2012; Stachowska, et al., 2012; Thuillier at al., 2013; Koba and Yanagita, 2014; Malinska, et al., 2015). In addition, it was reported as an antioxidant compound in animal and *in vitro* models (Bergamo at al., 2014). On the other hand, CLNA demonstrated anti-carcinogenic activity using *in vitro* and *in vivo* models and exerted cytotoxic effect on tumoural cells (Suzuki at al., 2001; Kohno et al., 2004; Pierre at al., 2013).

CLA refers to different isomers of linoleic acid (c9,c12-C18:2, LA) with a conjugated double bond, with the most biologically important isomers being c9,t11 and t10,c12 forms and the most prevalent in ruminant milk fat being c9,t11 and t7, c9. Likewise, there exist CLNA different isomers of the CLA (c9, c12, c15-C18:3, LNA).

CLA and CLNA are naturally present in milk and dairy products, being formed by ruminal microbial partial biohydrogenation of dietary linoleic (LA) or α-linolenic (LNA) acid to stearic acid metabolism pathway in the rumen (Kuhl and Lindner,

2016); CLA can also be produced by desaturation of trans vaccenic acid. Even if they are also present in some plant seed oils, the most important sources of CLA and CLNA for the human diet are milk and dairy products. In addition, many studies evidenced higher CLA content in fermented milk and cheese (Collomb et al., 2002; Siragusa et al., 2007) and the capability of LAB and bifidobacterial strains to synthesise CLA and CLNA (Abd El-Salam at al., 2010; Villar-Tajadura et al., 2014). Thus, the selection of LAB able to form conjugated fatty acids during the fermentation process represents an attractive target for further study in order to increase human intake.

Different studies provide evidence of the CLA production by different species of propionibacteria, LAB and Bifidobacteria. Up to now, *Propionibacterium freudenreichii, Lb. reuteri, Lb. rhamnosus, Lb. plantarum, Lb. brevis* (Huang at al., 2007*), Lb. acidophilus, Lc. lactis, Bifidobacterium* spp., *Streptococcus* spp. have been reported (Abd El-Salam et al., 2010; Kim at al., 2002; Van Nieuwenhove et al., 2007).

The production of CLNA has been demonstrated for *Lb. plantarum* and bifidobacteria. The CLA and CLNA production is strain specific, and it is influenced by the substrate concentration and culture conditions (incubation time and temperature as well as pH). Moreover, the isomers formed vary among the strains.

14.5 LOW-CALORIE SWEETENERS

LAB fermentation has a great deal of success in production of low-calorie sweeteners. These substances represent valid alternatives to sugar as food ingredients, having similar structural and sweetening effects. They combine several advantages, including low calorific value, low glycaemic response, no cariogenic effect and health-promoting benefits (Patra et al., 2009). In particular, sugar alcohols or polyols are hydrogenated carbohydrates in which the carbonyl group of the precursor sugar is reduced to produce the corresponding primary or secondary alcohol. Their main applications in the food, chemical and pharmaceutical industry and in medicine are as humectants, softeners, colour stabilisers, texturizing and anticaking agents. In addition, they do not undergo the Maillard reaction. Since polyols are incompletely digested and poorly absorbed in the small intestine, they reach the colon, and there they can be converted by gut microbiota into health-promoting short-chain fatty acids (Ortiz et al., 2013). The most well-known polyol produced by LAB is mannitol, a six-carbon sugar alcohol traditionally obtained by catalytic hydrogenation from fructose or a glucose/fructose (1:1) mixture at high temperatures and pressures. However, mannitol production by fermentation is attractive because it overcomes the problems connected to chemical synthesis and brings advances in terms of process yield, safety and versatility, such as the whole conversion of fructose into mannitol without co-formation of sorbitol, mild production conditions, and no requirement of highly purified substrates. When supplied with D-fructose in the medium, various heterofermentative LAB belonging to the genera *Leuconostoc* and *Lactobacillus* are efficient in converting D-fructose to D-mannitol, even if the optimum culture conditions for mannitol production are strain dependent (Patra et al., 2009; Ortiz et al., 2013).

These LAB use D-fructose as an alternative electron acceptor and divert a part of fructose-6-phosphate from the glycolytic pathway to mannitol by using mannitol-1-phosphate dehydrogenase. Increased mannitol yields have been reached by co-fermentation of fructose with glucose. Instead, homofermentative LAB usually produce only moderate levels of mannitol (Mazzoli et al., 2014). In addition to the general applications of polyols, mannitol has also been shown to have antioxidant effect by scavenging off free hydroxyl radicals, to guarantee osmoprotection to LAB cultures during drying and to extend the shelf life of various foodstuffs. Thanks to its low caloric value (1.6 kcal/g) and a relatively low glycaemic index, mannitol in the food industry is primarily used as a sweetener in sugar-free chewing gums (Patra et al., 2009). Erythritol is a polyol derived from erythrose, with very low caloric value (0.2 kcal/g). Besides industrial production based on fermentation by osmophilic yeasts, LAB offer an alternative to the cheap catalytic hydrogenation of erythrose. In particular, *Oenococcus oeni*, *Ln. mesenteroides*, and *Lb. sanfranciscensis* have been implicated up to now. Special properties differentiate erythritol from other polyols, in particular its small molecular size that allows this molecule to be absorbed (60–90%) but not systemically metabolized. Only small amounts are thus accessible to colonic bacteria, avoiding gastrointestinal side effects which usually come after the ingestion of the polyols currently used as sugar replacers. Like mannitol, erythritol has been recognized as a safe sweetener for food applications (Ortiz et al., 2013). Of dairy interest is tagatose. Tagatose is not a sugar alcohol, but an isomer of fructose, naturally occurring in dairy products. The hydrolysis of lactose present in milk generates glucose and galactose. The latter is then converted into tagatose under alkaline conditions. Microbial production of tagatose, as an alternative to a chemical method, is linked to biocatalysts, more exactly L-arabinose isomerase, that mediates the transformation of D-galactose into D-tagatose *in vitro*. LAB, that is, *Lb. plantarum* and *Lb. gayonii*, are biocatalyst sources and represent an economically feasible way for production of tagatose. In particular, the alginate immobilized *Lb. fermentum* CGMCC2921 cells were shown to be an efficient biocatalyst in D-tagatose production in the presence of borate (Patra et al., 2009; Xu et al., 2012). This functional sweetener has the advantages owned by polyols (tooth-friendly property and low caloric value) without having laxative effect as polyols. It is similar to sucrose in texture and taste, provides 92% of the sweetness of sucrose but with only 38% of the calories, and without aftertaste or cooling sensation. Consequently, it can be successfully used to replace the table sugar. Being malabsorbed in the small intestine (20% of the ingested amount), the unabsorbed fraction is fermented by the colonic microflora, leading to short-chain fatty acids which are rapidly absorbed and metabolized. In this sense, tagatose is a potential prebiotic. Tagatose fermentation is performed by a restricted number of intestinal microorganisms: *Lactobacillus* and *Enterococcus*. As confirmation of its prebiotic character, tagatose has been attested to stimulate the growth of these beneficial bacteria and increase the butyrate production both *in vitro* and *in vivo*. In other words, this sweetener is an outstanding substitute of sucrose as it emerges by looking at the long list of tagatose health benefits, including promotion of weight reduction and obesity treatment, and has anti-cariogenic, antiplaque, antihalitosis and anti-biofilm properties, and no glycaemic effect, anti-diabetic effect, and antioxidant activity (Patra et al., 2009).

14.6 EXOPOLYSACCHARIDES (EPS)

In the flourishing market of reduced-fat and reduced-calorie products, the improvement of the sensory quality is a priority, and functional attributes, such as firmness, syneresis, texture and mouth feel are basic, particularly in the dairy sector, hence the necessity to find fat and sugar substitutes in attempts to satisfy consumers' requirements. The microbial EPS represent one such type of substitute, and they have found applications from fermented milk to yoghurt and cheeses. Besides the provision of techno-functionality to food, a variety of further functions have been attributed to EPS, mainly therapeutic potential in terms of immunomodulation, anti-oxidation, hypocholesterolaemia, cholera toxin neutralization, inhibition of biofilm formation, anti-tumour and prebiotic activity. Furthermore, some EPS chemical structures confer bioactivity and pro-technological properties to the microbial producer and may have relevance in reclaiming an EFSA "probiotic" immunomodulatory health claim. EPS may be important for probiotic survival during the gastrointestinal transit and can also act as substrate for other organisms in complex ecosystems. They also have a key role in biofilm formation and surface adhesion, enabling the colonization of different environments. These abilities seem to be associated with probiotic bacteria. Moreover, EPS could interfere with adhesion to intestinal cells by a competitive inhibition mechanism, inhibiting the growth of pathogens. EPS synthesised by intestinal microbiota can represent fermentable substrates for microorganisms in the human gut, promoting shifts in short-chain fatty acids profiles and alterations in intestinal microbial populations. In this sense, they can be considered to be prebiotics, the "selectively fermented ingredient that results in specific changes in the composition and/or activity of the gastrointestinal microbiota, thus conferring benefit(s) upon host health". LAB-EPS may be considered an additional piece of the stress response machinery developed by LAB since they have a shielding effect around the cell wall. For instance, they have been reported to confer tolerance to lysozyme stress during technological processes upon microbial strains. Among scientific substantiation of EPS usefulness, the hypocholesterolemic *Lactobacillus mucosae* DPC 6426 was found to release in yoghurt more than half a gram per litre of EPS over the storage period, affecting significantly the technological and functional quality of yoghurt. The inclusion of an EPS-synthesising *Lc. lactis* starter in half-fat Cheddar cheese led to a quality similar to that of full-fat Cheddar. This was connected to higher primary proteolysis and diluted intact casein; in addition, the 9.5% increase in moisture content determined an 8.17% increase in yield. Similarly, a higher moisture level in Cheddar cheeses was obtained through addition of dextran-synthesising LAB (Ryan et al., 2015; Torino et al., 2015; Caggianello et al., 2016; Zannini et al., 2016). In the yoghurt sector, the consumer demand for stirred yoghurt with a smooth and creamy texture has led to the application of homogenization of the fermented milk coagulum. This technological step, however, negatively influences the rheology of the coagulum, enhancing serum separation. In order to improve the texture of fermented milk products and reduce syneresis, avoiding technological solution such as increasing milk solids (fat, proteins, sucrose or fructose) or adding stabilisers (pectin, starch, alginate, gelatine), EPS produced naturally by LAB used as starter culture in the fermentation have been exploited because of their ability to bind water, interact with proteins, and increase the viscosity of the milk serum phase. As valid viscosifiers, emulsifiers, gelling agents, texturizers and stabilisers, they avoid the use of food additives (Torino et al., 2015; Caggianello et al., 2016). The viscosity and the charge of the EPS determine largely the physical properties of the end product. EPS produced by LAB are taste-neutral, but the increased time of contact with the palate and taste receptors following the production of EPS implicates improved perception of taste. Encapsulated EPS from *St. thermophilus* were successfully used in part-skim Mozzarella cheese process affecting moisture level, yield, and melt properties without compromising the whey process recovery (Zannini et al., 2016). EPS are cell wall structural polysaccharides synthesised by LAB and either weakly or strongly bound to the bacterial cell surface. Some of them may form a loosely bound layer that can also be externally secreted.

EPS from LAB are diverse and can be classified according to different criteria. The most classical one considers their monomer composition, classifying them into two major groups: homopolysaccharides (HoPS) (e.g. cellulose, dextran, mutan, alternan, pullulan, levan and curdlan) and heteropolysaccharides (HePS) (e.g. gellan and xanthan). Depending on the linkage type and the position of the carbon involved in the bond, homopolysaccharides can be classified as α-D-glucans (dextran, mutan, reuteran, and alternan) and β-D-glucans, whereas those containing fructose are fructans (levan and inulin types). Glucans and fructans are found most frequently among the HoPS, and they are both applied as ingredients in the food industry. Polygalactans, composed of a pentameric repeating unit of galactose, have also been described. Long chains of HoPS and HePS consist of repeating units of sugars or sugar, which may be substituted with various chemical moieties. HoPS contain repeating units of one kind of monosaccharide, such as D-glucose or D-fructose; the most relevant are glucans and fructans, with molecular weights ranging from 10^5 to 10^6 Dalton (Da). HePS are composed commonly by glucose, galactose and rhamnose and in some cases by *N*-acetyl-D-glucosamine and *N*-acetyl- D-galactosamine, but may also contain phosphate or other moieties in their polymeric structure. The molecular mass is generally between 10^4 and 10^6 Da. The production of EPS by LAB has been correlated to specific gene clusters tagged as *eps* or *cps*, located mainly on the bacterial chromosome as in *St. thermophilus* or *Lb. plantarum*, or mainly on plasmids in species such as *Lc. lactis* and *Pediococcus damnosus*. The phenotype of EPS-producing strains varies, too, and this aspect can be quickly pointed out by the agar plate technique. The “ropy” phenotype forms a long filament when an inoculation loop is placed into the EPS-covered colony and then slowly withdrawn, while the “mucoid” one appears as shiny and smooth colonies growing on agar plate. The EPS yield and composition appear to be significantly influenced by culture and fermentation conditions (i.e. pH, temperature, incubation time, and substrate composition), while in some strains it is relatively constant under a variety of conditions. The yield of HePS is quite variable; largest producers are *Lb. rhamnosus* RW-9595 M (2.775 mg/l) and *Lactobacillus kefiranofaciens* WT-2B (2.500 mg/l), *Lc. lactis* subsp. *cremoris* (80–600 mg/l), *St. thermophilus* (50–350 mg/l), *Lb. delbrueckii* subsp. *bulgaricus* (60–150 mg/l), *Lb. casei* (50– 60 mg/l) and *Lb. plantarum* (140 mg/l). The benefits of EPS are detectable at extremely low concentrations (Ryan et al., 2015; Caggianello et al., 2016; Zannini et al., 2016). The higher EPS-producing strains can be investigated to resolve the problem of expensive production and recovery processes, making sustainable the EPS application in the food industry.

14.7 CONCLUSIONS

LAB strains offer the potential to transform food substrates into nutraceuticals. Several steps are still awaiting their introduction into the market, such as demonstration of claims through trials and tests of their efficacy on target population samples. At this stage, it is envisaged that the selection of strains with potential and high activity in the production of required metabolites in food fermentations will offer new clues for novel areas and niches of foods with beneficial or technological properties.

References

Abd El-Salam MH, El-Shafei K, Sharaf OM, Effat BA, Asem FM, El-Aasar M. 2010. Screening of some potentially probiotic lactic acid bacteria for their ability to synthesis conjugated linoleic acid. *Int J Dairy Technol* 63(1):62–69.

Adams CA. 2010. The probiotic paradox: live and dead cells are biological response modifiers. *Nutrition Res Rev* 23:37–46. DOI: 10.1017/S0954422410000090.

Adeghate E, Ponery AS. 2002. GABA in the endocrine pancreas: cellular localization and function in normal and diabetic rats. *Tissue Cell* 34(1):1–6.

Alzate A, Fernández-Fernández A, Pérez-Conde MC, Gutiérrez AM, Cámara C. 2008. Comparison of biotransformation of inorganic selenium by *Lactobacillus* and *Saccharomyces* in lactic fermentation process of yogurt and kefir. *J Agricul Food Chem* 56:8728–36. DOI: 10.1021/jf8013519.

Arena MP, Russo P, Capozzi V, López P, Fiocco D, Spano G. 2014. Probiotic abilities of riboflavin-overproducing *Lactobacillus* strains: a novel promising application of probiotics. *Appl Microbiol Biotechnol* 98:7569–81. DOI: 10.1007/s00253-014-5837-x.

Bailey RL, West JKP, Black RE. 2015. The epidemiology of global micronutrient deficiencies. *Annals Nutr Metab* 66: 22–33. DOI: 10.1159/000371618.

Bassaganya-Riera J, Hontecillas R, Horne WT, Sandridge M, Herfarth HH, Bloomfeld R, Isaacs KL. 2012. Conjugated linoleic acid modulates immune responses in patients with mild to moderately active Crohn's disease. *Clin Nutr* 31:721–27.

Beermann C, Hartung J. 2013. Physiological properties of milk ingredients released by fermentation. *Food Function* 4:185–99. DOI: 10.1039/c2fo30153a.

Beltrán-Barrientos LM, Hernández-Mendoza A, Torres-Llanez MJ, González-Córdova AF, Vallejo-Córdoba B. 2016. Invited review: fermented milk as antihypertensive functional food. *J Dairy Sci* 99:4099–10.

Bergamo P, Luongo D, Miyamoto J, Cocca E, Kishino S, Ogawa J, Rossi M. 2014. Immunomodulatory activity of a gut microbial metabolite of dietary linoleic acid, 10-hydroxy-cis-12-octadecenoic acid, associated with improved antioxidant/detoxifying defences. *J Function Foods* 11:192–202.

Bermudez-Brito M, Plaza-Díaz J, Muñoz-Quezada S, Gómez-Llorente C, Gil A. 2013. Probiotic mechanisms of action. *Annals Nutri Metab* 61:160–74. DOI: 10.1159/000342079.

Bhat ZF, Kumar S, Bhat HF. 2015. Bioactive peptides of animal origin: a review. *J Food Sci Technol* 52:5377–92. DOI: 10.1007/s13197-015-1731-5.

Bhushan B, Tomar SK, Manda S. 2016. Phenotypic and genotypic screening of human-originated lactobacilli for vitamin B12 production potential: process validation by micro-assay and UFLC. *Appl Microbiol Biotechnol* 100:6791–803. DOI: 10.1007/s00253-016-7639-9.

Bonjour JP, Benoit V, Atkin S, Walrand S. 2015. Fortification of yogurts with vitamin D and calcium enhances the inhibition of serum parathyroid hormone and bone resorption markers: A double blind randomized controlled trial in women over 60 living in a community dwelling home. *J Nutri Health Aging* 19:563–69. DOI: 10.1007/s12603-015-0498-8.

Bourdichon F, Casaregola S, Farrokh C, Frisvad JC, Gerds ML, Hammes WP, Harnett J, Huys G, Laulund S, Ouwehand A, Powell IB, Prajapati JB, Seto Y, Ter Schure E, Van Boven A, Vankerckhoven V, Zgoda A, Tuijtelaars S, Bech Hansen E. 2012. Food fermentations: microorganisms with technological beneficial use. *Int J Food Microbiol* 156:301–303. DOI: 10.1016/j.ijfoodmicro.2011.12.030.

Brasca M, Hogenboom JA, Morandi S, Rosi V, D'Incecco P, Silvetti T, Pellegrino L. 2016. Proteolytic activity and production of γ-aminobutyric acid by *Streptococcus thermophilus* cultivated in microfiltered pasteurized milk. *J Agri Food Chem* 64:8604–14.

Burgess CM, Smid EJ, Rutten G, van Sinderen D. 2006. A general method for selection of riboflavin-overproducing food grade micro-organisms. *Microbial Cell Factories* 5:24. DOI: 10.1186/1475-2859-5-24.

Caggianiello G, Kleerebezem M, Spano G. 2016. Exopolysaccharides produced by lactic acid bacteria: from health-promoting benefits to stress tolerance mechanisms. *App Microbiol Biotechnol* 100:3877–86. DOI: 10.1007/s00253-016-7471-2.

Capriotti AL, Cavaliere C, Piovesana S, Samperi R, Laganà A. 2016. Recent trends in the analysis of bioactive peptides in milk and dairy products. *Analytical Bioanal Chem* 408:2677–85. DOI: 10.1007/s00216-016-9303-8.

Chen SC, Lin YH, Huang HP, Hsu WL, Houng JY, Huang CK. 2012. Effect of conjugated linoleic acid supplementation on weight loss and body fat composition in a Chinese population. *Nutrition* 28:559–65.

Cho YR, Chang JY, Chang HC. 2007. Production of gamma-aminobutyric acid GABA by *Lactobacillus buchneri* isolated from kimchi and its neuroprotective effect on neuronal cells. *J Microbiol Biotechnol* 17(1):104–109.

Choi J, Sabikhi L, Hassan A, Anand S. 2012. Bioactive peptides in dairy products. *Int J Dairy Technol* 65:1–12. DOI: 10.1111/j.1471-0307.2011.00725.x.

Collomb M, Butikofer U, Sieber R, Jeangros B, Bosset JO. 2002. Composition of fatty acids in cow's milk fat produced in the lowlands, mountains and highlands of Switzerland using high-resolution gas chromatography. *Int Dairy J* 12:649–59.

Corbo MG, Bevilacqua A, Petruzzi L, Casanova FP, Sinigaglia M. 2014. Functional beverages: The emerging side of functional foods. Commercial trends, research, and health implications. *Compr Rev Food Sci Food Safety* 13:1192–1206. DOI: 10.1111/1541-4337.12109.

Cotter PD, Hill C. 2003. Surviving the acid test: responses of gram-positive bacteria to low pH. *Microbiol Mol Biol Rev* 67:429–53.

DeGuire JR, Makarem N, Vanstone CA, Morin S, Duque G, Weiler HA. 2012. Conjugated linoleic acid is related to bone mineral density but does not affect parathyroid hormone in men. *Nutr Res* 32:911–20.

del Valle MJ, Laiño JE, de Giori GS, LeBlanc JGJ. 2014. Riboflavin producing lactic acid bacteria as a biotechnological strategy to obtain bio-enriched soymilk. *Food Res Technol* 62:1015–19. DOI: 10.1016/j.foodres.2014.05.029.

Diplock AT, Aggett PJ, Ashwell M, Bornet F, Fern EB, Robertfroid MB. 2000. Scientific concepts of functional foods in Europe: consensus document. *Special Publication-Royal Society Chemistry* 248:8–60.

Divya JB, Nampoothiri KM. 2015. Folate fortification of skim milk by a probiotic *Lactococcus lactis* CM28 and evaluation of its stability in fermented milk on cold storage. *J Food Sci Technol* 52:3513–19. DOI: 10.1007/s13197-014-1406-7.

EFSA. 2010. Scientific opinion on the substantiation of health claims related to live yoghurt cultures improved lactose digestion ID 1143, 2976 pursuant to Article 131 of Regulation EC No 1924/20061. *EFSA J* 810:1763.

Eszenyi P, Sztrik A, Babka B, Prokisch J. 2011. Elemental, nano-sized 100–500 nm selenium production by probiotic lactic acid bacteria. *Int J Biosci Biochem Bioinforma* 1:148–52. DOI: 10.7763/IJBBB.2011.V1.27.

FAO/WHO Food and Agricultural Organization/World Health Organization. 2006. Probiotic in foods. Health and nutritional properties and guidelines for evaluation. In: FAO Food and Nutrition. Paper 85.

Gall JE, Boyd RS, Rajakaruna N. 2015. Transfer of heavy metals through terrestrial food webs: a review. *Environ Monitor Assessment* 187:201. DOI: 10.1007/s10661-015-4436-3.

Gardner-Fortier C, St-Gelais D, Champagne CP, Vuillemard J-C. 2013. Determination of optimal conditions for γ-aminobutyric acid production by *Lactococcus lactis* ssp. *lactis*. *Int Dairy J* 32:136–43.

Gille D, Schmid A. 2015. Vitamin B_{12} in meat and dairy products. *Nutri Rev* 73:106–15. DOI: 10.1093/nutrit/nuu011.

Gobbetti M, Di Cagno R, De Angelis M. 2010. Functional microorganisms for functional food quality. *Crit Rev Food Sci Nutrition* 50:716–27. DOI: 10.1080/10408398.2010.499770.

González-Arrojo A, Soldado A, Vicente F, Fernández Sánchez ML, Sanz-Medel A, de la Roza-Delgado B. 2016. Changes on levels of essential trace elements in selenium naturally enriched milk. *J Food Nutri Res* 4:303–308. DOI: 10.12691/jfnr-4-5-6.

Griffiths MW, Tellez AM. 2013. *Lactobacillus helveticus*: the proteolytic system. *Frontiers Microbiol* 4:30. DOI: 10.3389/fmicb.2013.00030.

Hati S, Mandal S, Prajapati JB. 2013. Novel starters for value added fermented dairy products. *Curr Res Nutr Food Sci* 1:83–91. DOI: http://dx.doi.org/10.12944/CRNFSJ.1.1.09.

Hayakawa K, Kimura M, Kasaha K, Matsumoto K, Sansawa H, Yamori Y. 2004. Effect of γ-aminobutyric acid enriched dairy product on the blood pressure of spontaneously hypertensive and normotensive Wista r-Kyoto rats. *Br J Nutr* 92:411–17.

Health Canada. 2009. Accepted claims about the nature of probiotic microorganisms in food. Health Canada. [online] http://www.hc-sc.gc.ca/fn-an/label-etiquet/claims-reclam/probiotics_claims-allegations_probiotiques-eng.php%20.

Hiraga K, Ueno YH, Oda KH. 2008. Glutamate decarboxylase from *Lactobacillus brevis*: activation by ammonium sulfate. *Biosci Biotechnol Biochem.* 72:1299–1306.

Hou JW, Yu RC, Chou CC. 2000. Changes in some components of soymilk during fermentation with bifidobacteria. *Food Res Int* 33:39–7. DOI: 10.1016/S0963-99690000061-2.

Huang J, Lehe M, Sheng Q, Shanjing YAO, Dongqiang LIN. 2007. Purification and characterization of glutamate decarboxylase of *Lactobacillus brevis* CGMCC 1306 isolated from fresh milk. *Chinese J Chemical Engineering* 152:157–61.

Hugenholtz J. 2008. The lactic acid bacterium as a cell factory for food ingredient production. *Int Dairy J* 18:466–75. DOI: 10.1016/j.idairyj.2007.11.015.

Hugenholtz J. 2013. Traditional biotechnology for new foods and beverages. *Curr Opin Biotechnol* 24:155–9. DOI: 10.1016/j.copbio.2013.01.001.

Inoue K, Shirai T, Ochiai H, Kasao M, Hayakawa K, Kimura M, Sansawa H. 2003. Blood-pressure-lowering effect of a novel fermented milk containing γ-aminobutyric acid GABA in mild hypertensives. *Eur J Clin Nutr* 57:490–95.

Jakobs C, Jaeken J, Gibson KM. 1993. Inherited disorders of GABA metabolism. *J Inher Metabol Dis* 164:704–15.

Kennedy DO. 2016. B vitamins and the brain: mechanisms, dose and efficacy—a review. *Nutrients* 8:68. DOI: 10.3390/nu8020068.

Kim YJ, Liu RH. 2002. Increase of conjugated linoleic acid content in milk by fermentation with lactic acid bacteria. *J Food Sci* 675:1731–7.

Koba K, Yanagita T. 2014. Health benefits of conjugated linoleic acid CLA. *Obesity Res Clin Practice* 86:e525–32.

Kohno H, Suzuki R, Yasui Y, Hosokawa M, Miyashita K, Tanaka T. 2004. Pomegranate seed oil rich in conjugated linolenic acid suppresses chemically induced colon carcinogenesis in rats. *Cancer Sci* 956:481–6.

Komatsuzaki N, Shima J, Kawamoto S, Momose H, Kimura T. 2005. Production of γ-aminobutyric acid GABA by *Lactobacillus paracasei* isolated from traditional fermented foods. *Food Microbiol* 226:497–504.

Kuhl GC, De Dea Lindner J. 2016. Biohydrogenation of linoleic acid by lactic acid bacteria for the production of functional cultured dairy products: A Review. *Foods* 51, 13.

Kurek E, Ruszczyńska A, Wojciechowski M, Łuciuk A, Michalska-Kacymirow M, Motyl I, Bulska E. 2016. Bio-transformation of selenium in Se-enriched bacterial strains of *Lactobacillus casei*. *Roczniki Państwowego Zakładu Higieny* 67:253–62.

Lacroix N, St-Gelais D, Champagne CP, Vuillemard JC. 2013. Gamma-aminobutyric acid-producing abilities of lactococcal strains isolated from old-style cheese starters. *Dairy Sci Technol.* 93:315–27.

Laiño JE, del Valle MJ, De Giori GS, LeBlanc JGJ. 2013. Development of a high folate concentration yoghurt naturally bio-enriched using selected lactic acid bacteria. *LWT – Food Sci Technol* 54:1–5. DOI: 10.1016/j.lwt.2013.05.035.

LeBlanc JG, Milani C, de Giori GS, Sesma F, van Sinderen D, Ventura M. 2013. Bacteria as vitamin suppliers to their host: a gut microbiota perspective. *Curr Opin Biotechnol* 24:160–68. DOI: 10.1016/j.copbio.2012.08.005.

Lecerf JM, Colin J, Hebel P, Bongard V, Ferrières J. 2016. Les consommateurs de produits laitiers frais: des consommateurs comme les autres? Analyse de leurs profils alimentaires et nutritionnels. Who are fresh dairy products consumers? Analysis of their dietary nutritional profiles. *Nutrition clinique et métabolisme* 30:11–21. DOI: 10.1016/j.nupar.2015.12.001.

Leroy F, De Vuyst L. 2004. Lactic acid bacteria as functional starter cultures for the food fermentation industry. *Trends Food Sci Technol* 15:67–78. DOI: 10.1016/j.tifs.2003.09.004.

Leskauskaite D, Jasutiene I, Malinauskyte E, Kersiene M, Matusevicius P. 2016. Fortification of dairy products with vitamin D-3. *Int J Dairy Technol* 69:177–83. DOI: 10.1111/1471-0307.12242.

Li H, Cao Y. 2010, Lactic bacterial cell factories for gamma-aminobutyric acid. *Amino Acids* 39:1107–16.

Malinska H, Hüttl M, Oliyarnyk O, Bratova M, Kazdova L. 2015. Conjugated linoleic acid reduces visceral and ectopic lipid accumulation and insulin resistance in chronic severe hypertriglyceridemia. *Nutrition* 31:1045–51.

Mazzoli R, Bosco F, Mizrahi I, Bayer EA, Pessione E. 2014. Towards lactic acid bacteria-based biorefineries. *Biotechnol Adv* 32:1216–36. DOI: 10.1016/j.biotechadv.2014.07.005.

Mogna L, Nicola S, Pane M, Lorenzini P, Strozzi G, Mogna G. 2012. Selenium and zinc internalized by *Lactobacillus buchneri* Lb26 DSM 16341 and *Bifidobacterium lactis* Bb1 DSM 17850: improved bioavailability using a new biological approach. *J Clin Gastroenterol* 46: S41–45. DOI: 10.1097/MCG.0b013e318268861d.

Mrvčić J, Šolić E, Butorac A, Stanzer D, Bačun-Družina V, Stehlik-Tomas V. 2011. The effect of metal ions supplementation on growth and binding capacity of lactic acid bacteria. Pp. 67. 7th International Congress of Food Technologists, Biotechnologists and Nutritionist/Medić, Helga ur. – Zagreb: Baris, Zaprešić, 2011.

Mrvčić J, Stanzer D, Šolić E, Stehlik-Tomas V. 2012. Interaction of lactic acid bacteria with metal ions: opportunities for improving food safety and quality. *World J Microbiol Biotechnol* 28:2771–82. DOI: 10.1007/s11274-012-1094-2.

Namdari A, Nejati F. 2016. Development of antioxidant activity during milk fermentation by wild isolates of *Lactobacillus helveticus*. *Appl Food Biotechnol* 3:178–86.

Nomura M, Kimoto H, Someya Y, Furukawa S, Suzuki I. 1998. Production of γ-aminobutyric acid by cheese starters during cheese ripening. *J Dairy Sci* 81:1486–91.

Okada T, Sugishita T, Murakami T, Murai H, Saikusa T, Hotorino T, Onoda A, Kajimoto O, Takahashi R, Takahashi T. 2000. Effect of the defatted rice germ enriched with GABA for sleeplessness, depression, autonomic disorder by oral administration. *J Jap Soc Food Sci Technol* 47:596–603.

Oh, SH, Oh CH. 2003. Brown rice extracts with enhanced levels of GABA stimulate immune cells. *Food Sci Biotechnol* 12:248–52.

Oh SH. 2006. Physiological characteristics of *Lactobacillus crispatus* RMK 567 isolated from milk and optimization of GABA producing conditions. Master thesis. Sungkyunkwan University, Suwon, South Korea.

Oh SH, Yu JJ. 2011. γ-Aminobutyric acid production and glutamate decarboxylase activity of *Lactobacillus sakei* OPK2-59 isolated from Kimchi. *Korean J Microbiol* 47:316–22.

Ortiz ME, Bleckwedel J, Raya RR, Mozzi F. 2013. Biotechnological and in situ food production of polyols by lactic acid bacteria. *Appl Microbiol Biotechnol* 11:4713–26. DOI: 10.1007/s00253-013-4884-z.

Palomo M, Gutiérrez AM, Pérez-Conde MC, Cámara C, Madrid Y. 2014. Se metallomics during lactic fermentation of Se-enriched yogurt. *Food Chemistry* 164:371–9. DOI: 10.1016/j.foodchem.2014.05.007.

Park S-Y, Kim K-S, Lee M-K, Lim S-D. 2013. Physiological characteristics and GABA production of *Lactobacillus plantarum* K255 isolated from Kimchi. *Korean J Food Sci Anim Res.* 33:595–602.

Park YW, Nam MS. 2015. Bioactive peptides in milk and dairy products: a review. *Korean J Food Sci Animal Res* 35:831–40.

Patel A, Shah N, Prajapati JB. 2013. Biosynthesis of vitamins and enzymes in fermenting foods by lactic acid bacteria and related genera. *Croatian J Food Sci Technol* 5:85–91.

Patel A, Shah N. 2016. Bioactive components of fermented foods. In: Montet D, Ray RC (Eds). *Fermented foods, Part I: Biochemistry and biotechnology*, Boca Raton: CRC Press, Taylor and Francis Group, pp 267–88.

Patel S. 2015. Emerging trends in nutraceutical applications of whey protein and its derivatives. *J Food Sci Technol* 52:6847–58. DOI: 10.1007/s13197-015-1894-0.

Patra F, Tomar SK, Arora S. 2009. Technological and functional applications of low-calorie sweeteners from lactic acid bacteria. *J Food Science* 74:16–23. DOI: 10.1111/j.1750-3841.2008.01005.x.

Pessione E, Cirrincione S. 2016. Bioactive molecules released in food by lactic acid bacteria: encrypted peptides and biogenic amines. *Frontiers Microbiol* 7:876. DOI: 10.3389/fmicb.2016.00876.

Pierre AS, Minville-Walz M, Fèvre C, Hichami A, Gresti J, Pichon L, Riall M. 2013. Trans-10, cis-12 conjugated linoleic acid induced cell death in human colon cancer cells through reactive oxygen species-mediated ER stress. *Biochim Biophys Acta* 18314:759–68.

Pophaly SD, Singh R, Pophaly SD, Kaushik JK, Tomar SK. 2012. Current status and emerging role of glutathione in food grade lactic acid bacteria. *Microbial Cell Factories* 11:114. DOI: 10.1186/1475-2859-11-114.

Pophaly SD, Poonam Singh P, Kumar H, Tomar SK, Sing R. 2014. Selenium enrichment of lactic acid bacteria and bifidobacteria: a functional food perspective. *Trends Food Sci Technol* 39:135–45. DOI: 10.1016/j.tifs.2014.07.006.

Punnagaiarasi A, Karthikeyan N, Elango A, Pandiyan C. 2016. Fortifications in dairy products – review. *Int J Adulteration* 1:52–57.

Rad AH, Khosroushahi AY, Khalili M, Jafarzadeh S. 2016. Folate bio-fortification of yoghurt and fermented milk: a review. *Dairy Sci Technol* 96:427–41. DOI: 10.1007/s13594-016-0286-1.

Ramesh V, Kumar R, Singh RRB, Kaushik J, Mann B. 2012. Comparative evaluation of selected strains of lactobacilli for the development of antioxidant activity in milk. *Dairy Sci Technol* 92:179–88. DOI: 10.1007/s13594-011-0048-z.

Ravyts F, De Vuyst L, Leroy F. 2012. Bacterial diversity and functionalities in food fermentations. *Engineering Life Sci* 12:356–67. DOI: 10.1002/elsc.201100119.

Rice BH, Quann EE, Miller GD. 2013. Meeting and exceeding dairy recommendations: effects of dairy consumption on nutrient intakes and risk of chronic disease. *Nutrition Rev* 71:209–223. DOI: 10.1111/nure.12007.

Russo P, López P, Capozzi V, Palencia P, Dueñas MT, Spano G, Fiocco D. 2012. Beta-glucans improve growth, viability and colonization of probiotic microorganisms. *Int J Mol Sci* 13:6026–39. DOI: 10.3390/ijms13056026.

Ryan PM, Ross RP, Fitzgerald GF, Caplicee NM, Stanton C. 2015. Sugar-coated: exopolysaccharide producing lactic acid bacteria for food and human health applications. *Food Function* 6:679. DOI: 10.1039/c4fo00529e.

Saubade F, Hemery YM, Guyot JP, Humblot C. 2016. Lactic acid fermentation as a tool for increasing the folate content of foods. *Critical Rev Food Sci Nutrition*. DOI: 10.1080/10408398.2016.1192986.

Schuller HM, Al-Wadei HA, Majidi M. 2008. Gamma-aminobutyric acid, a potential tumor suppressor for small airway-derived lung adenocarcinoma. *Carcinogenesis* 29(10):1979–85.

Shan Y, Man CX, Han X, Li L, Guo Y, Deng Y, Li T, Zhang LW, Jiang YJ. 2015. Evaluation of improved γ-aminobutyric acid production in yogurt using *Lactobacillus plantarum* NCD75017. *J Dairy Sci* 98:2138–49.

Shigwedha N, Sichel L, Jia L, Zhang L. 2014. Probiotical cell fragments PCFs as "Novel Nutraceutical Ingredients". *J Biosci Medicines* 2:43–55. DOI: 10.4236/jbm.2014.23007.

Shortt C, O'Brien J. 2004. Handbook of functional dairy products. Ames, IA: Wiley-Blackwell.

Siragusa S, De Angelis M, Di Cagno R, Rizzello CG, Coda R, Gobbetti M. 2007. Synthesis of γ-aminobutyric acid by lactic acid bacteria isolated from a variety of Italian cheeses. *Appl Environ Microbiol* 73(22):7283–90.

Siró I, Kápolna E, Kápolna B, Lugasi A. 2008. Functional food product development, marketing and consumer acceptance. a review. *Appetite* 51:456–67. DOI: 10.1016/j.appet.2008.05.060.

Somkuti GA, Renye JA Jr, Steinberg DH. 2012. Molecular analysis of the glutamate decarboxylase locus in *Streptococcus thermophilus* ST110. *J Ind Microbiol Biotechnol* 39:957–63.

Stachowska E, Siennicka A, Ba´skiewcz-Hałasa M, Bober J, Machalinski B, Chlubek D. 2012. Conjugated linoleic acid isomers may diminish human macrophages adhesion to endothelial surface. *Int J Food Sci Nutr* 63:30–35.

Sun TS, Zhao SP, Wang HK, Cai CK, Chen YF, Zhang HP. 2008. ACE-inhibitory activity and gamma-aminobutyric acid content of fermented skim milk by *Lactobacillus helveticus* isolated from Xinjiang koumiss in China. *Eur Food Res Technol* 228:607–12.

Suzuki R, Noguchi R, Ota T, Abe M, Miyashita K, Kawada T. 2001. Cytotoxic effect of conjugated trienoic fatty acids on mouse tumor and human monocytic leukemia cells. *Lipids* 365:477-82.

Sybesma W, Starrenburg M, Kleerebezem M, Mierau I, de Vos W M, Hugenholtz J. 2003. Increased production of folate by metabolic engineering of *Lactococcus lactis*. *Appl Environ Microbiol* 69:3069–76. DOI: 10.1128/AEM.69.6.3069-3076.2003.

Tamang JP, Shin DH, Jung SJ, Chae SW. 2016. Functional properties of microorganisms in fermented foods. *Frontiers Microbiol* 7:578. DOI: 10.3389/fmicb.2016.00578.

Thakur K, Tomar SK. 2015. Exploring indigenous *Lactobacillus* species from diverse niches for riboflavin production. *J Young Pharmacists* 7:122–7.

Thakur K, Tomar SK, De S. 2016. Lactic acid bacteria as a cell factory for riboflavin production. *Microbial Biotechnology* 9:441–51. DOI: 10.1111/1751-7915.12335.

Thuillier P, Pande NT, Ghena A, Song S, Lawrence Y, Shridhar V, Akkari Y, Pejovic T, Olson S. 2013. Dietary conjugated linoleic acids arrest cell cycle progression and prevent ovarian cancer xenografts growth suggesting a trans-10 cis-12 isoform specific activity. *J Cancer Ther* 4:33–42.

Torino MI, Font de Valdez G, Mozzi F. 2015. Biopolymers from lactic acid bacteria. Novel applications in foods and beverages. *Frontiers Microbiol* 6:834. DOI: 10.3389/fmicb.2015.00834.

Tujioka K, Ohsumi M, Horie K, Kim M, Hayas K, Yokogoshi H. 2009. Dietary γ-aminobutyric acid affects the brain protein synthesis rate in ovariectomized female rats. *J Nutr Sci Vitamin* 55:75–80.

Ueno H. 2000. Enzymatic and structural aspects of glutamate decarboxylase. *J Mol Catal B Enzym* 10:67–79.

Van Nieuwenhove CP, Oliszewski R, Gonzalez SN, Perez Chaia AB. 2007 Conjugated linoleic acid conversion by dairy bacteria cultured in MRS broth and buffalo milk. *Lett Appl Microbiol* 445:467–4.

Villar-Tajadura MA, Rodríguez-Alcalá LM, Martín V, Gómez de Segura A, Rodríguez JM, Requena T, Fontecha J. 2014. Production of conjugated linoleic and conjugated α-linolenic acid in a reconstituted skim milk-based medium by bifidobacterial strains isolated from human breast milk. *BioMed Res Int* Article ID 2014:725406. DOI: 10.1155/2014/725406.

Varzakas TH, Arvanytoiannis IS, Berzitzoglou E. 2010. Functional dairy foods and flora modulation. In: Yildiz F (Ed). Development and manufacture of yogurt other functional dairy products. Boca Raton: CRC Press Taylor and Francis Group, pp 339–74.

Walther B, Philip KJ, Booth SL, Boyaval PK. 2013. Menaquinones, bacteria, and the food supply: the relevance of dairy and fermented food products to vitamin K requirements. *American Soc Nutri Advances Nutr* 4:463–73. DOI: 10.3945/an.113.003855.

Xu Z, Li S, Fu F, Li G, Feng X, Xu H, Ouyang P. 2012. Production of d-tagatose, a functional sweetener, utilizing alginate immobilized *Lactobacillus fermentum* CGMCC2921 cells. *Appl Biochem Biotechnol* 166:961–73. DOI: 10.1007/s12010-011-9484-8.

Zannini E, Waters DM, Coffey A, Arendt EK. 2016. Production, properties, and industrial food application of lactic acid bacteria-derived exopolysaccharides. *Appl Microbiol Biotechnol* 100:1121–35. DOI: 10.1007/s00253-015-7172-2.

15 Dairy technologies in yogurt production

Panagiotis Sfakianakis and Constantina Tzia

Food Chemistry and Technology, School of Chemical Engineering, National Technical University of Athens, Zografou, Greece

15.1 INTRODUCTION

Yogurt and the fermented milk products in general, date back to 9000–8000 BC from ancient Egypt and Mesopotamia. The first version of fermented milk had occurred from a spontaneous and accidental lactic acid fermentation which led to the acidification and coagulation of the milk due to the temperature conditions in effect at those areas. The result was a gel-like, edible product, more resistant to spoilage than the raw milk and simultaneously acceptable for its sensorial characteristics (Chandon, 2006; Tamime and Robison, 2007a; Robinson, 2011). Over the centuries, the know-how of making yogurt and fermented dairy products spread all over the world. In addition to the acceptance that yogurt achieved, due to the likable taste, flavor and texture, the health benefits presented by Nobel Laureate, E. Metchnikof, and the Pasteur institute in the 20th century, boosted its consumption and production even more (Corrieu and Beal, 2015; Metchnikoff, 1907).

Modern food science and international legislation describe yogurt as the product being manufactured from milk, with or without the addition of some natural components of milk, such as skim milk powder, whey concentrates and caseinates or cream (Gahruie et al., 2015), with a gel structure that is formed from the coagulation of the milk proteins, due to the lactic acid produced by defined species of bacteria cultures (mixed starter culture of *Streptococcus thermophilus* and *Lactobacillus delbrueckii* ssp. *Bulgaricus*). Furthermore, these bacteria must be "viable and abundant" at the time of consumption, and the titratable acidity must be at least 0.9%, expressed as lactic acid (FAO/WHO CODEX STAN 243-2003). The main principles of yogurt manufacture have changed little throughout the ages; the basic processes involved are heating, inoculating of milk with the starter culture, fermentation process and storage of final product. After the 20th century and the commercialization of yogurt, several new processes have been added, standardization of milk constitution (fat and solid non-fat content) and pasteurization. However the most important stage of the entire process is the stage of fermentation. The action of the microorganism involved is of paramount importance and has great scientific interest. During fermentation, several biochemical

Microbiology in Dairy Processing: Challenges and Opportunities, First Edition.
Edited by Palmiro Poltronieri.

Published 2018 by John Wiley & Sons Ltd and the Institute of Food Technologists.

reactions take place that change the composition of the milk and subsequently cause physicochemical changes that lead to the formation of the curd and the development of the yogurt flavor; therefore, the fermentation process and all the individual reactions that are involved have significant effect on the final product and its characteristics (Sfakianakis and Tzia, 2014).

15.2 YOGURT TYPES

The late decades of the 20th century had changed drastically the production of yogurt due to increased consumer demand and the implementation of industrial practices, in addition to the recent developments of dairy science and nutrition. Quite important is the discrimination of yogurt products, based on the origins of the milk, the chemical composition, method of production, flavor additives used, the post-incubation processing and the contribution of other species of microorganism in the fermentation process (Sfakianakis and Tzia, 2014).

Yogurt can be manufactured using the milk of most domestic mammal species, with the most common being the bovine milk. Also yogurt from milk of the goat, sheep, buffalo, camel, ass and horse is available; however, these products present variety in their chemical composition, in particular goat-, sheep-, buffalo- and camel-derived yogurts have higher fat content, while ass- and horse-derived have lower (Shah, 2003).

Regarding the composition, yogurts are divided into full fat (fat content over 3.5%), reduced fat (fat content varies 0.5 – 2%) and low fat (fat content lower than 0.5%).

Based on the manufacturing process, yogurt products are classified as set type and stirred. Specifically, set-type yogurts are fermented in the retail container, resulting in continuous semisolid curd; stirred yogurts, in contrast, are not allowed to form a strong gel structure, and the curd is disturbed during fermentation, resulting in a creamy texture. A variation of stirred yogurt is the greek style yogurt, that requires the removal of whey from the product. The result is a more viscous and firmer yogurt product.

Another discrimination of yogurt products is based on the flavor. In addition to the plain yogurt with the distinctive acidic-sweet taste, yogurt-type products with different flavors exist. At the late stages or after the fermentation process, fruit fragments, puree or jam, sweetening agents or flavor additives are added in the curd, resulting in yogurt-type desserts with different flavors.

Last but not least is the differentiation of yogurt products that occurs with further processing after the incubation. Processes like drying, freezing and additional pasteurization have been implemented on yogurt production, resulting in the formation of products like frozen yogurt, yogurt powder and ultra-high temperature (UHT)/pasteurized yogurt. Frozen yogurt is a product similar to ice cream in terms of texture, while it maintains the chemical composition and flavor characteristics of yogurt. The manufacturing process is identical to conventional yogurt, with the addition of the freezing process. Yogurt in powder form, also known as dried yogurt, is made by putting the yogurt curd under a drying process, such as freeze, spray or sun drying. These processes remove the water trapped in the casein matrix and cause the loss of several flavor components. UHT and pasteurized yogurt manufacture involves the additional heating of the yogurt curd after its formation, thus killing any microorganisms, along with the starter culture in addition to reduce the concentration of several yogurt flavor components (Corrieu and Beal, 2015).

Although not typically yogurts, some fermented milk products that include other microbial strains are worth being mentioned. These products are known as probiotics, products that contain probiotic microorganisms. Probiotic microorganisms are defined as live microbes, which, when ingested, benefit the health of the host through their effect on the intestinal microflora (Metchnikoff, 1907; Salminen et al., 2011). In order to be labeled as probiotic, food products must have 10^7–10^8 CFU/g or ml (Kneifel et al., 1999; Vinderola et al., 2011). The most common probiotic bacteria are strains and species of Lactobacilli, Bifidobacteria, Enterococci and Lactococci. The species most thoroughly studied are *Lactobacillus acidophilus, Lactobacillus casei, Lactobacillus lactis, Lactobacillus helveticus, Bifidobacterium longum, Bifidobacterium lactis, Bifidobacterium animalis* ssp. *lactis* and *Bifidobacterium bifidum,* and *Bifidobacterium bifidus* (Elizaquível et al., 2011; Vinderola et al., 2011). These species have been proven to cause health benefits to human consumers, in particular to improve the functions of the gastrointestinal system and digestion. Also, in the last 20 years probiotic fermented milk products are available in retail markets, with high acceptance from consumers (Cruz et al., 2013).

15.3 YOGURT MANUFACTURING PROCESS

The manufacturing process of yogurt starts with the milking of the mammal and ends with the package and distribution or storage of the final product (Tamime and Robinson, 2007b). (See Figure 15.1.) Throughout the manufacturing process, the milk undergoes severe treatments in order to achieve the desired characteristics of the final product, including the preliminary treatment of milk right after the milking, the standardization of the milk composition in terms of fat and solid non-fat (SNF) content, (Kristo et al., 2003) the homogenization of milk, the heat treatment, the inoculation with the starter culture, the fermentation, the packaging and the storage. The most important parameter, throughout the entire manufacturing process is the temperature, due to the impact it has on the milk itself, the microbial content and the growth of the starter culture during fermentation (Sfakianakis and Tzia, 2014; Walstra et al., 2006a).

15.3.1 Initial treatment of milk

Milk is prone to contamination, especially during and soon after its secretion from the mammary glands. The milk can become contaminated in the interior of the udder, on the exterior of the udder, and in the milking or in the storage equipment. Contaminants in the newly produced milk are somatic cells of the mammal, dust, hair and other solid impurities from the environment, increased concentration of fatty acids and a wide variety of microorganisms (Shah, 2003). The microbial pollutants that can be found in the fresh milk are mostly pathogens, more specifically Gram-positive bacteria (*Bacillus cereus, Bacillus anthracis, Clostridium perfringens, Staphylococcus aureus, Streptococcus agalactiae, Streptococcus pyogenes, Listeria monocytogenes, Mycobacterium tuberculosis* and others), Gram-negative bacteria (*Escherichia coli, Salmonella, Yersinia enterocolitica, Campylobacter jejuni, Aeromonas hydrophila,* and others), rickettsia (*Coxiella burnetii*), viruses (enteroviruses, foot and mouth disease virus and others), fungi and protozoa

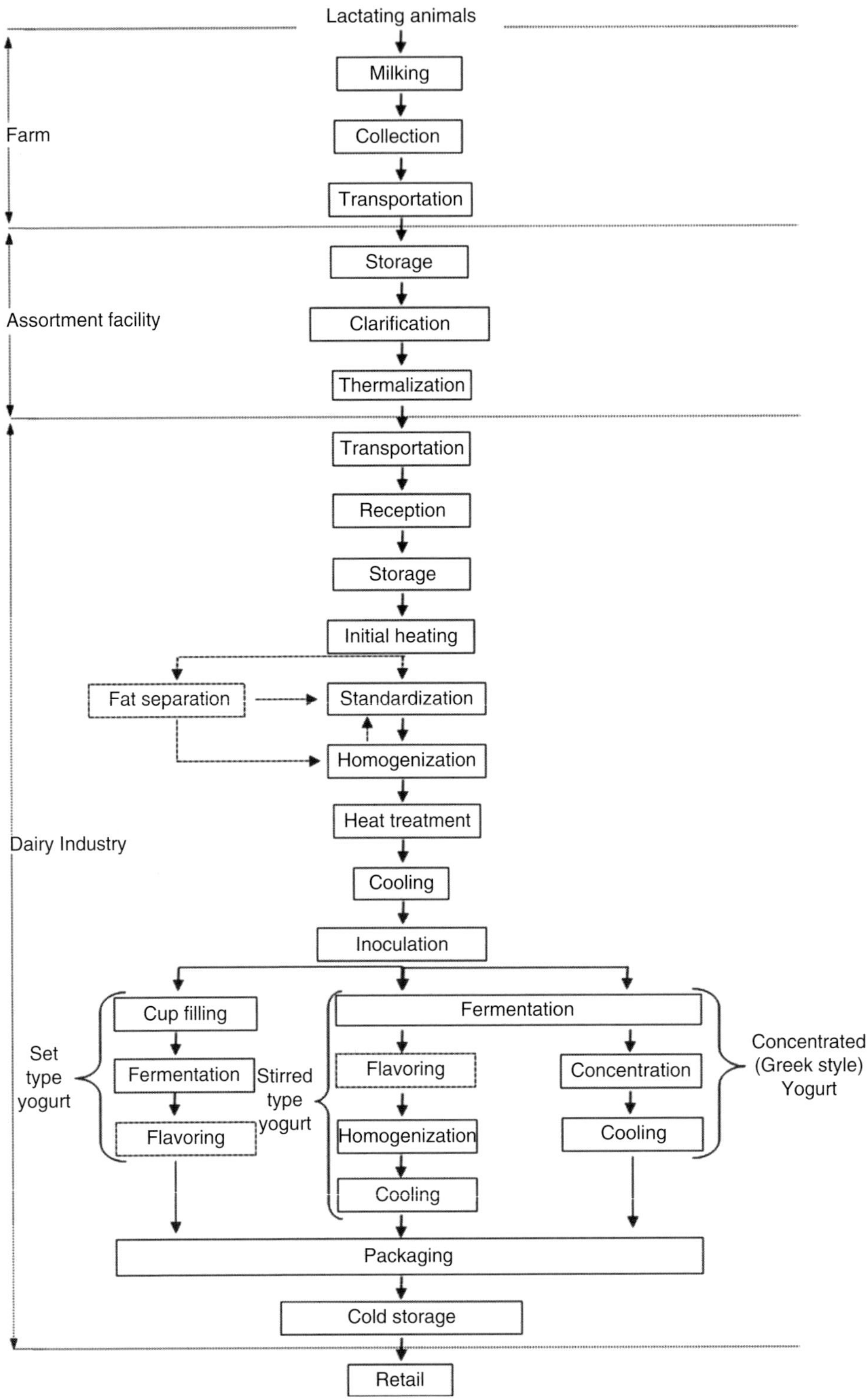

Figure 15.1 Flow chart of yogurt manufacturing process

(*Entamoeba histolytica, Giardia lamblia* and *Toxoplasma gondii*) (Shah, 2003; Corrieu and Beal, 2015). The solid impurities are removed via centrifugal clarification or filtration. The pathogens are eliminated with a mild heat treatment, known as thermalization, at a temperature range of 60–69 °C for 20–30 s and subsequently cooling at temperatures lower than 5 °C or inoculated with lactic acid bacteria or other microfloras (Tamime and Robison, 2007; Sfakianakis and Tzia, 2014). Free fatty acid molecules in milk are formed in the udder of the mammals, by the hydrolysis of triglycerides by the enzyme lipoprotein lipase. However, this reaction can continue after the milking and further increase the concentration of the fatty acids, thus resulting in rancid flavors. Thermalization hinders the lipase's activity and prevents further increase of the fatty acids. Apart from those post-milking processes, the quality and safety of the milk is affected by the health of the mammals, the feed they consume, the conditions of the milking, the quality and hygiene of the milking equipment and the good handling practices throughout the procedure (Shaker et al., 2000; Sfakianakis and Tzia, 2014; Walstra et al., 2006c).

15.3.2 Standardization of milk components – fat and SNF content

The composition of milk varies, depending on the producing species, the time of the year and the breeding and feeding of the mammal. The average composition of milk used for yogurt manufacture is presented in the Table 15.1 (Fox, 2011b).

The commercialization of yogurt indicates that yogurt products available in the market ought to have standard composition and characteristics. In order to achieve that, the major components, the fat and SNF content of milk, must present a not significant variance. In bovine milk, the most commonly used raw material for yogurt production, the fat content varies between 3.2 and 4.2% w/w; during the manufacturing process, the fat content of milk is standardized at percentages lower than 0.5 for skimmed milk, at 1.5–2% for semi-skimmed milk and at 3.5% for full fat milk (Fox, 2011b; Tamime and Robison, 2007b). These percentages lead to the production of yogurts with 0.1–10% fat content. The SNF components of milk mainly consist of lactose, protein and minerals; SNF content of milk varies from 11 to 14% of the total weight of the milk, leading to yogurt product with SNF content between 9 and 16% (Sfakianakis and Tzia, 2014). The desired percentages of fat and SNF content in milk is achieved by either the addition of skim milk or milk fat or the separation of fat from milk via centrifuge or separation and

Table 15.1 Composition of milk of selected species.

Species	Total solids	Fat	Protein	Lactose	Ash
Cow	12.7	3.7	3.4	4.8	0.7
Goat	12.3	4.8	2.9	4.1	0.8
Sheep	19.3	7.4	4.5	4.8	1.0
Camel	15.0	5.4	3.8	5.2	0.7
Buffalo	16.8	7.4	3.8	4.8	0.8
Horse	11.2	1.9	2.5	6.2	0.5
Ass	11.7	1.4	2.0	7.4	0.5

mixing milk fat with skimmed milk in the desired ratio (Shah, 2003). The main reason behind the standardization, in yogurt manufacture, is the fact that the fat and SNF content has significant effect on the fermentation process and the texture and the viscosity of the final product. In particular, high fat and protein content of milk leads to yogurt with high viscosity and high texture characteristics in addition to the effect on the milk fermentation process, the maximum pH decrease rate and the lag phase (Corrieu and Beal, 2015; Tamime and Robison, 2007a).

15.3.3 Homogenization

Milk is a complex solution that includes mainly water, fats and proteins. The fat in milk exists in milk fat globules (MFGs). MFGs are spherical formations, with diameter varying 0.1–20 μm in raw milk (Taylor and Mac Gibbons, 2011). The outer layer of MFG, also known as MFG membrane, consists of protein molecules (20–60%) and lipids (triacyglycerols, diacylglycerols, monacylglycerols, sterols, sterol-esters and phospholipids) in lower percentages. The inner layer of MFG consists fully of fatty molecules. This system is a typical oil-in-water (o/w) emulsion, with the MFG being the dispersed phase and the MFG membrane acting as emulsifier. However, milk is not a stable emulsion and tends to separate into two phases with the fat rising to the surface of the volume because MFGs collide and coalesce. The main reasons responsible for the MFG coalition and the separation phenomenon are the Brownian motion of the MFG, the reaction of agglutinins, the interfacial tension and the fact that the pressure is greater inside the smaller globules than inside the larger, and hence, large fat globules tend to grow at the expense of the smaller (Laplace principle) (Fox, 2011a). To prevent the separation phenomenon, milk undergoes homogenization. During this process the MFGs are submitted to high pressure and temperature gradient, and the MFG membrane collapses, unleashing the contained fat in the milk volume. Subsequently the protein molecules of the previous membrane form a new membrane in the fat–serum interface, and smaller MFGs with a richer membrane in protein are created. During the formation of the new membrane, protein molecules, mostly caseins, from the milk serum are absorbed in the membrane in order to emulsify the newly formed globules, since the fat surface area increases (Cano-Ruiz and Richter, 1997). The newly formed MFGs, due to their smaller size and the composition of the membrane, are less keen to aggregate and separate from the volume of the milk, thus increasing the emulsion stability of the milk (Aguilera and Kessler, 1988). The exposed casein surface area, due to increased number of casein-covered fat globules, can then participate in the acid-coagulated milk gel (Cho and Lucey, 1999; Walstra et al., 2011).

The methods available for homogenization are application of high pressure, ultrasound, high-speed mixing, microfluidization and membrane emulsification (Sfakianakis and Tzia, 2014).

High-pressure homogenization is the most commonly used method of homogenization in dairy industry (Fox, 2011b). The typical high-pressure homogenizer includes a system of two valves—the first applies higher pressure, 10–30 Mpa, and the second lower, 5–15 MPa—and a number of pistons to ensure the generation of enough driving pressure. The main principle behind high-pressure homogenization is the application of 10–30 MPa pressure on the milk at the first valve, which forces it to accelerate at high velocity (200–300 m*s^{-1}) through the narrow space of the valves (10 μm) until it reaches the second valve, where lower pressure (5–10 MPa) is a applied. It finally exits to environmental conditions. During the milk's voyage in the homogenizer, the high-velocity movement causes turbulence in the serum of the milk and shear stress is developed. The stress causes the MFGs to collapse and reform into smaller particles (Fox, 2011b).

Ultrasound (US) treatment is considered as an alternative method for reducing MFG size and can be effectively applied to homogenize milk. Several studies have proven that the application of US in milk can reduce the average MFG diameter to 0.1–0.6 μm (Demirdöven and Baysal, 2009; Dolatowski et al., 2007). Furthermore, US treatment has been referred to alter the composition of the MFG membrane, increasing the concentration of protein molecules. In effect the changes that US treatment causes on milk are similar with conventional, high-pressure homogenization. However US homogenization has several additional effects on the milk and its components; in particular, high-intensity US has been proven to cause denaturation of milk proteins; unfold the peptide chains of whey proteins (β-lactoglobulin, serum albumin) and subsequently facilitates the formation of whey–whey and whey–casein aggregates via disulfide bonds (Chandrapala et al., 2011; Madadlou et al., 2009; Wu et al., 2009; Morand et al., 2011; Sfakianakis et al., 2015). Furthermore, high-amplitude US has been reported to reduce the microbial content of milk and cause off-flavors due to the emission of certain volatiles (Bermúdez-Aguirre et al., 2008; Riener et al., 2009a). Flavor components like aldehydes, ketones and some aliphatic hydrocarbons with rubbery and burned aroma have been detected in ultrasonicated milk (Riener et al., 2009b).

Yogurt produced from milk treated by high-intensity US have shown improved physical properties and high value of viscosity and texture characteristics (firmness, cohesiveness) (Vercet et al., 2002; Nguyen and Anema, 2010). The higher the amplitude level of the US treatment, the higher the values of viscosity and firmness; moreover, the same treatment leads to yogurt microstructure similar to a honeycomb (Wu et al., 2001). The increased viscosity and texture of yogurt derived from ultrasonicated milk can be attributed to the denaturation of whey proteins and the formation of casein–whey aggregates; during acidification, the denatured whey proteins associated with casein micelles act as bridging material between the casein micelles facilitating the formation of the yogurt matrix and increasing the strength of the yogurt coagulum (Krešić et al., 2008; Sfakianakis et al., 2015).

High-speed mixing is the most commonly used method for emulsifying artificial emulsions and has very limited application for milk. The main principle is that the high-speed rotation generates velocity gradients in liquids, which result in the disruption of the MFG membrane and, subsequently, homogenization. For high-speed mixing, several types of equipment are available: blades, propellers and turbine mixers. The design of the stirrer determines to a large extent the efficiency of the emulsification/homogenization process. Additionally the efficiency of the homogenization is affected by the duration of the process and the rotational speed of the homogenizer (Hupperts, 2011)

Microfluidization is a method of homogenization via shear stress, turbulence and cavitation. The process involves the acceleration of the fluid and separates the flow into two microstreams that intersect in a chamber and collide. The impact causes intense turbulence and cavitation, the MFG membrane collapses and the homogenization is achieved (Kasaai et al., 2003). MFG diameter after microfluidization is less than 2 μm

(Ciron et al., 2010). The application of microfluidization in yogurt manufacture leads to yogurt curds with increased syneresis, reduced viscosity and lower firmness compared to conventionally manufactured yogurt. In addition, microfluidization has significant effect on the yogurt microstructure, giving more interconnectivity in the protein networks with embedded fat globules, but with similar texture profiles and water retention compared to yogurt prepared from conventionally homogenized milk (Hupperts, 2011; Ronkart et al., 2010; Skurtys and Aguilera, 2008).

15.3.4 Heat treatment

The next stage of yogurt manufacture after homogenization is the heat treatment. Heat treatment aims at reducing the milk's microbial content to safe levels for the consumer and causes certain modification to protein molecules that encases the formation of the yogurt curd, as well as improve the latter's characteristics (Kilara, 2006). The pathogenic microorganisms present in raw milk are, mainly, *M. tuberculosis, C. burnetii, St. aureus, Salmonella* species, *L. monocytogenes* and *C. jejuni*. A safe milk product is easy to produce, due to the fact that the above-mentioned pathogens are either killed by milk heat treatment or because other, high-heat-resistant pathogens do not occur in milk (e.g., *Bacillus anthracis*) or are outnumbered by other native microorganisms (e.g., *Clostridium perfringens*), or cause spoilage before their quantity is enough to cause health issues (e.g., *Bacillus cereus*) (Rukke et al., 2011; Sfakianakis and Tzia, 2014).

Apart from the effect on the pathogenic microorganism, heat treatment causes additional changes in milk physicochemical properties. The exposure of milk to heat that causes the emission of CO_2 and O_2, also increases the concentration of insoluble colloidal calcium phosphate and decreases the concentration of calcium cations. Furthermore, it induces the isomerization of lactose and causes a Maillard reaction, with significant effect on the pH and flavor of milk (Fox, 2011; Sfakianakis and Tzia, 2014).

Milk proteins, both whey and caseins, suffer radical changes during heat treatment. These changes can be exploited in the yogurt manufacture and enhance the formation of the yogurt curd as well as improve the latter's rheological and textural characteristics (Tamime and Robison, 2007b). In their native state, casein molecules exist in the form of micelles or aggregates of submicelles; these formations consist of αs1-, αs2- and β-casein molecules held together by the κ-casein, calcium and calcium phosphate molecules. Casein micelles are quite stable and require a high amount of energy to be disrupted (Horne, 2011). In comparison, whey proteins are fairly stable molecules and exist in solution in the serum. Over 80 °C, the peptide chain of whey proteins (β-lactoglobulin, serum albumin) are denatured irreversibly (Brew, 2011; Creamer et al., 2011). The unfolding of the peptide chains exposes the thiol groups and enables them to interact with other molecules, forming disulfide bonds. Denatured whey proteins can form bonds with other whey proteins, caseins (κ- and αs1- caseins, mostly) and can be incorporated at the MFG membrane. Such interaction between whey proteins and caseins increases the hydrophilic properties of the casein and, in addition, allows the whey proteins to contribute in the yogurt casein matrix and strengthen the yogurt curd. Also, if the thiol groups of the whey proteins are exposed, an interaction between casein and whey protein molecules occur during fermentation and casein-whey bonds are formed. Thus, whey proteins are incorporated into the curd matrix, reducing the propensity of the gel to syneresis and strengthening the latter, resulting in a firmer yogurt. Therefore, the heat-induced denaturation of whey protein, with the subsequent interaction with caseins, is beneficial for the texture development of yogurt and is exploited during the yogurt manufacturing process (Ion-Titapiccolo et al., 2013; Jaros and Rohm, 2003).

The heat treatments applied in dairy processing and in yogurt manufacture vary; the temperature range and the duration of the process differ in accordance with the desired result. The most commonly used heat treatments utilized in yogurt manufacture are thermalization, low and high pasteurization, sterilization and ultra-heat treatment (UHT) (Lewis, 2003; Sfakianakis and Tzia, 2014).

Thermalization is a mild heat treatment with low impact on the milk components. The milk is exposed to a temperature range of 60–69 °C for 20–30 s. The effects of thermalization are the killing of most vegetative microorganisms and the partial deactivation of several enzymes, with no additional, irreversible changes on the milk (Rukke et al., 2011).

Low pasteurization is a heat treatment, slightly more intense than thermalization and more effective. During low pasteurization, milk is heated at 63–65 °C for 20 min or at 72–75 °C for 15–20 s (also referred as HTST, high temperature short time). With this process, most pathogens, vegetative bacteria, yeast and molds are eliminated from the milk, in addition to the deactivation of several enzymes. However, low pasteurization does not cause any off-flavors in milk, little or no serum proteins are denatured, and the cold agglutination and bacteriostatic properties of the milk are not affected (Walstra et al., 2006d).

High-temperature pasteurization is an intense heat treatment that is performed at a combination of temperature and time of 85 °C for 20–30 min or 90–95 °C for 5 min. The effect that high-temperature pasteurization has on milk is the elimination of most vegetative microorganisms except those from spores. Also, at those temperatures most enzymes included in milk are deactivated except milk proteinase, plasmin and some bacterial proteinases and lipases. Additional milk components affected are whey proteins, which are denatured, and several milk lipids, which are oxidized and subsequently produce ketones, aldehydes and several sulfides, leading to the development of a "cooked" aroma. Thermalization, low and high pasteurization is carried out indirectly, in tubular or plate heat exchangers (Deeth and Datta, 2011; Sfakianakis and Tzia, 2014).

Sterilization is the most intense and most effective heat treatment in milk processing. The milk is heated at 110 °C for 30 min or at 130 °C for 40 s. This treatment causes the complete extermination of all the microbial flora of milk, including the bacterial spores and the inactivation of most milk enzymes (except several bacterial lipases). In addition, it facilitates Maillard reaction, and therefore the milk color darkens. Also during sterilization, the lipids of the milk are oxidized, but no off-flavor is caused because flavor volatiles, at those temperatures, are evaporated. Last, sterilization causes a high degree of denaturation and irreversible damage to all milk proteins, even caseins. Sterilization is carried out in tubular or plate heat exchangers (indirect) or by steam injection or infusion (direct) (Deeth and Datta, 2011; Hinrichs and Atamar, 2011).

UHT is also an effective treatment in terms of microbial content reduction, but causes considerable damage. UHT involves the heating of the milk at 145 °C for 1–2 s. Ninety-nine percent of milk microflora is eliminated, and several whey proteins are denatured (β-lactoglobulin, serum albumin, and some immunoglobulins). UHT causes lipid oxidation and the emission of several volatiles in milk, such as: 2-pentanone, 2-heptanone, 2-nonanone, 2-undecanone, 2,6-dimethylpyrazine, 2-ethylpyrazine, 2-ethyl-3-methylpyrazine, methional pentanoic acid, benzothiazole, vanillin, hexanal, decalactone, H2S, methanethiol, dimethylsulfide and carboxylsulfide. These sulfur containing molecules are responsible for the "cooked" off-flavor developed during UHT and high-temperature pasteurization (Boelrijk et al., 2003). UHT is carried out directly with injection or infusion of supersaturated steam in the milk, and indirectly in tubular or plate heat exchangers or with the combination of direct and indirect methods (Deeth and Datta, 2011).

In industrial yogurt manufacture and dairy processing, high-temperature pasteurization is the most commonly applied heat treatment (Walstra et al., 2006b).

15.3.5 Fermentation process

The most important stage of yogurt making is the fermentation process. During this stage, several bio-physicochemical reactions occur that lead to the formation of the curd and the development of the distinct flavor and the texture (Tamime and Robison, 2007a). The main factor in the fermentation process is the starter culture; the bacterial strains in effect define the entire process, the conditions under which the process takes place as well as the outcome. The species involved in milk fermentation are quite a few, *Lb. acidophilus, Lb. casei, Lactobacillus delbrueckii* subsp. *lactis, Lb. helveticus* and biotype *jogurti, B. animalis* ssp. *lactis, B. longum, B. bifidus* and *Bifidobacterium infantis, Streptococcus salivarius* subsp. *thermophilus*. However, in order to label a fermented milk product as "yogurt," international and regional legislations dictate that it should contain the two bacterial strains of *Streptococcus salivarius* subsp. *thermophilus* (ST) and *Lactobacillus delbrueckii* subsp. *bulgaricus* (LB) live and in abundance. ST is the only species in the streptococcus genus that is used in dairy starter cultures (Salminen et al., 2011; Sfakianakis and Tzia, 2014; Tamime and Robison, 2007b).

ST is a Gram-positive bacterial strain with an optimum growth temperature at 35–53 °C, classified as thermophilic. The ST cells change their morphology throughout their lives: in their early life they are spheres and form chains, and as they mature their morphology resembles a rod. LB are Gram-positive, rod-shaped, anaerobic bacteria with optimum growth temperature at 40–44 °C. Additionally, LB can produce very high amounts of lactic acid by metabolizing lactose. The metabolic pathway of both LB and ST leads to the reduction of lactose in milk and the increase of lactic acid, thus the pH of milk drops (Vedamuthu, 2006; Walstra et al., 2006e). The presence of LB in the milk facilitates the growth of ST and vice versa, therefore the combined action of both strains leads to higher lactose metabolism and lactic acid production compared to each one acting individually. The synergy of those species is based in several characteristics that each species has and enhances the other's growth. In particular ST is more "aerotolerant" than LB, lacks good proteolytic ability in comparison to LB, but possesses greater peptidase activity. When grown together in milk, ST grows vigorously at first, whereas LB grows slowly. ST, because of its proteolytic activity, creates an abundance of peptides to stimulate the growth of LB. The optimum growth temperature for the symbiotic culture of ST and LB is 45 °C (Corrieu and Beal, 2015; Shah, 2003).

The fermentation stage of yogurt manufacture initiates with the inoculation of the milk bulk with the symbiotic starter culture. The inoculation mix contains 10^6–10^8 CFU of bacteria. After the inoculation the mix, milk and starter culture, it is transferred to the fermentation tanks (for stirred, drinking, or concentrated yogurt manufacture) or distributed in the retail cups (for set-type yogurt and greek style yogurt manufacture). The temperature of the milk during inoculation and throughout the fermentation process is 45 °C. The effect of LB and ST in milk is mainly the consumption of 1 mole of lactose in order to produce 1 mole of galactose and 2 moles of lactic acid plus intracellular energy in the form of 2 moles of ATP. The metabolic pathway of the glycolysis of lactose to lactic acid initiates with lactose molecule entering the cell with the aid of a lactose permease, energized by a proton gradient. Afterward the lactose molecule is hydrolyzed into glucose and galactose by the enzyme β-galactosidase. ST produces significant levels of lactase, which catalyzes the hydrolysis of lactose to glucose and galactose. Glucose is catabolized to pyruvate via the glycolytic pathway and galactose is expelled from the cell. Subsequently, pyruvate, by the action of lactate dehydrogenase, is reduced into lactic acid. Each microorganism synthesizes different isomers of lactic acid: ST produces L(+) lactic acid and LB lactic acid D(−). Afterward, the intracellular lactate is excreted from the cell via a symport with protons and causes acidification of the extracellular environment. In the initial stage of the fermentation process, only the cells of ST are in effect, and as they grow ST, because of their great proteolytic activity, they create an abundance of peptides to stimulate the growth of LB. Through coordinated tandem activities, both bacteria accelerate the entire fermentation, which none of them would be able to achieve individually. When the pH of the yogurt approaches 5.0, activity of ST subsides and LB gradually dominates the overall fermentation process until the target value of pH (4.8–4.5) is reached and the fermentation process ceases (Corrieu and Beal, 2015; Shah, 2003; Sfakianakis and Tzia, 2014; Vedamuthu, 2006). In order to hinder the action of the symbiotic culture and stop the fermentation process, the system's temperature is lowered to 4 °C; at this temperature the culture survives, but its activity is drastically lowered.

Other bacterial species involved in milk fermentation, may follow heterofermentative pathways to produce their intracellular energy and lactic acid. In particular, other Lactobacilli (*Lactobacillus rhamnosus, Lb. casei,* and *Lactobacillus paracasei*) produce 1 mole of each lactic acid, ethanol, CO_2, and ATP from 1 mole of glucose while Bifidobacteria synthesize 3 moles of acetic acid and 2 moles of L(+) lactic acid and ATP, without generation of CO_2 from 2 moles of glucose. The difference in the metabolic pathways of different species changes the acidification rates and, consequently, the flavor and texture of the final product (Corrieu and Beal, 2015).

After the fermentation process of milk by ST and LB, the composition of the system has suffered drastic changes. In particular lactose, milk proteins and microbial content, as well as several carbon compounds, suffer major changes, whereas minor changes occur for vitamins and minerals. The concentration of lactose is reduced by 30% and produces double the molar amount of lactic acid. Proteins (caseins and whey) aggregate, creating the yogurt curd. Due to proteolysis caused by starter culture, amino acids (mainly proline and glycine) are released into the yogurt, even during storage at 4 °C (Vedamuthu, 2006). The microbial content increases from 10^8 to 10^{10} CFU/g. The concentration of several carbonyl components (lactic acid, acetaldehyde, dimethyl sulfide, 2,3-butanedione, 2,3-pentanedione, 2-methylthiophene, 3-methyl-2-butenal, 1-octen-3-one, dimethyl trisulfide, 1-nonen-3-one, acetic acid, methional, (cis,cis)-nonenal, 2-methyl tetrahydrothiophen-3-one, 2-phenylacetaldehyde, 3-methylbutyric acid, caproic acid and benzothiazole).

These molecules are by-products of the metabolism of the symbiotic culture, and they contribute to the distinctive yogurt flavor (Boelrijk et al., 2003). Among them, acetaldehyde is the major flavor compound of yogurt that gives the pleasant fresh aroma. Acetaldehyde's concentration in yogurt is measured to be 5–40 mg/kg. During fermentation, acetaldehyde is synthesized from pyruvate with the aid of the enzyme pyruvate decarboxylase or indirectly from acetyl coenzyme A through the action of pyruvate dehydrogenase and aldehyde dehydrogenase. Additionally, LB synthesizes acetaldehyde by converting threonine into acetaldehyde and glycine, with the aid of the enzyme serine hydroxyl-methyl transferase (Shah, 2003). Vitamin B increases throughout the fermentation process and storage. Finally, the quantity of minerals in yogurt remain the same as in milk, the only change is that, due to pH lowering, these minerals are in ionic rather than colloidal form (Corrieu and Beal, 2015; Shah, 2003; Sfakianakis and Tzia, 2014).

The most important change that milk proteins suffer during fermentation is the formation of the protein matrix that makes the coagulum. The decrease in pH leads to coagulation as a result of destabilization of the casein micelles. The mechanism relies on two concomitant phenomena. The carboxyl groups dissociate, serine phosphate is ionized, and the negative charge between casein micelles is increased. Simultaneously, the colloidal calcium–phosphate complex is solubilized, which results in the depletion of calcium in the micelles. Then, electrostatic and casein-casein attractions increase due to enhanced hydrophobic interactions. When the isoelectric point of caseins (pH 4.6) is achieved, coagulation occurs as a result of the formation of a three-dimensional network consisting of clusters and chains of caseins, which leads to the formation of the yogurt gel (Cho et al., 1999; Horne, 1999; Shaker et al., 2000; Sfakianakis and Tzia, 2014).

15.3.5.1 Monitoring of fermentation process – prediction of fermentation evolution

The milk fermentation process of yogurt can be described adequately by the evolution of pH and viscosity with respect to time; the model that expresses the evolution of pH during fermentation time is the modified Gompertz models of de Brabandere and de Baerdemaeker (1999).

$$pH = pH_0 + \left(pH_0 - pH_\infty\right) - \left\{-exp\left[\frac{\mu_{pH} \cdot e}{\left(pH_0 - pH_\infty\right)} \cdot \left(\lambda_{pH} - t\right) + 1\right]\right\} \tag{15.1}$$

where

pH_0, pH_∞ = initial and end values of pH, respectively,
μ_{pH} (min^{-1}) = maximum rate of pH decrease,
λ_{pH} (min) = duration of pH lag phase.

The μ_{pH} is the key factor in the fermentation process and the latter's duration. It is influenced by various factors, such as the composition and activity of the starter culture starter, the preliminary treatment of the milk, and, most importantly, by fermentation temperature. However, the coagulation phenomena occurring at pH lower than 5.1 make the accurate temperature control very difficult. Therefore, the variation of μ_{pH} is quite high, and subsequently the duration of the fermentation process varies from 3 to 8 h (De Brabandere and De Baerdemaeker, 1999).

The λ_{pH} parameter of the equation is also a factor of importance. It signifies the ability of the starter culture to acclimatize in the milk environment and start to thrive. As mentioned above, ST plays a major role because it is the first microorganism that acclimatizes in the system and initiates the fermentation process. A key factor for λ_{pH} is the fermentation temperature and the pre-fermentation treatment. The effect that the fermentation temperature has on ST and subsequently on the λ_{pH} is that the closer the temperature is to the optimum growth temperature, the shorter is λ_{pH}. The preliminary heat treatment facilitates the interaction of the ST cells with the needed molecules for their growth; therefore, the more adequate the heat treatment, the shorter the λ_{pH} (De Brabandere and De Baerdemaeker, 1999).

Furthermore, the model that describes the evolution of viscosity during fermentation is the modified Gomperz model of Soukoulis, et al. (2007).

$$\mu_\alpha = \mu_{\alpha 0} + \left(\mu_{\alpha 0} - \mu_{\alpha\infty}\right) - \left\{-exp\left[\frac{\mu_v \cdot e}{\left(\mu_{\alpha 0} - \mu_{\alpha\infty}\right)} \cdot \left(\lambda_v - t\right) + 1\right]\right\} \tag{15.2}$$

where

$\mu\alpha_0$, $\mu\alpha_\infty$ (Pa·s) = initial and end values of viscosity, respectively
μ_v (min^{-1}) = maximum rate of viscosity decrease,
λ_v (min) = duration of viscosity lag phase.

A typical graph of the evolution of pH and viscosity during fermentation process is illustrated by Figure 15.2. Both curves are sigmoidal and, in most cases, have good regression with the equations 1 and 2.

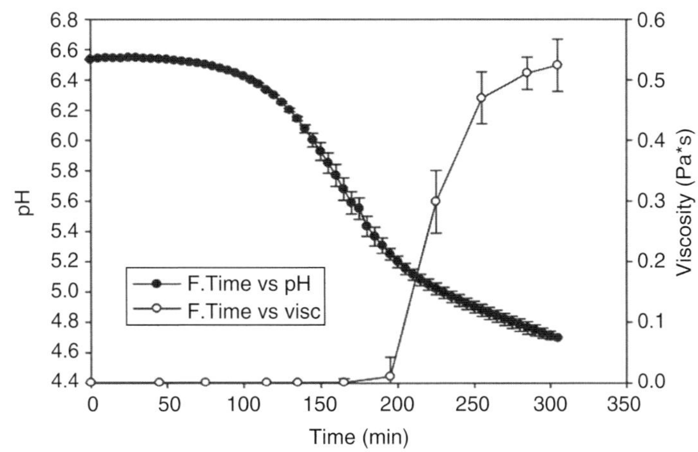

Figure 15.2 Typical graph of yogurt fermentation, pH vs time and viscosity vs time

On the pH versus time curve can be observed 3 distinct areas: the first area, from $t=0$ to the first turning point, the second area, between the two turning points, and the third from the second turning point to end of the curve. The first area ends when the fermentation time reaches the $\lambda_{pH.}$ During that time, the ST is mainly in effect, and the starter culture is acclimatizing in the milk environment and the pH reduction rate is low. The second area of the pH versus time graph signifies the time of the fermentation when both bacterial strains are in the peak of their growth and production of lactic acid. During that time, the pH reduction rate is equal to μ_{pH}. After the second turning point, around pH values of 5.2–5.1, the pH reduction rate drops along with the effectiveness of ST. During that time, the LB cells are mostly in effect and the formation of the casein matrix is taking place (Soukoulis et al., 2007).

The graph of viscosity versus time is almost the reverse of the graph of pH versus time: three distinct phases separated by two turning points. The first phase, from $t=0$ to λ_{visc}, is with a low viscosity increase rate, the second is between the two turning points, with high viscosity increase range and the third is after the second turning point with low viscosity increase rate. The first phase of the viscosity versus time curve ends simultaneously with the second phase of the pH versus time curve. In literature this specific point is referred to as geletation time. The entire evolution of viscosity during fermentation is mainly affected by the interaction between the milk proteins and the formation of the bonds between them; therefore, the key factors are the production rate of acids by the bacteria and the state of the protein molecules. Subsequently, the most important role in the evolution of viscosity is played by the preliminary heat treatment and the latter's effect on the protein molecules (Soukoulis et al., 2007).

15.3.6 Post-fermentation processing

15.3.6.1 Cooling – addition of additives

The fermentation process is stopped when the system's pH reaches the desired value. Depending on the manufacturer and the consumer's demands the pH end value varies from 4.7 to 4.3. After this point the yogurt's temperature must drop, as fast as possible, to lower than 20 °C in order to inhibit the action of the starter culture and stop the acidification (Tamime and Robison, 2007a). The methods used for cooling depend on the yogurt type produced. Set-type yogurts are cooled with a single-stage method and stirred with two stages. Set yogurts are cooled within 1 or 2 h to 4 or 5 °C using cold air in ventilated cabinets, cooling rooms, or tunnels, while stirred yogurts are cooled in an external heat exchanger reaching intermediate temperature (between 15 and 20 °C) for less than 1 h (20–60 min for industrial tanks) and then gradually reach the storage temperature of 4 or 5 °C (Chandan and O'Rell, 2006; Shah, 2006). At this first stage of cooling are added the additional aroma compounds, sweeteners and fruits (jam, pulp, and pieces) (Corrieu and Beal, 2015; Sfakianakis and Tzia, 2014).

15.3.6.2 Addition of fruit

In the case of yogurt products with fruit additives, the stage when the addition is made depends on the yogurt type. For stirred yogurt, the fruits are added at the first stage of cooling, while for set-type yogurt the fruits are added before the fermentation stage. In the case of stirred yogurt with fruit additives, after the yogurt goes through the heat exchanger and has reached temperature lower than 20 °C, it is mixed with fruits either in a mixing tank coupled with a dynamic stirrer or in a filling line coupled with a static mixer. In the latter case, yogurt and fruits are pumped into filling line from two separate dosing units (Chandon and O'Rell, 2006; Shah, 2006)

In set-type fruit yogurt production, the fruits are added to the milk prior to inoculation with the starter culture, and the mix is fermented afterward. This practice has a significant effect on the fermentation kinetics, since fruit sugar may slow down the metabolic activity of yogurt starter bacteria (see Figure 15.3). Also, the presence of the fruit in the curd increases the risk of syneresis. In order to prevent that, the polysaccharide-producing strains of yogurt starter bacteria are preferred or stabilizers are added. The optimum concentrations of stabilizers vary between 0.25 and 0.75%, depending on the type of stabilizer. Pectin and carboxymethylcellulose (CMC) are the most commonly used stabilizers in fruit yogurts. When sodium benzoate or potassium sorbate is used as a preservative, the concentrations of such preservatives should not exceed 0.03%.

Fruit fragments added in yogurts, whether after or prior to fermentation, must have been pasteurized beforehand. However the pasteurization heat treatment causes losses in texture and aroma of these fruits. In order to counter this, a vacuum infusion technique was developed. In this technique, fruit aroma is concentrated prior to heat treatment (Behare et al., 2016; Tamime and Robison, 2007a).

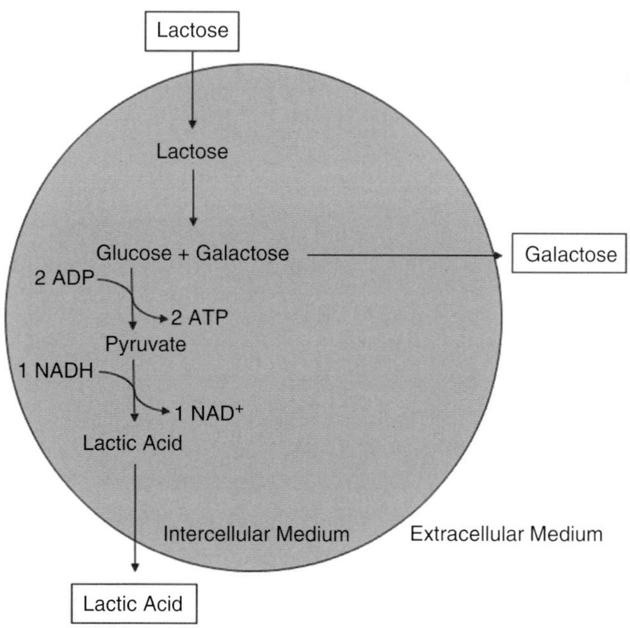

Figure 15.3 Simplified representation of the metabolic route in yogurt starter culture

15.3.6.3 Packaging

In order for yogurt products to reach the consumer in a state that is safe and that consumers like, the products must be adequately packaged and preserved. The material of industrial yogurt's packages include plastic, ceramic and glass. In the case of set-type yogurt, the cups are filled and sealed right after the inoculation with the starter culture, and the fermentation takes place in the retail containers. In the case of stirred yogurt, the cups are filled after the fermentation process. The technology of packaging is referred as 'form–fill–seal' technology. The process involves the following: the thermo-formation of the containers at 150–200 °C using multilayer thermoplastic materials, the filling of the preformed containers under a closed environment and sterile air overpressure, and the thermos-sealing of the filled containers with an aluminum lid labeled to deliver product information. After packaging, the packed yogurts are stored at low temperature (4 or 5 °C), which is maintained during transportation and commercialization. This low temperature ensures the safety and preserves the quality of the products as well as hinders the action of the starter culture, preventing post-acidification (Alvarez and Pascall, 2011; Corrieu and Beal, 2015; Shah, 2006).

15.3.7 Quality control of yogurt production

Yogurt quality requires control of the raw materials and the processes during the course of the manufacturing as well as on the final product. The microbiological content of all the raw material involved, fresh milk, powder milk, fruits, sweeteners, and starter culture, must be controlled in advance. Also, the presence of somatic cells on fresh milk must be dealt with. In addition, many physicochemical properties must be checked: titratable acidity, and fat and protein contents of the fresh milk; the absence of antibiotics, solubility, moisture, and fat content of the milk powder; and pH, viscosity, and Brix value of the added fruits. For the starter culture, the acidification activity and the viability of each strain must be assessed (Corrieu and Beal, 2015).

During the processing of yogurt manufacture, control over each individual process must be maintained in order to ensure minimum variation of the production and maximum levels of quality and food safety of the product. The main parameters controlled are the temperature (in fermentation tanks, heat exchangers, incubation rooms, and cooling systems), pH (by sampling either in the fermentation tanks or directly in cups) and the duration of each step of the manufacturing process. In addition to these controls, modern legislation forces the application of ISO 22000 or International Food Standards in order to ensure and verify the quality and safety of the product. Hazard Analysis and Critical Control Point (HACCP) systems are applied, as well, to yogurt manufacturing process, and critical control points are established for the pasteurization of milk, the refrigerated packaging, and cold storage. In addition, packaging is also related to physical hazards (Corrieu and Beal, 2015).

Finally, at the end of the manufacturing process and during the shelf life, several tests are performed on the final products. The frequency of sampling is defined by each manufacturer, as stated by its good hygiene practices (GMPs). The CFU of the two bacterial strains (ST and LB) must be measured in order to verify that they are within the range of the target value of 10^7 CFU*g$-^1$. Additionally, microbiological tests must ensure that the product is free of spoilage and pathogenic microorganisms, including *Listeria monocytogenes*, *Salmonella* spp., coliforms, yeasts and molds. Also, physicochemical properties such as the fat and total solid contents, the titratable acidity or the pH must be at specific levels and therefore continuously controlled. Last, the sensory profiles of the samples must be assessed, in particular properties such as appearance (syneresis and color), texture (palatability, firmness and consistency), aroma and odor, taste and aftertaste (freshness, acidity and persistency). Another important characteristic related to yogurt quality is syneresis occurring during storage as whey separates from the curd; this is considered a defect, thus reducing the product quality level and acceptance. The value of each physicochemical and sensory property of yogurt varies depending on the type: set-type yogurts have different thresholds for discarding compared to stirred type (Corrieu and Beal, 2015).

15.4 CONCLUSIONS

Yogurt is the best-known and most widely consumed fermented milk product, being manufactured and consumed by humans for more than 8000 years. Since the industrialization of the manufacturing process and the development of dairy science, yogurt making has improved quite a lot. All the processes involved have been thoroughly studied and improved, leading to production of standardized and high-quality yogurt products. Preliminary treatment of milk, homogenization, heat treatment, fermentation and post-fermentation process all contribute to the manufacture of safe and high-quality yogurt products. However, current innovations of food and dairy science are constantly applied to yogurt manufacture and continue to improve the product and the production process, making the classic process a field of significant scientific and industrial interest.

References

Aguilera JM, Kessler HG. 1988. Physicochemical and rheological properties of milk-fat globules with modified membranes. *Milchwissenschaft* 43:411–15.

Alvarez VB, Pascall MA. 2011. Packaging. In: Fuquay J, Fox PF, McSweeney PLH (Eds.). *Encyclopedia of dairy sciences*, vol. 4, 2nd ed. London: Elsevier Ltd. Academic Press, pp 16–23.

Behare P, Kumar H, Mandal S. 2016. Yogurt: yogurt based products. In: Finglas P, Toldrá F (Eds.). *Encyclopedia of food and health*. London: Academic Press, pp 625–35

Bermúdez-Aguirre D, Corradini MG, Mawson R, Barbosa-Canova GV. 2008. Modeling the inactivation of Listeria innocua in raw whole milk treated under thermo-sonication. *Innov Food Sci Emerg Technol* 10:172–8.

Boelrijk AEM, de Jong C, Smit G. 2003. Flavour generation in dairy products. In: Smith G. (Ed.). *Dairy processing*, chapter 7. Cambridge, UK: Woodhead Publishing Ltd., pp 128–53.

Brew K. 2011. α-Lactalbumin. In: Fuquay J, Fox PF, McSweeney PLH (Eds.). *Encyclopedia of dairy sciences*, 2nd ed. London, UK: Elsevier Ltd. Academic Press, pp 780–6.

Cano-Ruiz ME, Richter RL. 1997. Effect of homogenization pressure on the milk fat globule membrane proteins. *J Dairy Sci* 11:2732–39.

Chandan RC. 2006. History and consumption trends, chapter 1. In: Chandan RC (Ed.). *Manufacturing yogurt and fermented milks*. Ames, IA: Blackwell Publishing, pp 3–17.

Chandan RC, O'Rell KR. 2006. Principles of yogurt processing, chapter 12. In: Chandan RC (Ed.). *Manufacturing yogurt and fermented milks*. Ames, IA: Blackwell Publishing, pp 195–211.

Chandrapala J, Zisu B, Palmer M, Kentish S, Ashokkumar M. 2011. Effects of ultrasound on the thermal and structural characteristics of proteins in reconstituted whey protein concentrate. *Ultrason Sonochem* 18:951–57.

Cho YH, Lucey JA, Singh H. 1999. Rheological properties of acid milk gels as affected by the nature of the fat globule surface material and heat treatment of milk. *Int Dairy J* 9:537–45.

Ciron CIE, Gee VL, Kelly AL, Auty MAE. 2010. Comparison of the effects of high-pressure microfluidization and conventional homogenization of milk on particle size water retention and texture of non-fat and low-fat yoghurts. *Int Dairy J* 20:314–20.

Corrieu G, Beal C. Yogurt: the product and its manufacture. 2015. In: Caballero B, Finglas P, Toldrá F. (Eds). *Encyclopedia of food and health*. London: Elsevier Ltd. Academic Press pp 617–24.

Creamer LK, Loveday SM, Sawyer L. 2011. β-Lactoglobulin. In: Fuquay J, Fox PF, McSweeney PLH. (Eds.). *Encyclopedia of dairy sciences*, 2nd ed. London: Elsevier Ltd. Academic Press, pp 787–94.

Cruz AG, Cadena RS, Castro WF, Esmerino EA, Rodrigues JB, Gaze L, Faria JAF, Freitas MQ, Deliza R, Bolini HMA. 2013. Consumer perception of probiotic yogurt: performance of check all that apply (CATA) projective mapping sorting and intensity scale. *Food Res Int* 54:601–10.

De Brabandere AG, de Baerdemaeker JG. 1999. Effects of process conditions on the pH development during yogurt fermentation. *J Food Eng* 41:221–7.

Deeth HC, Datta N. 2011. Ultra-high temperature treatment (UHT): heating systems. In: Fuquay J, Fox PF, McSweeney PLH (Eds.). *Encyclopedia of dairy sciences*, 2nd ed. London: Elsevier Ltd. Academic Press. pp 699–707.

Demirdöven A, Baysal T. 2009. The use of ultrasound and combined technologies in food preservation. *Food Rev Int* 25:1–11.

Doganis P, Sarimveis H. 2008. Optimal production scheduling for the dairy industry. *Ann Oper Res* 159(1):315–31.

Dolatowski ZJ, Stadnik J, Stasiak D. 2007. Applications of ultrasound in food technology. *Acta Sci Pol Technol Aliment* 6:89–99.

Elizaquível P, Sánchez G, Salvador A, Fiszman S, Dueñas MT, López P, Fernández de Palencia P, Aznar R. 2011. Evaluation of yogurt and various beverages as carriers of lactic acid bacteria producing 2-branched (13)-β-D-glucan. *J Dairy Sci* 94:3271–78.

FAO/WHO Codex Alimentarius. 2011. CODEX STAN 243-2003 for Fermented Milks. Rome, Italy.

Fox PP. 2011a. Fat globules in milk. In: Fuquay J., Fox PF, McSweeney PLH (Eds.). *Encyclopedia of dairy sciences*, 2nd ed. London: Elsevier Ltd. Academic Press, pp 1564–48.

Fox PP. 2011b. Milk. In: Fuquay J, Fox PF, McSweeney PLH. (Eds). *Encyclopedia of dairy sciences*, 2nd ed. London: Elsevier Ltd. Academic Press, pp 458–631.

Gahruie HH, Eskandari MH, Mesbahi G, Hanifpour MA. 2015. Scientific and technical aspects of yogurt fortification: a review. *Food Sci Human Wellness* 4:1–8.

Hinrichs J, Atamar Z. 2011. Sterilization of milk and other products. In: Fuquay J, Fox PF, McSweeney PLH (Eds.). *Encyclopedia of dairy sciences*, 2nd ed. London: Elsevier Ltd. Academic Press, pp 761–64.

Horne DS. 1999. Formation and structure of acidified milk gels. *Int Dairy J* 9:261–8.

Horne DS. 2011. Casein micellar structure. In: Fuquay J, Fox PF, McSweeney PLH (Eds.). *Encyclopedia of dairy sciences*, 2nd ed. London: Elsevier Ltd. Academic Press, pp 772–79.

Hupperts T. 2011. Other types of homogenizer (high-speed mixing ultrasonics microfluidizers membrane emulsification). In: Fuquay J, Fox PF, McSweeney PLH (Eds.). *Encyclopedia of dairy sciences*, 2nd ed. London: Elsevier Ltd. Academic Press. pp, 761–64.

Ion-Titapiccolo G, Alexander M, Corredig M. 2013. Heating of milk before or after homogenization changes its coagulation behavior during acidification. *Food Biophys* 8:81–89.

Jaros D, Rohm H. 2003. Controlling the texture of fermented dairy products: the case of yoghurt, chapter 8. In: Smith G. (Ed.). *Dairy processing*. Cambridge, UK: Woodhead Publishing Ltd, pp 88–155.

Kasaai MR, Charlet G, Paquin P, Arul J. 2003. Fragmentation of chitosan by microfluidization process. *Innov Food Sci Emerg Technol* 4:403–13.

Kelly AL, Zeece M. 2009. Applications of novel technologies in processing of functional foods. *Aust J Dairy Technol* 64:12–16.

Kilara A. 2006. Basic dairy processing principles, chapter 5. In: Chandan RC (Ed.). *Manufacturing yogurt and fermented milks*. Ames, IA: Blackwell Publishing, pp 73–89.

Kneifel W, Mattila-Sandholm T, von Wright A. 1999. Probiotic bacteria detection and estimation in fermented and non-fermented dairy products. *Encycl Food Microbiol* 3:1783–89.

Krešić G, Lelas L, Jambrak AR, Herceg Z, Brnčić SR. 2008. Influence of novel food processing technologies on the rheological and thermophysical properties of whey proteins. *J Food Eng* 1:64–73.

Kristo E, Biliaderis CG, Tzanetakis N. 2003. Modelling of the acidification process and rheological properties of milk fermented with a yogurt starter culture using response surface methodology. *Food Chem* 83:437–46.

Lewis MJ. 2003. Improvements in the pasteurisation and sterilisation of milk, chapter 5. In: Smith G (Ed.). *Dairy processing*. Cambridge, UK: Woodhead Publishing Ltd, pp 79–102.

Madadlou A, Mousavi ME, Emam-Djomeh Z. 2009. Comparison of pH-dependent sonodisruption of re-assembled casein micelles by 35 and 130 kHz ultrasounds. *J Food Eng* 95:505–509.

Metchnikoff E. 1907. Essais Optimistes. Paris. In: Mitchell PC (Ed.). *The prolongation of life. Optimistic studies*. London: Heinemann.

Morand M, Guyomarc'h F, Famelart M. 2011. How to tailor heat-induced whey protein/κ-casein complexes as a means to investigate the acid gelation of milk – a review. *Dairy Sci Technol* 91:97–126.

Nguyen HA, Anema SG. 2010. Effect of ultrasonication on the properties of skim milk used in the formation of acid gels. *Innov Food Sci Emerg Technol* 11:616–22.

Riener J, Noci F, Cronin DA. 2009a. The effect of thermosonication of milk on selected physicochemical and microstructural properties of yoghurt gels during fermentation. *Food Chem* 114:905–11.

Riener J, Noci F, Cronin DA, Morgan D, Lyng JG. 2009b. Characterisation of volatile compounds generated in milk by high intensity ultrasound. *Int Dairy J* 19:269–72.

Robinson RK. 2011. Yogurt: types and manufacture. In: Fuquay J, Fox PF, McSweeney PLH (Ed.). *Encyclopedia of dairy sciences*, 2nd ed. London: Elsevier Ltd., pp 525–28.

Ronkart SN, Paquot M, Deroanne C, Fougnies C, Besbes S, Blecker CS. 2010. Development of gelling properties of inulin by microfluidization. *Food Hydrocoll* 24:318–24.

Rukke EO, Sørhaug T, Stephaniak L. 2011. Thermization of milk In: Fuquay J, Fox PF, McSweeney PLH (Eds.). *Encyclopedia of dairy sciences*, 2nd ed. London: Elsevier Ltd. Academic Press, pp 693–98.

Salminen S, Kenifel W, Ouwhand AC. 2011. Bacteria beneficial. Probiotics applications in dairy products. In: Fuquay J, Fox PF, McSweeney PLH (Eds.). *Encyclopedia of dairy sciences*, 2nd ed. London: Elsevier Ltd. Academic Press, pp 412–19.

Shah NP. 2003. Yogurt: the product and its manufacture. In: Caballero B, Trugo L, Finglas P (Eds.). *Encyclopedia of food sciences and nutrition*, 2nd ed. Amsterdam: Elsevier Science BV, pp 6252–59.

Shah NP. 2006. Health benefits of yogurt and fermented milks. In: Chandan R (Ed.). *Manufacturing yogurt and fermented milks*. Ames, IA: Blackwell Publishing, pp 322–40.

Shaker RR, Jumah RY, Abu-Jdayil B. 2000. Rheological properties of plain yogurt during coagulation process: impact of fat content and preheat treatment of milk. *J Food Eng* 44:175–80.

Skurtys O, Aguilera JM. 2008. Applications of microfluidic devices in food engineering. *Food Biophys* 3:1–15.

Soukoulis C, Panagiotidis P, Koureli R, Tzia C. 2007. Industrial yogurt manufacture: monitoring of fermentation process and improvement of final product quality. *J Dairy Sci* 90:2641–54.

Sfakianakis P, Topakas E, Tzia C. 2015. Comparative study on high-intensity ultrasound and pressure milk homogenization: Effect on the kinetics of yogurt fermentation process. *Food Bioprocess Technol* 8(3):548–57.

Sfakianakis P, Tzia C. 2014. Conventional and innovative processing of milk for yogurt manufacture development of texture and flavor: a review. *Foods* 3(1):176–93.

Tamime AY, Robison RK. 2007a. Historical background chapter 1. In: Tamime AY, Robison RK. (Eds). *Yogurt: science and technology*, 3rd ed. Cambridge UK: Woodhead Publishing Ltd, pp 1–10.

Tamime AY, Robison RK. 2007b. Backround to manufacturing practice, chapter 2. In: Tamime AY, Robison RK (Eds.). *Yogurt: science and technology*, 3rd ed. Cambridge UK: Woodhead Publishing Ltd, pp 11–118.

Taylor MW, Mac Gibbon AKH. 2011. Milk lipids. In: Fuquay J, Fox PF, McSweeney PLH (Eds.). *Encyclopedia of dairy sciences*, 2nd ed. London, UK: Elsevier Ltd., pp 649–740.

Vedamuthu ER. 2006. Starter cultures for yogurt and fermented milks, chapter 6. In: Chandan RC (Ed.). Manufacturing yogurt and fermented milks. Ames, IA: Blackwell Publishing, pp 89–117.

Vercet A, Oria R, Marquina P, Crelier S, López-Buesa P. 2002. Rheological properties of yoghurt made with milk submitted to manothermosonication. *J Agric Food Chem* 50:6165–71.

Vinderola G, Binetti A, Burns P, Reinheimer J. 2011. Cell viability and functionality of probiotic bacteria in dairy products. *Front Microbio* 2:70.

Walstra P, Wouters JTM, Geurts TJ. 2006a. Milk components, chapter 2. In: *Dairy science and technology*. Boca Raton: Taylor & Francis Group LLC, pp 17–108.

Walstra P, Wouters JTM, Geurts TJ. 2006b. Heat treatment, chapter 7. In: *Dairy science and technology*. Boca Raton: Taylor & Francis Group LLC, pp. 225–72.

Walstra P, Wouters JTM, Geurts TJ. 2006c. Homogenization, chapter 9. In: *Dairy science and technology*. Boca Raton: Taylor & Francis Group LLC, pp 276–79.

Walstra P, Wouters JTM, Geurts TJ. 2006d. Cooling and freezing, chapter 11. In: *Dairy science and technology*. Boca Raton: Taylor & Francis Group LLC, pp 297–307.

Walstra P, Wouters JTM, Geurts TJ. 2006e. Fermented milks, chapter 22. In: *Dairy science and technology*. Boca Raton: Taylor & Francis Group LLC, pp 551–73.

Wu H, Hulbert GJ, Mount JR. 2001. Effects of ultrasound on milk homogenization and fermentation with yoghurt starter. *Innov Food Sci Emerging Technol* 1:211–18.

Wu H, Hulbert GJ, Mount JR. 2009. Effects of ultrasound on milk homogenization and fermentation with yogurt starter. *Innov Food Sci Emerg Technol* 3:211–218.

16 Milk protein composition and sequence differences in milk and fermented dairy products affecting digestion and tolerance to dairy products

Maria Gabriella Giuffrida[1], Marzia Giribaldi[2], Laura Cavallarin[1] and Palmiro Poltronieri[3]

[1]CNR-ISPA, Institute of Sciences of Foods Productions, Dipartimento di Scienze Veterinarie dell'Università di Torino, Grugliasco, TO, Italy
[2]CREA-IT, Research Centre for Engineering and Agro-Food Processing, Turin, Italy
[3]Institute of Sciences of food Productions (CNR-ISPA), CNR, National Research Council of Italy, Lecce, Italy

16.1 INTRODUCTION

Milk, from a physico-chemical point of view, is an emulsion of lipid globules and a colloidal suspension of proteins and mineral aggregates, in a solution of carbohydrates (mainly lactose) and vitamins. Milk components have a great potential for use as health-promoting food and for novel applications, such as prebiotics, special foods and specific aliment to recover from disease states. Research in foods derived from milk constituents include (i) utilization of biological models to assess the molecular mechanisms and physiological effects of bioactive food components, (ii) development of bioprocessing techniques to isolate bioactive compounds or produce novel health-promoting foods that maintain desirable quality and safety over shelf-life, and (iii) characterisation of the molecular properties of food with health-promoting activities.

Milks of different origins have long been used and processed to dairy products for their longer shelf life and contribution to diets with macro- and micronutrients. Ruminant and non-ruminant milks, that possess a variable content in protein, lactose and fats (Figure 16.1) (Salimei et al., 2004; Pieszka et al., 2011, Sanz Ceballos et al., 2009), are the main sources available for humans to manufacture dairy products and fermented milks (Raynal-Ljutovac et al., 2008). Beside cow's milk and milk from other small ruminants (such as goat and sheep), research on milk from other species is still poorly exploited, despite the distribution of consumption of milk from these species (e.g. buffalo milk is the second in diffusion in the world) (FAO, 2013). More recently local producers have established a niche for the application of donkey milk, due to its peculiar

Microbiology in Dairy Processing: Challenges and Opportunities, First Edition.
Edited by Palmiro Poltronieri.

Published 2018 by John Wiley & Sons Ltd and the Institute of Food Technologists.

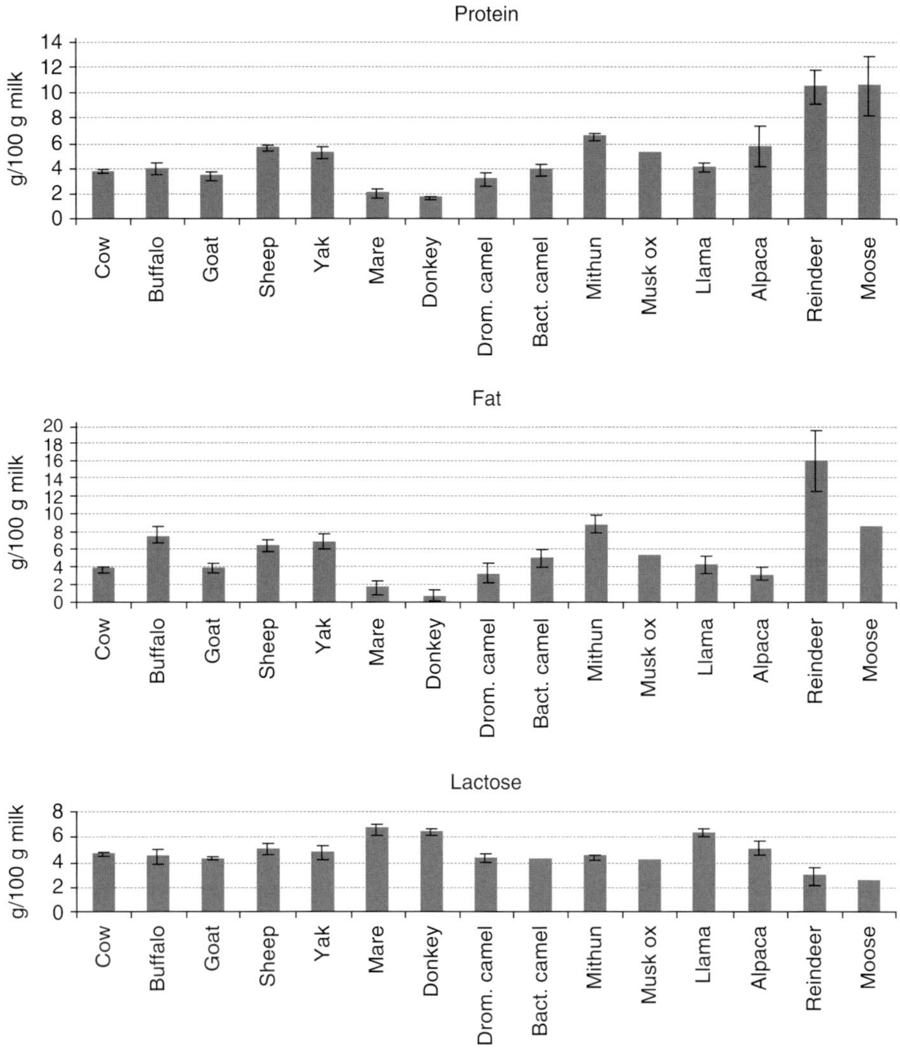

Figure 16.1 Differences in milk proteins, lactose and fats in milks of 15 species. Adapted from: FAO, 2013.

protein composition, which is more similar than bovine milk to that of human milk (Cavallarin et al., 2015; Martini et al., 2014).

Proteins are involved in most of the technological processes responsible for the wide range of dairy products worldwide. Milk proteins, independent from the origin, are grouped into caseins and whey proteins (whose proportions vary significantly according to the species, feeding type and lactation stage) and milk fat globule membrane (MFGM) proteins. In general, ruminant milk is relatively rich in caseins (nearly 80% of the total proteins), while non-ruminant milk has proportionally a higher whey protein content (Claeys et al., 2014). The proportion of non-protein nitrogen (free amino acids, peptides) is particularly relevant for milk of small ruminants, similar to human milk (Silanikove et al., 2010).

16.2 CASEINS

Caseins form a micellar colloidal suspension together with nanoclusters of amorphous calcium, and this fraction is involved in cheese-making and fermentation-associated modifications. From a nutritional point of view, the specific conformation of casein micelles is thought to have a role in the prolonged release of nutrients during digestion (Holt et al., 2013). The casein fraction of milk is characterised by a high degree of variability between species and individuals. Cow's milk contains α_{s1}-casein, α_{s2}-casein, β-casein and κ-casein phosphoproteins. Differentiation between α_S-,β-, and κ-caseins is based upon their primary structures (amino acid sequences), while α_S-casein classes are named based on differences in the degree of phosphorylation. Human, equine and goat milk contain higher concentrations of β-casein, which is more susceptible to peptic hydrolysis than α_S-caseins, particularly α_S1, which conversely predominates in cow and in sheep milk (Claeys et al., 2014). α-s1-casein plays a functional role in cheese curd formation, being characterised by a greater solubility in the presence of calcium.

Caseins are hydrophobic on their surface and are organised in milk as a suspension of casein micelles, with the interior side highly hydrated. The caseins in the micelles are held together by calcium ions that bind to phosphate groups of adjacent caseins and by hydrophobic interactions.

Casein contains in high quantity proline (17% in β-casein), contributing to the low presence of α helices in secondary structure. As for the essential amino acids, tryptophan, methionine and cysteine are present, especially in κ-casein (Sindayikengera and Xia, 2006; Salmen et al., 2012).

There are no disulfide bridges in α_{S1}- and β-casein, while α_{s2}- and κ-casein contain four cysteine residues forming disulfide bonds. α_{s2}-Casein exists as a disulfide-linked dimer, while up to 10 κ-casein molecules may be linked by disulfide bridges. K-casein is glycosylated, and contain galactose, *N*-acetyl galactosamine, *N*-acetylneuraminic acid (sialic acid, NANA), which is repeated every three or tetra saccharides, with a maximum of nine glycoside bonds attached to threonine. All caseins are phosphorylated on serines, from 1 to 2 groups in κ-casein (12% of total bovine caseins), 4 to 5 groups on β-casein (35% of total caseins), 8 or 9 in α_{s2}-casein (10% of total caseins), to 10 up to 13 in α_{S1}-casein (38% of total caseins).

Identification of milk proteins α_{s1}-, β-, and κ-caseins is based on their amino acid sequences. Caseins have been named on the basis of differences within these proteins due to posttranslational modification. α_{s0}-casein is identical to α_{s1}-casein, and α_{s3}-, α_{s4}-, and α_{s6}-caseins are identical to α_{s2}-casein with differences in degree of phosphorylation. Additionally, proteose peptone components 5, 8-slow and 8-fast, and γ_1-, γ_2-, and γ_3-caseins are N-terminal and C-terminal fragments, respectively, of β-casein formed during proteolysis by plasmin.

Rennet proteases cleave the Phe_{105}–Met_{106} bond at the carboxy terminal domain on κ-casein, containing the phosphorylated amino acids, thus releasing caseinomacropeptide (CMP), which causes milk clotting. CMP content is higher in cow, sheep and goat whey, reflecting the higher casein content. CMP is very low in equine milk, either due to a low casein/whey protein ratio, and to the low content of κ-casein (since other caseins support the precipitation of casein coagulum), and it contains 68 amino acids, compared to 63 amino acids in bovine CMP and 65 in human CMP. It is also named glycomacropeptide (GMP) and is a hydrophilic, thermostable peptide, with a net negative charge. CMP comprises from 10 to15% of whey proteins, thus is the third most abundant peptide in rennet-based cheese whey. CMP is glycosylated up to 36% in goat and to 30% in ovine milk (Moreno et al., 2000). The goat CMP contains 25 μg sialic acid per milligram of dry weight, twofold less than in bovine CMP (Moreno et al., 2001). CMP contains several phosphorylated serines and glycosylated threonines, and sialic acid, *N*-acetylgalactosamine and other mucin-like glucans (Thomä-Worringer et al., 2006; Moreno et al., 2000). The region of the major site of phosphorylation is well conserved in the caseins of other species, and it is strongly resistant to digestive degradation, thus increasing permanence in the gut and the risk of sensitization (Tsabouri et al., 2014). Recently, equine milk proteins were shown to be rapidly digested by gastric juices (Inglingstad et al., 2010).

Milk protein genetic polymorphism has received considerable research interest in recent years because of possible associations between milk protein and economically important traits in livestock. Four casein genes (*CSN1S1*, *CSN2*, *CSN1S2* and *CSN3*, respectively) are located as a cluster on chromosome 6 in cattle, sheep and goats (Ferretti et al., 1990; Hayes et al., 1993; Popescu et al., 1996). Polymorphism of casein genes have been well studied in cattle and goats (Caroli et al., 2007; Marletta et al., 2007; Rando et al., 2000), and the variants have been found to be associated with the composition and biological properties of milk (Martin et al., 2002; Vinesh et al., 2013; Johansson et al., 2015).

16.2.1 Gene polymorphisms in *κ-casein* genes

The κ-casein (*CSN3* gene) protein is a single-chain polypeptide of 169 amino acids with a molecular weight of 19.2 kDa. The κ-casein protein plays an important role in milk chemistry by providing colloidal stability to the casein micelle. In the micelle, κ-casein is mostly located at the periphery, with its hydrophilic C-terminal sequence protruding into the solvent. The *κ-casein* gene has been assigned to chromosome 6 (6q31) in cattle (Threadgill and Womack, 1990). Up to 16 genetic variants of bovine *κ-casein* have been identified, corresponding to proteins with different posttranslational modifications (PTMs) and levels of phosphorylation and glycosylation, in most cases due to the presence of serine (O'Riordan et al., 2014).

Two κ-casein protein variants have been described in detail as κ-casein A and B. The difference between A and B lies in single amino acid substitutions at positions 136 and 148 (Eigel et al., 1984). Attention has been drawn to the mutations in bovine κ-casein, Arg_{10} (variant F2) and Arg_{97} (variants C, D and G1) and to Pro_{130}, resulting either in the loss (position 10 and 97) or in the appearance (position 130) of a site of cleavage by pancreatic trypsin in the digestive tract. In addition, the Ser_{104}Ala modification occurring in the I variant altering the chymosin-sensitive site Ser-Phe-Met-Ala affecting the rennetability and clotting of milk (Martin et al., 2011).

In *Bos indicus* cattle, the predominant genotype is *CSN3 AA*, as in European breeds. The *κ-casein* gene *BB* genotype showed a greater influence on monthly milk yield, 305-days milk yield, monthly soluble nitrogen fraction (SNF) yield and monthly protein yield. The study was an aid in the selection of superior animals, that is, those with genotype *BB*, in order to increase the production of milk and its constituents (Rachagani and Gupta, 2008).

Five polymorphic sites have been described at the *κ-casein* (*CSN3*) locus in the Indian domestic Gaddi goat (*Capra hircus*). Sequence analysis of the *κ-casein* gene in sheep showed three novel nucleotide changes in malpura sheep when compared with the exotic sheep. These results highlight the importance of taking into consideration the *CSN3* SNPs (single nucleotide polymorphisms) when performing selection for milk composition in dairy livestock breeds. There are also studies on goat milk differences in European breeds (Martin et al., 2011; Martini et al., 2010).

The *κ-casein B* allele and *β-lactoglobulin (β-LG) B* variant are associated in cows of "Brown Italian" breed (Boettcher et al., 2004). This milk shows better technological performance, higher protein content, and is at the base of dairy products labelled as "made of Brown Italian milk".

16.2.2 Gene polymorphisms in *β-casein* gene

The most polymorphic milk protein gene is *β-casein*; it can give rise to 13 protein variants in cattle (Singh et al., 2015). The studies explored the genetic polymorphisms in exon 7 of *β-casein* and exon 4 of *κ-casein* genes in Arunachali yaks (*Bos grunniens*), Sahiwal (*Bos indicus*) cattle, malpura sheep (*Ovis aries*) and Gaddi goat (*Capra hircus*). Results of the study revealed presence of 11 SNP variants in all livestock species. Four SNPs were observed in *Bos indicus*, two SNPs in *Bos grunniens*, three SNPs in *Ovis aries* and three SNPs in *Capra hircus*. These variations are found to be synonymous in nature as these variations do not result in changes in corresponding amino acids.

The principal differences between β-casein protein variants are: A1/A2: $His_{67}Pro$; A2/A3: $His_{106}Gln$; A1/B: $Ser_{122}Arg$; A1/C: $Glu_{37}Lys/SepP_{35}Ser$; A2/D: $SepP_{18}Lys$; A2/E: $Glu_{36}Lys$.

In the A2 protein variant of β-casein a proline residue occurs at position 67, whereas the A1 and B variants of β-casein have a histidine residue at this position. In the case of the variants containing proline, the enzymatic hydrolysis of the Ile_{66}-Pro_{67} bond does not occur or occurs at a very low rate. A1 and A2 are processed differently by digestive enzymes, and a seven-amino peptide, β-casomorphin-7 (BCM-7), (β-casein f(60-66)), can be released by digestion of A1-β-casein (Hartwig, 1997; Jinsmaa et al., 1999).

Since individuals raised some concern on the safety of A1-β-casein and the dairy products containing it, especially for the potential delayed intestinal transit linked to gut opioid receptors for the bioactive peptide β-casomorphin (BCM-7), EFSA released a safety study which assessed there is no impact on health by consuming A1-β-casein and on lack of correlation between production of BCM-7 and risk factors for diabetes, cardiovascular disease and atherosclerosis (EFSA, 2009).

The A1-β-casein is the most common β-casein type protein found in cow's milk in Europe (excluding France), the United States, Australia and New Zealand. The other types of dairy cows are smaller, brownish and white in colour. Brown Swiss, Jersey and Guernsey cows produce lesser volumes of milk, are naturally resistant to diseases, and convert grass to milk quite efficiently. The level of A1 casein in these animals is very low, and they have higher levels of A2. Their milk is similar to that of other animals, including goat, sheep, buffalo, yak, donkey and camel: the milk from these animals contain mostly A2 and little A1.

16.3 PROTEOLYTIC RELEASE OF BIOACTIVE PEPTIDES IN FERMENTED MILK AND CHEESE

There is a difficulty in identifying and quantifying bioactive peptides in complex food matrices containing many peptides as a result of milk protein hydrolysis. Recent studies with 21 *Lactobacillus* strains isolated from fermented milks have demonstrated that these lactic acid bacteria (LAB) were able to degrade 80–90% of the β-casein following 72–96 h *in vitro* incubation of a sodium caseinate solution (Tzvetkova et al., 2007). LAB proteinases from *Lactococcus lactis* subsp. *cremoris* Wg2 hydrolysed 40% of the peptide bonds in β-casein, resulting in more than 100 oligopeptides (Mierau et al., 1997; Juillard et al., 1995).

The formation/degradation of BCM-7 has been studied mainly using purified caseins or synthetic peptides as substrates and single bacterial strains or their associated proteinase/peptidase systems.

A number of studies have reported on the formation and fate of BCM-7 in fermented dairy products, including cheese, fermented milk and yoghurt, with varying information on the level of BCM-7 in dairy products (Sabikhi and Mathur, 2001). The formation of casomorphins in dairy products fermented with LAB is temporary, since dairy-associated bacteria have a high intracellular X-prolyldipeptidylaminopeptidase (PepX) activity (Gobbetti et al., 2002). PepX is the best characterised proline-specific enzyme produced by LAB, and it releases X-Pro dipeptides from the N-terminus of peptides. Hydrolysis with PepX removes the X-Pro sequence central for the bioactivity of BCM-7. Cell-free extracts of wild-type or PepX-mutant strains of *Lactobacillus helveticus* L89 totally or partially hydrolysed synthetic BCM-7, forming BCM-4 after incubation at 37 °C for 120 min (Matar and Goulet, 1996). Formation of BCM-4 but not BCM-7 in a pasteurized (65 °C for 30 min) milk fermented with a PepX-deficient mutant of *Lb. helveticus* L89 was shown (Matar and Goulet, 1996). Inhibition by BCM-7 on *in vitro* activity of PepO and PepN from *Lc. lactis* ssp. *lactis* MG1363 was hypothesised, but PepX was able to hydrolyse BCM-7 (Stepaniak et al., 1995).

The type of starter used seems to affect the nature of the bioactive peptides released in fermented milks. *Lactobacillus delbrueckii* subsp. *bulgaricus* and *S. thermophilus,* used in combination as starters in yoghurt manufacture, release many peptides. In general, due to the short fermentation time, lysis of cells of LAB is unlikely and intracellular peptidase activity is unlikely to significantly contribute to proteolysis during yoghurt manufacture (Meisel and Bockelmann, 1999). The proteolytic activity in yoghurt containing *Lb. delbrueckii* ssp. *bulgaricus* Lb1466 and *Streptococcus thermophilus* St1342 occurs during the first 24 h following manufacture and subsequently continues to increase at a slower rate during storage (28 d) at 4 °C (Donkor et al., 2007). The storage at 4 °C for 3weeks resulted in proteolysis of milk proteins in yoghurt made from skimmed milk heated to 90 °C and fermented (44 °C for 3 h) using the above two LAB strains (Schieber and Brückner, 2000). In this product, the majority of the peptides arose from β-casein A1 breakdown, but only peptides, including BCM-containing sequences (i.e. β-casein 57-68 and β-casein 57-72), were identified by means of HPLC-MS and peptide sequencing.

The formation and fate of BCM-7 have been studied in different cheese varieties which are considered a very complex food system. The extent and the pattern of proteolysis in cheese varies as a function of heat treatment of milk (unprocessed or pasteurized), type of coagulant (animal or vegetable), curd handling (cooking, salting, pH at draining), starter culture and ripening conditions (pH, time, temperature, humidity, secondary microbiota). Most of the proteolysis occurs during ripening, and as a result caseins are hydrolysed by endogenous proteinases, coagulant, and starter and non-starter proteinases. All these factors influence the release and degradation of bioactive peptides and must be determined under real cheese-making conditions.

16.4 MINOR MILK PROTEINS

Milk has important biological activities in terms of human health (Gill and Cross 2002; Clare et al. 2003; Smolenski et al. 2007). Minor bovine milk glycoproteins include many MFGM proteins in addition to α-lactalbumin (αLA) (Teschemacher, 2003), lactoferrin (Lf), lactoperoxidase (LP) (Severin and Wenshui, 2005), folate-binding protein (FBP) and glycolactin (O'Riordan et al., 2014; Martín-Ortiz et al., 2016).

Nonglycosylated proteins are whey acidic protein (Nukumi et al., 2007), lysozyme (Jirillo and Magrone, 2014), and β-lactoglobulin (Hernandez-Ledesma et al., 2008). High mobility group (HMG) proteins (Yamamura et al., 1999), milk basic protein (Matsuoka et al., 2002; Weinberg, 2003; Wakabayashi et al., 2006); ceruloplasmin (Sokolov et al., 2006) are also included in the minor fraction of milk proteins.

Whey protein composition differs in milks of different mammal species, thus it may range from 6 to 10 g/l of proteins. The casein/whey protein ratio differs between mammalian species, ranging from 3.2:1 up to 4.7:1 in goat, sheep and cow, while it is 0.4:1 in human, 1.1:1 in horse and 1.3:1 in donkey milk (Uniacke-Lowe et al., 2010). Whey proteins contain a high amount of sulfured amino acids, able to enhance the immune function of the organism. The major whey proteins undergo denaturation upon heating to a temperature above 60° C, forming inter- and intramolecular bonds that are either physical (hydrophobic, electrostatic, etc.) or chemical (disulfide bonds). The presence of protein variants differing in amino acid substitutions codified by alleles in different varieties or in different species increase the variability in heat stability of each whey protein.

16.4.1 Lactoferrin

Lf is a natural defence protein with iron-binding ability, present in exocrine secretions and in body fluids that are commonly exposed to normal flora. Lf content in whey varies according to the species of origin: from high content, i.e. 1-2 g/l in human milk (20% of total N), to 600 mg/l in equine whey, but only 100-200 mg/l in bovine, goat and sheep whey. Cow, buffalo, goat and sheep Lfs share over 90% sequence identity and form an extremely closely related group. Lf binds iron ions on a wide range of pH (10% of Lf is saturated with Fe2+) and other essential elements (vanadium, manganese, molybdenum and zinc ions). Lf structure is bilobular, each lobe binding and retaining one Fe2+, even at low pH (except the C-terminal lobe of camel Lf) (Baker et al., 2000; Baker and Baker, 2005).

Lf is highly glycosylated, and shows a wide variability in glycosylation sites and types among different species, but also among individuals. Some of the peculiar features of Lf, such as resistance to proteolysis, bifidogenic potential and eventual allergenicity, may be dependent on differential glycosylation (O'Riordan et al., 2014; Gallina et al., 2016).

Most of the glycosylation sites are highly exposed on the protein surface, but there are glycosylation sites, such as Asn_{545} in cow, buffalo, sheep and goat Lfs, in a surface cleft between the two domains of the C-lobe, providing additional interactions with both domains, that may help stabilize the closed state (Jabeen et al., 2005; Albar et al., 2014). Lf resistance to proteolysis has been demonstrated by recovery of partially digested proteins in the stool of nursing infants (Lonnerdal, 2003).

16.4.2 β-Lactoglobulin (β-LG)

β-Lactoglobulin is a globular protein, not present in human milk, and it is the most abundant protein in the whey of other mammalian species: 3.7 g/l in sheep, 74% of whey proteins, from 3 to 3.2 g/l in goat and cow whey, respectively (>50% of whey proteins and 12% of total milk proteins), and 2.5 g/l in equine whey. Yak milk contains almost twice as much β-LG than bovine milk. The sequence of bovine, ovine and equine β-LGs contains several amino acid substitutions, and in equine β-LG an additional glycine produces a longer protein. There is 60% sequence homology between β-LG from equine milk and cow milk. Dromedary camel and Bactrian camel milks do not contain measurable amounts of β-LG, similarly to human milk (Omar et al., 2016; Felfoul et al., 2017).

Bovine β-LG has a molecular weight of 18 kDa and contains two intramolecular bonds with five cysteines. Porcine and β-LG in other species lack one cysteine. The cysteine residue is important during milk thermal denaturation: it reacts with cysteines in κ-casein, and affects rennet coagulation and heat stability of milk.

Bovine β-LG is quite resistant to digestion by pepsin. This resistance may contribute to the intolerance to cow milk. Resistance to digestion is not uniform among species, with ovine β-LG being more sensitive to pepsin proteolysis (El-Zahar et al., 2005). Horse and donkey β-LG are well tolerated and digested.

16.4.3 α-Lactalbumin (α-LA)

α-Lactalbumin is the second most abundant whey protein and the major whey protein in human and camel milk. Its content is 2 g/l in goat (27% of whey proteins), 1.2 g/l in sheep and cow (15% of total whey proteins), and 2.4 g/l in equine whey. It binds and transports retinol, fatty acids and calcium ions. α-LA is rich in branched amino acids isoleucine, leucine and valine and contains several cysteines, forming four intra-molecular disulfide bonds. α-LA contains two major domains: the α-domain, which forms four α-helices, and the β-domain, which contains a β-sheet and loop regions (Wu et al., 1996). In α-LA, metal binding resides in the calcium-binding loop (residues DKFLDDDITDDI in human protein) connecting the two domains (Permyakov et al., 2016). In addition to four disulfide bonds, the native fold of α-lactalbumin is stabilized by binding of calcium. Although in various mammals, α-LAs are poorly conserved, possessing an overall 16% sequence identity, the positions of all eight cysteines and a calcium-binding site are strongly conserved. The conserved calcium-binding loop is located within a region with high structural flexibility. α-LA interacts with other metal ions, such as zinc (for which it has a specific binding site). The binding of Mg^{2+} to the Ca^{2+} site increases α-lactalbumin stability against action of heat and denaturing agents, with high stabilization effect. A feature of α-LA structure is the acidic conformational transition occurring between pH 3 and 4, due to competition between calcium ions and protons for the carboxyl side chains (Permyakov et al., 2016). The functionality of α-LA is probably due to being a rich source of tryptophan, which promotes the synthesis of serotonin, an important regulator of appetite and mood (Beulens et al., 2004). A study on consumption of a diet enriched with α-LA showed positive results (Markus et al., 2000), with an increase in the plasma tryptophan and reduction of depressive mood in stress-vulnerable subjects under acute stress.

16.5 PROTEINS WITH BIOACTIVE ROLES

Ceruloplasmin is a copper-binding glycoprotein, also known as ferroxidase, with a molecular weight of 126 kDa, binding six atoms of Cu^{2+} per molecule. Prosaposin is a high molecular weight glycoprotein in bovine milk, a necrotrophic factor with a role in development and maintenance of nervous tissue (Calcutt et al., 1999).

β_2-Microglobulin (β2-MG) consists of 98 amino acids, with an estimated mass of 11.636 Da, probably originated by proteolysis of a larger protein. β2-MG is highly conserved among different species, and overall structures are virtually identical, implying a universality of β2-MG.

A whey protein with a molecular weight of 10 kDa was shown to stimulate the proliferation and differentiation of murine osteoblastic MC3T3-E1 cells *in vitro*. The amino-terminal sequence of this 10-kDa protein was identical to bovine high mobility group-1 (HMG-1) protein. This 10-kDa protein is a basic protein with a HMG box consensus sequence motif (Yamamura et al., 1999).

Osteopontin (OPN) is a highly phosphorylated acidic glycoprotein of 60 kDa with 50 potential calcium-binding sites. An acidic whey fraction containing OPN was found to reduce bone loss in ovariectomized rats (O'Mahony et al., 2013). OPN was shown to bind lactoperoxidase and lactoferrin and may work as a carrier for these proteins.

Whey acidic protein (WAP) has a molecular weight of 14–30 kDa, varying with differences in glycosylation, and contains three to four disulfide bonds. In humans and ruminants, the *WAP* gene contains a frameshift with inactivation of translation. WAP functions as protease inhibitor and has antibacterial activity (Martin et al., 2011). Proline-rich polypeptide (PRP) was studied for immunotropic functions, including promotion of T cell activation and inhibition of autoimmune disorders such as multiple sclerosis (Camfield et al., 2011).

Lysozyme (Lyz) is an enzyme conferring antibacterial property to saliva and body fluids (Moatsou, 2010). Its content in mammalian milks varies from less than 1 mg/l in bovine milk to 300 mg/l in human milk and 790 mg/l in equine milk (Fox and Kelly, 2006) so that antimicrobial activity of human, horse and donkey milk is mainly determined by lysozyme and lactoferrin, whereas lactoperoxidase and immunoglobulins are the main defence proteins in bovine milk. In literature, values of specific activity in raw donkey milk range from about 4000 to 162,000 U/ml of milk (Ragona et al., 2016).

Folate-binding protein (FBP) is a glycoprotein with about 22% carbohydrates (Rubinoff et al., 1977; Hansen et al, 1979; Chen et al., 2006) and binds 9.2 mg folic acid/g of protein. Glycosylation seems to have a role in determining its resistance to digestion, which facilitates the vitamin-carrier role of the protein (Lonnerdal, 2003; Nygren-Babol et al., 2004).

16.6 MFGM-ASSOCIATED PROTEINS

The milk fat globule membrane (MFGM) core proteins (Fong et al., 2007; Cavaletto et al., 2008; Spertino et al., 2012; Pisanu et al., 2012; Lu et al., 2016) include: mucins (MUC) 1 and 5; adipophilin (ADPH); lactadherin or periodic acid–Schiff glycoproteins (PAS) 6 and 7 (PAS-6/7) (O'Riordan et al., 2014); fatty acid binding protein (FABP); cluster differentiation 36 (CD36), a 78-kDa *N*-glycosylated protein; and butyrophilin (BTN), with the role of anchoring xanthine oxidoreductase (XOR), a metal (Mo, Fe) binding protein, to the MFGM (Cavaletto et al., 2008; Spertino et al., 2012).

Proteose peptone component 3 (PP3) is loosely associated to MFGM. It is a 28-kDa phosphorylated glycoprotein also known as lactophorin (O'Roirdan et al., 2014), with a lipase inhibition activity. PP3 contains two distinct domains. The first covers the N-terminal part (residues 1–97) and is negatively charged. The second domain in the C-terminal, lactophoricin, is positively charged, amphipathic, and adopts an α-helical structure (Campagna et al., 2001, 2004). Proteose peptone (PP) is abundant in bovine (300 mg/l) and goat milk and whey. This fraction has been characterised as a mixture of heat-stable, acid-soluble (at pH 4.6), phosphorylated glycoprotein of 135 amino acids in length.

16.7 COW'S MILK PROTEIN ALLERGY (CMPA)

CMPA affects 2–7.5% of children; persistence in adulthood is uncommon since a tolerance develops in 51% of cases within 2 years and 80% within 3–4 years. CMPA is an immunological reaction to one or more milk proteins: bovine serum albumin (BSA), bovine insulin, β-lactoglobulin, α-lactalbumin and β-casein (Karjalainen et al., 1992; Vaarala et al., 1996; Monetini et al., 2002; Luopajärvi et al., 2008). IgE or non-IgE-associated responses have been linked to immediate or late onset symptoms (Solinas et al., 2010; Pal et al., 2015). Studies have suggested α_{S1}-casein to be a major milk allergen, causing strong allergic reactions (Tsabouri et al, 2014). Sensitization is very frequent against α_{S1}-casein and k-casein, as well as against β-LG. Some individuals show cross-reactivity to milks of related species (sheep, goat): this is less frequent with equine milk. Milk from some goat breeds, lacking α_{S1}-casein, was shown to be less allergenic (El-Agamy, 2007). The World Allergy Organization states that goat milk should not be used as a substitute milk for children with cow milk allergy (Fiocchi et al., 2010). Patients with positive skin prick tests (SPT) tolerate well camel, donkey and (to a lesser extent) goat milk (Ehlayel et al., 2011).

β-lactoglobulin is involved in allergic reactions to cow's milk (Sabikhi, 2007). Enhanced humoral and cellular immune response to β-LG was shown in some diabetic patients compared with control subjects (Savilahti et al., 1993; Saukkonen et al., 1995; Vaaralaet al., 1996). A recent study points out to a possible cross-reaction between anti-β-LG and the human protein glycodelin (PP14), a T cell modulator, undermining T cell regulation of beta cells in infancy (Goldfarb, 2008).

Owing to the similarity of the milk protein profile in equine and human milk, equine milks have been suggested for use in children with severe IgE-mediated CMPA (Businco et al., 2000; Monti et al., 2007, 2012). A recent study tested the milk from donkeys of a unique native donkey breed (Amiata donkey) in 32 children with CMPA, showing that 78.8% resulted in negative response to skin tests, and 97% of them tolerated donkey milk (Sarti et al., 2016). The good tolerability of donkey milk in children suffering from CMPA could thus be due to the levels of its major allergenic milk components. Its low casein content and casein/whey protein ratio may play a role in the sensitization capacity of the milk. Other factors may help to explain the good tolerability of donkey milk, such as the number of casein fractions, the primary structure of the milk proteins, and the differences in digestibility of potential milk allergens, factors which have not yet been analysed in depth (Martini et al., 2014).

16.8 CONCLUSIONS

The possibility to exploit milks of different origin may lead to the consumption of dairy products with beneficial effects to specific groups of individuals, thanks to the availability of protein variants with characteristics such as resistance to proteolytic enzymes or ease of processing by LAB and dairy-associated species during dairy fermentation.

Up to now, EFSA has declared that there is no significant risk for consumption of β-casomorphins and related peptides deriving from β-casein variant A1.

Other polymorphisms and their association in certain types of milk were shown to have favourable effects on milk technological traits such as yield and rennetability of these types of milk.

References

Albar AH, Almehdar HA, Uversky VN, Redwan EM. 2014. Structural heterogeneity and functionality of lactoferrin. *Curr Prot Peptide Sci* 15:778–97.

Baker EN, Baker HM. 2005. Molecular structure, binding properties and dynamics of lactoferrin. *Cell Mol Life Sci* 62:2531–39.

Baker HM, Anderson BF, Kidd RD, Shewry SC, Baker EN. 2000. Lactoferrin three dimensional structure: a framework for interpreting function. In: Shimazaki K, Tomita M, Tsuda H (Eds.). *Lactoferrin: structure, function and applications*. Amsterdam: Elsevier Science, pp 3–15.

Beulens JWJ, Bindels JG, de Graaf C, Alles MS, Wouters-Wesseling W. 2004. Alpha-lactalbumin combined with a regular diet increases plasma Trp-LNAA ratio. *Physiol Behav* 81:585–93.

Boettcher PJ, Caroli A, Stella A, Chessa S, Budelli E, Canavesi F, Ghiroldi S, Pagnacco G. 2004. Effects of casein haplotypes on milk production traits in Italian Holstein and Brown Swiss cattle. *J Dairy Sci* 87(12):4311–17.

Businco L, Giampietro PG, Lucenti P, Lucaroni F, Pini C, Di Felice G, Lacovacci P, Curadi C, Orlandi M. 2000. Allergenicity of mare's milk in children with cow's milk allergy. *J Allergy Clin Immunol* 105(5):1031–34.

Calcutt NA, Campana WM, Eskeland NL, Mohiuddin L, Dines KC, Mizisin AP, O'Brien JS. 1999. Prosaposin gene expression and the efficacy of a prosaposin-derived peptide in preventing structural and functional disorders of peripheral nerve in diabetic rats. *J Neuropathol Exp Neurol* 58(6):628–36.

Camfield DA, Owen L, Scholey AB, Pipingas A, Stough C. 2011. Dairy constituents and neurocognitive health in ageing. *Brit J Nutrition* 22:1–17.

Campagna S, Cosette P, Molle G, Gaillard JL. 2001. Evidence for membrane affinity of the C-terminal domain of bovine milk PP3 component. *Biochim Biophys Acta* 1513:217–22.

Campagna S, Mathot AG, Fleury Y, Girardet JM, Gaillard JL. 2004. Antibacterial activity of Lactophoricin, a synthetic 23-residues peptide derived from the sequence of bovine milk component-3 of protease peptone. *J Dairy Sci* 87:1621–26.

Caroli A, Chiatti F, Chessa S, Rignanese D, Ibeagha-Awemu EM, Erhardt G. 2007. Characterization of the casein gene complex in West African goats and description of a new αs1-casein polymorphism. *J Dairy Sci* 90:2989–96.

Cavaletto M, Giuffrida MG, Conti A. 2008. Milk fat globule membrane components – a proteomic approach. *Adv Exp Med Biol* 606:129–41.

Cavallarin L, Giribaldi M, Soto-Del Rio MdlD, Valle E, Barbarino G, Gennero MS, Civera T. 2015. A survey on the milk chemical and microbiological quality in dairy donkey farms located in NorthWestern Italy. *Food Control* 50:230–35.

Chen W, Lee PJ, Stapels M, Gebler JC. 2006. The use of mass spectrometry to determine location and extent of N-glycosylation on folate binding protein from bovine milk. *Rapid Commun Mass Spectrom* 20:313–16.

Claeys WL, Verraes C, Cardoen S, De Block J, Huyghebaert A, Raes K, Dewettinck K, Herman L. 2014. Consumption of raw or heated milk from different species: An evaluation of the nutritional and potential health benefits. *Food Control* 42:188–201.

Clare DA, Catignani GL, Swaisgood HE. 2003. Biodefense properties of milk: the role of antimicrobial proteins and peptides. *Curr Pharm Des* 9:1239–55.

Donkor ON, Henriksson A, Singh TK, Vasiljevic T, Shah NP. 2007. ACE inhibitory activity of probiotic yoghurt. *Int Dairy J* 17(11):1321–31.

EFSA. 2009. Scientific Report of EFSA prepared by a DATEX Working Group on the potential health impact of β-casomorphins and related peptides. EFSA Scientific Report 231, 1-107. *EFSA J* 7(2).

Ehlayel M, Bener A, Abu Hazeima K, Al-Mesaifri F. 2011. Camel milk is a safer choice than goat milk for feeding children with cow milk allergy. *ISRN Allergy* 391641.

Eigel WN, Butler JE, Ernstrom CA, Farrell HM Jr, Harwalkar VR, Jenness R, Whitney RMcL. 1984. Nomenclature of proteins of cow's milk: 5th rev. *J Dairy Sci* 67:1599–631

El-Agamy EI. 2007. The challenge of cow milk protein allergy. *Small Ruminant Res* 68(1–2): 64–72.

El-Zahar K, Sitohy M, Choiset Y, Métro F, Haertlé T, Chobert J-M. 2005. Peptic hydrolysis of ovine b-lactoglobulin and a-lactalbumin. Exceptional susceptibility of native ovine b-lactoglobulin to pepsinolysis. *Int Dairy J* 15:17–27.

FAO. 2013. Muehlhoff E, Bennett A, McMahon D. Milk and dairy products in human nutrition. Rome. Available at: www.fao.org/docrep/018/i3396e/i3396e.pdf.

Felfoul I, Jardin J, Gaucheron F, Attia H, Ayadi MA. 2017. Proteomic profiling of camel and cow milk proteins under heat treatment. *Food Chem* 216:161–69.

Ferretti L, Leone P, Sgaramella V. 1990. Long range restriction analysis of the bovine casein genes. *Nucleic Acids Res* 18:6829–33

Fiocchi A, Brozek J, Schünemann H, Bahna SL, Berg AV, Beyer K, Bozzola M, Bradsher JB, Compalati E, Ebisawa M, Guzmán MA, Li H, Heine RG, Keith P, Lack G, Landi M, Martelli A, Rancé F, Sampson H, Stein A, Terracciano L, Vieths S. 2010. World Allergy Organization (WAO) diagnosis and rationale for action against cow's milk allergy (DRACMA) guidelines. *Pediatr Allergy Immunol* 21(suppl. 21):1–125.

Fong BY, Norris CS, McGibbon AKH. 2007. Protein and lipid composition of bovine milk-fat globule membrane. *Int Dairy J* 17:275–88.

Fox PF, Kelly AL. 2006. Indigenous enzymes in milk: overview and historical aspects – Part 2. *Int Dairy J* 16:517–32.

Gallina S, Saletti R, Cunsolo V, Muccilli V, Foti S, Roepstorff P, Rasmussen MI. 2016. Site-specific glycosylation of donkey milk lactoferrin investigated by high-resolution mass spectrometry. *Amino Acids* 48(7):1569–80.

Gill HS, Cross ML. 2000. Anticancer properties of bovine milk. *Brit J Nutri* 84:161–66.

Gobbetti M, Stepaniak L, De Angelis M, Corsetti A, Di Cagno R. 2002. Latent bioactive peptides in milk proteins: Proteolytic activation and significance in dairy processing. *Crit Rev Food Sci Nutr* 42(3):223–39.

Goldfarb MF. 2008. Relation of time of introduction of cow milk protein to an infant and risk of type-1 diabetes mellitus. *J Proteome Res* 7(5):2165–67.

Hansen SI, Holm J, Lyngbye J. 1979. Change in binding properties of folate-binding protein in cow's whey due to removal of a cofactor during affinity chromatographic purification. *Biochim Biophys Acta* 579(2):479–82.

Hartwig A, Teschemacher H, Lehmann W, Gauly M, Erhardt G. 1997. Influence of genetic polymorphisms in bovine milk on the occurrence of bioactive peptides. Seminar on milk protein polymorphism, International Dairy Federation, special issue 9702, 459–460.

Hayes H, Petit E, Bouniol C, Popescu P. 1993; Localisation of the alpha-S2-casein gene (CASAS2) to the homologous cattle, sheep and goat chromosome 4 by in situ hybridisation. *Cytogenet Cell Genet* 64:282–85.

Hernandez-Ledesma B, Recio I, Amigo L. 2008. β-Lactoglobulin as source of bioactive peptides. *Amino Acids* 35:257–65.

Holt C, Carver JA, Ecroyd H, Thorn DC. 2013. Invited review: caseins and the casein micelle: Their biological functions, structures, and behavior in foods. *J Dairy Sci* 96:6127–46.

Inglingstad RA, Devold TG, Eriksen EK, Holm H, Jacobsen M, Liland KH, Rukke EO, Vegarud GE. 2010. Comparison of the digestion of caseins and whey proteins in equine, bovine, caprine and human milks by human gastrointestinal enzymes. *Dairy Sci Technol* 90(5):549–3.

Jabeen T, Sharma S, Singh N, Bhushan A, Singh TP. 2005. Structure of the zinc-saturated C-terminal lobe of bovine lactoferrin at 2.0 A resolution. *Acta Chrystallogr. D Biol Crystallogr* 61:1107–15.

Jinsmaa M, Yoshikawa M. 1999. Enzymatic release of neocasomorphin and bcasomorphin from bovine b-casein. *Peptides* 20:957–62.

Jirillo F, Magrone T. 2014. Anti-inflammatory and anti-allergic properties of donkey's and goat's milk. *Endocr Metab Immune Disord Drug Targets* 14:27–37.

Johansson M, Högberg M, Andrén A. 2015. Relation between α-casein content and coagulation properties of milk from Swedish dairy goats. *The Open Food Sci J* 9:1–4.

Juillard V, Laan H, Kunji ER, Jeronimus-Stratingh CM, Bruins AP, Konings WN. 1995. The extracellular PI-type proteinase of *Lactococcus lactis* hydrolyzes beta-casein into more than one hundred different oligopeptides. *J Bacteriol* 177(12):3472–78.

Karjalainen J, Martin JM, Knip M, Ilonen J, Robinson BH, Savilahti E, Akerblom HK, Dosch HM. 1992. A bovine albumin peptide as a possible trigger of insulin dependent diabetes mellitus. *New Engl J Med* 327(5):302–307.

Lönnerdal B. 2003. Nutritional and physiologic significance of human milk proteins. *Am J Clin Nutr* 77(6): 1537S–43S.

Lu J, Liu L, Pang X, Zhang S, Jia Z, Ma C, Zhao L, Lv J. 2016. Comparative proteomics of milk fat globule membrane in goat colostrum and mature milk. *Food Chem* 209:10–6.

Luopajärvi K, Savilahti E, Virtanen SM, Ilonen J, Knip M, Akerblom HK, Vaarala O. 2008. Enhanced levels of cow's milk antibodies in infancy in children who develop type 1 diabetes later in life. *Acta Diabetologica* 44:S30.

Markus CR, Olivier B, Panhuysen GEM, Van Der Gugten J, Alles MS, Tuiten A, Westenberg HGM, Fekkes D, Koppeschaar HF, De Haan E. 2000. The bovine protein α-lactalbumin increases the plasma ratio of tryptophan to the other large neutral amino acids, and in vulnerable subjects raises brain serotonin activity, reduces cortisol concentration, and improves mood under stress. *Am J Clin Nutr* 71:1536–44.

Marletta D, Criscione A, Bordonaro S, Guastella AM, D'Urso G. 2007. Casein polymorphism in goat's milk. *Lait* 87:491–504.

Martin P, Cebo C, Miranda G. 2011. Interspecies comparison of milk proteins: quantitative variability and molecular diversity. In: Fuquay JW, Fox PF, McSweeney PLH (Eds). *Encyclopedia of dairy sciences*. London: Academic Press, pp 821–42.

Martin P, Szymanowska M, Zwierzchowski L, Leroux C. 2002. The impact of genetic polymorphisms on the protein composition of ruminant milks. *Reprod Nutr Dev* 42:433–59.

Martini M, Altomonte I, Salari F, Caroli AM. 2014. Monitoring nutritional quality of Amiata donkey milk: Effects of lactation and productive season. *J Dairy Sci* 97(11):6819–22.

Martini M, Salari F, Altomonte I, Rignanese D, Chessa S, Gigliotti C, Caroli A. 2010. The Garfagnina goat: a zootechnical overview of a local dairy population. *J Dairy Sci* 93:4659–67.

Martín-Ortiz A, Salcedo J, Barile D, Bunyatratchata A, Moreno FJ, Martin-García I, Clemente A, Sanz ML, Ruiz-Matute AI. 2016. Characterization of goat colostrum oligosaccharides by nano-liquid chromatography on chip quadrupole time-of-flight mass spectrometry and hydrophilic interaction liquid chromatography-quadrupole mass spectrometry. *J Chromatogr A* 1428:143–53.

Matar C, Goulet J. 1996. ß-casomorphin 4 from milk fermented by a mutant of *Lactobacillus helveticus*. *Int Dairy J* 6(4):383–97.

Matsuoka Y, Serizawa A, Yoshioka T, Yamamura J, Morita Y, Kawakami H, Toba Y, Takada Y, Kumegawa M. 2002. Cystatin C in milk basic protein (MBP) and its inhibitory effect on bone resorption in vitro. *Biosci Biotechnol Biochem* 66 (12):2531–36.

Meisel H, Bockelmann W. 1999. Bioactive peptides encrypted in milk proteins: proteolytic activation and thropho-functional properties. *Ant van Leeuwenhoek* 76 (1):207–15.

Mierau I, Kunji ERS, Venema G, Kok J. 1997. Casein and peptide degradation in lactic acid bacteria. *Biotechnol Gen Engin Rev* 14:279–301.

Moatsou G. 2010. Indigenous enzymatic activities in ovine and caprine milks. *Int J Dairy Technol* 63:16–31.

Monetini L, Cavallo MG, Manfrini S, Stefanini L, Picarelli A, Di Tola M, Petrone A, Bianchi M, La Presa M, Di Giulio C, Baroni MG, Thorpe R, Walker BK, Pozzilli P. IMDIAB Group. 2002. Antibodies to bovine beta-casein in diabetes and other autoimmune diseases. *Hormone Metab Res* 34(8):455–59.

Monti G, Bertino E, Muratore MC, Coscia A, Cresi F, Silvestro L, Fabris C, Fortunato D, Giuffrida MG, Conti A. 2007. Efficacy of donkey's milk in treating highly problematic cow's milk allergic children: an in vivo and in vitro study. *Pediatr Allergy Immunol* 18(3):258–64.

Monti G, Viola S, Baro C, Cresi F, Tovo PA, Moro G, Ferrero MP, Conti A, Bertino E. 2012. Tolerability of donkey's milk in 92 highly-problematic cow's milk allergic children. *J Biol Regul Homeost Agents* 26:75–82.

Moreno JF, Recio I, Olano A, Lopez-Fandino R. 2000. Chromatographic characterization of ovine κ-casein macropeptide. *J Dairy Res*, 67, 349–59.

Moreno JF, Recio I, Olano A, Lopez-Fandino R. 2001. Heterogeneity of caprine κ-casein macropeptide. *J Dairy Res* 68:197–208.

Nukumi N, Iwamori T, Kano K, Naito K, Tojo H. 2007. Whey acidic protein (WAP) regulates the proliferation of mammary epithelial cells by preventing serine protease from degrading laminin. *J Cell Physiol* 213(3):793–800.

Nygren-Babol L, Sternesjo Å, Bjorck L. 2004. Factors influencing levels of folate-binding protein in bovine milk. *Int Dairy J* 14(9):761–65.

O'Mahony JA, Fox PF, Kelly AL. 2013. Indigenous milk enzymes. In: McSweeney PLH, Fox PF (Eds). *Advanced dairy chemistry*, 4th ed. New York: Springer, pp 337–85.

O'Riordan N, Kane M, Joshi L, Hickey RM. 2014. Structural and functional characteristics of bovine milk protein glycosylation. *Glycobiology* 24:220–36.

Omar A, Harbourne N, Oruna-Concha MJ. 2016. Quantification of major camel milk proteins by capillary electrophoresis. *Int Dairy J* 58:31–35.

Pal S, Woodford K, Kukulian S, Ho S. 2015. Milk intolerance, beta-casein and lactose. *Nutrients* 7(9):7285–97.

Permyakov EA, Permyakov SE, Breydo L, Redwan EM, Almehdar HA, Uversky VN. 2016. Disorder in milk proteins: α-Lactalbumin. Part C. Peculiarities of metal binding. *Curr Protein Peptide Sci* 17(8):735–45.

Pieszka M, Huszczyński J, Szeptalin A. 2011. Comparison of mare's milk composition of different breeds. *Nauka Przyroda Technologie* 5:112.

Pisanu S, Ghisaura S, Pagnozzi D, Falchi G, Biosa G, Tanca A, Roggio T, Uzzau S, Addis MF. 2012. Characterization of sheep milk fat globule proteins by two-dimensional polyacrylamide gel electrophoresis/mass spectrometry and generation of a reference map. *Int Dairy J* 24:78–86.

Popescu CP, Long S, Riggs P, Womack J, Schmutz S, Fries R, Gallagher DS. 1996. Standardization of cattle karyotype nomenclature: report of the committee for the standardization of the cattle karyotype. *Cytogenet Cell Genet* 74:259–61.

Rachagani S, Gupta ID. 2008. Bovine kappa-casein gene polymorphism and its association with milk production traits. *Genet Mol Biol (São Paulo)* 31(4):893–7.

Ragona G, Corrias F, Benedetti M, Paladini I, Salari F, Altomonte I, Martini M. 2016. Amiata donkey milk chain: animal health evaluation and milk quality. *IJFS* 5:173–78.

Rando A, Ramunno L, Masina P. 2000. Mutations in casein genes. *Zoot Nut. Anim* 26:105–14.

Raynal-Ljutovac K, Lagriffoul G, Paccard P, Guillet I Chilliard Y. 2008. Composition of goat and sheep milk products: an update. *Small Ruminant Res* 79(1):57–72.

Rubinoff M, Schreiber C, Waxman S. 1977. The isolation and characterization of the folate binding protein from goat milk. *FEBS Letters* 75(1):244–8.

Sabikhi L. 2007. Designer milk. *Adv Food Nutr Res* 53:161–98.

Sabikhi L, Mathur BN. 2001. Qualitative and quantitative analysis of b-casomorphins in Edam cheese. *Milchwissenschaft* 56(4):198–200.

Salimei E, Fantuz F, Coppola R, Chiofalo B, Polidori P, Varisco G. 2004. Composition and characteristics of ass's milk. *Anim Res* 53:67–78.

Salmen SH, Abu-Tarboush HM, Al-Saleh AA, Metwalli AA. 2012. Amino acids content and electrophoretic profile of camel milk casein from different camel breeds in Saudi Arabia, *Saudi J Biol Sci* 19:177–83.

Sanz Ceballos L, Ramos Morales E, de la Torre Adarve G, Díaz Castro J, Pérez Martínez L, Sanz Sampelayo MR. 2009. Composition of goat and cow milk produced under similar conditions and analyzed by identical methodology. *J Food Compos Anal* 22:322–9.

Sarti L, Martini M, Ragona G, Casati D, Belli F, Salari F, Altomonte I, Barni S, Mori F, Pucci N, Novembre E. 2016. Donkey milk of Amiatina race in the treatment of children with allergy to cow milk proteins. Poster. Proceedings of the Xth Forum of Pratical Nutrition, Milano. 21-22/04/2016.

Saukkonen T, Savilahti E, Landin-Olsson M, Dahlquist G. 1995. IgA bovine serum albumin antibodies are increased in newly diagnosed patients with insulin-dependent diabetes mellitus, but the increase is not an independent risk factor for diabetes. *Acta Paediatr* 84:1258–61.

Savilahti E, Saukkonen TT, Virtala ET, Tuomilehto J, Akerblom HK. 1993. Increased levels of cow's milk and beta-lactoglobulin antibodies in young children with newly diagnosed IDDM. The Childhood Diabetes in Finland Study Group. *Diabetes Care* 16(7):984.

Schieber A, Brückner H. 2000. Characterization of oligo-and polypeptides isolated from yoghurt. *Eur Food Res Technol* 210(5):310–13.

Séverin S, Wenshui X. 2005. Milk biologically active components as nutraceuticals: review. *Crit Rev Food Sci Nutr* 45(7-8):645–56.

Silanikove N, Leitner G, Merin U, Prosser CG. 2010. Recent advances in exploiting goat's milk: Quality, safety and production aspects. *Small Ruminant Research* 89:110–124.

Sindayikengera S, Xia W. 2006. Nutritional evaluation of caseins and whey proteins and their hydrolysates from protamex. *J Zhejiang Univ Sci B* 7:90–98.

Singh LV, Jayakumar S, Sharma A, Gupta SK, Dixit SP, Gupta N, Gupta SC. 2015. Comparative screening of single nucleotide polymorphisms in β-casein and κ-casein gene in different livestock breeds of India. *Meta Gene* 4:85–91.

Smolenski G, Haines S, Kwan FY, Bond J, Farr V, Davis SR, Stelwagen K, Wheeler TT. 2007. Characterisation of host defence proteins in milk using a proteomic approach. *J Proteome Res* 6(1):207–15.

Sokolov AV, Pulina MO, Zakharova ET, Susorova AS, Runova OL, Kolodkin NI, Vasilyev VB. 2006. Identification and isolation from breast milk of ceruloplasmin-lactoferrin complex. *Biochemistry (Mosc)* 71:160–66.

Solinas C, Corpino M, Maccioni R, Pelosi U. 2010. Cow's milk protein allergy. *J Matern Fetal Neonatal Med* 23(Suppl 3):76–79.

Spertino S, Cipriani V, De Angelis C, Giuffrida MG, Marsano F, Cavaletto M. 2012. Proteome profile and biological activity of caprine, bovine and human milk fat globules. *Mol BioSyst* 8:967–74.

Stepaniak L, Fox PF, Sorhaug T, Grabska J. 1995. Effect of peptides from the sequence 58-72 of beta-casein on the activity of endopeptidase, aminopeptidase, and X-prolyl-dipeptidyl aminopeptidase from *Lactococcus lactis* ssp. *lactis* MG1363. *J Agricul Food Chem* 43 (3):849–53.

Teschemacher H. 2003. Opioid receptor ligands derived from food proteins. *Curr Pharm Des* 9:1331–44.

Thomä-Worringer C, Sørensen J, López-Fandiño R. 2006. Health effects and technological features of caseinomacropeptide. *Int Dairy J* 16:1324–33.

Threadgill DW, Womack JE. 1990. Genome analysis of the major bovine milk protein genes. *Nucleic Acids Res* 18:6935–42.

Tsabouri S, Douros K, Priftis KN. 2014. Cow's milk allergenicity. *Endocr Metab Immune Disord Drug Targets* 14(1):16–26.

Tzvetkova I, Dalgalarrondo M, Danova S, Iliev I, Ivanova I, Chobert JM, Haertle T. 2007. Hydrolysis of major dairy proteins by lactic acid bacteria from Bulgarian yogurts. *J Food Biochem* 31(5):680–702.

Uniacke-Lowe T, Huppertz T, Fox PF. 2010. Equine milk proteins: chemistry, structure and nutritional significance. *Int Dairy J* 20(9):609–29.

Vaarala O, Klemetti P, Savilahti E, Reijonen H, Ilonen J, Akerblom HK. 1996. Cellular immune response to cow's milk beta-lactoglobulin in patients with newly diagnosed IDDM. *Diabetes* 45(2):178–82.

Vinesh PV, Brahma B, Kaur R, Datta TK, Goswami SL, De S. 2013. Characterization of β-casein gene in Indian riverine buffalo. *Gene* 527:683–88.

Wakabayashi H, Yamauchi K, Takase M. 2006. Lactoferrin research, technology and applications. *Int Dairy J* 16:1241–51.

Weinberg ED. 2003. The therapeutic potential of lactoferrin. *Expert Opin Investig Drugs* 12:841–51.

Wu LC, Schulman BA, Peng ZY, Kim PS. 1996. Disulfide determinants of calcium-induced packing in α-lactalbumin. *Biochemistry* 35:859–63.

Yamamura J, Takada Y, Goto M, Kumegawa M, Aoe S. 1999. High mobility group like protein in bovine milk stimulates the proliferation of osteoblastic MC3T3-E1 cells. *Biochem Biophys Res Commun* 1261:113–17.

Index

Microbiology in Dairy Processing: Challenges and Opportunities, First Edition.
Edited by Palmiro Poltronieri.

Published 2018 by John Wiley & Sons Ltd and the Institute of Food Technologists.